STUDENT SOLUTIONS MANUAL FOR LAROSE'S

Discovering Statistics

CHRISTINA MORIAN
LINCOLN UNIVERSITY

W. H. FREEMAN AND COMPANY / NEW YORK

Printed in the United States of America

ISBN-10: 1-4292-2753-2
ISBN-13: 978-1-4292-2753-7

Second printing

W. H. Freeman and Company
41 Madison Avenue
New York, NY 10010
Houndmills, Basingstoke
RG21 6XS England

www.whfreeman.com

Contents

SOLUTIONS TO EXERCISES

Chapter 1

The Nature of Statistics

Section 1.1

1. Answers will vary.

3. If you were writing a news story that sought to display the Katrina survivors in the most sympathetic light, the reasons you might emphasize are "I did not have a car or a way to leave," "I had to care for someone who was physically unable to leave," "I was physically unable to leave," "I worried that my possessions would be stolen or damaged if I left," and "I didn't want to leave my pet."

5. Answers will vary.

7. Answers will vary. The large differences in the heights of the rectangles in the graph for men and women for the emotions of sadness, anger, and disbelief provide evidence to support this point of view.

9. The Palm Beach County votes for Buchanan were unusually high. The scatterplot shows that there is a clear trend of a linear relationship except the point corresponding to Palm Beach County votes. This point lies a lot higher than it would if it followed the trend. This indicates that the Palm Beach County votes for Buchanan were unusually high.

11. Buchanan actually received about 3500 votes.

Section 1.2

1. Statistics is the *art* and *science* of collecting, analyzing, presenting, and interpreting data.

3. False. Statistical inference consists of methods for drawing conclusions about population characteristics based on the information contained in a subset (sample) of that population.

5. A *qualitative variable* is a variable that does not assume a numerical value, but is usually classified into categories. A *quantitative variable* is a variable that takes on numerical values.

7. True.

9. A *statistic* is a characteristic of a sample.

11. We call the process by which we learn about the characteristics of a population by studying the characteristics of a sample *statistical inference*.

13. It may be impossible to sample the entire population, such as a researcher interested in the quality of water in Lake Erie. It may be too expensive or take too much time to sample the whole population. It may be impossible to sample the whole population without destroying it, such as testing the lifetime of light bulbs or crash-testing cars.

15. **(a)** Quantitative. **(b)** The temperature (in degrees Fahrenheit) in the room you are in right now represents interval data. There is no natural zero (since 0 degrees Fahrenheit does not represent no heat). Also, division (70 / 35) does not make sense in terms of degrees Fahrenheit because 70 degrees Fahrenheit does not represent twice as much heat as 35 degrees Fahrenheit;

the data are not ratio data. However, subtraction does make sense, because 70 degrees Fahrenheit is 70 − 50 = 20 degrees Fahrenheit warmer than 50 degrees Fahrenheit.

17. **(a)** Quantitative. **(b)** The price of tea in China represents ratio data. There is a natural zero ($0.00 per pound or $0.00 per box). Here, division does make sense. That is, tea that costs $10.00 per pound costs twice as much as tea that costs $5.00 per pound.

19. **(a)** Quantitative. **(b)** The winning score in next year's Super Bowl represents ratio data. There is a natural zero because it is possible for a team to score 0 points. Here division makes sense because a score of 28 points represents twice as many points as a score of 14 points.

21. **(a)** Qualitative. **(b)** The rank of the winning Super Bowl team in their division represents ordinal data because the ranks may be arranged in a particular order and no arithmetic may be performed on them.

23. **(a)** Qualitative. **(b)** Your favorite video game represents nominal data because there is no natural or obvious way that the data may be ordered. Also, no arithmetic can be carried out on your favorite video game.

25. **(a)** Qualitative. **(b)** Your favorite topping on pizza represents nominal data because there is no natural or obvious way that the data may be ordered. Also, no arithmetic can be carried out on your favorite pizza topping.

27. **(a)** Quantitative. **(b)** How old your car is represents ratio data. Since a car that was just purchased is 0 years old, there is a natural zero. A car that is 6 years old is twice as old as a car that is 3 years old, so division makes sense in this case.

29. The population is all home sales in Tarrant County, Texas. The sample is the 100 home sales selected.

31. The population is all 4-H clubs in Utah. The sample is the ten 4-H clubs selected.

33. The population is all students at Portland Community College. The sample is the 50 Portland Community College students that were selected.

35. Descriptive statistics. The average price of homes sold in Newington, Connecticut is a descriptive statistic because it describes a sample. But no inference is made regarding a larger population.

37. Statistical inference. A sample was taken, and the sample average percentage of people in which the cholesterol level was lowered by daily exercise was calculated. Then this percentage was used to make an inference about how much daily exercise can lower everyone's cholesterol level.

39. Descriptive statistics. The goal-against average for the Charlestown Chiefs hockey team is a descriptive statistic because it describes the sample. But no inference was made regarding a larger population.

41. Descriptive statistics. The average on the first statistics test in Ms. Reynolds' class is a descriptive statistic because it describes a sample. But no inference was made regarding a larger population.

43. These four species represent a sample because there are a total of 11 species that are listed as endangered in Minnesota.

45. **(a)** The elements are the companies City of Santa Monica, St. John's Health Center, The Macerich Company, Fremont General Corporation, and Entravision Communications Corporation. **(b)** The two variables are Employees and Industry. **(c)** The variable Employees is quantitative. **(d)** The variable Industry is qualitative. **(e)** The observation for St. John's Health Center is that it has 1755 employees and it is a health services industry.

Company	Employees	Industry
St. John's Health Center	1755	Health services

47. **(a)** The elements are the crime types Robberies, Aggravated Assaults, Burglaries, Larceny/Thefts, and Motor Vehicle Thefts. **(b)** The four variables are 2005 Total; Per 100,000 People; National per 100,000 People; and Compared to National Average. **(c)** The variables 2005 Total; Per 100,000 People; and National per 100,000 People are all quantitative variables. **(d)** The variable Compared to National Average is qualitative. **(e)** The observation for motor vehicle thefts is that there were a total of 55 automobile thefts in Stillwater, Oklahoma in 2005; the rate of automobile thefts in Stillwater, Oklahoma in 2005 was 134.1 per 100,000 people; and the national rate of automobile thefts in 2005 was 526.5 per 100,000 people. So the rate of automobile thefts in Stillwater, Oklahoma in 2005 was better than the national rate of automobile thefts in 2005.

Crime type	2005 total	Per 100,000 people	National per 100,000 people	Compared to national average
Motor vehicle thefts	55	134.1	526.5	Better

49. **(a)** The elements are the hospitals Hardy Wilson Memorial Hospital, Humphreys County Memorial Hospital, Jefferson County Hospital, Lackey Memorial Hospital, Leake Memorial Hospital, Madison County Medical Center, Monfort Jones Memorial Hospital, Rankin Medical Center, and University of Mississippi Medical Center—Holmes County. **(b)** The three variables are Beds, City, and Zip. **(c)** The variable Beds is quantitative. **(d)** The variables City and Zip are qualitative. **(e)** The observation for Monfort Jones Memorial Hospital is that it has 72 beds and it is in the city of Kosciusko with a Zip Code of 39090.

Hospital name	Beds	City	Zip
Monfort Jones Memorial Hospital	72	Kosciusko	39090

51. The four variables are:

- ERA: Earned-Run Average, earned runs allowed per nine innings.
- HR: Total number of homeruns given up by the team's pitchers.
- BB: Total number of walks given up by the team's pitchers.
- SO: Total number of strikeouts made by the team's pitchers.

All four variables are quantitative.

53. The observation for the New York Yankees is that the earned-run average is 4.41, the total number of homeruns given up by the team's pitchers is 170, the total number of walks given up by the team's pitchers is 496, and the total number of strikeouts made by the team's pitchers is 1019.

Team	ERA	HR	BB	SO
New York Yankees	4.41	170	496	1019

55. The two variables are Births and Average Maternal Age. Both variables are quantitative.

57. **(a)** The elements are the different types of music that all radio listeners listened to in the winter of 2004. The only elements that we can list are Mainstream AC, Hot AC, Mod AC, and Soft AC. We cannot list the rest of the elements because they are not given. **(b)** The only variable is the Percentage of listeners of each type of music. It is a quantitative variable. **(c)** The percentage of all listeners that are Mainstream AC listeners is 7.8%. The percentage of AC listeners that are Mainstream AC listeners is $(7.8/14)100\% = 55.71\%$.

59. They tested a sample of their own light bulbs, found the average lifetime of the sample, compared it to the average lifetimes of other current models of light bulbs, and found the average lifetime of their sample to be longer than the reported average lifetimes of other current models of light bulbs.

61. **(a)** The elements are the years 1910, 1986, 2004, 1981, 1962, 1996, 1972, 1980, 1963, and 1951. **(b)** The variable is Deaths. It is a numerical variable. **(c)** This sample could not be considered a random sample of the annual number of tornado deaths of all years. The 10 years selected were not selected randomly, but were selected according to which 10 years had the fewest tornado deaths.

Section 1.3

1. An example of a situation where the information from a bad sample led to negative consequences is the *Literary Digest* poll. The *Literary Digest* used lists of people who owned cars and had telephones, which resulted in the exclusion of millions of poor and underprivileged people who largely supported Roosevelt.

The sample was therefore highly biased toward the richer people who were more likely to support Alf Landon. Thus, the results of the poll incorrectly indicated that Alf Landon would win. The paper later declared bankruptcy.

3. The *Literary Digest* could have decreased the bias in their poll by choosing a random sample of houses and apartments and surveying the people door to door. They would have been more likely to include people who were poor or underprivileged by using this method and thus their sample would have been more representative of the population.

5. A *random sample* is a sample for which every element has an equal chance of being selected.

7. Answers will vary.

9. In an *observational study*, the researcher observes whether the subjects' differences in the predictor variable are associated with differences in the response variable. No attempt is made to create differences in the predictor variable. In an *experimental study*, researchers investigate how varying the predictor variable affects the response variable. Subjects are randomly placed into treatment and control groups.

11. Stratified sampling is represented.

13. Systematic sampling is represented.

15. Random sampling is represented.

17. This is an example of a leading question.

19. An observational study is involved. The response variable is stock price. The predictor variable is whether the company gives large bonuses to their CEOs (at least $1 million per year).

21. Answers will vary. Your sample is different from the sample in the text because both samples were randomly selected.

23. The target population is all high schools in New England, and the potential population is all high schools in greater Boston. The potential for selection bias is that the sample is not a random sample of all high schools in New England. The drop-out rate for all of New England high school students may be different than the drop-out rate for those 15 high schools in greater Boston.

25. The question may be interpreted in more than one way. Some people might think that the question is asking for a choice to be made between rap and hip-hop music. Others may think the answer to the question is either yes or no. This is because it is actually two questions in one.

27. No, the researcher would not be justified in reporting "Two-thirds of women support abortion." The women responded to the question "Do you support the right of a woman to terminate a pregnancy when her life is in danger?" and not the question "Do you support abortion?" The women may have answered each question differently.

29. **(a)** No, we don't know how many employees the largest employer in the sample will have before we select the sample. Since the sample is chosen randomly, we don't know which three employers will be selected before we select the sample. Different samples may contain different largest employers. **(b)** Answers will vary. **(c)** No, another sample of size 3 is not likely to comprise the same three employers. Since the samples are selected randomly, the samples will probably contain different employers. **(d)** Answers will vary.

31. **(a)** No, we do not know what the lowest price in the sample will be before we select the sample. Since the sample is randomly selected, we don't know which two stocks will be selected before we select our sample. Different samples may contain different lowest stock prices. **(b)** Answers will vary. **(c)** No, if we take another sample of size 2, it is not likely to comprise the same two companies. Since the samples are randomly selected, they will probably contain different employers. **(d)** Answers will vary.

33. The response variable is the risk for a second heart attack and the predictor variable is whether the patient followed a Mediterranean diet or a Western diet.

35. **(a)** The predictor variable is whether the subject had the placebo bracelet or the ionized bracelet. **(b)** The treatment is wearing the ionized bracelet. **(c)** The response variable is the measure of pain.

37. Answers will vary. One possible answer is given below.

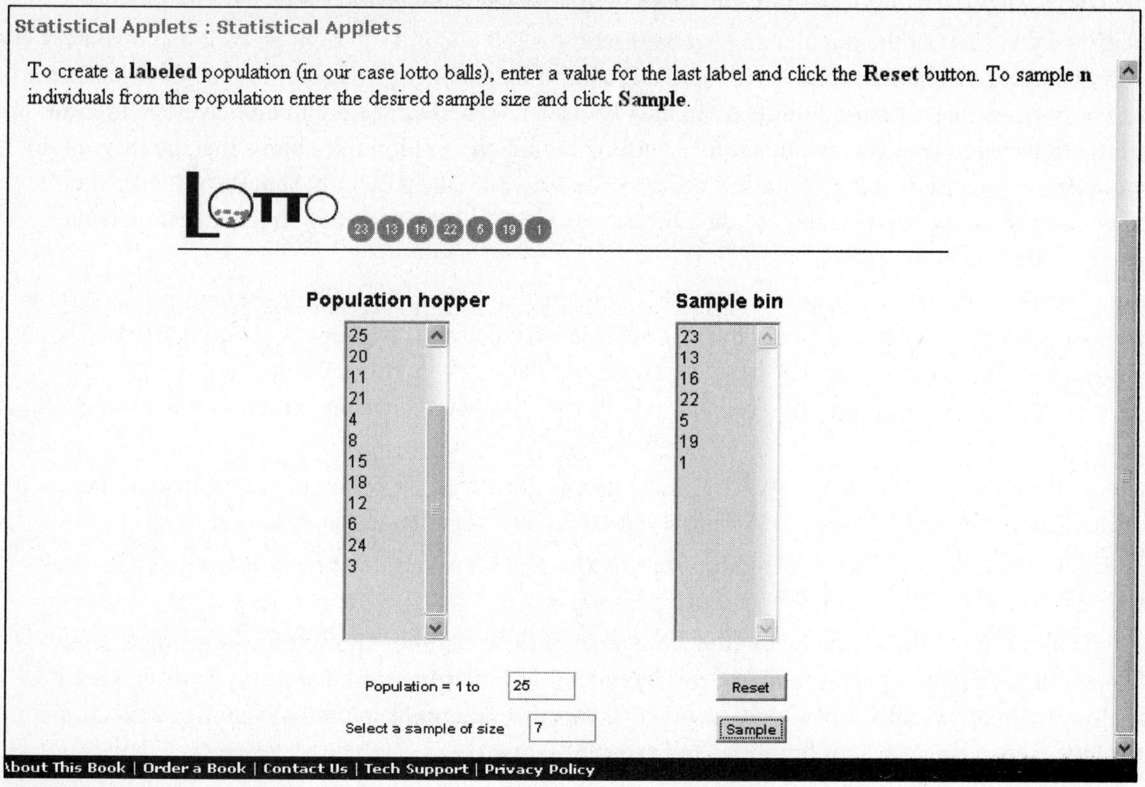

Statistical Applets : Statistical Applets

To create a **labeled** population (in our case lotto balls), enter a value for the last label and click the **Reset** button. To sample **n** individuals from the population enter the desired sample size and click **Sample**.

LOTTO 23 13 16 22 5 19 1

Population hopper	Sample bin
25	23
20	13
11	16
21	22
4	5
8	19
15	1
18	
12	
6	
24	
3	

Population = 1 to 25 Reset

Select a sample of size 7 Sample

About This Book | Order a Book | Contact Us | Tech Support | Privacy Policy

This represents the sample:

23. Orange County, California
13. Washington, DC, metro area
16. Nassau-Suffolk, NY
22. Miami-Hialeah, FL
5. West Palm Beach, FL
19. Austin, TX
1. Atlanta, GA

39. **(a)** Answers will vary. **(b)** There is no way of telling for certain in advance whether this city will appear in the random sample. **(c)** Answers will vary.

Chapter 1 Review Exercises

1. **(a)** The ten elements are the countries Iraq, United States, Pakistan, Canada, Madagascar, North Korea, Chile, Bulgaria, Afghanistan, and Iran. **(b)** The variables are Continent, Climate, Water Use (Per Capita Gallons per Day), and Main Use. **(c)** The observation for Canada is that it is on the continent of North America, it has a temperate climate, its water use is 1268 gallons per capita per day, and its main use of water is industry.

Country	Continent	Climate	Water use (per capita gallons per day)	Main use
Canada	North America	Temperate	1268	Industry

(d) Iraq uses the most water per capita in the world. **(e)** "Per capita" means per person.

3. The two temperate countries that share the same continent, climate, and main use of water are the United States and Canada. **(a)** Their water use is not the same. The United States uses 1565 gallons per capita per day, and Canada uses 1268 gallons per capita per day. **(b)** No, it is not likely that their water use would be exactly the same. The United States has more farming and more industry than Canada does, so it uses more water.

5. **(a)** The only way to find out the population average lifetime of all one million light bulbs in the inventory is to turn on all one million light bulbs and leave them all on until they burn out, measuring the time it takes for each light bulb to burn out. All of these lifetimes can then be used to calculate the population average lifetime of all one million light bulbs. **(b)** This would require burning out all one million light bulbs that are in stock so that there would be no good light bulbs left to sell. It would be better to take a random sample of the light bulbs, find the average lifetime of the sample, and use the sample average lifetime of the light bulbs to estimate the population average lifetime of the light bulbs.

7. **(a)** The population is all statistics students. **(b)** The sample is the random sample of students selected from the statistics class. **(c)** The variable is whether the student is left-handed. It is a categorical variable. **(d)** The sample proportion is not likely to be exactly the same as the population proportion. But it is not likely to be very far away from the population proportion because which statistics class a person enrolls in is not based on whether the person is left-handed.

9. **(a)** You would use an observational study. **(b)** Since people are already enrolled in their statistics classes, it would be impractical to randomly reassign people to a statistics class after classes have started.

11. **(a)** The experimental factor violated is randomization. **(b)** If a sample is not a random sample, the results may be biased.

13. If the results of the study showed that children from single-parent families showed lower average cognitive skills than children from two-parent families, this would not necessarily mean that living in a one-parent family causes lower cognitive skills. There may be other factors that determine a child's cognitive skills besides whether they come from a single-parent family or a two-parent family.

Chapter 1 Quiz

1. False. Statistical inference consists of methods for estimating and drawing conclusions about *population* characteristics based on the information contained in a *sample.*

2. True.

3. False. A parameter is a characteristic of a *population.*

4. collecting

5. observation

6. equal chance

7. sample

8. A sample survey is an example of an observational study.

9. Convenience sampling is represented.

10. An experimental study is involved.

11. The predictor variable is whether an elderly patient with Alzheimer's is given the new drug or the placebo. The response variable is whether the patient's Alzheimer's symptoms are reduced.

12. The elements are the sectors of the information technology economy: Hardware, Software, and Communications.

13. The variables are:

- Gross Domestic Product in 1996
- Gross Domestic Product in 1997
- Gross Domestic Product in 1998
- Gross Domestic Product in 1999
- Gross Domestic Product in 2000

- Gross Domestic Product in 2001
- Gross Domestic Product in 2002
- Gross Domestic Product in 2003

The variables are all quantitative.

14. Hardware production fell sharply in 2001.

15. The response variable is the amount of insect damage to crops. The predictor variable is whether the new form of pesticide is used or if the traditional pesticide is used.

16. The researchers would use an experimental study.

17. The treatment is the new pesticide and the control is the traditional pesticide.

18. Different people have different interpretations of the words often, occasionally, sometimes, and seldom.

19. No, the researcher did not contact every single smoker in the country. It would have been too time-consuming and too expensive.

20. The favorite brands should be considered statistics because they most likely came from a sample.

21. According to the survey, 60.6% of all African-Americans living in the Northeast prefer Kool cigarettes.

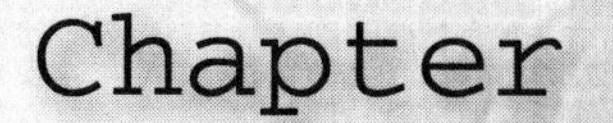

Describing Data Using Graphs and Tables

Section 2.1

1. We use graphical and tabular form to summarize data in order to organize it in a format where we can better assess the information. If we just report the raw data, it may be extremely difficult to extract the information contained in the data.

3. True.

5.

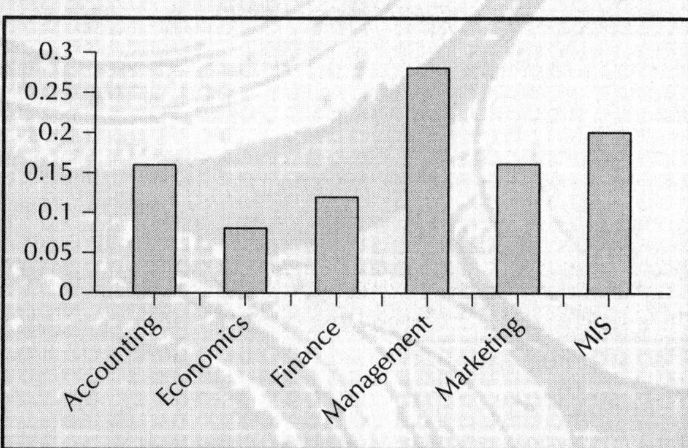

7. **(a)** and **(b)**

Climate	Frequency	Relative frequency
Arid	5	5/10 = 0.50
Temperate	4	4/10 = 0.40
Tropical	1	1/10 = 0.10

(c)

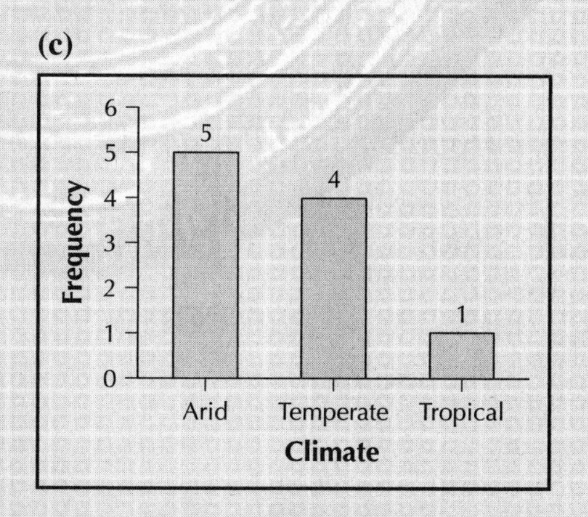

(d)

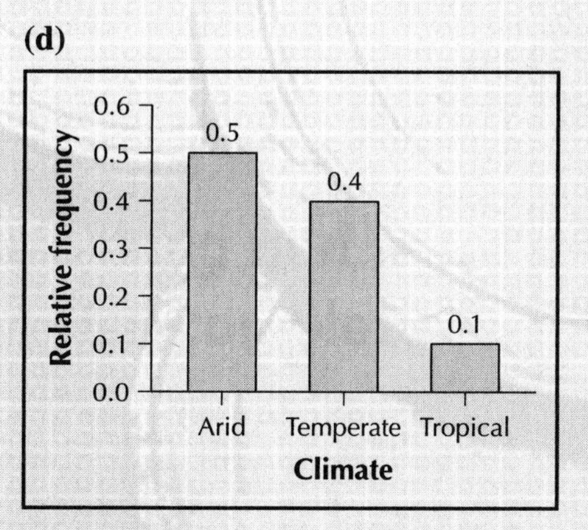

(e)

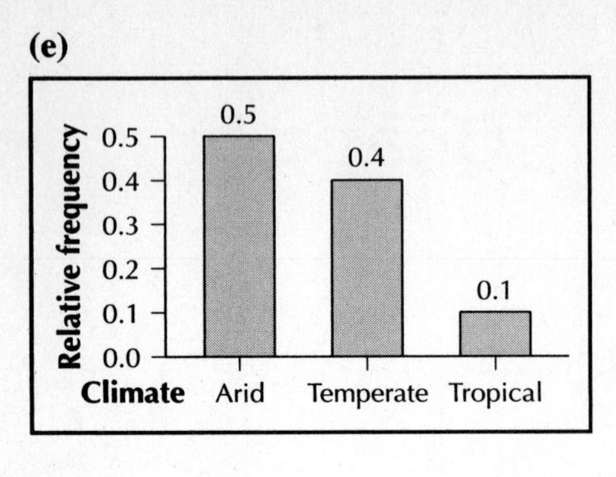

(f)

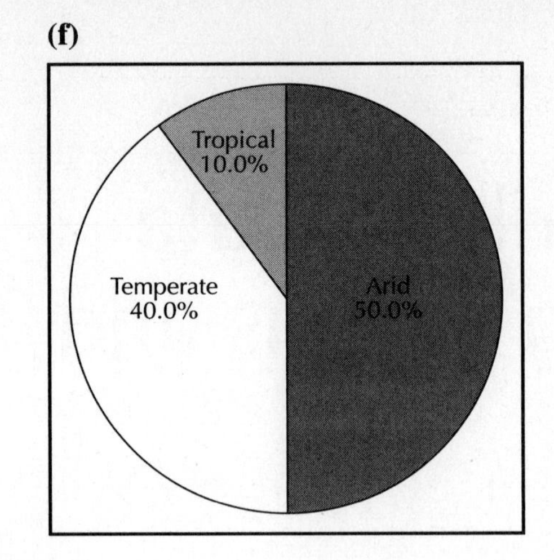

9. (a)

End-use sector	Relative frequency
Residential	1213.9/5923.2 ≈ 0.2049
Commercial	1034.1/5923.2 ≈ 0.1746
Industrial	1736.0/5923.2 ≈ 0.2931
Transportation	1939.2/5923.2 ≈ 0.3274

(b)

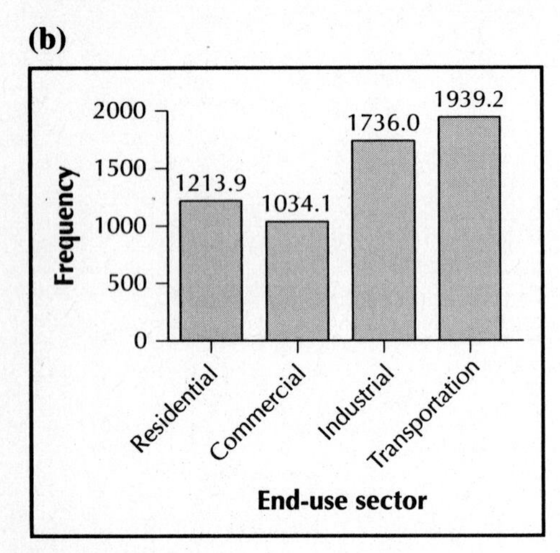

(c)

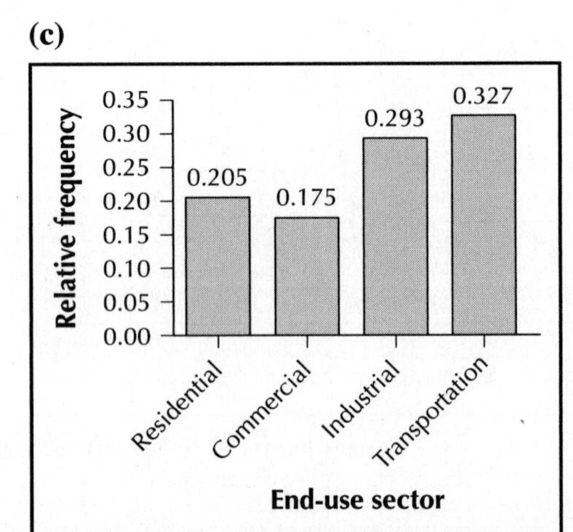

(d)

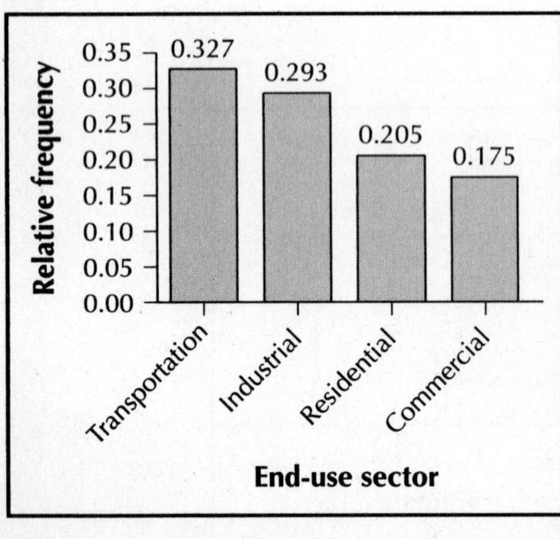

(e) NOTE: In the chart, the relative frequencies are expressed as percentages.

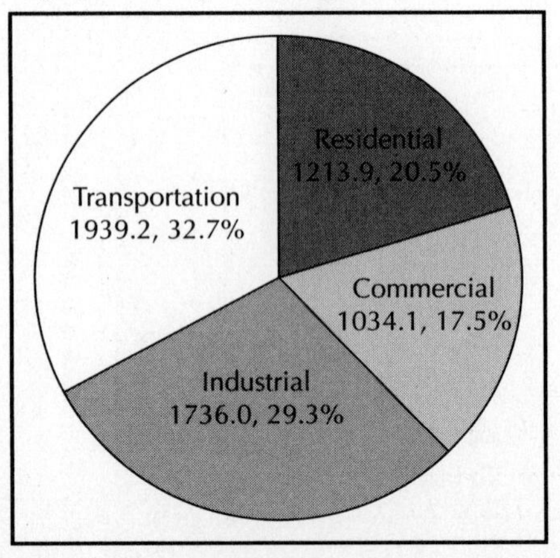

11. **(a)** and **(b)**

Type	Frequency	Relative frequency
F	5	5/10 = 0.5
NF	5	5/10 = 0.5

(c)

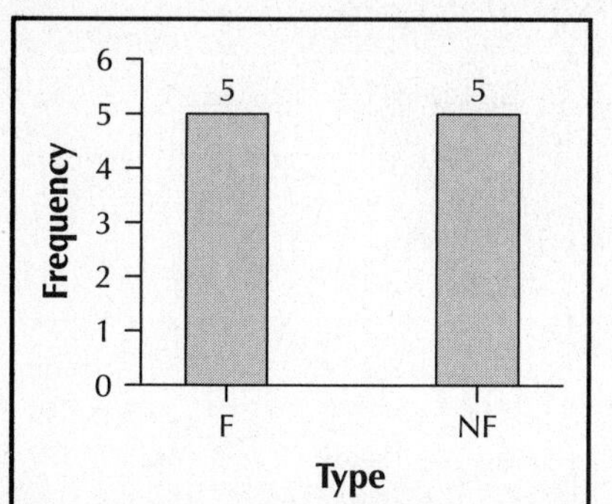

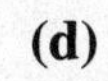

(d)

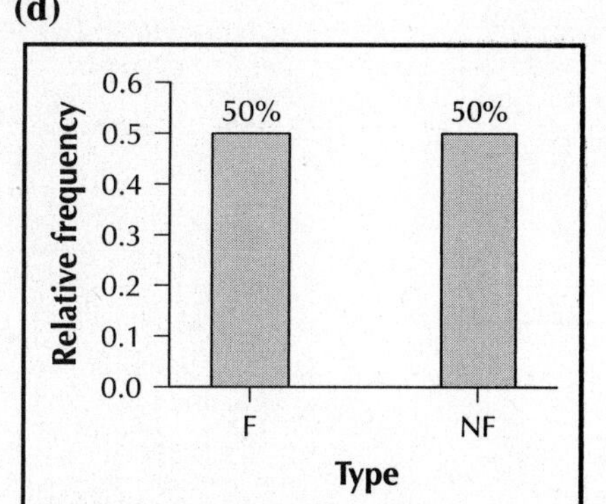

(e)

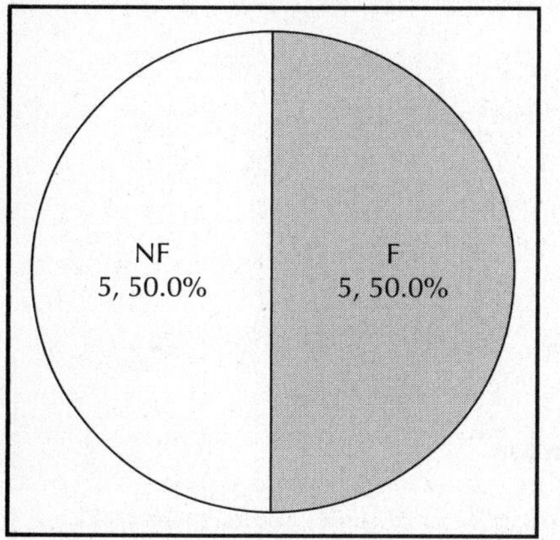

13. **(a)** and **(b)**

Gender	Frequency	Relative frequency
Female	5	5/10 = 0.5
Male	5	5/10 = 0.5

(c)

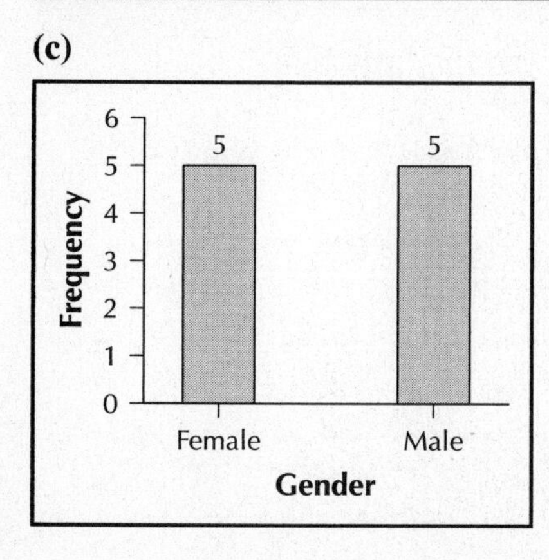

(d)

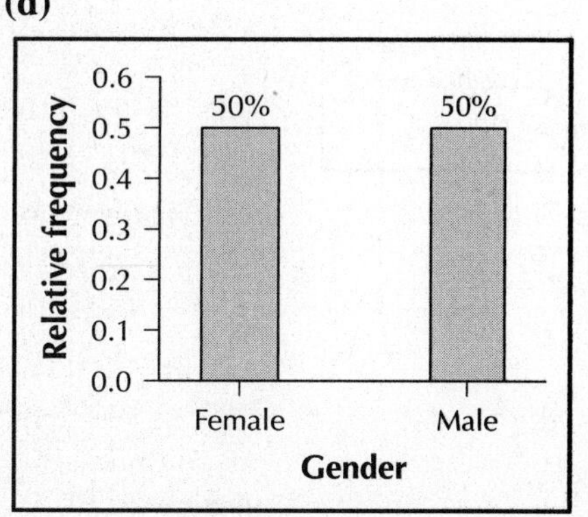

(e)

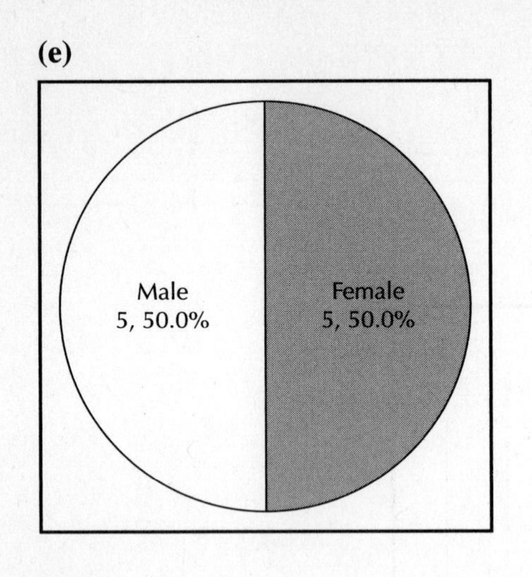

15.

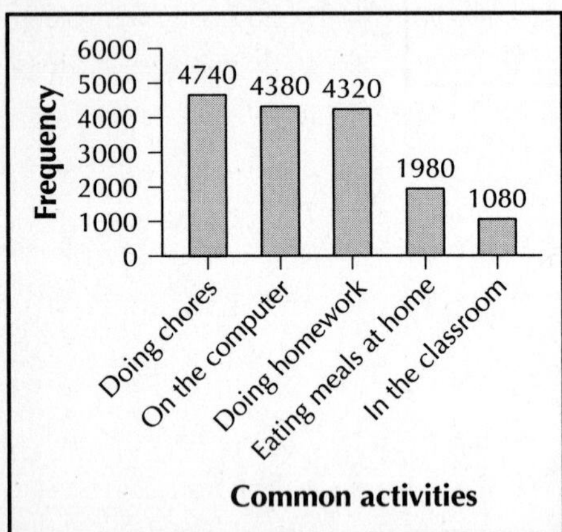

17. Yes. Two other possible answers could be "I don't know" or "Other factors."

19.

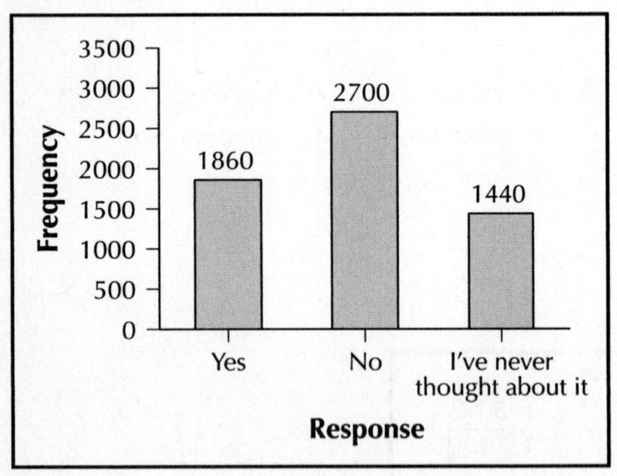

21. **(a)**

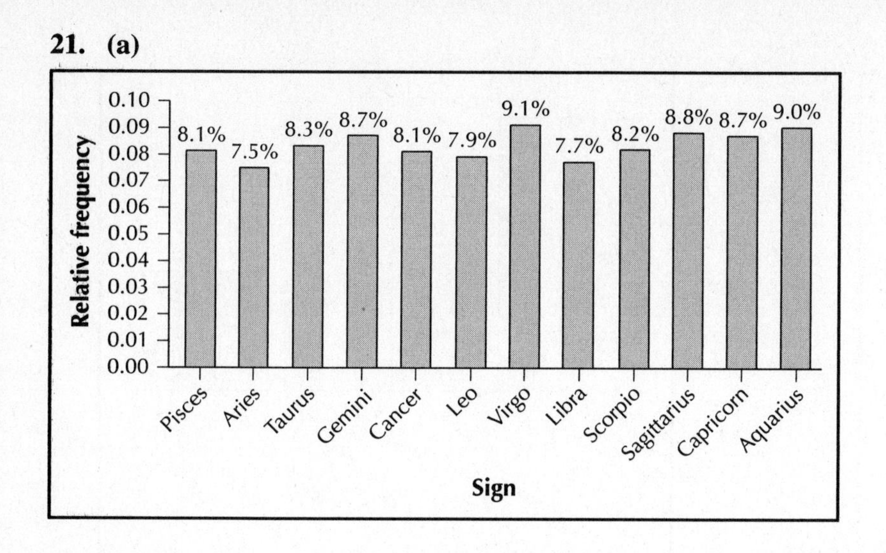

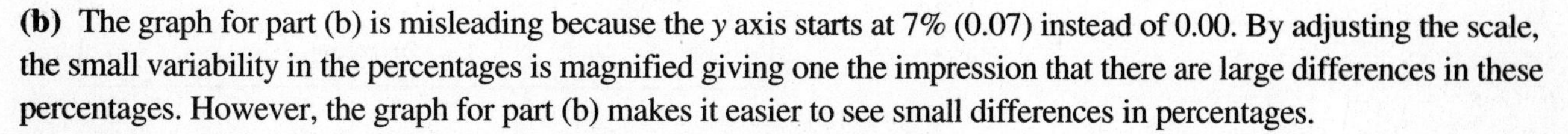
(b) The graph for part (b) is misleading because the y axis starts at 7% (0.07) instead of 0.00. By adjusting the scale, the small variability in the percentages is magnified giving one the impression that there are large differences in these percentages. However, the graph for part (b) makes it easier to see small differences in percentages.

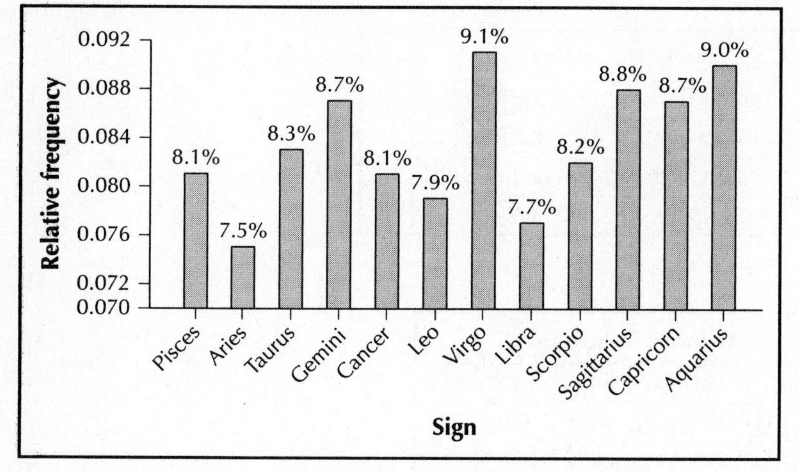

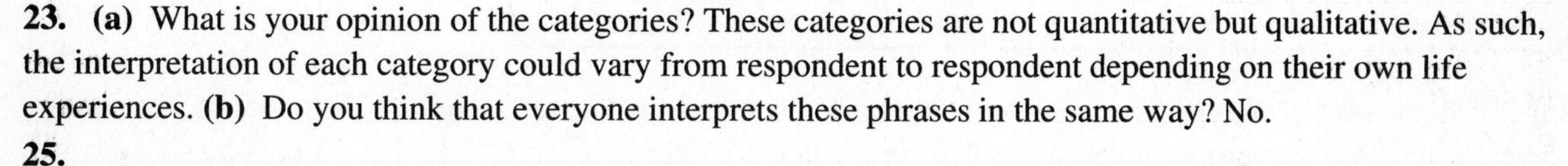
23. **(a)** What is your opinion of the categories? These categories are not quantitative but qualitative. As such, the interpretation of each category could vary from respondent to respondent depending on their own life experiences. **(b)** Do you think that everyone interprets these phrases in the same way? No.

25.

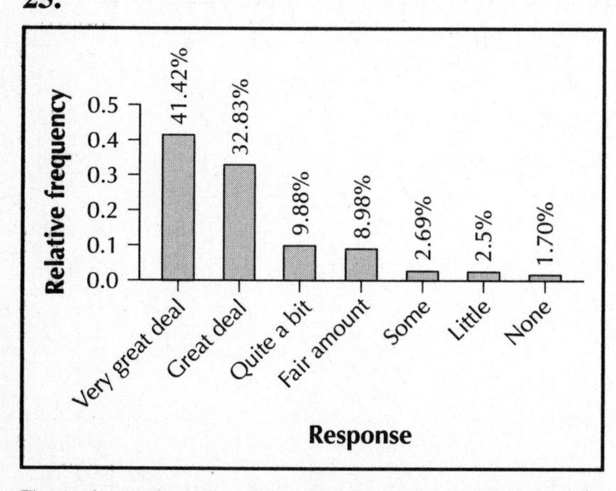

Based on the graphs presented in problems 24 and 25, one can see that the categories of "Very Great Deal" and "Great Deal" have the tallest bars, which reinforces the positive message in the data.

27. (a)

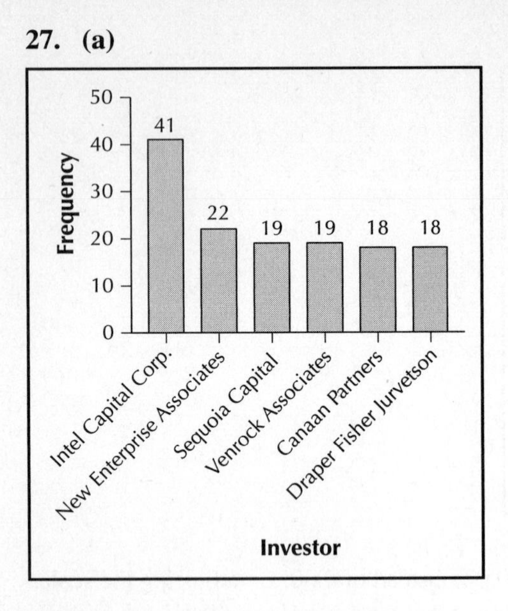

(b)

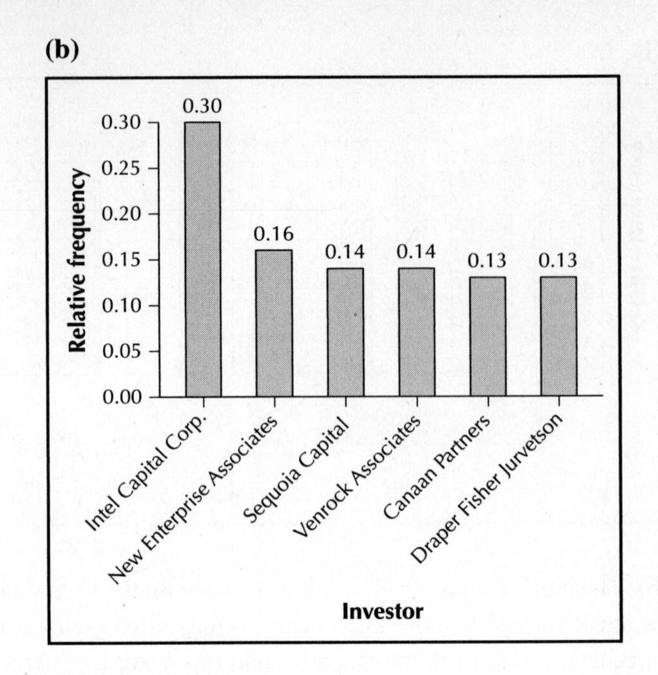

29. Distribution for under 65 years.

Type of insurance	Frequency	Relative frequency
Private	126,800,000	126,800,000/182,000,000 ≈ 0.6967
Medicaid	13,700,000	13,700,000/182,000,000 ≈ 0.0753
Other	5,700,000	5,700,000/182,000,000 ≈ 0.0313
Uninsured	35,800,000	35,800,000/182,000,000 ≈ 0.1967

31. Distribution for over 65 years.

Type of insurance	Frequency	Relative frequency
Private	20,800,000	20,800,000/34,800,000 ≈ 0.5977
Medicaid	11,700,000	11,700,000/34,800,000 ≈ 0.3362
Other	2,100,000	2,100,000/34,800,000 ≈ 0.0603
Uninsured	200,000	200,000/34,800,000 ≈ 0.0057

33. From the graph, one can observe a very large difference between the uninsured for the under 65 years and 65 years and over. The proportion of uninsured for the under 65 group is larger than that for the 65 and over.

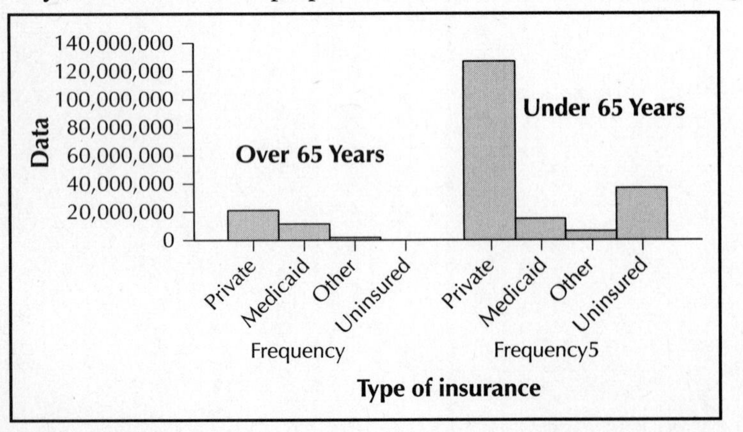

35.

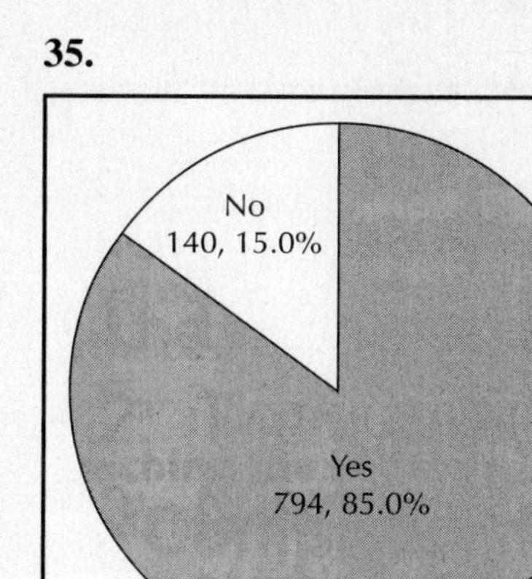
No
140, 15.0%
Yes
794, 85.0%

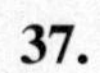

37.

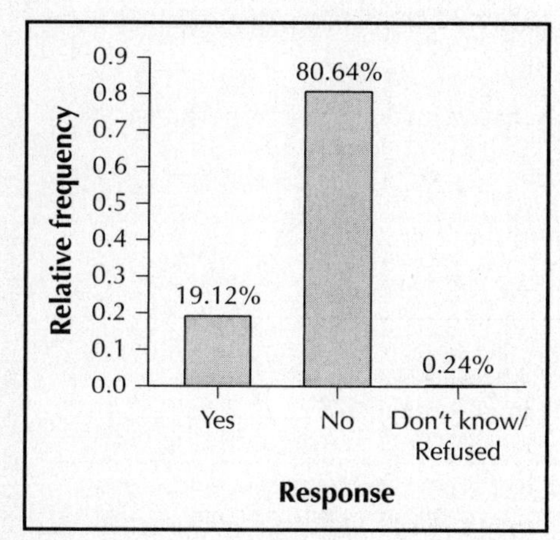
Relative frequency
0.9
0.8
0.7
0.6
0.5
0.4
0.3
0.2
0.1
0.0
80.64%
19.12%
0.24%
Yes
No
Don't know/
Refused
Response

39.

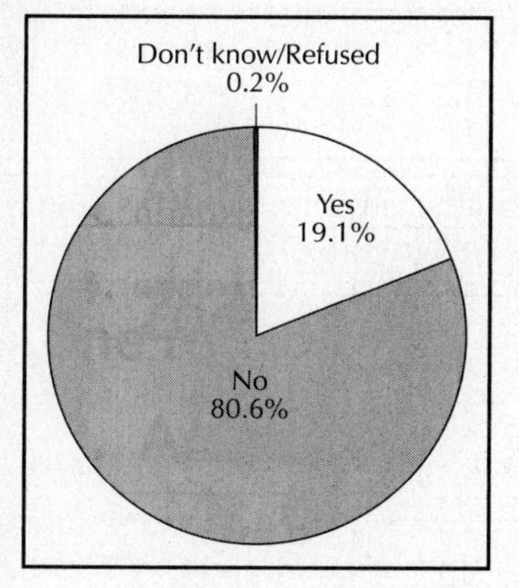
Don't know/Refused
0.2%
Yes
19.1%
No
80.6%

41.

Food store	Frequency	Relative frequency
Supermarkets	73,357	73,357/198,514 ≈ 0.3695
Convenience food stores	30,748	30,748/198,514 ≈ 0.1549
Convenience food/gasoline	23,035	23,035/198,514 ≈ 0.1160
Delicatessens	6,123	6,123/198,514 ≈ 0.0308
Meat and fish markets	8,941	8,941/198,514 ≈ 0.0450
Retail bakeries	20,418	20,418/198,514 ≈ 0.1029
Fruit and vegetable markets	2,971	2,971/198,514 ≈ 0.0150
Candy, nuts, confectionery	5,029	5,029/198,514 ≈ 0.0253
Dairy products stores	2,340	2,340/198,514 ≈ 0.0118
Other food stores	25,552	25,552/198,514 ≈ 0.1287

43.

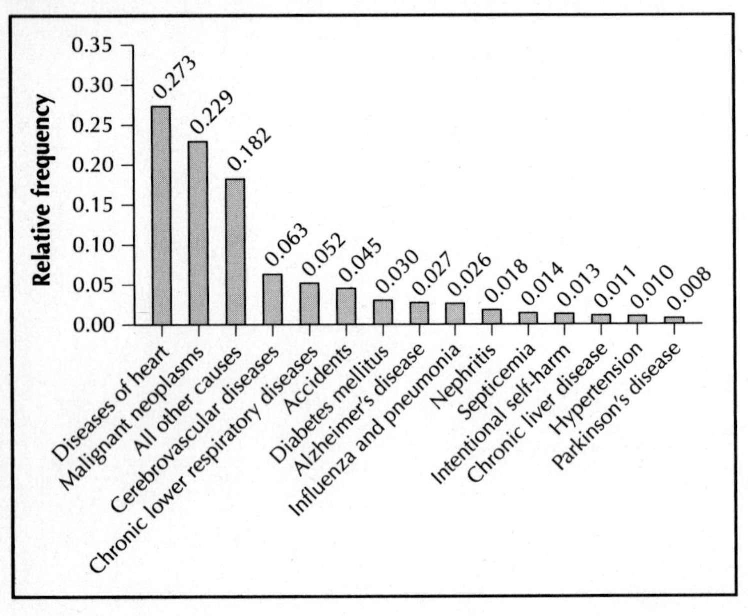

45. Pie chart with categories combined that are 3% or less. It is helpful to combine some of the categories with the lower frequencies.

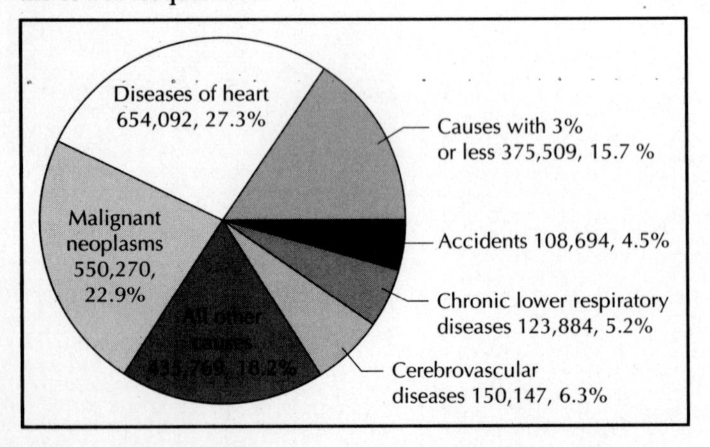

47.

Gender	Frequency	Relative frequency
Boy	26	$26/44 \approx 0.5909$
Girl	18	$18/44 \approx 0.4091$

49.

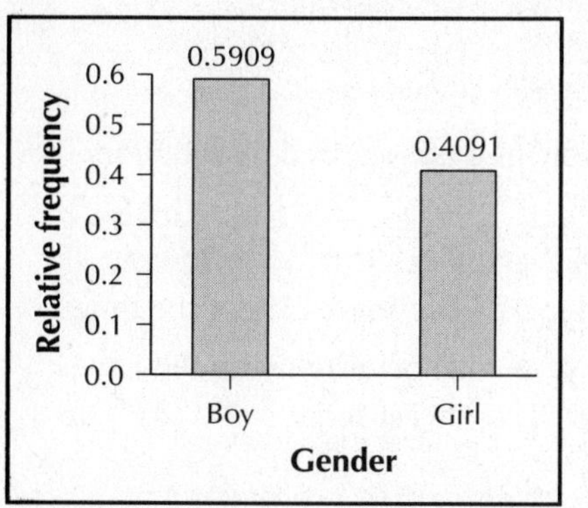

No, we only need to multiply the scale on the y axis by the sample size of $n = 44$.

51. (a)

Tally for Discrete Variables: GENDER

GENDER	Percent
boy	47.49
girl	52.51

(b)

Tally for Discrete Variables: GOALS

GOALS	Percent
Grades	51.67
Popular	29.50
Sports	18.83

53. There is only one qualitative variable: state.
There are seven quantitative variables: Tot_hhld, Fam_tpc, Fam_mpc, Fam_fpc, Nfm-tpc, Nfm_lpc, and Ave_size

55. (a) and **(b)**

Class	Frequency	Relative frequency
Freshmen	5	$5/20 = 0.25$
Sophomores	5	$5/20 = 0.25$
Juniors	5	$5/20 = 0.25$
Seniors	5	$5/20 = 0.25$

57. Answers will vary.

Section 2.2

1. frequency distribution—can be used for both qualitative and quantitative data; relative frequency distribution—can be used for both qualitative and quantitative data; histograms—can be used for quantitative data only; frequency polygons—can be used for quantitative data only; stem-and-leaf displays—can be used for quantitative data only; dot plots—can be used for quantitative data only.

3. Between 5 and 20.

5. Dot plots may be useful for comparing two or more variables.

7. Example of a right-skewed distribution—price of precious gems. Example of a left-skewed distribution—lifetime of electrical bulbs.

9. **(a)** The value of 4 occurs with the highest frequency. **(b)** The values of 1 and 6 occur with the lowest frequency. **(c)** The value of 3 occurred with a frequency of 15. **(d)** The value of 3 occurred 15% of the times.

11. **(a)** The score that occurred with the highest frequency is 46. **(b)** The score that occurred with the lowest frequency is 33 (not including a frequency of 0). **(c)** The highest score is 49. The lowest score is 33. **(d)** The distribution is left-skewed.

13. **(a)** The greatest number of items sold is 13. **(b)** Five children sold the smallest number of items—1 (one). **(c)** The number sold with the greatest frequency is 5. **(d)** The distribution is right-skewed.

15. **(a)** 8 **(b)** 8/100 = 0.08 **(c)** 6 **(d)** 6/100 = 0.06

17. **(a)** You can change to a relative frequency distribution by dividing the frequency values by the total frequency. The classes will not be affected. **(b)** The relative frequency histogram can be converted into a frequency histogram by changing the scale along the relative frequency axis (vertical axis). This is done by multiplying the relative frequency values by the total frequency. This change in scale will not affect the shape of the distribution. **(c)** The sample size is 19.

19. **(a)** 0 **(b)** 0 **(c)** The class $25 to $27.50 has the largest relative frequency. This value is $4/19 \approx 0.2105$. **(d)** 3 **(e)** 0

21. Data set: 23 24 25 26 27 28 28 29 30 31 31 32 32 32 33 35 36 37 39 40

23. Histogram with five classes.

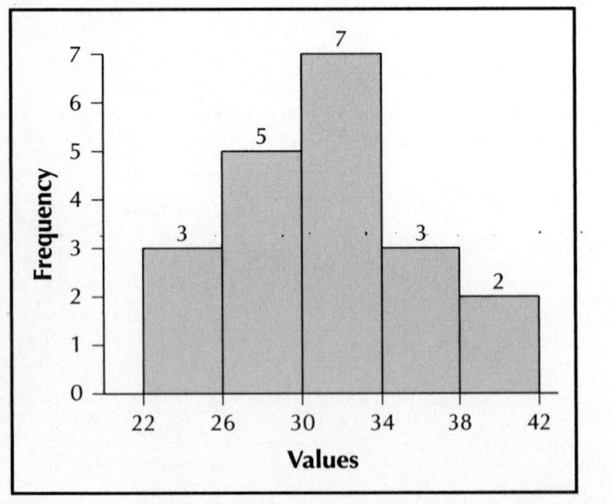

25. **(a)** 15 **(b)** 37.5 **(c)** 52.5 **(d)** 67.5 to 82.5 **(e)** 22.5 to 37.5

27. **(a)** 2000 **(b)** 1000 **(c)** The class with the highest frequency is 1000 to 3000. **(d)** The class with the lowest frequency is 17,000 to 19,000.

29. **(a)** and **(b)**

Continent	Frequency	Relative frequency
Africa	1	1/10 = 0.1
Asia	5	5/10 = 0.5
Europe	1	1/10 = 0.1
North America	2	2/10 = 0.2
South America	1	1/10 = 0.1

(c) Cannot construct a frequency histogram for the variable *continent* because it is a qualitative variable. **(d)** Cannot construct a dot plot for the variable *continent* because it is a qualitative variable. **(e)** Cannot construct a stem-and-leaf display for the variable *continent* because it is a qualitative variable.

31. 

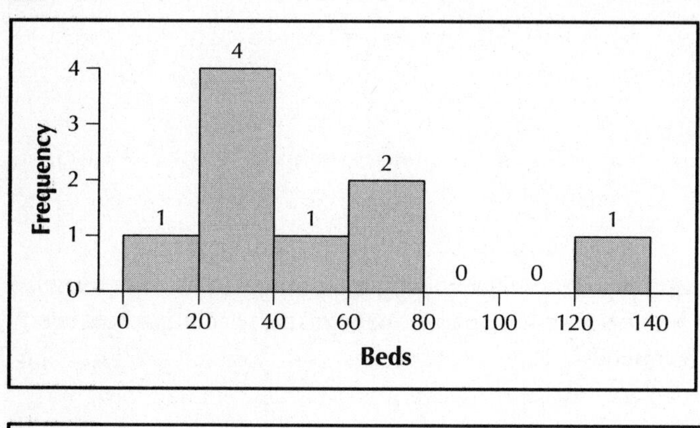

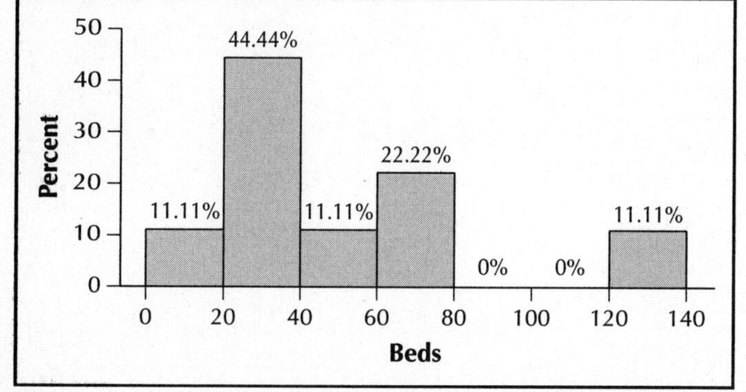

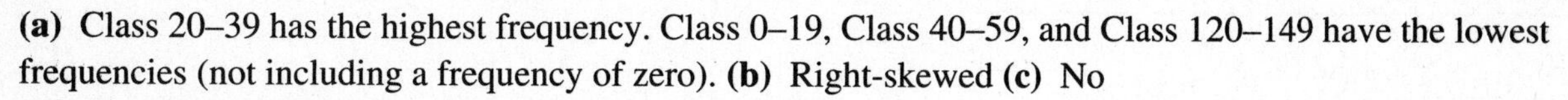
(a) Class 20–39 has the highest frequency. Class 0–19, Class 40–59, and Class 120–149 have the lowest frequencies (not including a frequency of zero). **(b)** Right-skewed **(c)** No

33. Stem-and-leaf display for the number of beds.

Stem	Leaf
1	5
2	55
3	04
4	9
5	
6	7
7	2
8	
9	
10	
11	
12	
13	4

35.

Classes	Frequency	Relative frequency
0900–1099	4	4/10 = 0.4
1100–1299	2	2/10 = 0.2
1300–1499	0	0/10 = 0.0
1500–1699	2	2/10 = 0.2
1700–1899	2	2/10 = 0.2

37.

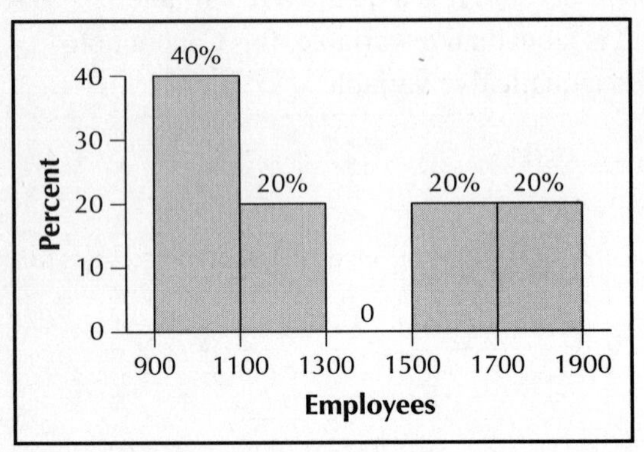

The only difference between the frequency histogram and the relative frequency histogram is the scale used for the y axis. In the frequency histogram the scale used for the y axis is the frequency; in the relative frequency histogram the scale used for the y axis is the relative frequency.

39.

Classes	Frequency	Relative frequency
550–599	1	$1/12 \approx 0.0833$
600–649	1	$1/12 \approx 0.0833$
650–699	1	$1/12 \approx 0.0833$
700–749	0	$0/12 = 0.0000$
750–799	3	$3/12 = 0.2500$
800–849	3	$3/12 = 0.2500$
850–899	1	$1/12 \approx 0.0833$
900–949	2	$2/12 \approx 0.1667$

41.

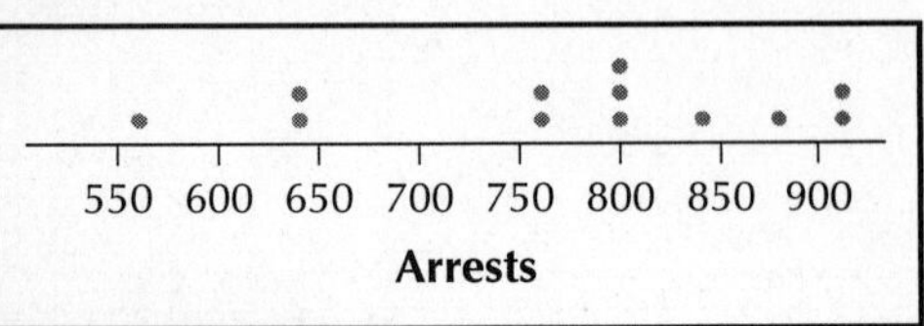

43. From the relative frequency distribution, it is difficult to determine what is the most common number of shutouts because we have intervals of values.

Interval	Frequency	Relative frequency
2–3	1	1/14 = 0.071
4–5	1	1/14 = 0.071
6–7	2	2/14 = 0.143
8–9	8	8/14 = 0.571
10–11	0	0/14 = 0
12–13	2	2/14 = 0.143

45. From the histogram given in this solution, **(a)** Proportion of teams with at least 8 shutouts = $10/14 \approx 0.7143$. **(b)** Proportion of teams with less than 2 shutouts = 0. **(c)** Proportion of teams with between 12 and 14 shutouts = $2/14 \approx 0.1429$. **(d)** Proportion of teams with at most 10 shutouts = $12/14 \approx 0.8571$. **(e)** Proportion of teams with more than 10 shutouts = $2/14 \approx 0.1429$.

47. A "typical" value would lie between 56% and 64%. An approximate "typical" value would be 62%. This value occurs approximately in the middle of the distribution.

49. **(a)** and **(b)**

Exam Scores	Frequency	Relative frequency
50–59	2	2/20 = 0.10
60–69	4	4/20 = 0.20
70–79	6	6/20 = 0.30
80–89	5	5/20 = 0.25
90–99	3	3/20 = 0.15

(c)

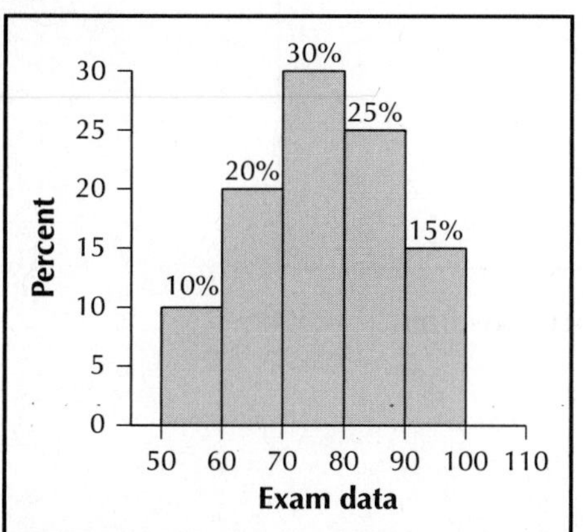

51.

	Dotplot	Histogram	Stem-and-leaf
(a) Symmetry and skewness	Appropriate to use for small ranges of data	Appropriate to use	Appropriate to use for small ranges of data
(b) Construct using pencil and paper	Easily done for small ranges of data	Easily done for small ranges of data	Easily done for small ranges of data
(c) Presentation in front of non-statisticians	Appropriate	Appropriate	May not be appropriate
(d) Maximum leeway for presentation	Appropriate	Appropriate	Appropriate

53. There are 51 observations in this data set. There are 8 variables. The only qualitative variable is **State**. The quantitative variables are tot_hhld, fam_tpc, fam_mpc, fam_fpc, nfm_tpc, nfm_lpc, ave_size.

55. From the Minitab display the maximum number of households is 10,381,206. From the original data set this is the number of households in the state of California. The histogram indicates that this is an unusually large number of households.

57. From the histogram there is one state where the average household size for the state is unusually small, and there are two states where the average household size is unusually large. From the original data the state where the average household size is unusually small is the District of Columbia with 2.3, and the two states where the average household size is unusually large is Hawaii with 3.0 and Utah with 3.2.

59. There are 961 observations in the data set and there are 22 variables.

61. Yes; fats and oils

63. From Minitab the highest cholesterol value is 2053 grams. From the original data set one whole cheesecake is the food with the highest cholesterol level.

65.

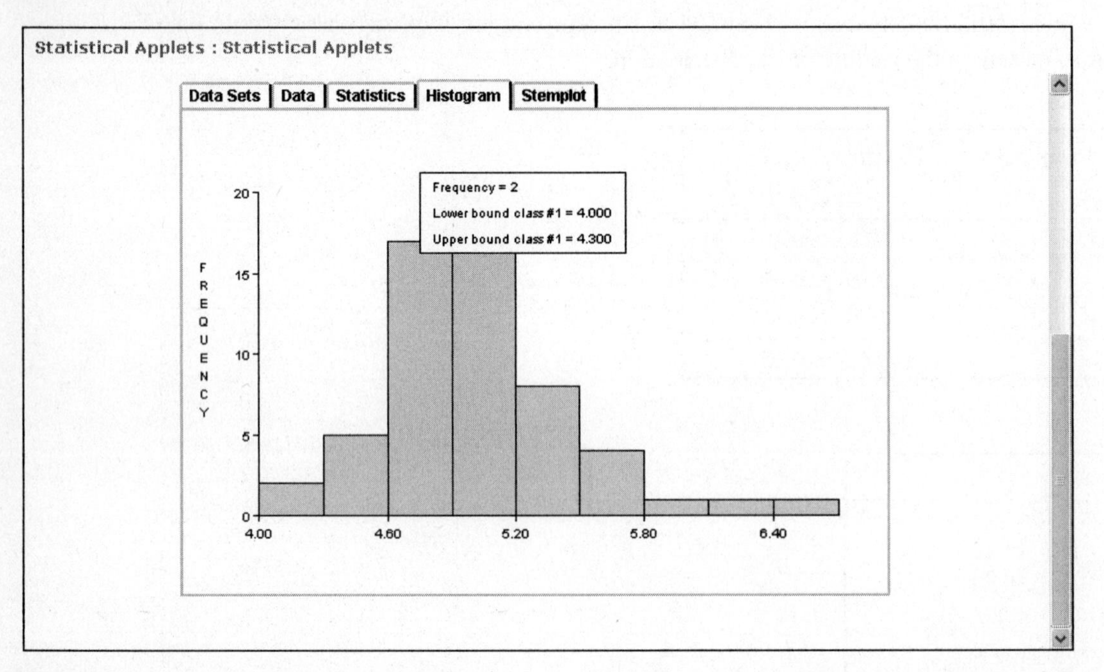

(a) The frequency for the class is 2. **(b)** Lower-class limit = 4.000; Upper class limit = 4.300

67.

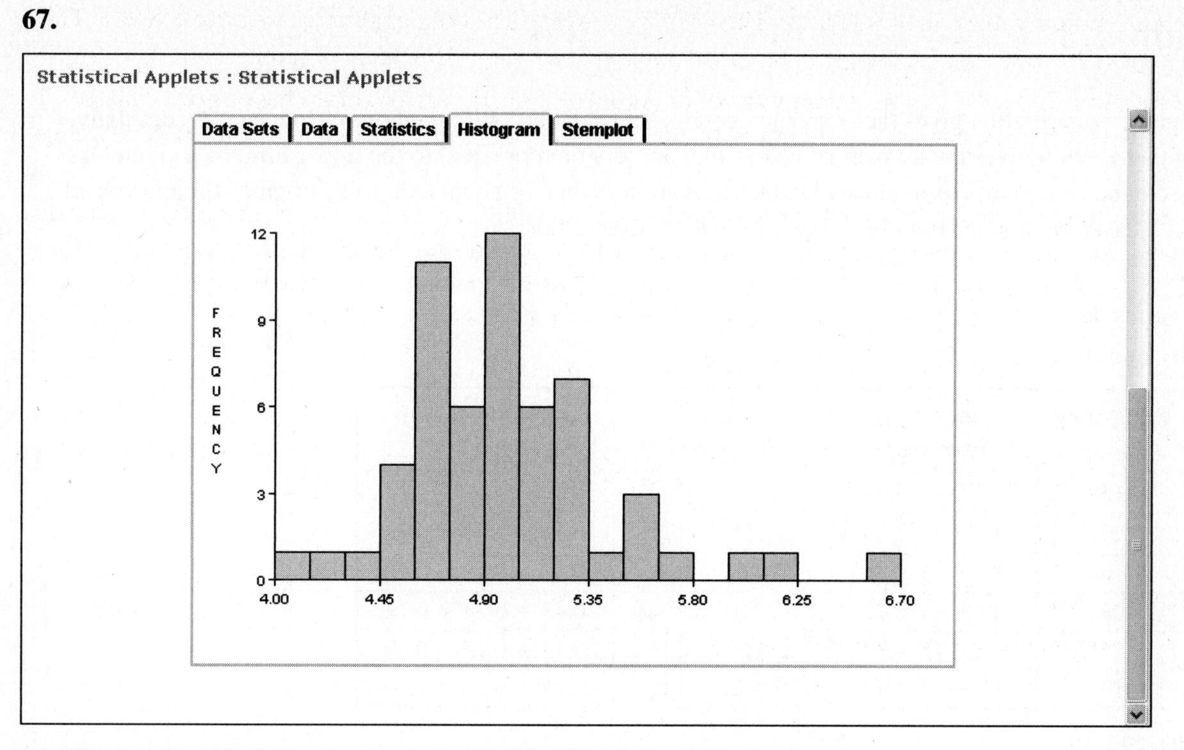

(a) The number of classes increases to 18. **(b)** The class width decreases.

69.

Statistical Applets : Statistical Applets

Data Sets | Data | Statistics | Histogram | Stemplot

☑ Split stems

```
4 | 023
4 | 5555666666666777888888999999
5 | 00000011111122222334
5 | 5558
6 | 12
6 | 6
```

(a) Now there are 6 stems. **(b)** There are still 57 leaves. **(c)** The split stem.

71. Answers will vary.

Section 2.3

1. A frequency distribution gives the frequency counts for each class (grouped or ungrouped). A cumulative frequency distribution gives the number of values that are less than or equal to the upper limit of a given class for grouped data, or it gives the number of values that are less than or equal to a given number for ungrouped data. In both cases we assume that the values are in ascending order.

3. Ogive.

5. Time series data.

7. **(a)**, **(b)**, and **(c)**

Value	Frequency	Cumulative frequency	Relative frequency	Cumulative relative frequency
1	13	13	13/100 = 0.13	0.13
2	20	13 + 20 = 33	20/100 = 0.20	0.13 + 0.20 = 0.33
3	15	33 + 15 = 48	15/100 = 0.15	0.33 + 0.15 = 0.48
4	24	48 + 24 = 72	24/100 = 0.24	0.48 + 0.24 = 0.72
5	15	72 + 15 = 87	15/100 = 0.15	0.72 + 0.15 = 0.87
6	13	87 + 13 = 100	13/100 = 0.13	0.87 + 0.13 = 1.00

9. **(a)**, **(b)**, and **(c)**

Stock prices	Frequency	Cumulative frequency	Relative frequency	Cumulative relative frequency
6.25	1	1	1/19 ≈ 0.0526	0.0526
8.75	1	1 + 1 = 2	1/19 ≈ 0.0526	0.0526 + 0.0526 = 0.1052
11.25	2	2 + 2 = 4	2/19 ≈ 0.1053	0.1052 + 0.1053 = 0.2105
13.75	1	4 + 1 = 5	1/19 ≈ 0.0526	0.2105 + 0.0526 = 0.2631
16.25	2	5 + 2 = 7	2/19 ≈ 0.1053	0.2631 + 0.1053 = 0.3684
18.75	0	7 + 0 = 7	0/19 ≈ 0.0000	0.3684 + 0.0000 = 0.3684
21.25	3	7 + 3 = 10	3/19 ≈ 0.1579	0.3684 + 0.1579 = 0.5263
23.75	3	10 + 3 = 13	3/19 ≈ 0.1579	0.5263 + 0.1579 = 0.6842
26.25	4	13 + 4 = 17	4/19 ≈ 0.2105	0.6842 + 0.2105 = 0.8947
28.75	2	17 + 2 = 19	2/19 ≈ 0.1053	0.8947 + 0.1503 = 1.0000

11. **(a)**

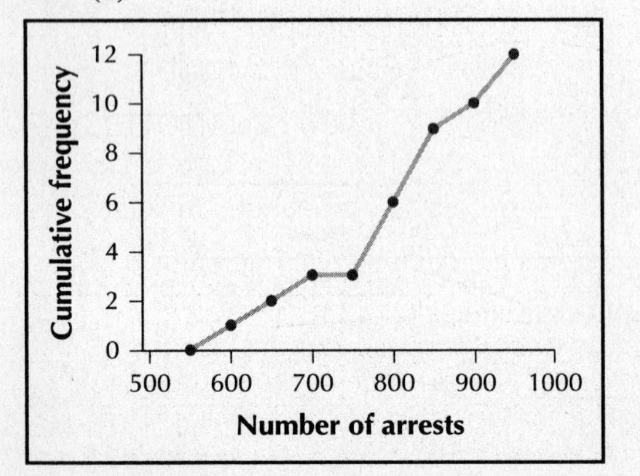

(b)

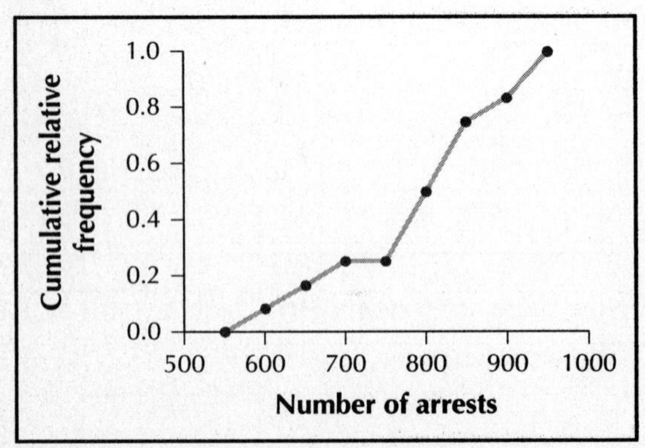

13. **(a)** The class width is the difference between the lower-class limits of successive classes. If we use the lower-class limits of the two leftmost classes, then the class width = 3.19 – 2.39 = 0.8. **(b)** 2.39

(c) The lower-class limit of the leftmost class = 2.39 – 0.8 = 1.59. The class midpoint of the leftmost class is the average of the lowest two consecutive lower-class limits, so the class midpoint of the leftmost class = (1.59 + 2.39)/2 = 3.98/2 = 1.99.

15. To change to a relative frequency ogive, just divide the frequency values along the *y* axis by 367. This will not change the points or the line segments along the graph.

17. **(a)** approximately 45% **(b)** approximately 75% **(c)** approximately 75% – 45% = 30%

19. **(a)** The class with the highest frequency is \$1.5 – \$2.99. Nine states belong to this class. **(b)** The class with the lowest frequency (not including a frequency of zero) is \$10.5 – \$11.99. This class represents the state of California.

Exports	Frequency	Cumulative frequency
0–1.49	4	4
1.5–2.99	9	4 + 9 = 13
3–4.49	6	13 + 6 = 19
4.5–5.99	0	19 + 0 = 19
6–7.49	0	19 + 0 = 19
7.5–8.99	0	19 + 0 = 19
9–10.49	0	19 + 0 = 19
10.5–11.99	1	19 + 1 = 20

(c) The states in the leftmost class include Georgia, Pennsylvania, Michigan, and South Dakota.

21.

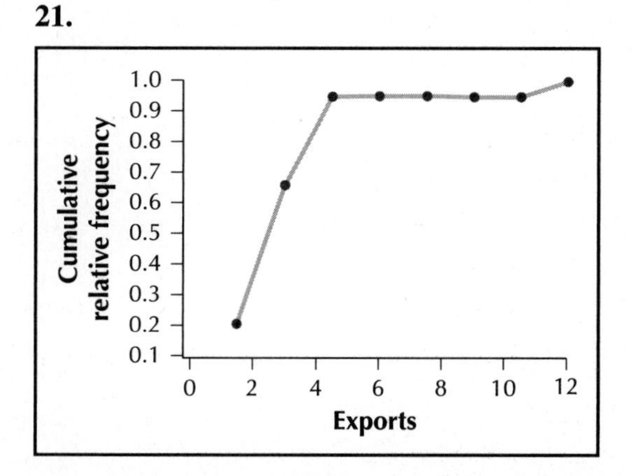

23. **(a)** 5.26% **(b)** 100% − 5.26% = 94.74%

Magnitude	Relative frequency	Cumulative relative frequency
3.7–4.09	1/57 ≈ 0.0175	0.0175
4.1–4.49	2/57 ≈ 0.0351	0.0175 + 0.0351 = 0.0526
4.5–4.89	21/57 ≈ 0.3684	0.0526 + 0.3684 = 0.4210
4.9–5.29	23/57 ≈ 0.4035	0.4210 + 0.4035 = 0.8245
5.3–5.69	6/57 ≈ 0.1053	0.8245 + 0.1053 = 0.9298
5.7–6.09	1/57 ≈ 0.0175	0.9298 + 0.0175 = 0.9473
6.1–6.49	2/57 ≈ 0.0351	0.9473 + 0.0351 = 0.9824
6.5–6.89	1/57 ≈ 0.0175	0.9824 + 0.0175 = 0.9999 ≈ 1

25. **(a)** 82.45% **(b)** 100% − 82.45% = 17.55%

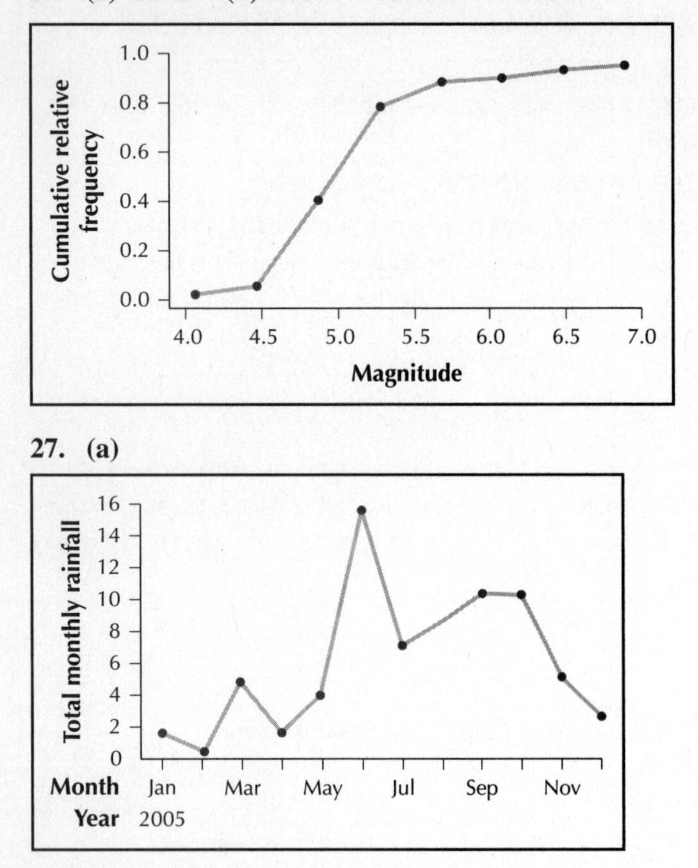

27. **(a)**

(b) Yes. It is wetter in the summer than in the winter (in Forth Lauderdale).

29. **(a)**

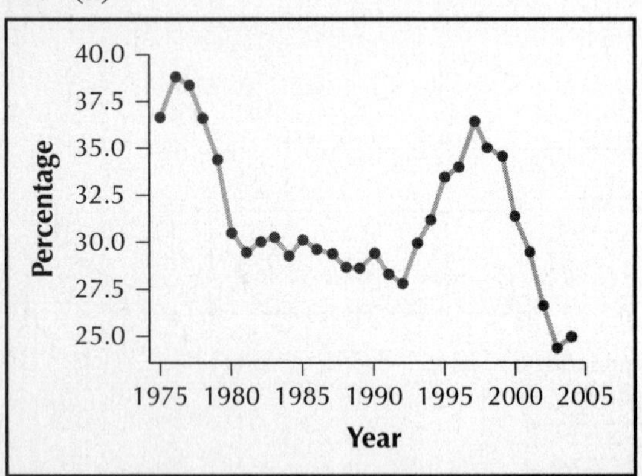

(b) The trend was downward until 1992, when it started to trend upward until 1997 and then again downward.

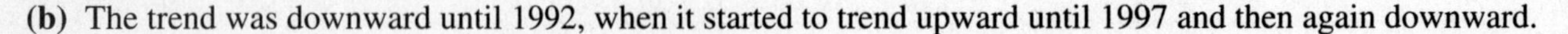

Section 2.4

1. In order to have a critical view of the presented data. If one has a critical approach one would be able to recognize the misleading aspects of misleading graphs.

3. Figure 2.31 is more effective at convincing the general public that a problem exists because the graphical representation visually reinforces the magnitude of the differences.

5. The insurance company would prefer to use Table 2.26 rather than Table 2.27. Table 2.26 gives the actual number of cars stolen.

7. **(a)** Biased distortion or embellishment, omitting the zero on the relevant scales, inaccuracy in relative lengths of bars in a bar chart. **(b)** One may use a Pareto chart or pie chart.

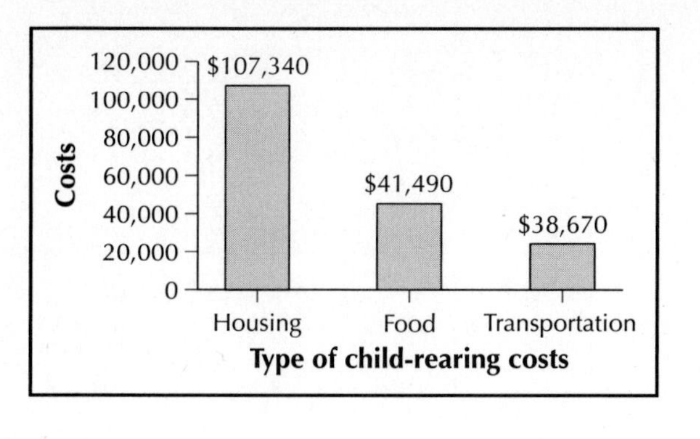

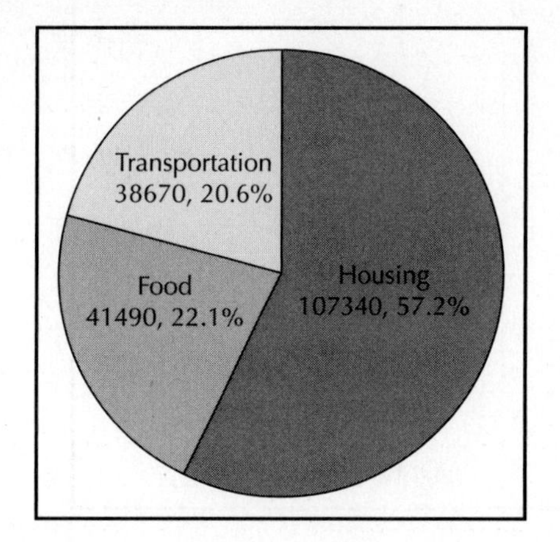

9. **(a)** The point of the graph is to indicate that the percentage (or number) of doctors who are devoted to family practice is decreasing over the years. **(b)** Inaccuracy in relative lengths of bars in the bar chart. **(c)** One may use a bar chart.

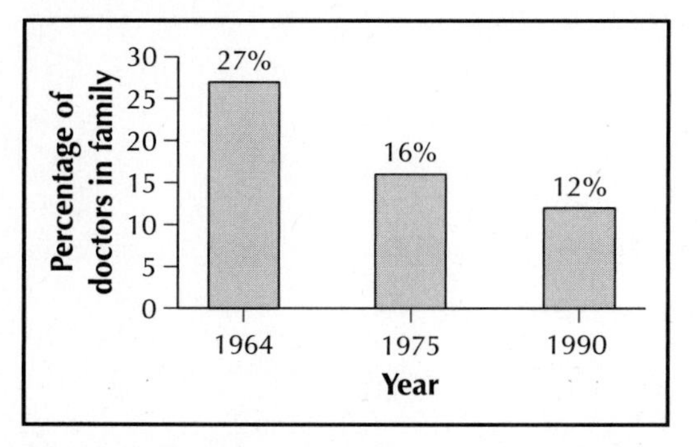

11. **(a)** Omitting the zero on the vertical scale. **(b)** Bar graph that is not misleading.

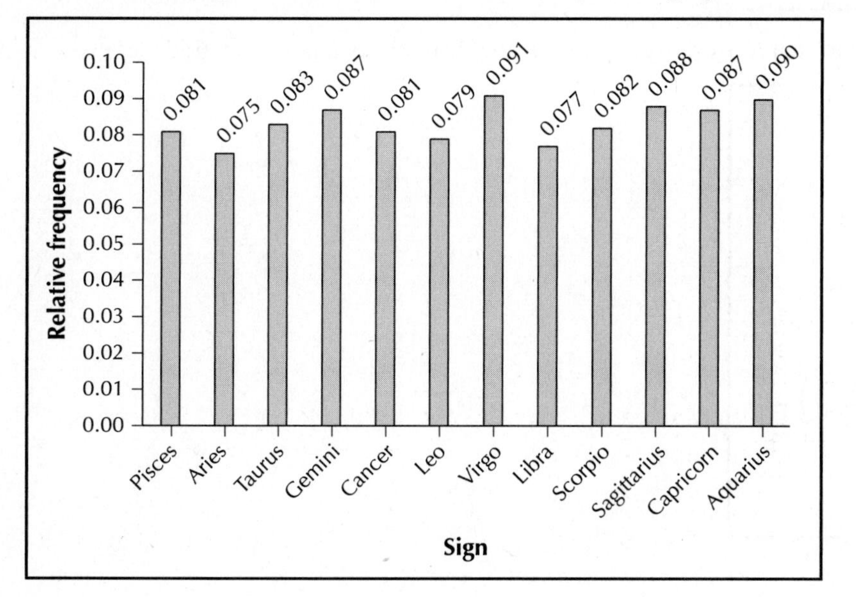

13. **(a)** Bar chart that overemphasizes the differences among the grade levels.

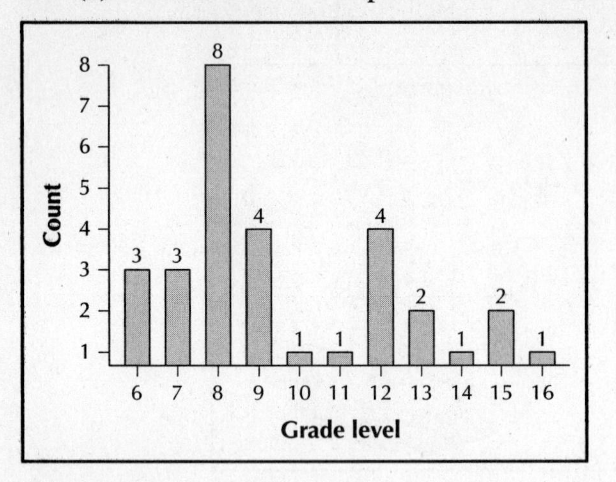

(b) Manipulating the scale. **(c)** Bar chart that underemphasizes the differences among the grade levels.

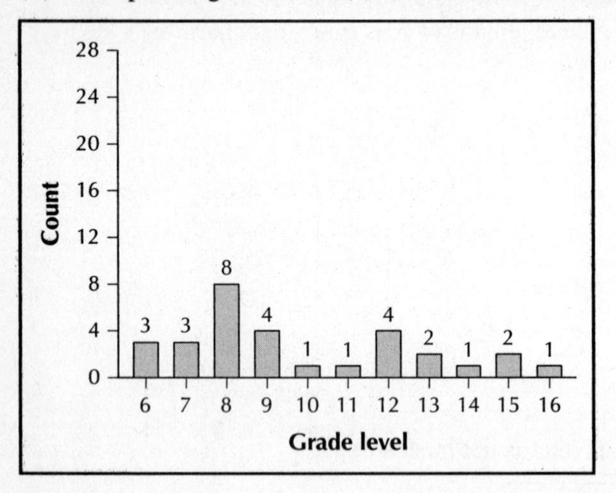

(d) Manipulating the scale. **(e)** Bar chart that most fairly represents the differences among the grade levels.

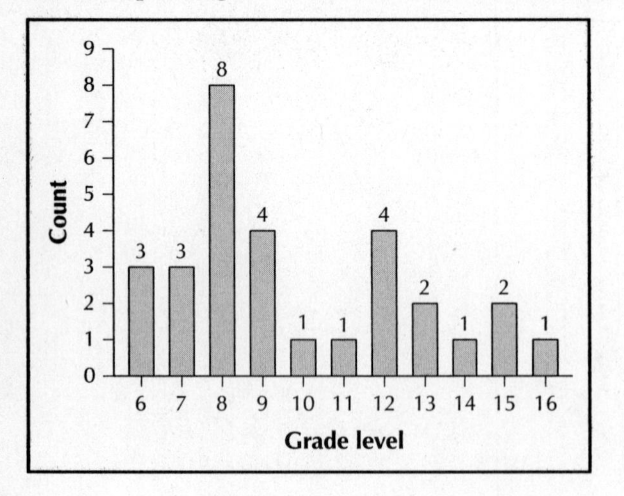

15.

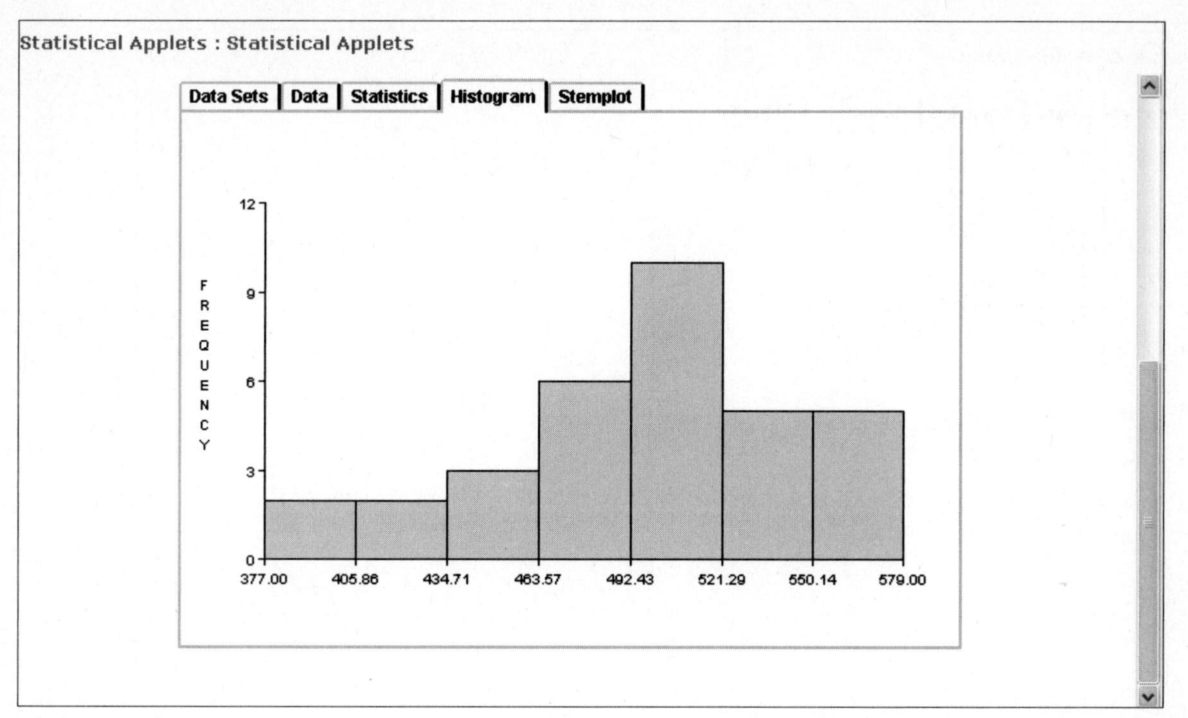
Statistical Applets : Statistical Applets
Data Sets
Data
Statistics
Histogram
Stemplot
FREQUENCY
0
3
6
9
12
377.00
405.86
434.71
463.57
492.43
521.29
550.14
579.00

17. (a)

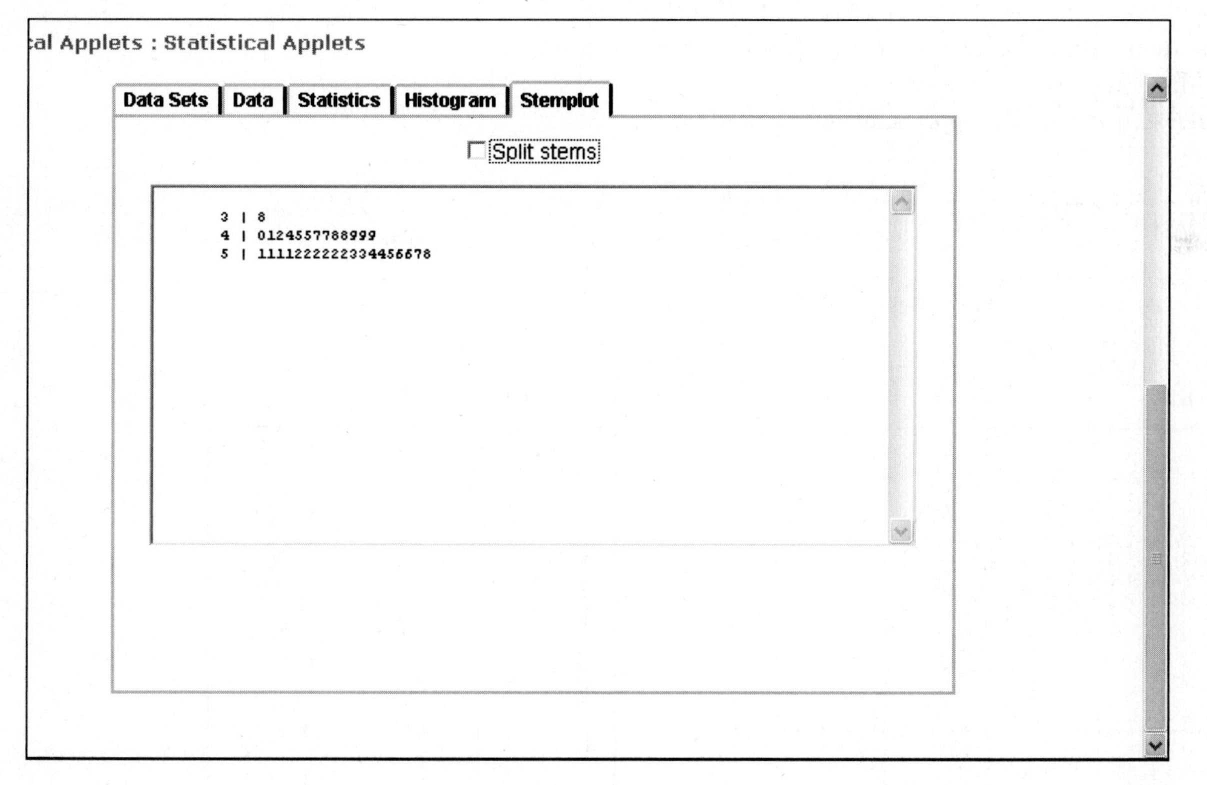
al Applets : Statistical Applets
Data Sets
Data
Statistics
Histogram
Stemplot
Split stems
3 | 8
4 | 0124557788999
5 | 11112222222334456678

(b)

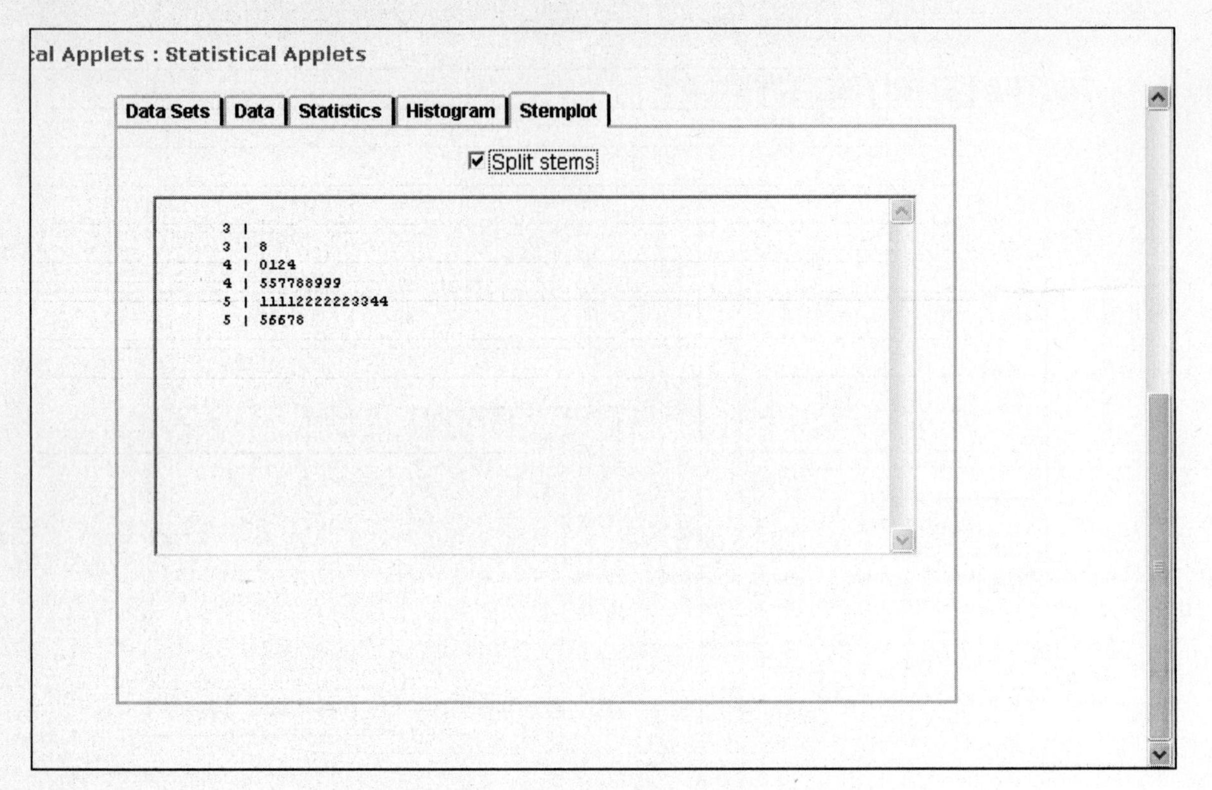

The split-stem stem-and-leaf display is preferable for this data set.

(c) The distribution is left-skewed.

(d) The description in (c) supports the left-skewed histogram.

Chapter 2 Review

1. No, because the variable is categorical.

3. The relative frequencies are expressed as percentages.

(a) and **(b)**

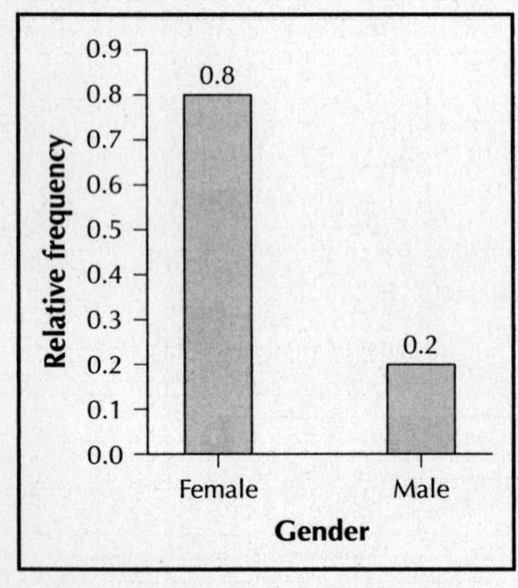

(c) and **(d)**

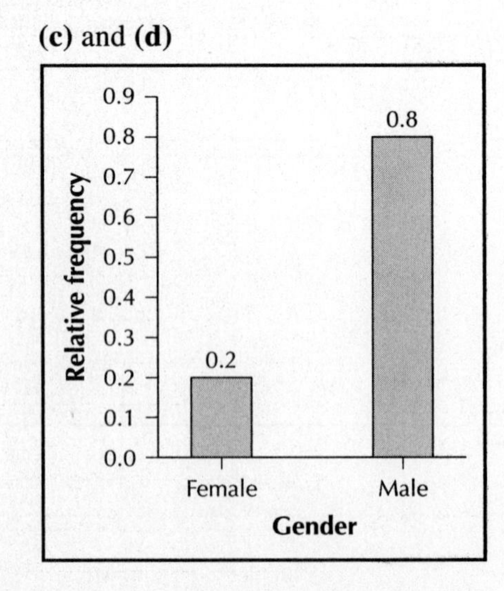

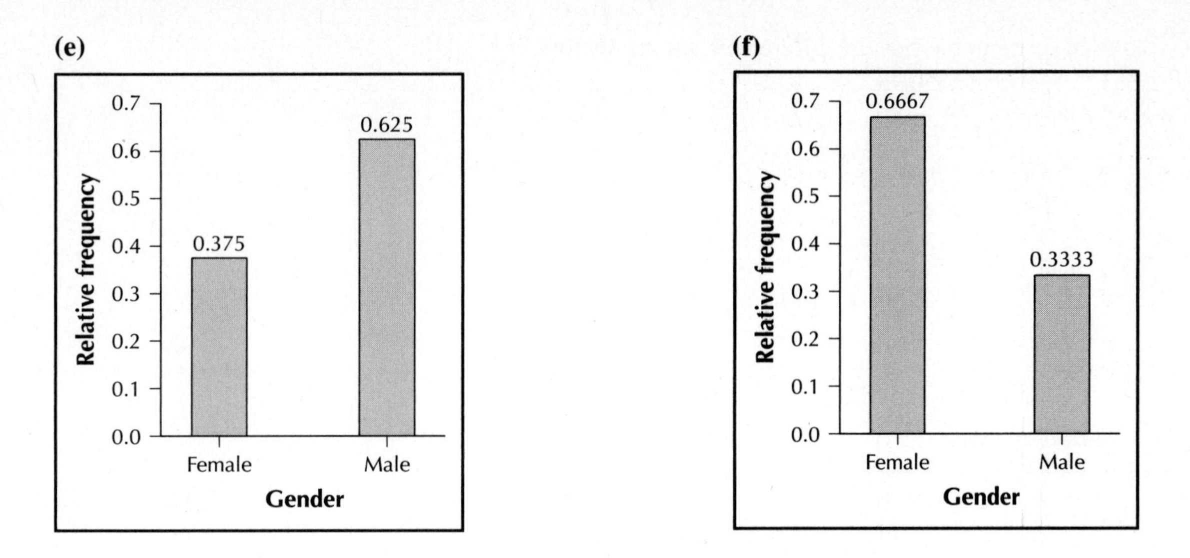

5. No, because the variable is categorical and not quantitative.

7. No. From the relative frequency histogram it will not be possible to extract the frequency counts from the class relative frequencies. This will only be possible if the original sample size is known.

9. Around a value of 62.

11. **(b)** Tending to be right-skewed.

13. The shape of the graph will not be affected. The scale will just move 0.5 units to the right.

15. **(a)** and **(b)**

Beds	Frequency	Cumulative frequency	Relative frequency	Cumulative relative frequency
0–19	1	1	$1/9 \approx 0.1111$	0.1111
20–39	4	$1 + 5 = 5$	$4/9 \approx 0.4444$	$0.1111 + 0.4444 = 0.5555$
40–59	1	$5 + 1 = 6$	$1/9 \approx 0.1111$	$0.5555 + 0.111 = 0.6666$
60–79	2	$6 + 2 = 8$	$2/9 \approx 0.2222$	$0.6666 + 0.2222 = 0.8888$
80–99	0	$8 + 0 = 8$	$0/9 = 0.0000$	$0.8888 + 0 = 0.8888$
100–119	0	$8 + 0 = 8$	$0/9 = 0.0000$	$0.8888 + 0 = 0.8888$
120–139	1	$8 + 1 = 9$	$1/9 \approx 0.1111$	$0.8888 + 0.1111 = 0.9999 \approx 1$

17.

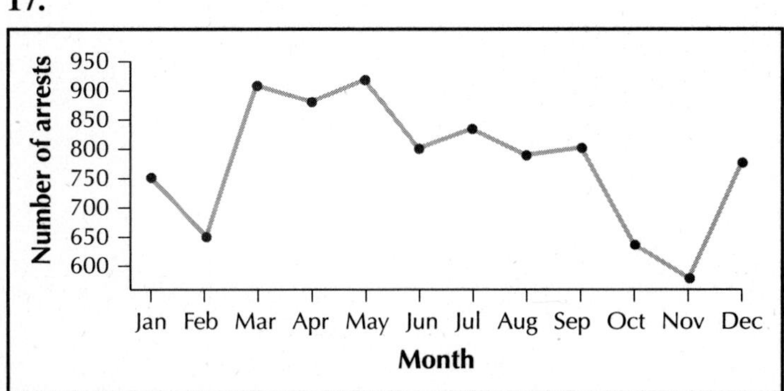

19. **(a)** Bar chart that overemphasizes the differences among sectors.

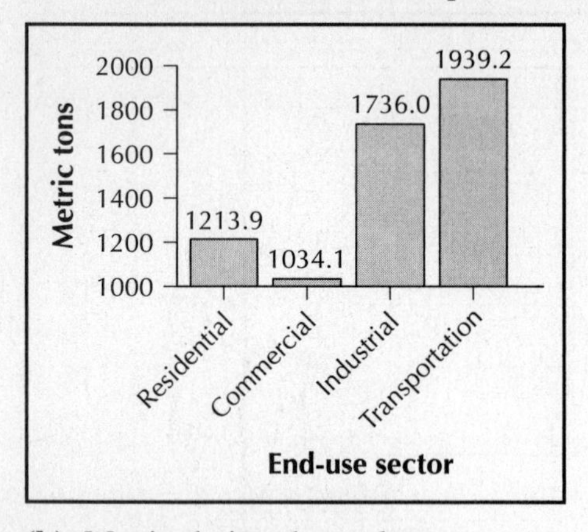

(b) Manipulating the scale.
(c) Chart that underemphasizes the differences among the responses.

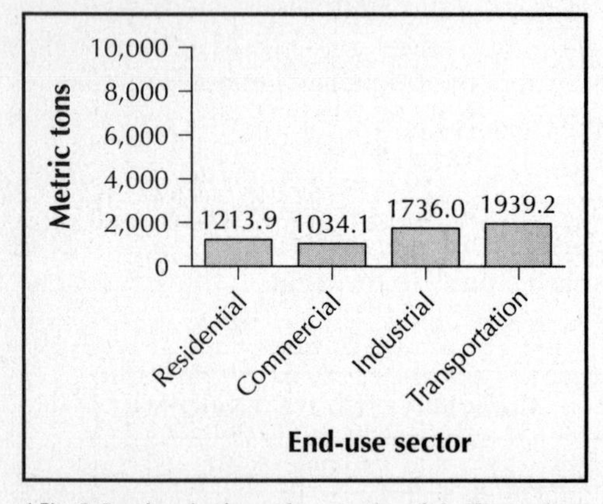

(d) Manipulating the scale. **(e)** Bar chart that most fairly represents the data.

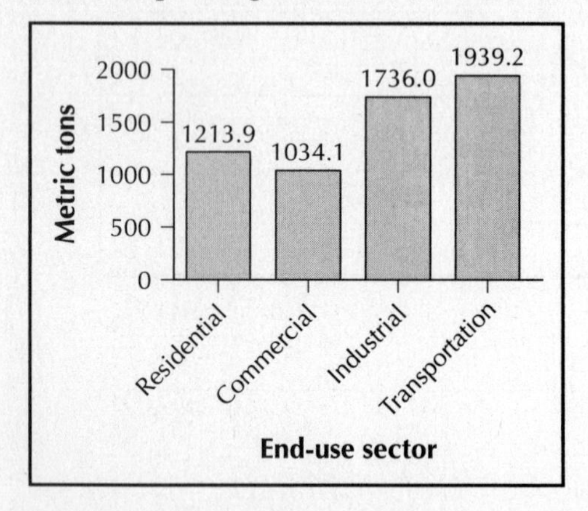

21. **(a)**

Tally for Discrete Variables: SCHOOL

SCHOOL	Percent
Brent El	14.02
Brent Mid	17.57
Brown Mid	10.88
Elm	4.39
Main	14.23
Portage	12.76
Ridge	10.04

(b) Brent Middle School, Elm

(c) **(d)**

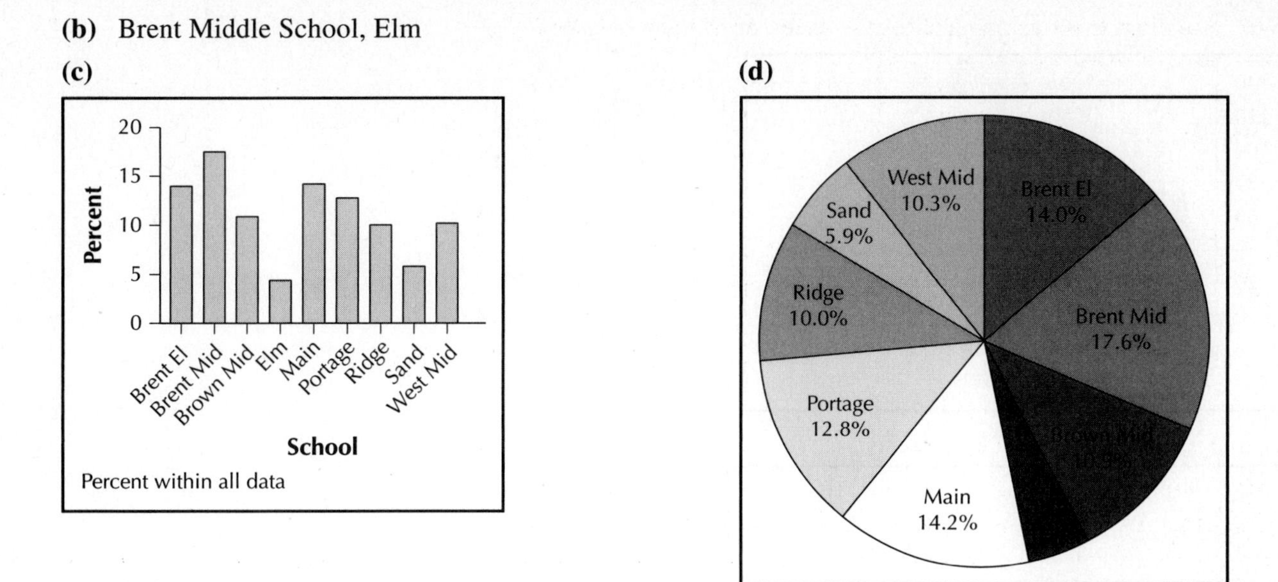

(e) Answers will vary. Answers will vary. Pie chart. Pie chart.

Chapter 2 Quiz

1. False. Stem-and-leaf displays retain the information contained in the data set.
2. True
3. True
4. Sample size
5. Bar chart
6. Frequency distribution
7. The sum of the relative frequencies must equal one.
8. The image is symmetric.
9. The distribution is right skewed.

10–13.

Vowels	Frequency	Cumulative frequency	Relative frequency	Cumulative relative frequency
a	73	73	$73/378 \approx 0.1931$	0.1931
e	130	$73 + 130 = 203$	$130/378 \approx 0.3439$	$0.1931 + 0.3439 = 0.5370$
i	74	$203 + 74 = 277$	$74/378 \approx 0.1958$	$0.5370 + 0.1958 = 0.7328$
o	74	$277 + 74 = 351$	$74/378 \approx 0.1958$	$0.7328 + 0.1958 = 0.9286$
u	27	$351 + 27 = 378$	$27/378 \approx 0.0714$	$0.9286 + 0.0714 = 1.0000$

14.

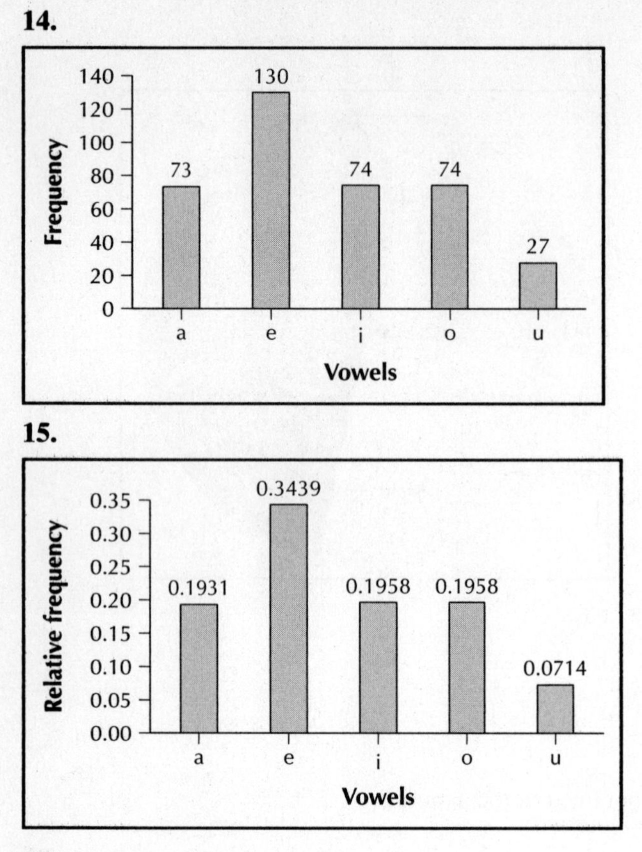

15. (chart above)

16.

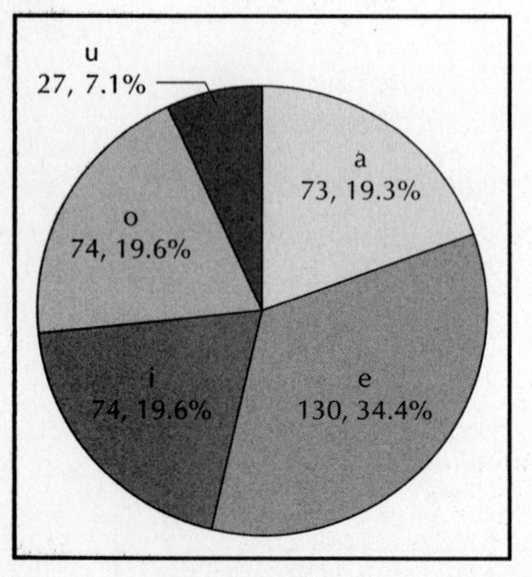

17. and **18.** Cannot be done because the variable is qualitative.

19.

Response	Frequency	Relative frequency
More serious than Pearl Harbor	412	412/618 ≈ 0.6667
Equal to Pearl Harbor	154	154/618 ≈ 0.2492
Not as serious as Pearl Harbor	29	29/618 ≈ 0.0469
Not sure	23	23/618 ≈ 0.0372

20.

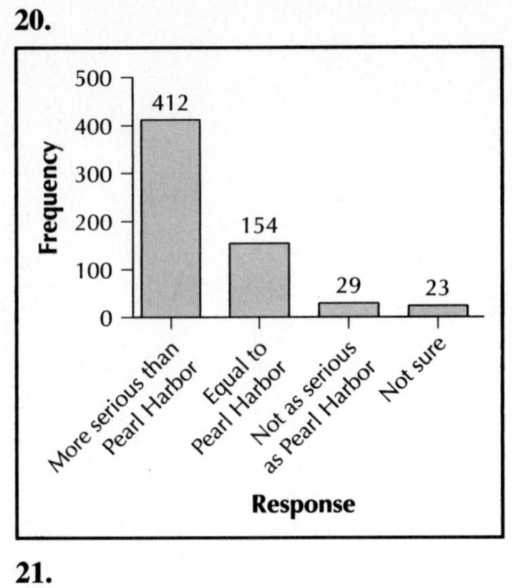

21.

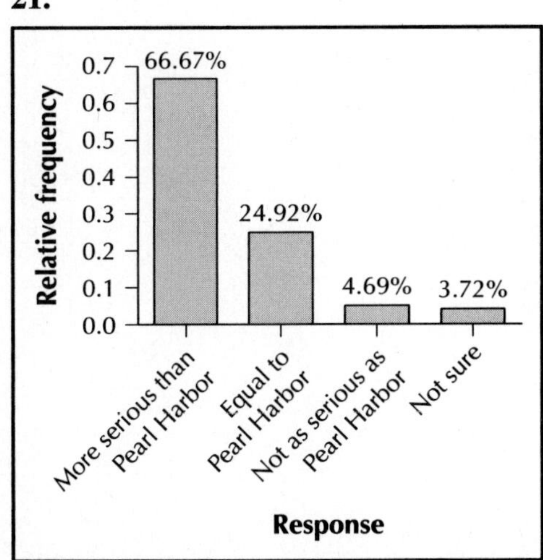

If we wanted to change the relative frequency bar chart to a frequency bar chart, the bars will not have to be redrawn. All one would need to do is to multiply the values along the *y* scale by the sample size to get the frequencies.

22.

Type of music	Relative frequency	Frequency
Hip-hop/rap	0.27	$6000 \times 0.27 = 1620$
Pop	0.23	$6000 \times 0.23 = 1380$
Rock/punk	0.17	$6000 \times 0.17 = 1020$
Alternative	0.07	$6000 \times 0.07 = 420$
Christian/gospel	0.06	$6000 \times 0.06 = 360$
R & B	0.06	$6000 \times 0.06 = 360$
Country	0.05	$6000 \times 0.05 = 300$
Techno/house	0.04	$6000 \times 0.04 = 240$
Jazz	0.01	$6000 \times 0.01 = 60$
Other	0.04	$6000 \times 0.04 = 240$

23.

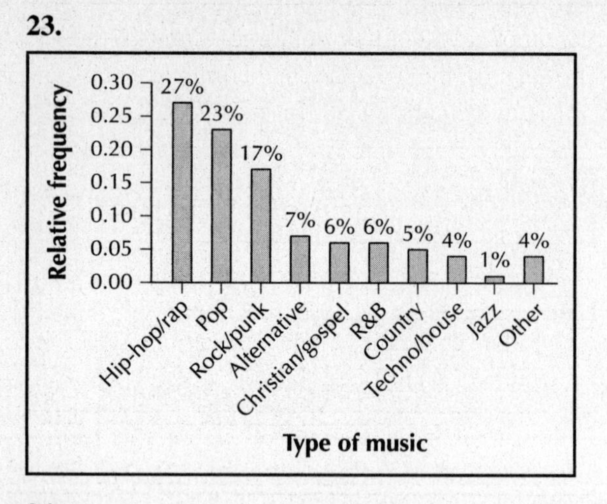

24.

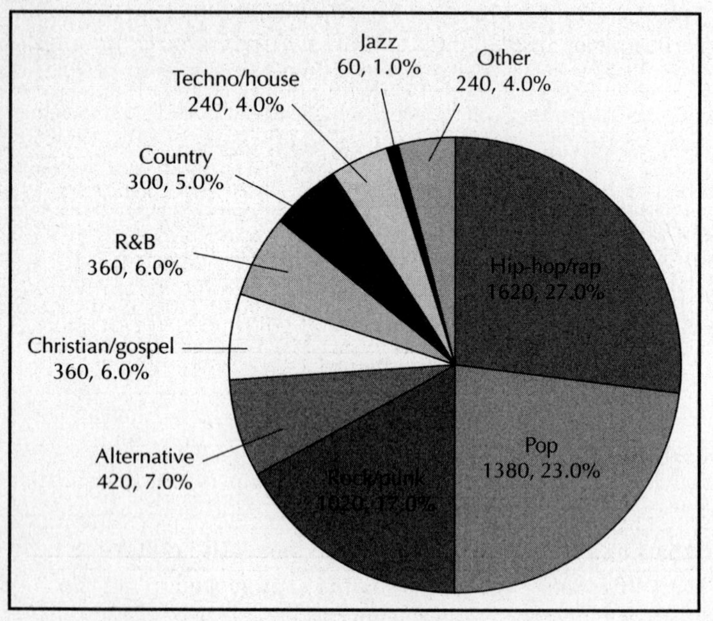

25. A frequency and relative frequency bar chart are different in that the frequencies and the relative frequencies respectively are plotted along the y axis. They are the same in that the heights of the bars are the same for both graphs.

Chapter 3

Describing Data Numerically

Section 3.1

1. A measure of center is a value that locates the center of the data set.

3. The mean is sensitive to outliers because each value in the data set is used to compute the mean. If there is an outlier in the data it will skew the value of the mean. The median is not sensitive to outliers because it is the middle value when the data are put in ascending order.

5. Data set arranged in ascending order: 17, 18, 20, 20, 24, 36

(a) $n = 6$. The sample mean is

$$\bar{x} = \frac{\sum x}{n} = \frac{17 + 18 + 20 + 20 + 24 + 36}{6} = \frac{135}{6} = 22.5.$$

(b) Since $n = 6$ is even, the median is the mean of the two data values that lie on either side of the $\left(\frac{n+1}{2}\right)^{\text{th}} = \left(\frac{6+1}{2}\right)^{\text{th}} = 3.5\text{th}$ position. That is, the median is the mean of the 3rd and 4th data values, 20 and 20.

$$\text{median} = \frac{20 + 20}{2} = \frac{40}{2} = 20.$$

(c) The mode is the value that occurs with the greatest frequency. There are two 20s. No other number occurs more than once. Therefore, the mode is 20.

7. Data set in ascending order: 5, 5, 15, 15, 16, 18, 19, 20, 24, 29

(a) $n = 10$. The sample mean is

$$\bar{x} = \frac{\sum x}{n} = \frac{5 + 5 + 15 + 15 + 16 + 18 + 19 + 20 + 24 + 29}{10}$$

$$= \frac{166}{10} = 16.6.$$

(b) Since $n = 10$ is even, the median is the mean of the two data values that lie on either side of the $\left(\frac{n+1}{2}\right)^{\text{th}} = \left(\frac{10+1}{2}\right)^{\text{th}} = 5.5\text{th}$ position. That is, the median is the mean of the 5th and 6th data values, 16 and 18.

$$\text{median} = \frac{16 + 18}{2} = \frac{34}{2} = 17.$$

(c) The mode is the data value that occurs with the greatest frequency. There are two 5s and two 15s. No other number occurs more than once. Therefore there are two modes, 5 and 15.

9. Data set in ascending order: 3, 4, 4, 5, 7, 8, 9, 9, 10, 11

(a) $n = 10$. The sample mean is

$$\bar{x} = \frac{\sum x}{n} = \frac{3 + 4 + 4 + 5 + 7 + 8 + 9 + 9 + 10 + 11}{10}$$

$$= \frac{70}{10} = 7.$$

(b) Since $n = 10$ is even, the median is the mean of the two data values that lie on either side of the $\left(\frac{n+1}{2}\right)^{\text{th}} = \left(\frac{10+1}{2}\right)^{\text{th}} = 5.5$th position. That is, the median is the mean of the 5th and 6th data values, 7 and 8.

$$\text{median} = \frac{7+8}{2} = \frac{15}{2} = 7.5.$$

(c) The mode is the value that occurs with the greatest frequency. There are two 4s and two 9s. No other numbers occur more than once. Therefore there are two modes, 4 and 9.

11. Data set in ascending order: 25, 26, 28, 30, 33, 34, 35, 35, 39, 49

(a) $n = 10$. The sample mean is

$$\bar{x} = \frac{\sum x}{n} = \frac{25 + 26 + 28 + 30 + 33 + 34 + 35 + 35 + 39 + 49}{10}$$
$$= \frac{334}{10} = 33.4.$$

(b) Since $n = 10$ is even, the median is the mean of the two data values that lie on either side of the $\left(\frac{n+1}{2}\right)^{\text{th}} = \left(\frac{10+1}{2}\right)^{\text{th}} = 5.5$th position. That is, the median is the mean of the 5th and 6th data values, 33 and 34.

$$\text{median} = \frac{33+34}{2} = \frac{67}{2} = 33.5.$$

(c) The mode is the value that occurs with the highest frequency. There are two 35s. No other number occurs more than once. Therefore, the mode is 35.

13. Data set in ascending order: 497, 500, 501, 501, 505, 515, 518, 522

(a) $n = 8$. The sample mean is

$$\bar{x} = \frac{\sum x}{n} = \frac{497 + 500 + 501 + 501 + 505 + 515 + 518 + 522}{8}$$
$$= \frac{4059}{8} = 507.38.$$

(b) Since $n = 8$ is even, the median is the mean of the two data values that lie on either side of the $\left(\frac{n+1}{2}\right)^{\text{th}} = \left(\frac{8+1}{2}\right)^{\text{th}} = 4.5$th position. That is, the median is the mean of the 4th and 5th data values, 501 and 505.

$$\text{median} = \frac{501+505}{2} = \frac{1006}{2} = 503.$$

(c) The mode is the data value that occurs with the highest frequency. There are two 501s. No other number occurs more than once. Therefore, the mode is 501.

15. (a) Mode = English. This does not mean that most of the students are majoring in English. English occurs with the highest frequency but only accounts for 20% of the data set. **(b)** Since this is qualitative data, the idea of the mean or median does not make sense. **(c)** Economics could not be the most popular major because it does not occur with the highest frequency.

17. Data set times 5: \$84.75, \$109.75, \$130.00, \$134.75, \$139.95

(a) $n = 5$. The sample mean is

$$\bar{x} = \frac{\sum x}{n} = \frac{\$84.75 + \$109.75 + \$130.00 + \$134.75 + \$139.95}{5}$$
$$= \frac{\$599.20}{5} = \$119.84.$$

(b) This mean of \$119.84 is \$23.968 × 5. **(c)** If a set of values is multiplied by a positive constant, then the mean for this new data set will equal the mean of the original data set times the constant.

19. Since the age of the car in 2009 is 2009 Model year, the data values for the age of the car in ascending order are: 2, 2, 2, 2, 2, 2, 2, 3, 3, 3, 3, 4, 4, 4, 5, 5, 5, 5, 6, 6

(a) $n = 20$. The mode is the data value that occurs with the highest frequency. There are seven 2s. No other data value occurs more than four times. Therefore, the mode is 2.

(b) The sample mean is

$$\bar{x} = \frac{\sum x}{n} = \frac{7(2) + 4(3) + 3(4) + 4(5) + 2(6)}{20} = \frac{70}{20} = 3.5.$$

Since $n = 20$ is even, the median is the mean of the two data values that lie on either side of the $\left(\frac{n+1}{2}\right)^{\text{th}} = \left(\frac{20+1}{2}\right)^{\text{th}} = 10.5$th position. That is, the median is the mean of the 10th and 11th data values, 3 and 3.

$$\text{median} = \frac{3+3}{2} = \frac{6}{2} = 3.$$

21. **(a)** Mean < Median < Mode **(b)** Mode < Median < Mean **(c)** Mean = Median = Mode

23. Since the mean is larger than the median, the implication is that the distribution is right-skewed.

25. The mean, median, and mode will all be halved. Each statistic will be affected equally.

27. **(a)** Since $n = 65$ is odd, the median is the data value in the $\left(\frac{n+1}{2}\right)^{\text{th}} = \left(\frac{65+1}{2}\right)^{\text{th}} = 33$rd position. Therefore, the median pulse rate for males is 73. **(b)** Since $n = 65$ is odd, the median is the data value in the $\left(\frac{n+1}{2}\right)^{\text{th}} = \left(\frac{65+1}{2}\right)^{\text{th}} = 33$rd position. Therefore, the median pulse rate for females is 76. **(c)** Females have a higher median pulse rate. Yes it does agree.

29. **(a)** Since the mean is $\bar{x} = \frac{\sum x}{n}$, a decrease in the fastest (largest) pulse rate for males would result in a decrease in $\sum x$. Since the number of observations would not change, the mean pulse rate for males would decrease. **(b)** Since the median pulse rate is the middle value in the data set, a decrease in the fastest pulse rate for males would not affect the median. Thus, the median pulse rate for males would be unchanged. **(c)** Since the mode is not the fastest pulse rate or the second fastest pulse rate, a decrease in the fastest pulse rate would not affect the mode men's pulse rate. Thus, the modal pulse rate for males would be unchanged.

31. Total earnings for college graduate = $60,000 × 40 = $2,400,000. This as a percentage of Alex Rodriguez's 2004 salary = (2400000/2172688) ×100% = 11.0462%.

33. No. The measure of center for this sample will not be representative for all U.S. states because all values in the sample are from northeastern states and from states that have the highest participation rates.

35. **(a)** Since $n = 65$ is odd, the median is the data value in the $\left(\frac{n+1}{2}\right)^{\text{th}} = \left(\frac{65+1}{2}\right)^{\text{th}} = 33$rd position. Thus, the median body temperature for males is 98.1. **(b)** Since $n = 65$ is odd, the median is the data value in the $\left(\frac{n+1}{2}\right)^{\text{th}} = \left(\frac{65+1}{2}\right)^{\text{th}} = 33$rd position. Thus, median body temperature for females is 98.4.

(c) Females; yes.

37. **(a)** Decrease because all values are used in the computation of the mean. A smaller data value will thus decrease the value of the mean. **(b)** Remain unchanged because the median is not affected by an outlying value. **(c)** Remain unchanged because the mode is not affected by an outlying value.

39. Solutions will vary. The data set should be left-skewed.

41. Solutions will vary. The original data set should be symmetric. Then increasing the largest number in the data set will result in the median remaining the same and the mean increasing.

43. The mean is the sum of the three numbers divided by 3. When the largest number is increased, the sum of the three numbers is increased. Therefore, the mean is increased. However, the median is the middle value of the three numbers, which is unaffected by an increase in the largest value.

Section 3.2

1. The deviation for a data value is obtained by subtracting the mean of the data set from the value. That is, the deviation for a data value gives the distance the value is from the mean.

3. Less variability in a data set is better than more variability. Less variability will lead to more precise estimates, higher confidence in conclusions, and greater discerning power.

5. To calculate the variance using the computational formula we first need to find $\sum x$ and $\sum x^2$.

x	x^2
3	9
6	36
6	36
9	81
$\sum x = 24$	$\sum x^2 = 162$

$n = 4$. Therefore, the sample variance is

$$s^2 = \frac{\sum x^2 - \left(\sum x\right)^2/n}{n-1} = \frac{162 - (24)^2/4}{4-1} = 6.$$

To calculate the variance using the other formula we first need to find the sample mean $\bar{x}$.

$$\bar{x} = \frac{\sum x}{n} = \frac{3+6+6+9}{4} = \frac{24}{4} = 6$$

x	$(x - \bar{x})$	$(x - \bar{x})^2$
3	−3	9
6	0	0
6	0	0
9	3	9
		$\sum(x - \bar{x})^2 = 18$

Therefore, the variance is

$$s^2 = \frac{\sum(x - \bar{x})^2}{n-1} = \frac{18}{4-1} = 6.$$

7. Range = largest value − smallest value = 25 − 0 = 25.

9. The sample standard deviation is $s = \sqrt{s^2} = \sqrt{82.5} \approx 9.08$.

11. To calculate the variance using the computational formula we first need to find $\sum x$ and $\sum x^2$.

x	x^2
−5	25
−10	100
−15	225
−20	400
$\sum x = -50$	$\sum x^2 = 750$

$n = 4$. Therefore, the sample variance is

$$s^2 = \frac{\sum x^2 - \left(\sum x\right)^2/n}{n-1} = \frac{750 - (-50)^2/4}{4-1} \approx 41.67.$$

To calculate the variance using the other formula we first need to find the sample mean $\bar{x}$.

$$\bar{x} = \frac{\sum x}{n} = \frac{(-5) + (-10) + (-15) + (-20)}{4} = \frac{-50}{4} = -12.5$$

x	$(x - \bar{x})$	$(x - \bar{x})^2$
−5	7.5	56.25
−10	2.5	6.25
−15	−2.5	6.25
−20	−7.5	56.25
$\Sigma(x - \bar{x})^2 = 125$		

Therefore, the variance is

$$s^2 = \frac{\sum(x - \bar{x})^2}{n-1} = \frac{125}{4-1} \approx 41.67.$$

13. Range = largest value − smallest value = 49 − 25 = 24 years.

15. To calculate the variance we first need to find Σx and Σx^2.

x	x^2
36	1296
24	576
20	400
20	400
18	324
17	289
$\Sigma x = 135$	$\Sigma x^2 = 3285$

$n = 6$. Therefore, the sample variance is

$$s^2 = \frac{\sum x^2 - \left(\sum x\right)^2/n}{n-1} = \frac{3285 - (135)^2/6}{6-1} = 49.5 \text{ mpg squared.}$$

The sample standard deviation is $s = \sqrt{s^2} = \sqrt{49.5} \approx 7.04$ mpg.

17. To calculate the variance we first need to find Σx and Σx^2.

x	x^2
28.5	812.25
33.2	1102.24
33.9	1149.21
34.3	1176.49
34.4	1183.36
34.5	1190.25
34.6	1197.16
34.7	1204.09
$\Sigma x = 268.1$	$\Sigma x^2 = 9015.05$

$n = 8$. Therefore, the sample variance is

$$s^2 = \frac{\sum x^2 - \left(\sum x\right)^2/n}{n-1} = \frac{9015.05 - (268.1)^2/8}{8-1} \approx 4.336 \text{ years squared.}$$

The sample standard deviation is $s = \sqrt{s^2} = \sqrt{4.336} \approx 2.082$ years. The standard deviation is more easily understood because it has the same units as the variable *age*.

19.

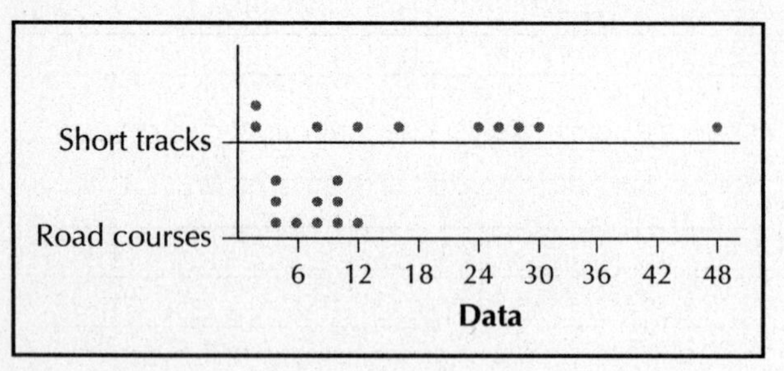

The Short Tracks variable has the greater variability.

21. Range for the Super Speedways variable = largest value − smallest value = 29 − 5 = 24. Range for the Road Courses variable = largest value − smallest value = 11 − 3 = 8.
Based on these ranges it would imply that the Super Speedways variable has the larger variability.

23. Range for the zooplankton variable = largest value − smallest value = 0.26 − (−6.60) = 6.86. Range for the phytoplankton variable = largest value − smallest value = 10.61 − 0.65 = 9.96. **(a)** The phytoplankton variable has the larger range. **(b)** The range says that the phytoplankton variable has the greater variability.

25. Mean number of cases sold

$$\bar{x} = \frac{\sum x}{n} = \frac{1929 + 1385 + 811 + 541 + 537 + 536 + 530 + 220 + 180 + 97}{10}$$
$$= \frac{6766}{10} = 676.6 \text{ million.}$$

Number of cases (in millions)	Deviations (in millions)
1929	1929 − 676.6 = 1252.4
1385	1385 − 676.6 = 708.4
811	811 − 676.6 = 134.4
541	541 − 676.6 = −135.6
537	537 − 676.6 = −139.6
536	536 − 676.6 = −140.6
530	530 − 676.6 = −146.6
220	220 − 676.6 = −456.6
180	180 − 676.6 = −496.6
97	97 − 676.6 = −579.6

27. **(a)** By removing the value of 1929 from the sample, it would affect (reduce) the values of the range, variance, and standard deviations. **(b)** New range for the number of cases = largest value − smallest value = 1385 − 97 = 1288. To calculate the variance we first need to find $\sum x$ and $\sum x^2$.

x	x^2
1385	1,918,225
811	657,721
541	292,681
537	288,369
536	287,296
530	280,900
220	48,400
180	32,400
97	9,409
$\sum x = 4837$	$\sum x^2 = 3,815,401$

$n = 9$. Therefore, the new sample variance is

$$s^2 = \frac{\sum x^2 - \left(\sum x\right)^2/n}{n-1} = \frac{3,815,401 - (4837)^2/10}{9-1} = 151,973.$$

The new standard deviation for the number of cases is

$$s = \sqrt{s^2} = \sqrt{151,973} \approx 390.$$

29. For Colony A:
To calculate the variance we first need to find $\sum x$ and $\sum x^2$.

x	x^2
109	11,881
120	14,400
94	8,836
61	3,721
72	5,184
134	17,956
94	18,836
113	2,769
111	12,321
106	11,236
$\sum x = 1014$	$\sum x^2 = 107,140$

$n = 10$. Therefore, the sample variance is

$$s^2 = \frac{\sum x^2 - \left(\sum x\right)^2/n}{n-1} = \frac{107,140 - (1014)^2/10}{10-1} \approx 480.04.$$

For Colony B:

To calculate the variance we first need to find $\sum x$ and $\sum x^2$.

x	x^2
148	21,904
110	12,100
110	12,100
97	9,409
136	18,496
115	13,225
101	10,201
158	24,964
67	4,489
114	12,996
$\sum x = 1156$	$\sum x^2 = 139{,}886$

$n = 10$. Therefore, the sample variance is

$$s^2 = \frac{\sum x^2 - \left(\sum x\right)^2/n}{n-1} = \frac{139{,}886 - (1156)^2/10}{10-1} \approx 694.49.$$

(a) Colony B has the larger variance. **(b)** Colony B has the greater variability according to the variance. Yes, it concurs. **(c)** The standard deviation for Colony B will be greater because it has the larger variance.

31. Range for SAT I Verbal = largest value − smallest value = 522 − 497 = 25.

Range for SAT I Math = largest value − smallest value = 523 − 499 = 24.

For SAT I Verbal: To calculate the variance we first need to find $\sum x$ and $\sum x^2$.

x	x^2
497	247,009
515	265,225
518	268,324
501	251,001
522	272,284
505	255,025
501	251,001
500	250,000
$\sum x = 4{,}059$	$\sum x^2 = 2{,}060{,}069$

$n = 8$. Therefore, the sample variance is

$$s^2 = \frac{\sum x^2 - \left(\sum x\right)^2/n}{n-1} = \frac{2{,}060{,}069 - (4059)^2/8}{8-1} = 90.55$$

The sample standard deviation is $s = \sqrt{s^2} = \sqrt{90.55} \approx 9.52$.

For SAT I Math:

To calculate the variance we first need to find $\sum x$ and $\sum x^2$.

x	x^2
510	260,100
515	265,225
523	273,529
574	264,196
521	271,441
501	251,001
502	252,004
499	249,001
$\sum x = 4{,}085$	$\sum x^2 = 2{,}086{,}497$

$n = 8$. Therefore, the sample variance is

$$s^2 = \frac{\sum x^2 - \left(\sum x\right)^2/n}{n-1} = \frac{2{,}086{,}497 - (4085)^2/8}{8-1} = 84.8392857143 \approx 84.84.$$

The sample standard deviation is $s = \sqrt{s^2} = \sqrt{84.8392857143} \approx 9.21$.

These values support the judgment in the previous exercise because they are similar in magnitude.

33. **(a)** The typical distance from the mean for the WMU team is about 6 inches. **(b)** An estimate for the typical distance from the mean for the NCU team would be 2 inches, the average for the absolute deviations from the mean. Note: the absolute deviations from the mean are 3, 2, 1, 1, 3.

35. **(a)** Changing the value of 66 to 62 in the data set for NCU will increase the range, variance, and standard deviation for the data set. **(b)** New range = largest value − smallest value = 72 − 62 = 10. To calculate the variance we first need to find $\sum x$ and $\sum x^2$.

x	x^2
62	3844
67	4489
70	4900
70	4900
72	5184
$\sum x = 341$	$\sum x^2 = 23{,}317$

$n = 5$. Therefore, the new sample variance is

$$s^2 = \frac{\sum x^2 - \left(\sum x\right)^2/n}{n-1} = \frac{23{,}317 - (341)^2/5}{5-1} = 15.2.$$

The new standard deviation is $s = \sqrt{s^2} = \sqrt{15.2} \approx 3.9$. Yes, the judgment in (a) was supported.

37. About 300.

39. To calculate the variance we first need to find $\sum x$ and $\sum x^2$.

x	x^2
462	213,444
621	385,641
104	10,816
907	822,649
293	85,849
186	34,596
470	220,900
136	18,496
675	455,625
114	12,996
$\sum x = 3968$	$\sum x^2 = 2{,}261{,}012$

$n = 10$. Therefore, the standard deviation is

$$s = \sqrt{\frac{\sum x^2 - \left(\sum x\right)^2/n}{n-1}} = \sqrt{\frac{2{,}261{,}012 - (3968)^2/10}{10-1}} \approx 276.2.$$

(a) Difference $= 300 - 276.2 = 23.8$. **(b)** The frequency counts for the syllables typically differ from the mean of 396.8 by 276.2.

41. **(a)** The same as given in the descriptive statistics output. **(b)** Standard deviation for female pulse rate $= 8.11$. **(c)** Very close with a difference of 0.11. **(d)** Standard deviation for male pulse rate $= 5.875$. **(e)** The same as given in the descriptive statistics output. **(f)** Female. This agreed with the earlier surmise as to the relative sizes of the standard deviations. **(g)** Females have the greater variability in pulse rates.

43. **(a)** Males: Mean body temperature $= 98.105$
Females: Mean body temperature $= 98.394$
The estimates in problem 43 for the means were very close.

(b) Males: Maximum body temperature $= 99.5$
Minimum body temperature $= 96.3$
Range for the body temperature $= 3.2$
Females: Maximum body temperature $= 100.8$
Minimum body temperature $= 96.4$
Range for the body temperature $= 4.4$
The estimates in problem 43 for the minimum, maximum, and range are the same as the actual values.

(c) Females: Standard deviation $= 0.743$
Males: Standard deviation $= 0.699$

(d) The standard deviation for the females is larger than that for the males, which indicates that the variability in body temperature is larger for the females.

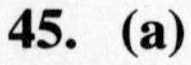

45. **(a)**

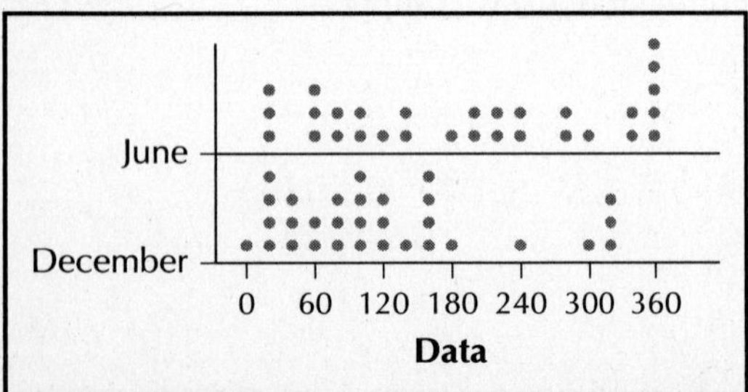

(b) Based on the above graphs it seems that the June data has more variability.

47. **(a)** The range will remain the same because the value of 360 is smaller than the maximum value and larger than the minimum value. **(b)** and **(c)** Both the variance and standard deviation will increase because these statistics use all the values in their computations. The value of 360 will contribute more than 180 to these statistics.

49. Answers will vary.

51. Answers will vary.

Section 3.3

1. These formulas will only provide estimates because we will not know the exact data values. We will estimate the data values in each class with the class midpoint. These midpoints will then be used in the computations.

3. The quantity that $\hat{\mu}$ and $\bar{x}$ estimate is the population mean.

5. $$\bar{x}_w = \frac{\sum w_i x_i}{\sum w_i} = \frac{w_1x_1 + w_2x_2 + w_3x_3}{w_1 + w_2 + w_3} = \frac{(0.25)(50) + (0.40)(80) + (0.35)(70)}{0.25 + 0.40 + 0.35} = \frac{69}{1.0} = 69$$

7. $$\bar{x}_w = \frac{\sum w_i x_i}{\sum w_i} = \frac{w_1x_1 + w_2x_2 + w_3x_3 + w_4x_4 + w_5x_5}{w_1 + w_2 + w_3 + w_4 + w_5}$$
$$= \frac{(3)(3.00) + (3)(2.5) + (3)(3.5) + (3)(4.0) + (4)(3.0)}{3 + 3 + 3 + 3 + 4} = \frac{51}{16} \approx 3.2$$

9. $$\bar{x}_w = \frac{\sum w_i x_i}{\sum w_i} = \frac{w_1x_1 + w_2x_2 + w_3x_3 + w_4x_4 + w_5x_5}{w_1 + w_2 + w_3 + w_4 + w_5}$$
$$= \frac{(14)(1.0) + (14)(1.5) + (15)(2.5) + (15)(3.0) + (16)(3.5)}{14 + 14 + 15 + 15 + 16} = \frac{173.5}{74} = 2.3446$$

11.

Class limits	Midpoints
10–12.49	$\frac{10 + 12.50}{2} = 11.25$
12.50–14.99	$\frac{12.50 + 15}{2} = 13.75$
15–17.49	$\frac{15 + 17.50}{2} = 16.25$
17.50–19.99	$\frac{17.50 + 20}{2} = 18.75$

13. $\sum m_i f_i = (5)(10) + (10)(20) + (15)(20) + (20)(10) + (25)(10) = 1000$

$N = \sum f_i = 10 + 20 + 20 + 10 + 10 = 70$

$$\text{Estimated mean } \hat{\mu} = \frac{\sum m_i f_i}{N} = \frac{1000}{70} \approx 14.2857$$

15. $\sum m_i f_i = (50)(20) + (150)(30) + (250)(30) + (450)(40) + (750)(30) + (1150)(20) = 76{,}500$

$N = \sum f_i = 20 + 30 + 30 + 40 + 30 + 20 = 170$

$$\text{Estimated mean } \hat{\mu} = \frac{\sum m_i f_i}{N} = \frac{76{,}500}{170} = 450$$

17. From Exercise 14, $\hat{\mu} = 6.25$ and $N = 40$.

Midpoint (m_i)	Frequency (f_i)	$\hat{\mu}$	$m_i - \hat{\mu}$	$(m_i - \hat{\mu})^2 \cdot f_i$
−10	3	6.25	−16.25	792.1875
−5	2	6.25	−11.25	253.125
0	5	6.25	−6.25	195.3125
5	12	6.25	−1.25	18.75
10	8	6.25	3.75	112.5
15	10	6.25	8.75	765.625
				$\Sigma(m_i - \hat{\mu})^2 \cdot f_i = 2137.5$

Thus the estimated variance is

$$\hat{\sigma}^2 = \frac{\sum(m_i - \hat{\mu})^2 \cdot f_i}{N} = \frac{2137.5}{40} = 53.4375$$

and the estimated standard deviation is

$$\hat{\sigma} = \sqrt{\hat{\sigma}^2} = \sqrt{53.4375} \approx 7.3101.$$

19. (a)

Age	Frequency	Midpoints
0–4.99	63,422	$\frac{0+5}{2} = 2.5$
5–17.99	240,629	$\frac{5+18}{2} = 11.5$
18–64.99	540,949	$\frac{18+65}{2} = 41.5$

(b) $\sum m_i f_i = (2.5)(63{,}422) + (11.5)(240{,}629) + (41.5)(540{,}949) = 25{,}375{,}172$

$N = \sum f_i = 63{,}422 + 240{,}629 + 540{,}949 = 845{,}000$

$$\text{Estimated mean } \hat{\mu} = \frac{\sum m_i f_i}{N} = \frac{25{,}375{,}172}{845{,}000} \approx 30.0298$$

(c) From (b), $\hat{\mu} \approx 30.0298$ and $N = 845{,}000$.

Midpoint (m_i)	Frequency (f_i)	$\hat{\mu}$	$m_i - \hat{\mu}$	$(m_i - \hat{\mu})^2 \cdot f_i$
2.5	63,422	30.0298	−27.5298	48,066,892.48
11.5	240,629	30.0298	−18.5298	82,620,806.47
41.5	540,949	30.0298	11.4702	71,170,219.19
				$\Sigma(m_i - \hat{\mu})^2 \cdot f_i = 201{,}857{,}918.10$

Thus the estimated variance is

$$\hat{\sigma}^2 = \frac{\sum(m_i - \hat{\mu})^2 \cdot f_i}{N} = \frac{201{,}857{,}918.10}{845{,}000} \approx 238.8851$$

and the estimated standard deviation is

$$\hat{\sigma} = \sqrt{\hat{\sigma}^2} = \sqrt{238.8851} \approx 15.4559.$$

21. (a)

Dollar value	Housing units (frequency)	Midpoints
\$0–\$49,999	5,430	$\frac{\$0 + \$50{,}000}{2} = \$25{,}000$
\$50,000–\$99,000	90,605	$\frac{\$50{,}000 + \$100{,}000}{2} = \$75{,}000$
\$100,000–\$149,999	90,620	$\frac{\$100{,}000 + \$150{,}000}{2} = \$125{,}000$
\$150,000–\$199,999	54,295	$\frac{\$150{,}000 + \$200{,}000}{2} = \$175{,}000$
\$200,000–\$299,000	34,835	$\frac{\$200{,}000 + \$300{,}000}{2} = \$250{,}000$
\$300,000–\$499,999	15,770	$\frac{\$300{,}000 + \$500{,}000}{2} = \$400{,}000$
\$500,000–\$999,999	5,595	$\frac{\$500{,}000 + \$1{,}000{,}000}{2} = \$750{,}000$

(b) $\sum m_i f_i = (\$25{,}000)(5{,}430) + (\$75{,}000)(90{,}605) + (\$125{,}000)(90{,}620) + (\$175{,}000)(54{,}295) + (\$250{,}000)(34{,}835) + (\$400{,}000)(15{,}770) + (\$750{,}000)(5{,}595) = \$46{,}973{,}250{,}000$

$$N = \sum f_i = 5{,}430 + 90{,}605 + 90{,}620 + 54{,}295 + 34{,}835 + 15{,}770 + 5{,}595 = 297{,}150$$

$$\text{Estimated mean } \hat{\mu} = \frac{\sum m_i f_i}{N} = \frac{\$46{,}973{,}250{,}000}{297{,}150} \approx \$158{,}079.25$$

(c) From (b), $\hat{\mu} \approx \$158{,}079.25$ and $N = 297{,}150$.

Midpoint (m_i)	Frequency (f_i)	$\hat{\mu}$	$m_i - \hat{\mu}$	$(m_i - \hat{\mu})^2 \cdot f_i$
\$25,000	5,430	\$158,079.25	−133,079.25	$9.616577122 \times 10^{13}$
\$75,000	90,605	\$158,079.25	−83,079.25	$6.253703681 \times 10^{14}$
\$125,000	90,620	\$158,079.25	−33,079.25	$9.915973705 \times 10^{13}$
\$175,000	54,295	\$158,079.25	16,920.75	$1.554529813 \times 10^{13}$
\$250,000	34,835	\$158,079.25	91,920.75	$2.943356948 \times 10^{14}$
\$400,000	15,770	\$158,079.25	241,920.75	$9.22949484892 \times 10^{14}$
\$750,000	5,595	\$158,079.25	591,920.75	$1.960321125 \times 10^{15}$
				$\sum(m_i - \hat{\mu})^2 \cdot f_i = 4.013847479 \times 10^{15}$

Thus the estimated variance is

$$\hat{\sigma}^2 = \frac{\sum(m_i - \hat{\mu})^2 \cdot f_i}{N} = \frac{4.013847479 \times 10^{15}}{297{,}150} \approx 13{,}507{,}815{,}950 \text{ dollars}$$

squared and the estimated standard deviation is

$$\hat{\sigma} = \sqrt{\hat{\sigma}^2} = \sqrt{13{,}507{,}815{,}950} \approx \$116{,}223.13.$$

23. $$\bar{x}_w = \frac{\sum w_i x_i}{\sum w_i} = \frac{w_1x_1 + w_2x_2 + w_3x_3 + w_4x_4}{w_1 + w_2 + w_3 + w_4}$$

$$= \frac{(25)(70) + (32)(80) + (33)(90) + (10)(100)}{25 + 32 + 33 + 10} = \frac{8280}{100} = 82.8$$

Final course grade = 82.8%.

25. $$\bar{x}_w = \frac{\sum w_i x_i}{\sum w_i} = \frac{w_1x_1 + w_2x_2 + w_3x_3}{w_1 + w_2 + w_3}$$

$$= \frac{(130)(89.66) + (111)(100.00) + (120)(92.16)}{130 + 111 + 120}$$

$$= \frac{33{,}815}{361} \approx 93.67.$$

Weighted mean size of all ants = 93.67 milligrams.

27. If $w_i = 1$ for all i then the weighted mean formula will be equivalent to the formula for the sample mean.

Section 3.4

1. The 95th percentile of a data set is the data value at which 95 percent of the values in the data set are less than or equal to this value.

3. The term *z*-score indicates how many standard deviations a particular value is from the mean. The *z*-score can provide us with information about data values even when we do not fully understand the original data set.

5. It is possible for the 1st percentile to equal the 99th percentile if all of the data values are the same.

Data set for Exercises 7, 9, 11 arranged in ascending order:
17, 18, 20, 20, 24, 36
$n = 6$

7. **(a)** For the 50th percentile $p = 50$. Thus

$$i = \left(\frac{p}{100}\right) \cdot n = \left(\frac{50}{100}\right) \cdot 6 = 3.$$

Since i is an integer, the 50th percentile is the mean of the data values in positions 3 and 4. Thus the 50th percentile is $\frac{20 + 20}{2} = 20$ mpg.

(b) For the 75th percentile $p = 75$. Thus

$$i = \left(\frac{p}{100}\right) \cdot n = \left(\frac{75}{100}\right) \cdot 6 = 4.5.$$

Since i is not an integer, round i up to 5. The 75th percentile is the data value in the fifth position. Thus the 75th percentile is 24 mpg.

(c) For the 25th percentile $p = 25$. Thus

$$i = \left(\frac{p}{100}\right) \cdot n = \left(\frac{25}{100}\right) \cdot 6 = 1.5.$$

Since i is not an integer, round i up to 2. The 25th percentile is the data value in the second position. Thus the 25th percentile is 18 mpg.

9. **(a)** For the Toyota Camry, $x = 24$ mpg. Thus the *z*-score for the Toyota Camry is

$$z = \frac{x - \bar{x}}{s} = \frac{24 - 22.5}{7.0356} = 0.21.$$

(b) For the Lincoln Town Car, $x = 17$ mpg. Thus the z-score for the Lincoln Town Car is

$$z = \frac{x - \bar{x}}{s} = \frac{17 - 22.5}{7.0356} = -0.78.$$

(c) For the Jaguar X-type, $x = 18$ mpg. Thus the z-score for the Jaguar X-type is

$$z = \frac{x - \bar{x}}{s} = \frac{18 - 22.5}{7.0356} = -0.64.$$

11. Since the z-score for the Honda Civic is 1.92 and $-2 < 1.92 < 2$, the mileage for the Honda Civic does not represent an outlier.

Data set for Exercises 13, 15 arranged in ascending order:

90, 90, 100, 110, 110, 110, 110, 110, 110, 120, 120, 120, 130 $\quad n = 12$

13. (a) For the 25th percentile $p = 25$. Thus

$$i = \left(\frac{p}{100}\right) \cdot n = \left(\frac{25}{100}\right) \cdot 12 = 3.$$

Since i is an integer, the 25th percentile is the mean of the data values in positions 3 and 4. Thus the 25th percentile is $\frac{100 + 110}{2} = 105$ calories.

(b) For the 95th percentile $p = 95$. Thus

$$i = \left(\frac{p}{100}\right) \cdot n = \left(\frac{95}{100}\right) \cdot 12 = 11.4.$$

Since i is not an integer, round i up to 12. The 95th percentile is the data value in the 12th position. Thus the 95th percentile is 130 calories.

For Exercise 15: The sample mean is

$$\bar{x} = \frac{\sum x}{n} = \frac{90 + 90 + 100 + 110 + 110 + 110 + 110 + 110 + 110 + 120 + 120 + 130}{12}$$

$$= \frac{1310}{12} \approx 109.1667 \text{ calories.}$$

To calculate the standard deviation we first need to find $\sum x$ and $\sum x^2$.

x	x^2
90	8,100
90	8,100
100	10,000
110	12,100
110	12,100
110	12,100
110	12,100
110	12,100
110	12,100
120	14,400
120	14,400
130	16,900
$\sum x = 1{,}310$	$\sum x^2 = 144{,}500$

$n = 12$. Therefore, the standard deviation is

$$s = \sqrt{\frac{\sum x^2 - \left(\sum x\right)^2/n}{n-1}} = \sqrt{\frac{144{,}500 - (1{,}310)^2/12}{12-1}} \approx 11.6450 \text{ calories.}$$

15. **(a)** For Bran Flakes, $x = 90$ calories. Thus the z-score for the Bran Flakes is

$$z = \frac{x - \bar{x}}{s} = \frac{90 - 109.1667}{11.6450} = -1.65.$$

(b) For Cap'n Crunch, $x = 120$ calories. Thus the z-score for the Cap'n Crunch is

$$z = \frac{x - \bar{x}}{s} = \frac{120 - 109.1667}{11.6450} = 0.93.$$

For Exercises 17, 19:

Data set arranged in ascending order:
2.0, 2.1, 2.4, 2.8, 3.1, 3.5, 3.8, 4.2, 4.3, 4.4, 5.2, 7.1, 7.7, 8.8, 14.7

17. **(a)** For the 50th percentile $p = 50$. Thus

$$i = \left(\frac{p}{100}\right) \cdot n = \left(\frac{50}{100}\right) \cdot 15 = 7.5.$$

Since i is not an integer, round i up to 8. The 50th percentile is the data value in the 8th position. Thus the 50th percentile is 4,200,000.
(b) For the 75th percentile $p = 75$. Thus

$$i = \left(\frac{p}{100}\right) \cdot n = \left(\frac{75}{100}\right) \cdot 15 = 11.25.$$

Since i is not an integer, round i up to 12. The 75th percentile is the data value in the 12th position. Thus the 75th percentile is 7,100,000.

For Exercise 19:
The sample mean is

$$\bar{x} = \frac{\sum x}{n} = \frac{2.0 + 2.1 + 2.4 + 2.8 + 3.1 + 3.5 + 3.8 + 4.2 + 4.3 + 4.4 + 5.2 + 7.1 + 7.7 + 8.8 + 14.7}{15}$$

$$= \frac{76.1}{15} \approx 5.0733 \text{ million.}$$

To calculate the standard deviation we first need to find $\sum x$ and $\sum x^2$.

x	x^2
2.0	4.00
2.1	4.41
2.4	5.76
2.8	7.84
3.1	9.61
3.5	12.25
3.8	14.44
4.2	17.64
4.4	19.36
5.2	27.04
7.1	50.41
7.7	59.29
8.8	77.44
14.7	216.09
$\sum x = 76.1$	$\sum x^2 = 544.07$

$n = 15$. Therefore, the standard deviation is

$$s = \sqrt{\frac{\sum x^2 - (\sum x)^2/n}{n-1}} = \sqrt{\frac{544.07 - (76.1)^2/15}{15-1}} \approx 3.3593 \text{ million.}$$

19. (a) For Ginseng, $x = 8.8$ million. Thus the z-score for Ginseng is

$$z = \frac{x - \bar{x}}{s} = \frac{8.8 - 5.0733}{3.3593} = 1.11.$$

(b) For Garlic supplements, $x = 7.1$ million. Thus the z-score for Garlic supplements is

$$z = \frac{x - \bar{x}}{s} = \frac{7.1 - 5.0733}{3.3593} = 0.60.$$

21. (a) **(b)**

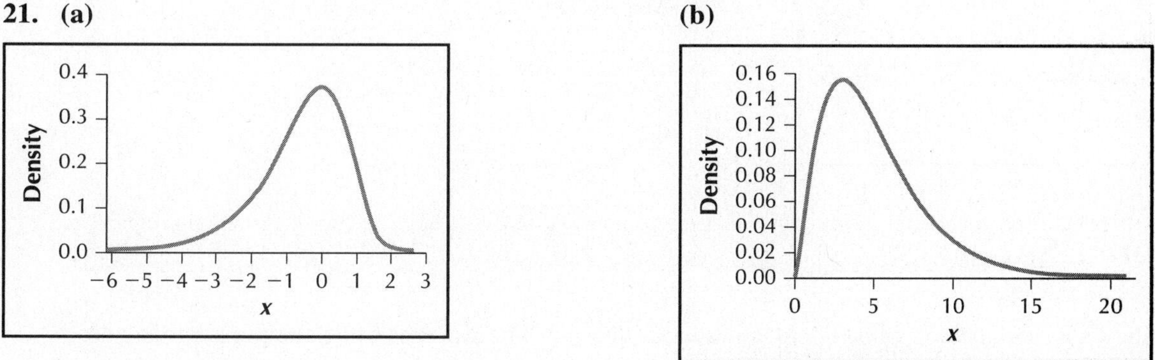

(c) For a bell-shaped distribution, the 50th percentile equals the mean. Thus Brandy's score was the mean, which was 503.

For Exercise 23, the data is already arranged in ascending order in the table and $n = 50$.

23. (a) For the 5th percentile $p = 5$. Thus

$$i = \left(\frac{p}{100}\right) \cdot n = \left(\frac{5}{100}\right) \cdot 50 = 2.5.$$

Since i is not an integer, round i up to 3. The 5th percentile is the data value in the 3rd position. Thus the 5th percentile is \$48,422.

(b) For the 95th percentile $p = 95$. Thus

$$i = \left(\frac{p}{100}\right) \cdot n = \left(\frac{95}{100}\right) \cdot 50 = 47.5.$$

Since i is not an integer, round i up to 48. The 95th percentile is the data value in the 48th position. Thus the 95th percentile is \$78,312.

(c) For the 1st percentile $p = 1$. Thus

$$i = \left(\frac{p}{100}\right) \cdot n = \left(\frac{1}{100}\right) \cdot 50 = 0.5.$$

Since i is not an integer, round i up to 1. The 1st percentile is the data value in the 1st position. Thus the 1st percentile is \$47,550.

(d) For the 99th percentile $p = 99$. Thus

$$i = \left(\frac{p}{100}\right) \cdot n = \left(\frac{99}{100}\right) \cdot 50 = 49.5.$$

Since i is not an integer, round i up to 50. The 99th percentile is the data value in the 50th position. Thus the 99th percentile is \$82,406.

25. **(a)** The difference between the 5th percentile and the 50th percentile is $|8998-6381| = 2617$ and the difference between the 95th percentile and the 50th percentile is $|17{,}188-8998| = 8190$. Since the 50th percentile is closer to the 5th percentile than it is to the 95th percentile the distribution of expenditures is right-skewed.
(b) Since the distribution of expenditures it right-skewed, we would expect the mean expenditure per pupil to be greater than the 50th percentile which is \$8998.

(c)

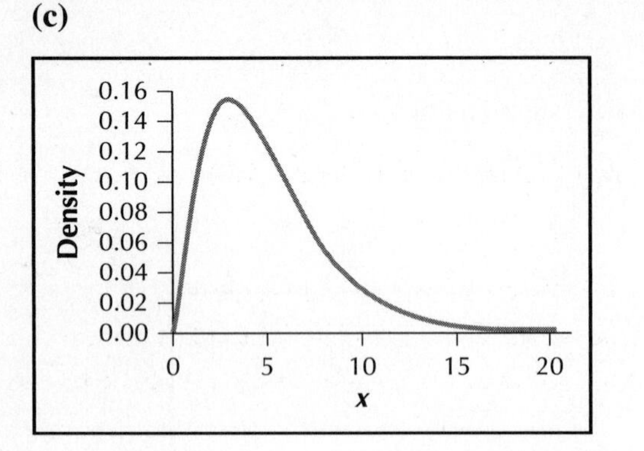

Section 3.5

1. Chebyshev's Rule may be applied to any continuous data set. The value of k must be greater than one.

3. The phrase "within two standard deviations of the mean" means the interval between the mean minus two standard deviations and the mean plus two standard deviations.

5. True.

7. Yes. Chebyshev's Rule may be used for any shaped distribution as long as it is continuous.

9. Since we don't know whether the distribution is bell-shaped, we will use Chebyshev's Rule.

(a) First we need to find k.

$$k = \left|\frac{\text{data value} - \text{mean}}{\text{standard deviation}}\right| = \left|\frac{80 - 100}{10}\right| = 2 \text{ and}$$

$$k = \left|\frac{\text{data value} - \text{mean}}{\text{standard deviation}}\right| = \left|\frac{120 - 100}{10}\right| = 2.$$

Therefore, at least $\left[1 - \left(\frac{1}{k}\right)^2\right] 100\% = \left[1 - \left(\frac{1}{2}\right)^2\right] 100\% = 75\%$ of the data values will fall within two standard deviations of the mean.

(b) First we need to find k.

$$k = \left|\frac{\text{data value} - \text{mean}}{\text{standard deviation}}\right| = \left|\frac{70 - 100}{10}\right| = 3 \text{ and}$$

$$k = \left|\frac{\text{data value} - \text{mean}}{\text{standard deviation}}\right| = \left|\frac{130 - 100}{10}\right| = 3.$$

Therefore, at least $\left[1 - \left(\frac{1}{k}\right)^2\right] 100\% = \left[1 - \left(\frac{1}{3}\right)^2\right] 100\% \approx 88.9\%$ of the data values will fall within three standard deviations of the mean.

11. Since we don't know whether the distribution is bell-shaped, we will use Chebyshev's Rule.

(a) First we need to find k.

$$k = \left|\frac{\text{data value} - \text{mean}}{\text{standard deviation}}\right| = \left|\frac{300 - 500}{100}\right| = 2 \text{ and}$$

$$k = \left|\frac{\text{data value} - \text{mean}}{\text{standard deviation}}\right| = \left|\frac{700 - 500}{100}\right| = 2.$$

Therefore, at least $\left[1 - \left(\frac{1}{k}\right)^2\right] 100\% = \left[1 - \left(\frac{1}{2}\right)^2\right] 100\% = 75\%$ of the data values will fall within two standard deviations of the mean.

(b) First we need to find k.

$$k = \left|\frac{\text{data value} - \text{mean}}{\text{standard deviation}}\right| = \left|\frac{100 - 500}{100}\right| = 4 \text{ and}$$

$$k = \left|\frac{\text{data value} - \text{mean}}{\text{standard deviation}}\right| = \left|\frac{900 - 500}{100}\right| = 4.$$

Therefore, at least $\left[1 - \left(\frac{1}{k}\right)^2\right] 100\% = \left[1 - \left(\frac{1}{4}\right)^2\right] 100\% \approx 93.75\%$ of the data values will fall within four standard deviations of the mean.

13. **(a)** Here we have to work backward. According to the Empirical Rule, 95% is associated with $k = 2$. So, about 95% of the data values lie within $k = 2$ standard deviations of the mean. The mean plus 2 standard deviations is $100 + 2(10) = 120$, and the mean minus 2 standard deviations is $100 - 2(10) = 80$. Therefore, 95% of all values lie between 80 and 120.

(b) 120 **(c)** 80

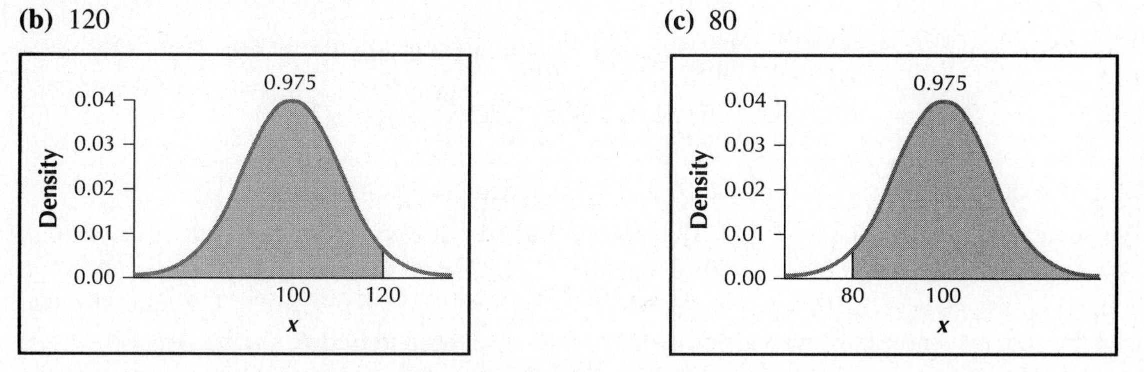

15. Since we don't know whether the distribution is bell-shaped, we will use Chebyshev's Rule.

(a) First we need to find k.

$$k = \left|\frac{\text{data value} - \text{mean}}{\text{standard deviation}}\right| = \left|\frac{4100 - 4500}{200}\right| = 2 \text{ and}$$

$$k = \left|\frac{\text{data value} - \text{mean}}{\text{standard deviation}}\right| = \left|\frac{4900 - 4500}{200}\right| = 2.$$

Therefore, at least $\left[1 - \left(\frac{1}{k}\right)^2\right] 100\% = \left[1 - \left(\frac{1}{2}\right)^2\right] 100\% = 75\%$ of the years have between 4100 and 4900 heating degree-days.

(b) First we need to find k.

$$k = \left|\frac{\text{data value} - \text{mean}}{\text{standard deviation}}\right| = \left|\frac{3900 - 4500}{200}\right| = 3 \text{ and}$$

$$k = \left|\frac{\text{data value} - \text{mean}}{\text{standard deviation}}\right| = \left|\frac{5100 - 4500}{200}\right| = 3.$$

Therefore, at least $\left[1 - \left(\frac{1}{k}\right)^2\right] 100\% = \left[1 - \left(\frac{1}{3}\right)^2\right] 100\% \approx 88.9\%$ of the years have between 3900 and 5100 heating degree-days.

(c) First we need to find k.

$$k = \left|\frac{\text{data value} - \text{mean}}{\text{standard deviation}}\right| = \left|\frac{4300 - 4500}{200}\right| = 1 \text{ and}$$

$$k = \left|\frac{\text{data value} - \text{mean}}{\text{standard deviation}}\right| = \left|\frac{4700 - 4500}{200}\right| = 1.$$

Since $k = 1$, we can't find the percentage of years with between 4300 and 4700 heating degree-days.

17. Since the distribution is bell-shaped we will use the Empirical Rule.

(a) First we need to find k.

$$k = \left|\frac{\text{data value} - \text{mean}}{\text{standard deviation}}\right| = \left|\frac{4100 - 4500}{200}\right| = 2 \text{ and}$$

$$k = \left|\frac{\text{data value} - \text{mean}}{\text{standard deviation}}\right| = \left|\frac{4900 - 4500}{200}\right| = 2.$$

Therefore, about 95% of the years have between 4100 and 4900 heating degree-days.

(b) First we need to find k.

$$k = \left|\frac{\text{data value} - \text{mean}}{\text{standard deviation}}\right| = \left|\frac{3900 - 4500}{200}\right| = 3 \text{ and}$$

$$k = \left|\frac{\text{data value} - \text{mean}}{\text{standard deviation}}\right| = \left|\frac{5100 - 4500}{200}\right| = 3.$$

Therefore, about 99.7% of the years have between 3900 and 5100 heating degree-days.

(c) First we need to find k.

$$k = \left|\frac{\text{data value} - \text{mean}}{\text{standard deviation}}\right| = \left|\frac{4300 - 4500}{200}\right| = 1 \text{ and}$$

$$k = \left|\frac{\text{data value} - \text{mean}}{\text{standard deviation}}\right| = \left|\frac{4700 - 4500}{200}\right| = 1.$$

Therefore, about 68% of the years have between 4300 and 4700 heating degree-days.

19. **(a)** **(i)** A z-score of -3 corresponds to $k = 3$. The number of shares of stock traded that is three standard deviations below the mean is $1{,}602{,}000{,}000 - 3(500{,}000{,}000) = 102{,}000{,}000$ shares. **(ii)** He is very truthful. **(b)** A z-score of -1 corresponds to the 16th percentile, which is less than the 25th percentile. Therefore the day corresponding to the 25th percentile had more shares traded. **(c)** First we need to find k.

$$k = \left|\frac{\text{data value} - \text{mean}}{\text{standard deviation}}\right| = \left|\frac{602 - 1602}{500}\right| = 2 \text{ and}$$

$$k = \left|\frac{\text{data value} - \text{mean}}{\text{standard deviation}}\right| = \left|\frac{2602 - 1602}{500}\right| = 2.$$

Therefore, at least $\left[1 - \left(\frac{1}{k}\right)^2\right] 100\% = \left[1 - \left(\frac{1}{2}\right)^2\right] 100\% = 75\%$ of the days had between 602 million and 2602 million shares traded.

21. Since we don't know whether the distribution is bell-shaped, we will use Chebyshev's Rule.

(a), (b), and (c) First we need to find k.

$$k = \left|\frac{\text{data value} - \text{mean}}{\text{standard deviation}}\right| = \left|\frac{7.6 - 13.6}{6}\right| = 1 \text{ and}$$

$$k = \left|\frac{\text{data value} - \text{mean}}{\text{standard deviation}}\right| = \left|\frac{19.6 - 136}{6}\right| = 1.$$

Since $k = 1$, we can't estimate the proportion of times that the wind speed in July is between 7.6 mph and 19.6 mph. We also can't estimate the proportion of times that the wind speed in July is either less than 7.6 mph or greater than 19.6 mph.

23. Since the distribution is bell-shaped we will use the Empirical Rule.

(a) First we need to find k.

$$k = \left|\frac{\text{data value} - \text{mean}}{\text{standard deviation}}\right| = \left|\frac{1.6 - 13.6}{6}\right| = 2 \text{ and}$$

$$k = \left|\frac{\text{data value} - \text{mean}}{\text{standard deviation}}\right| = \left|\frac{25.6 - 13.6}{6}\right| = 2.$$

Therefore, the proportion of times that the wind speed in July is between 1.6 mph and 25.6 mph is about 95%.

(b) From (a), about 95% of the time the wind speed in July is between 1.6 mph and 25.6 mph. Thus, the wind speed in July is either less than 1.6 mph or greater than 25.6 mph about 100% – 95% = 5% of the time. Since the distribution is symmetric, the wind speed in July is less than 1.6 mph about $\left(\frac{1}{2}\right)5\% = 2.5\%$ of the time.

25. The proportion of calm days in July is about 2.5%, and the proportion of calm days in January is between 16% and 50%.

27. Since we don't know whether the distribution is bell-shaped, we will use Chebyshev's Rule.

(a) First we need to find k.

$$k = \left|\frac{\text{data value} - \text{mean}}{\text{standard deviation}}\right| = \left|\frac{2.1 - 2.6}{0.5}\right| = 1 \text{ and}$$

$$k = \left|\frac{\text{data value} - \text{mean}}{\text{standard deviation}}\right| = \left|\frac{3.1 - 2.6}{0.5}\right| = 1.$$

Since $k = 1$, we can't estimate the proportion of students with a GPA between 2.1 and 3.1.

(b) First we need to find k.

$$k = \left|\frac{\text{data value} - \text{mean}}{\text{standard deviation}}\right| = \left|\frac{1.6 - 2.6}{0.5}\right| = 2 \text{ and}$$

$$k = \left|\frac{\text{data value} - \text{mean}}{\text{standard deviation}}\right| = \left|\frac{3.6 - 2.6}{0.5}\right| = 2.$$

Therefore, at least $\left[1 - \left(\frac{1}{k}\right)^2\right]100\% = \left[1 - \left(\frac{1}{2}\right)^2\right]100\% = 75\%$ of the students had a GPA between 1.6 and 3.6.

(c) From (b), at least 75% of the students had a GPA between 1.6 and 3.6. Therefore, at most 100% – 75% = 25% of the students had a GPA less than 1.6 or more than 3.6. Since we don't know whether the distribution is symmetric, the best we can say is that at most 25% of the students had a GPA more than 3.6. We can't be more precise because we don't know whether or not the distribution is symmetric.

29. The answers in Exercise 28 that were found using the Empirical Rule are more precise. The extra information with which we are paying for this increased precision is that the distribution is bell-shaped.

31. At the lower-performing high school the z-score for a 4.0 GPA is $z = \frac{x - \mu}{\sigma} = \frac{4.0 - 3.6}{0.5} = 0.8$. Thus $-2 < z\text{-score} < 2$, so a GPA of 4.0 is not considered unusual. At the higher-performing high school the z-score for a 4.0 GPA is $z = \frac{x - \mu}{\sigma} = \frac{4.0 - 2.6}{0.5} = 2.8$. Thus $2 \le z\text{-score} < 3$, so a GPA of 4.0 is considered moderately unusual.

For Exercise 33, the TI-84 calculator gives $\bar{x} \approx 2.8251$ mg per gram and $s \approx 2.6755$ mg per gram.

33. **(a)**

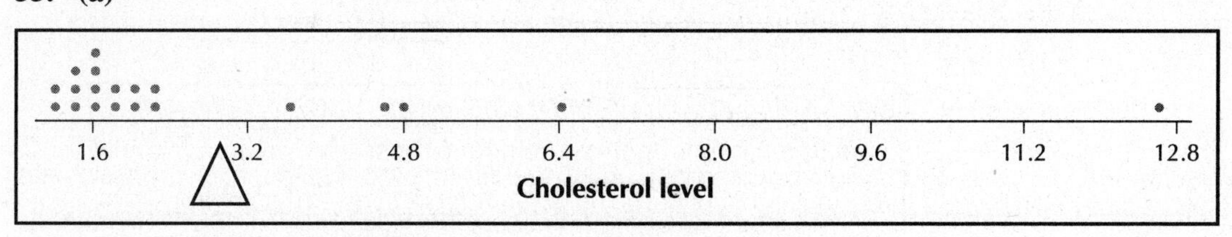

The mean is 2.8251 mg per gram.

(b) The dot plot shows evidence that the distribution of food items is not bell-shaped. **(c)** Since the distribution of food items is not bell-shaped we should not apply the Empirical Rule.

Section 3.6

1. We need robust measures because we need statistics that can summarize a data set while being less sensitive to the presence of outliers. The mean and the standard deviation are sensitive to extreme values and therefore might not be very representative to the center and the spread of the data.

3. The 3rd quartile is the 75th percentile, so it is the data value at which 75% of all of the values in the data set are less than or equal to this value. The 3rd percentile is the data value at which 3% of all of the values in the data set are less than or equal to this value.

5. (a) The data set would have all of the numbers between Q1 and Q2 be the same. Since Q2 is the median, the line for the median would be the same line as the line for Q1.
Example:

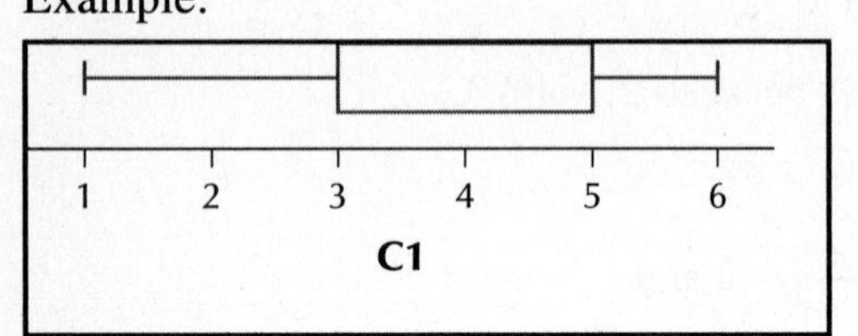

(b) All numbers in the data set would have to be the same. The boxplot would be a single line.
Example:

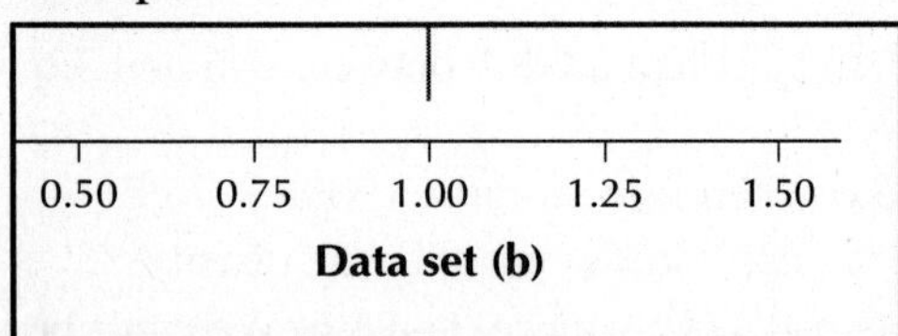

(c) The data set would have to be right-skewed with a few values much larger than the rest of the data set. The boxplot will have the line for the median closer to the line for Q3 than the line for Q1.
Example: Data $-4, -3, -2, -1, -1, -1, 0, 20$

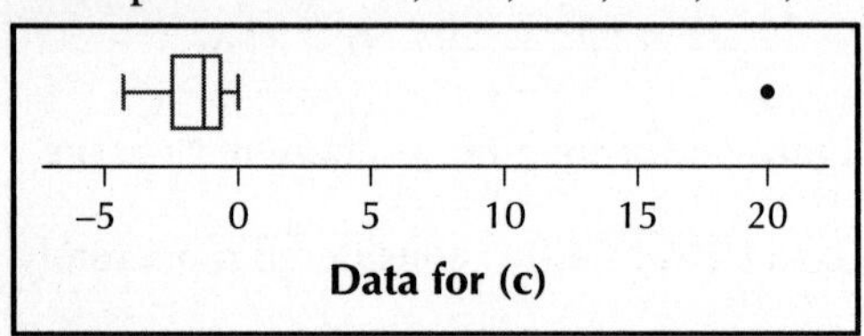

(d) The data set would have to be left-skewed with a few values much smaller than the rest of the data set. The boxplot would have the line for the median closer to the line for Q1 than the line for Q3.
Example: Data $-20, 0, 1, 1, 1, 2, 3, 4$

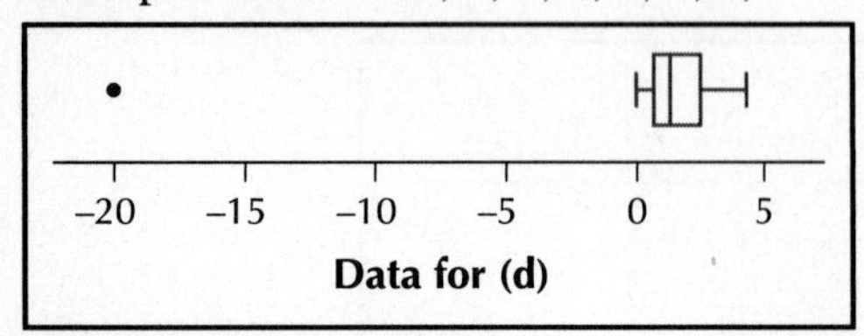

(e) Can't do. The 50th percentile of a data set can never be smaller than the 25th percentile.

(f) The numbers between Q1 and Q3 can't all be the same.
Example:

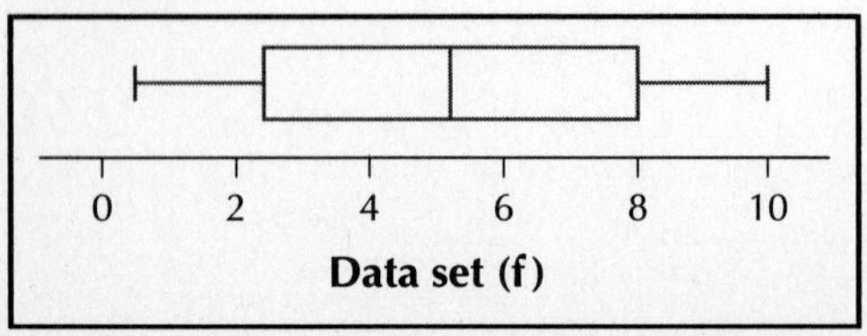

7. False. The five-number summary consists of the following: Minimum, Q1, *Median*, Q3, Maximum.

9. Data set arranged in ascending order:
25, 26, 28, 30, 33, 34, 35, 35, 39, 49 $n = 10$

(a) Here, Q1 is the 25th percentile, so $p = 25$. Therefore,

$$i = \left(\frac{p}{100}\right)n = \left(\frac{25}{100}\right)10 = 2.5.$$

Since i is not an integer, round i up to 3. We know that Q1 is the number in the 3rd position, so Q1 = 28.

(b) Here, Q2 is the 50th percentile, so $p = 50$. Therefore,

$$i = \left(\frac{p}{100}\right)n = \left(\frac{50}{100}\right)10 = 5.$$

Since i is an integer, we know that Q2 is the mean of the numbers in the 5th and 6th positions. Thus

$$Q2 = \frac{33 + 34}{2} = 33.5.$$

(c) Here, Q3 is the 75th percentile, so $p = 75$. Therefore,

$$i = \left(\frac{p}{100}\right)n = \left(\frac{75}{100}\right)10 = 7.5.$$

Since i is not an integer, round i up to 8. We know that Q3 is the number in the 8th position, so Q3 = 35.

11. Minimum = 25, Q1 = 28, Median = 33.5, Q3 = 35, Maximum = 49

13.

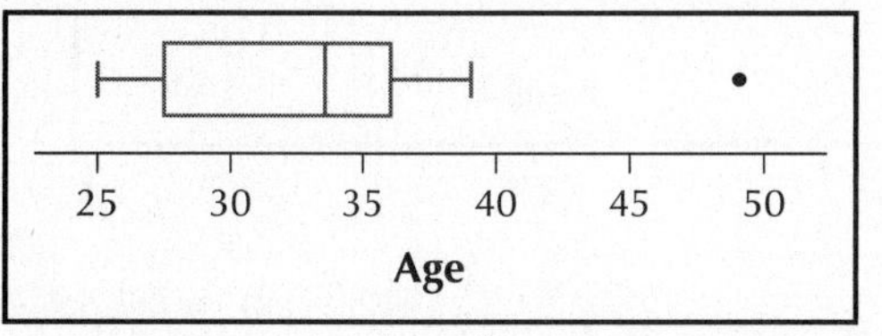

15. IQR = Q3 − Q1 = 85.5 − 68 = 17.5

17. Lower fence = Q1 − 1.5 * IQR = 68 − 1.5(17.5) = 41.75
Upper fence = Q3 + 1.5 * IQR = 85.5 +1.5(17.5) = 111.75
There are no values in the data set that lie outside of this interval so there are no outliers.

19. For WMU: Data set in ascending order: 60, 70, 70, 70, 75 $n = 5$

Here, Q1 is the 25th percentile, so $p = 25$. Therefore,

$$i = \left(\frac{p}{100}\right)n = \left(\frac{25}{100}\right)5 = 1.25.$$

Since i is not an integer, round i up to 2. We know that Q1 is the number in the 2nd position, so Q1 = 70.

Here, Q2 is the 50th percentile, so $p = 50$. Therefore,

$$i = \left(\frac{p}{100}\right)n = \left(\frac{50}{100}\right)5 = 2.5.$$

Since i is not an integer, round i up to 3. We know that Q2 is the number in the 3rd position, so Q2 = 70.

Here, Q3 is the 75th percentile, so $p = 75$. Therefore,

$$i = \left(\frac{p}{100}\right)n = \left(\frac{75}{100}\right)5 = 3.75.$$

Since i is not an integer, round i up to 4. We know that Q3 is the number in the 4th position, so Q3 = 70.

Therefore, the five-number summary for WMU is:

Minimum = 60, Q1 = 70, Median = 70, Q3 = 70, Maximum = 75

For NCU: Data set in ascending order: 66, 67, 70, 70, 72 $n = 5$

Here, Q1 is the 25th percentile, so $p = 25$. Therefore,

$$i = \left(\frac{p}{100}\right)n = \left(\frac{25}{100}\right)5 = 1.25.$$

Since i is not an integer, round i up to 2. We know that Q1 is the number in the 2nd position, so Q1 = 67.

Here, Q2 is the 50th percentile, so $p = 50$. Therefore,

$$i = \left(\frac{p}{100}\right)n = \left(\frac{50}{100}\right)5 = 2.5.$$

Since i is not an integer, round i up to 3. We know that Q2 is the number in the 3rd position, so Q2 = 70.

Here, Q3 is the 75th percentile, so $p = 75$. Therefore,

$$i = \left(\frac{p}{100}\right)n = \left(\frac{75}{100}\right)5 = 3.75.$$

Since i is not an integer, round i up to 4. We know that Q3 is the number in the 4th position, so Q3 = 70. Therefore, the five-number summary for WMU is:

Minimum = 66, Q1 = 67, Median = 70, Q3 = 70, Maximum = 72

A five-number summary for a data set of size five uses all five numbers in the data set, so a five-number summary probably shouldn't be used on a data set of size five.

21.

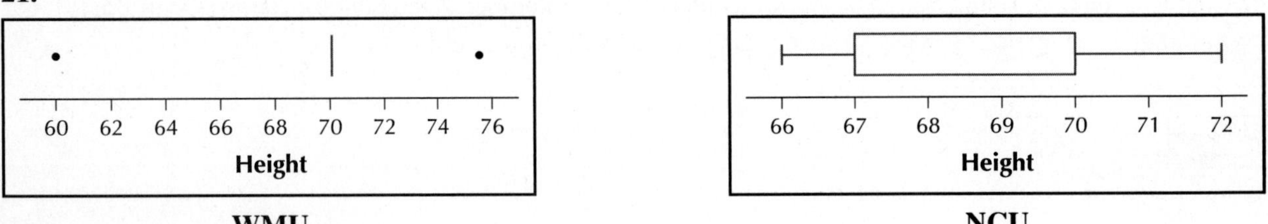

WMU **NCU**

The boxplots look so unusual because each data set only has five numbers.

23. IQR = Q3 − Q1 = 37.79 − 13.69 = 24.1. The IQR represents the spread of the middle 50% of the data. So for this problem the spread of the middle 50% of the data is 24.1.

25.

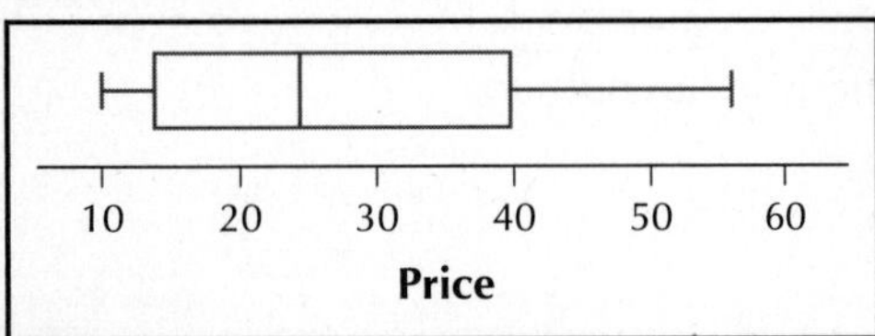

27. IQR = Q3 − Q1 = 0.08 − (−0.14) = 0.22. The IQR represents the spread of the middle 50% of the data values. So for this problem the spread of the middle 50% of the data is 0.22.

29.

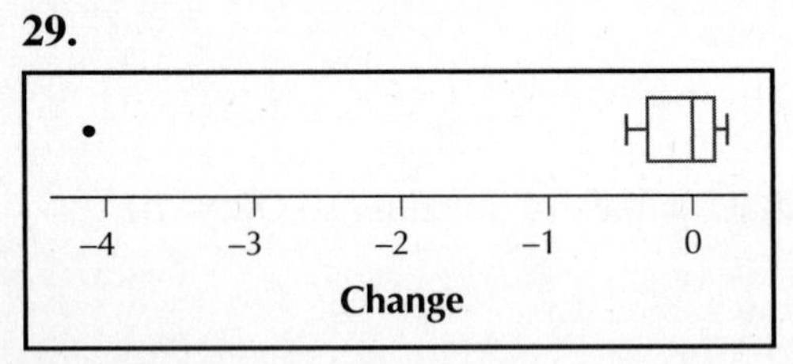

31. IQR = Q3 − Q1 = 115 − 105 = 10. For this data set the spread of the middle 50% of the data is 10.

33.

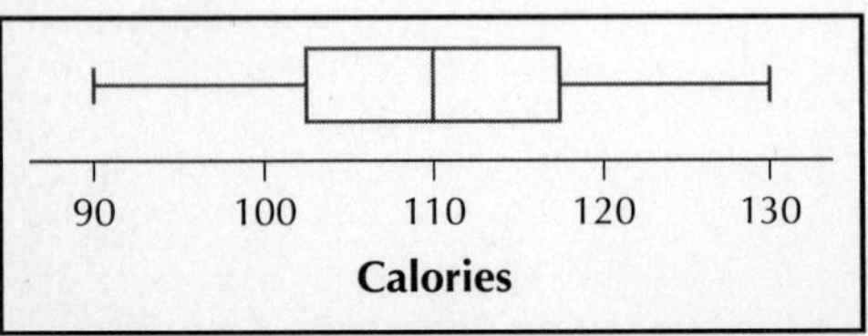

35. For Basic 4, $x = 130$ calories. Thus the z-score for Basic 4 is

$$z = \frac{x - \bar{x}}{s} = \frac{130 - 109.1667}{11.6450} = 1.79, \text{ so } -2 < z\text{-score} < 2. \text{ Therefore}$$

Basic 4 is not an outlier. From Exercise 32, there are no outliers when the robust method is used. Therefore, both methods have the same result.

37. IQR = Q3 − Q1 = 7,100,000 − 2,800,000 = 4,300,000. The spread of the middle 50% of this data set is 4,300,000.

39.

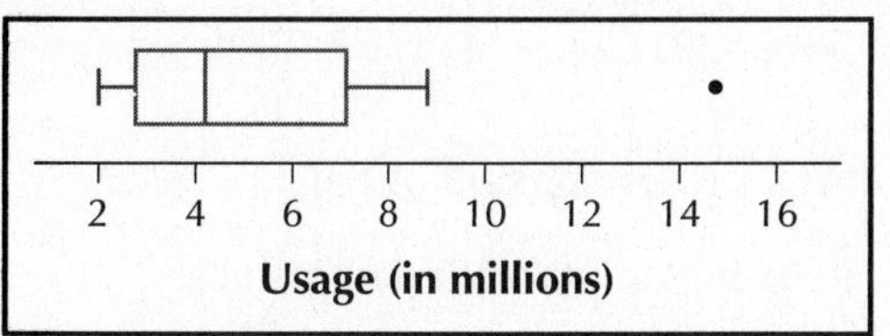

41. For Echinacea, $x = 14.7$ million. Thus the z-score for Echinacea is $z = \frac{x - \bar{x}}{s} = \frac{14.7 - 5.0733}{3.3593} = 2.87$, so $2 \le z\text{-score} < 3$. Thus the usage for Echinacea is considered moderately unusual but not an outlier. The robust method indicates that the usage for Echinacea is an outlier.

43. IQR = Q3 − Q1 = \$72,400 − \$67,600 = \$4,800. The spread of the middle 50% of the data is \$4,800.

45.

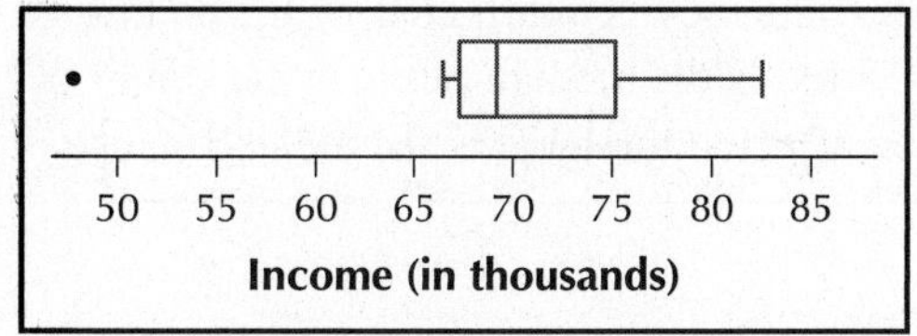

47. For West Virginia, $x = \$47.6$ thousand. Thus the z-score for West Virginia is

$$z = \frac{x - \bar{x}}{s} = \frac{47.6 - 70.1882}{7.867} = -2.87, \text{ so } -3 < z\text{-score} \le -2.$$

Therefore the median income of West Virginia is moderately unusual but not an outlier. The median income of West Virginia was found to be an outlier using the more robust method.

49. From Minitab:

Descriptive Statistics: IRON

```
Variable    N   N*   Mean  SE Mean  StDev  Minimum     Q1  Median     Q3
IRON      961    0  1.784    0.101  3.138    0.000  0.300   0.800  1.700

Variable  Maximum
IRON       37.600
```

Mean = 1.784 mg, standard deviation = 3.138 mg, Min = 0.000 mg, Q1 = 0.300 mg, Median = 0.800 mg, Q3 = 1.700 mg, Max = 37.600 mg

Range = Maximum − Minimum = 37.600 mg − 0.000 mg = 37.600 mg

IQR = Q3 − Q1 = 1.700 mg − 0.300 mg = 1.400 mg

51.

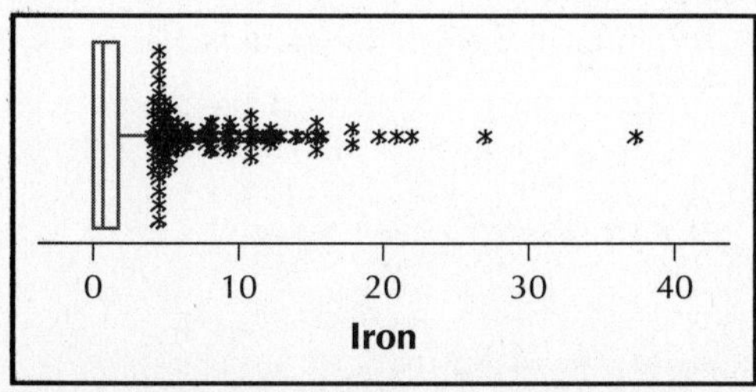

The boxplot is very right-skewed.

Chapter 3 Review

1. Mean number of cases sold

$$\bar{x} = \frac{\sum x}{n} = \frac{1929 + 1385 + 811 + 541 + 537 + 536 + 530 + 220 + 180 + 97}{10}$$

$$= \frac{6766}{10} = 676.6 \text{ million cases} = 676{,}600{,}000 \text{ cases.}$$

Data set arranged in ascending order (in millions): 97, 180, 220, 530, 536, 537, 541, 811, 1385, 1929
$n = 10$
Since $n = 10$ is even, the median is the mean of the two data values that lie on either side of the $\left(\frac{n+1}{2}\right)^{th} = \left(\frac{10+1}{2}\right)^{th} = 5.5$th position. That is, the median is the mean of the 5th and 6th data values, 536 and 537. Therefore,

$$\text{median} = \frac{536 + 537}{2} = 536.5 \text{ million cases} = 536{,}500{,}000 \text{ cases.}$$

Since median < mean, the distribution should be right-skewed.

3. The median because there are outlying values in the data set.

5. **(a)** The mode will not be affected because the value with the largest frequency is unaffected by the deletion of values less than 60 calories. **(b)** The measure that will be slightly affected is the median because the center of the data will be shifted slightly to the right. **(c)** The measure that will be affected the most is the mean because the mean uses all the values in its computation.

7. The mean, median, and mode will all be doubled.

9. Data set in ascending order after 4 inches has been added to the heights: 64, 74, 74, 74, 79 $n = 5$
(a) Range = largest value − smallest value = 79 − 64 = 15 inches.
To calculate the standard deviation we first need to find $\sum x$ and $\sum x^2$.

x	x^2
64	4096
74	5476
74	5476
74	5476
79	6241
$\sum x = 365$	$\sum x^2 = 26{,}765$

$n = 5$
Therefore, the standard deviation is

$$s = \sqrt{\frac{\sum x^2 - \left(\sum x\right)^2/n}{n-1}} = \sqrt{\frac{26{,}765 - (365)^2/5}{5-1}} \approx 5.48 \text{ inches.}$$

(b) Adding a positive constant to each value in a data set will not change the value of the original range or standard deviation.

(c) Note: In the dot plot the variable WMU4 represents the original variable values plus 4.

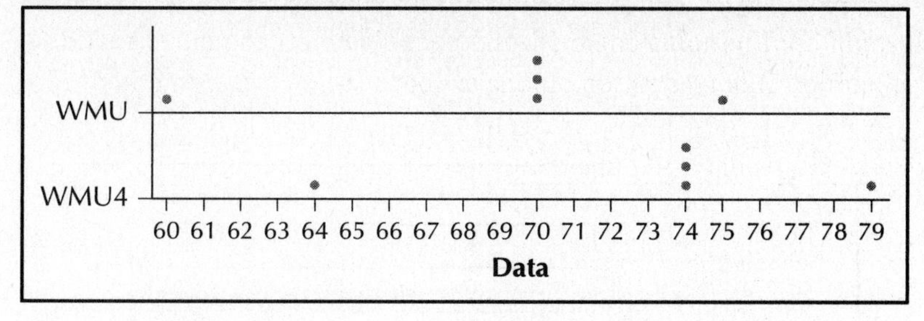

(i) The spread has not increased, it remains the same. However, the data points are shifted 4 units to the right.

(ii) Yes.

11. **(a)** The value of 7310 is a large outlying value relative to the rest of the data set, so if it is included in the computations, it will cause the variability to increase greatly. **(b)** Range without "the" = largest value − smallest value: 907 − 104 = 803.

Range with "the" = largest value − smallest value: 7310 − 104 = 7206.

Standard deviation without "the":

To calculate the standard deviation we first need to find Σx and Σx^2.

x	x^2
462	213,444
621	385,641
104	10,816
907	822,649
293	85,849
186	34,596
470	220,900
136	18,496
675	455,625
114	12,996
$\Sigma x = 3968$	$\Sigma x^2 = 2{,}261{,}012$

$n = 10$. Therefore, the standard deviation is

$$s = \sqrt{\frac{\Sigma x^2 - \left(\Sigma x\right)^2/n}{n-1}} = \sqrt{\frac{2{,}261{,}012 - (3968)^2/10}{10-1}} \approx 276.2.$$

Standard deviation with "the":

To calculate the standard deviation we first need to find $\sum x$ and $\sum x^2$.

x	x^2
462	213,444
621	385,641
104	10,816
907	822,649
293	85,849
186	34,596
470	220,900
136	18,496
675	455,625
114	12,996
7310	53,436,100
$\sum x = 11{,}278$	$\sum x^2 = 55{,}697{,}112$

$n = 11$. Therefore, the standard deviation is

$$s = \sqrt{\frac{\sum x^2 - \sum x^2/n}{n-1}} = \sqrt{\frac{55{,}697{,}112 - (11{,}278)^2/11}{11-1}} \approx 2101.$$

The above values bear out the intuition that the variability will increase.

13.

Class limits	Midpoints
0–1.99	$\frac{0+2}{2} = 1.0$
2–2.99	$\frac{2+3}{2} = 2.5$
3–3.99	$\frac{3+4}{2} = 3.5$
4–4.99	$\frac{4+5}{2} = 4.5$

$$\sum m_i f_i = (1.0)(126) + (2.5)(935) + (3.5)(2698) + (4.5)(3440) = 27{,}386.5$$

$$N = \sum f_i = 126 + 935 + 2698 + 3440 = 7199$$

$$\text{Estimated mean } \hat{\mu} = \frac{\sum m_i f_i}{N} = \frac{27{,}386.5}{7{,}199} \approx 3.8042$$

Midpoint (m_i)	Frequency (f_i)	$\hat{\mu}$	$m_i - \hat{\mu}$	$(m_i - \hat{\mu})^2 \cdot f_i$
1.0	126	3.8042	−2.8042	990.80574264
2.5	935	3.8042	−1.3042	1,590.3766934
3.5	2698	3.8042	−0.3042	249.66655272
4.5	3440	3.8042	0.6958	1,665.4334816
				$\sum(m_i - \hat{\mu})^2 \cdot f_i = 4{,}496.28247036$

Thus the estimated variance is

$$\hat{\sigma}^2 = \frac{\sum(m_i - \hat{\mu})^2 \cdot f_i}{N} = \frac{4496.28247036}{7{,}199} = 0.624570422331$$

and the estimated standard deviation is

$$\hat{\sigma} = \sqrt{\hat{\sigma}^2} = \sqrt{0.624570422331} \approx 0.7903.$$

15.

Class limits	Midpoints
0–12.99	$\frac{0+13}{2} = 6.5$
13–14.99	$\frac{13+15}{2} = 14$
15–24.99	$\frac{15+25}{2} = 20$
25–34.99	$\frac{25+35}{2} = 30$
35–44.99	$\frac{35+45}{2} = 40$
45–54.99	$\frac{45+55}{2} = 50$
55–64.99	$\frac{55+65}{2} = 60$
65–74.99	$\frac{65+75}{2} = 70$

$$\begin{aligned}\sum m_i f_i &= (6.5)(48) + (14)(60) + (20)(2{,}114) + (30)(9{,}361) + (40)(16{,}778) \\ &\quad + (50)(10{,}178) + (60)(3{,}075) + (70)(901) \\ &= 1{,}751{,}852\end{aligned}$$

$$N = \sum f_i = 48 + 60 + 2{,}114 + 9{,}361 + 16{,}778 + 10{,}178 + 3{,}075 + 901 = 42{,}515$$

Estimated mean $\hat{\mu} = \frac{\sum m_i f_i}{N} = \frac{1{,}751{,}852}{42{,}515} \approx 41.2055.$

Midpoint (m_i)	Frequency (f_i)	$\hat{\mu}$	$m_i - \hat{\mu}$	$(m_i - \hat{\mu})^2 \cdot f_i$
6.5	48	41.2055	−34.7055	57,814.643052
14	60	41.2055	−27.2055	44,408.353815
20	2,114	41.2055	−21.2055	950,609.208749
30	9,361	41.2055	−11.2055	1,175,397.39837
40	16,778	41.2055	−1.2055	24,382.2971345
50	10,178	41.2055	8.7945	787,199.397485
60	3,075	41.2055	18.7945	1,086,192.18302
70	901	41.2055	28.7945	747,040.030455
				$\sum(m_i - \hat{\mu})^2 . f_i = 4{,}873{,}043.51208$

Thus the estimated variance is

$$\hat{\sigma}^2 = \frac{\sum(m_i - \hat{\mu})^2 \cdot f_i}{N} = \frac{4{,}873{,}043.51208}{42{,}515} = 114.61939344$$

and the estimated standard deviation is

$$\hat{\sigma} = \sqrt{\hat{\sigma}^2} = \sqrt{114.61939344} \approx 10.7061.$$

For Exercises 16–18:

Data set arranged in ascending order: 1.2292, 1.2824, 1.3529, 1.3600, 1.4000, 1.5059, 1.5435, 1.5614, 1.6193, 1.7231, 1.8478, 1.9294, 1.9765, 2.2000, 2.2069, 3.5246, 4.5870, 4.8235, 6.300, 12.5294

$n = 20$

17. For the 50th percentile $p = 50$. Thus

$$i = \left(\frac{p}{100}\right)n = \left(\frac{50}{100}\right)20 = 10.$$

Since i is an integer, the 50th percentile is the mean of the data values in positions 10 and 11. Thus the 50th percentile is $\frac{1.7231 + 1.8478}{2} = 1.78545$ mg per gram.

19. For French toast, $x = 1.7231$ mg per gram. Thus the z-score is

$$z = \frac{x - \mu}{\sigma} = \frac{1.7321 - 2.8251}{2.6755} = -0.41.$$

21. For waffles, $x = 1.3600$ mg per gram. Thus the z-score is

$$z = \frac{x - \mu}{\sigma} = \frac{1.3600 - 2.8251}{2.6755} = -0.55.$$

23. **(a)** Zero standard deviations below the mean is $1.3 - 0(1.9) = 1.3$. One standard deviation below the mean is $1.3 - 1(1.9) = -0.6$. Thus, the values that lie between zero and one standard deviation below the mean are the values that lie between -0.6 triples and and 1.3 triples. There are 166 batters that lie between these two values. **(b)** One standard deviation below the mean is $1.3 - 1(1.9) = -0.6$. Two standard deviations below the mean is $1.3 - 2(1.9) = -2.5$. Thus, the values that lie between one and two standard deviations below the mean are the values that lie between -2.5 triples and -0.6 triples. There are zero batters that lie between these two values. **(c)** It is impossible to hit a negative number of triples, so there are zero batters in any interval that consists entirely of negative numbers. **(d)** There is no need to look for outliers on the low end of the triples distribution. It is impossible for a batter to hit a negative number of triples so there will be no batters in the low end of the triples distribution.

25. Since the distribution is bell-shaped, we will use the Empirical Rule.
First we need to find k.

$$k = \left|\frac{\text{data value} - \text{mean}}{\text{standard deviation}}\right| = \left|\frac{77 - 70}{5}\right| = 1.4.$$

Since $k = 1.4$ is between 1 and 2, the proportion of vehicle speeds that are over 77 mph is between 2.5% and 16%.

27. Since we don't know the distribution we can't find the 50th percentile.

29. The 36th percentile lies between the 16th percentile and the 50th percentile. Since the distribution is bell-shaped, the 16th percentile lies one standard deviation below the mean and the 50th percentile is the mean. So the 16th percentile is $70 - 1(5) = 65$ mph and the 50th percentile is 70 mph. Therefore the 36th percentile lies between 65 mph and 70 mph.

31. Data set arranged in ascending order: 1.2292, 1.2824, 1.3529, 1.3600, 1.4000, 1.5059, 1.5435, 1.5614, 1.6193, 1.7231, 1.8478, 1.9294, 1.9765, 2.2000, 2.2069, 3.5246, 4.5870, 4.8235, 6.300, 12.5294
$n = 20$

Here, Q1 is the 25th percentile, so $p = 25$. Therefore,

$$i = \left(\frac{p}{100}\right)n = \left(\frac{25}{100}\right)20 = 5.$$

Since i is an integer, we know that Q1 is the mean of the numbers in the 5th and 6th positions. Thus,

$$\text{Q1} = \frac{1.4000 + 1.5059}{2} = 1.45295.$$

Here, Q2 is the 50th percentile, so $p = 50$. Therefore,

$$i = \left(\frac{p}{100}\right)n = \left(\frac{50}{100}\right)20 = 10.$$

Since i is an integer, we know that Q2 is the mean of the numbers in the 10th and 11th positions. Thus

$$\text{Q2} = \frac{1.7231 + 1.8478}{2} = 1.78545.$$

Here, Q3 is the 75th percentile, so $p = 75$. Therefore,

$$i = \left(\frac{p}{100}\right)n = \left(\frac{75}{100}\right)20 = 15.$$

Since i is an integer, we know that Q3 is the mean of the numbers in the 15th and 16th positions. Thus,

$$Q3 = \frac{2.2069 + 3.5246}{2} = 2.86575.$$

Therefore, the five-number summary for cholesterol levels is:

Minimum = 1.2292, Q1 = 1.45295, Median = 1.78545, Q3 = 2.86575, Maximum = 12.5294

33. **(a)** For the yolk of a raw egg, $x = 12.5294$ mg per gram. Thus the z-score is

$$z = \frac{x - \mu}{\sigma} = \frac{12.5294 - 2.8251}{2.6755} = 3.63.$$

Since z-score > 3, the cholesterol level for the yolk of a raw egg may be considered an outlier.

For the chicken liver, $x = 6.3000$ mg per gram. Thus the z-score is

$$z = \frac{x - \mu}{\sigma} = \frac{6.3000 - 2.8251}{2.6755} = 1.30.$$

Since $-2 < z\text{-score} < 2$, the cholesterol level for chicken liver is not considered unusual or an outlier.

(b) IQR = Q3 − Q1 = 2.86575 − 1.45295 = 1.4128

Lower fence = Q1 − 1.5 * IQR =1.45295 − 1.5(1.4128) = −0.66625;

Upper fence = Q3 + 1.5 * IQR = 2.86575 + 1.5(1.4128) = 4.98495

The values that don't lie between −0.66625 and 4.98495 are 12.5294 and 6.3000. The z-score method only identified 12.5294 as an outlier.

35. IQR Q3 − Q1 = 48 − 25 = 23. The spread of the middle 50% of the data set is 23.

37. **(a)**

x	$z = \frac{x - \bar{x}}{s}$
8	$\frac{8 - 36.3}{16.51} = -1.71$
25	$\frac{25 - 36.3}{16.51} = -0.68$
25	$\frac{25 - 36.3}{16.51} = -0.68$
26	$\frac{26 - 36.3}{16.51} = -0.62$
31	$\frac{31 - 36.3}{16.51} = -0.32$
38	$\frac{38 - 36.3}{16.51} = 0.10$
43	$\frac{43 - 36.3}{16.51} = 0.41$
48	$\frac{48 - 36.3}{16.51} = 0.71$
59	$\frac{59 - 36.3}{16.51} = 1.37$
60	$\frac{60 - 36.3}{16.51} = 1.44$

None of the z-scores are larger than 3 or less than −3, so there are no outliers.

(b) Lower fence = Q1 − 1.5 * IQR = 25 − 1.5(23) = −9.5

Upper fence = Q3 + 1.5 * IQR = 48 + 1.5(23) = 82.5

All of the Ragweed Pollen Indices lie between −9.5 and 82.5.

So there are no outliers.

(c) The two methods concur.

Chapter 3 Quiz

1. True.

2. False. For example let data set A be 1, 6, 6, 6, 16 and data set B be 4, 6, 6, 6, 13. Then both data sets have a mean of 7, a median of 6, and a mode of 6, but they are not the same data sets.

3. False. The *standard deviation* is the square root of the *variance*.

4. False. The Empirical Rule only applies to data sets with bell-shaped distributions.

5. True.

6. 2

7. outlier

8. center

9. mean

10. average

11. deviation

12. squared deviations

13. median, Q2 – second quartile

14. robust measures

15. mode

16. zero

17. The class midpoint

18. **(a)** Left-skewed **(b)** Right-skewed **(c)** Symmetric unimodal

19. Data set arranged in ascending order: 31,054; 84,059; 98,008; 106,178; 117,964

(a) $n = 5$. The sample mean is

$$\bar{x} = \frac{\sum x}{n} = \frac{31{,}054 + 84{,}059 + 98{,}008 + 106{,}178 + 117{,}964}{5} = \frac{437{,}263}{5} \approx 87{,}453.$$

(b) Since $n = 5$ is odd, the median is the data value in the $\left(\frac{n+1}{2}\right)^{\text{th}} = \left(\frac{5+1}{2}\right)^{\text{th}} = 3\text{rd}$ position. Therefore, the median number of passengers is 98,008.

20. **(a)** Range = largest value − smallest value = 117,964 − 31,054 = 86,910. **(b)** To calculate the standard deviation we first need to find $\sum x$ and $\sum x^2$.

x	x^2
31,054	964,350,916
84,059	7,065,915,481
98,008	9,605,568,064
106,178	11,273,767,684
117,964	13,915,505,296
$\sum x = 437{,}263$	$\sum x^2 = 42{,}825{,}107{,}441$

$n = 5$. Therefore, the standard deviation is

$$s = \sqrt{\frac{\sum x^2 - \left(\sum x\right)^2/n}{n-1}} = \sqrt{\frac{42825107441 - (437263)^2/5}{5-1}} \approx 33{,}857.$$

21. Typically, the number of passengers differs from the mean of 87,453 by 33,857.

22.

Class limits	Midpoints
140–259.99	$\frac{140+260}{2}=200$
260–299.99	$\frac{260+300}{2}=280$
300–339.99	$\frac{300+340}{2}=320$
340–379.99	$\frac{340+380}{2}=360$
380–499.99	$\frac{380+500}{2}=440$

$$\Sigma m_i f_i = (200)(4) + (280)(5) + (320)(10) + (360)(6) + (440)(7) = 10{,}640$$

$$N = \Sigma f_i = 4 + 5 + 10 + 6 + 7 = 32$$

Estimated mean

$$\hat{\mu} = \frac{\sum m_i f_i}{N} = \frac{10{,}640}{32} \approx 332.5 \text{ points.}$$

Midpoint (m_i)	Frequency (f_i)	$\hat{\mu}$	$m_i - \hat{\mu}$	$(m_i - \hat{\mu})^2 \cdot f_i$
200	4	332.5	−132.5	70,225
280	5	332.5	−52.5	13,781.25
320	10	332.5	−12.5	1,562.5
360	6	332.5	27.5	4,537.5
440	7	332.5	107.5	80,893.75
				$\Sigma(m_i - \hat{\mu})^2 \cdot f_i = 171{,}000$

Thus the estimated variance is

$$\hat{\sigma}^2 = \frac{\sum (m_i - \hat{\mu})^2 \cdot f_i}{N} = \frac{171{,}000}{32} = 5343.75$$

and the estimated standard deviation is

$$\hat{\sigma} = \sqrt{\hat{\sigma}^2} = \sqrt{5343.75} \approx 73.1010 \text{ points.}$$

23.

Class limits	Midpoints
0–39.99	$\frac{0+40}{2}=20$
40–49.99	$\frac{40+50}{2}=45$
50–59.99	$\frac{50+60}{2}=55$
60–69.99	$\frac{60+70}{2}=65$
70–79.99	$\frac{70+80}{2}=75$
80–89.99	$\frac{80+90}{2}=85$

$$\Sigma m_i f_i = (20)(22) + (45)(31) + (55)(51) + (65)(47) + (75)(44) + (85)(44) = 14{,}735$$

$$N = \Sigma f_i = 22 + 31 + 51 + 47 + 44 + 44 = 239$$

Estimated mean

$$\hat{\mu} = \frac{\sum m_i f_i}{N} = \frac{14{,}735}{239} \approx 61.6527.$$

Midpoint (m_i)	Frequency (f_i)	$\hat{\mu}$	$m_i - \hat{\mu}$	$(m_i - \hat{\mu})^2 \cdot f_i$
20	22	61.6257	−41.6257	38,119.3758108
45	31	61.6257	−16.6257	8,568.83091519
55	51	61.6257	−6.6257	2,238.89492499
65	47	61.6257	3.3743	535.13732303
75	44	61.6257	13.3743	7,870.36362156
85	44	61.6257	23.3743	24,039.7476216
				$\Sigma(m_i - \hat{\mu})^2 \cdot f_i = 81{,}372.35022$

Thus the estimated variance is $\hat{\sigma}^2 = \frac{\sum(m_i - \hat{\mu})^2 \cdot f_i}{N} = \frac{81{,}372.35022}{239} = 340.4700846$, and the estimated standard deviation is $\hat{\sigma} = \sqrt{\hat{\sigma}^2} = \sqrt{340.4700846} \approx 18.4518.$

24. **(a)** $z = \frac{x-\mu}{\sigma} = \frac{308-308}{154} = 0$; **(b)** $z = \frac{x-\mu}{\sigma} = \frac{462-308}{154} = 1$; **(c)** $z = \frac{x-\mu}{\sigma} = \frac{616-308}{154} = 2$;

(d) $z = \frac{x-\mu}{\sigma} = \frac{154-308}{154} = -1$; **(e)** $z = \frac{x-\mu}{\sigma} = \frac{0-308}{154} = -2$

25. **(a)** $z = \frac{x-\bar{x}}{s} = \frac{120-60}{40} = 1.5$; **(b)** $z = \frac{x-\bar{x}}{s} = \frac{20-60}{40} = -1$; **(c)** $z = \frac{x-\bar{x}}{s} = \frac{100-60}{40} = 1$;

(d) $z = \frac{x-\bar{x}}{s} = \frac{0-60}{40} = -1.5$; **(e)** $z = \frac{x-\bar{x}}{s} = \frac{60-60}{40} = 0$

26. **(a)** Since we are assuming that the distribution is bell-shaped, we will use the Empirical Rule. From Exercise 24(e), $k = 2$, so $x = \$0$ is about the 2.5th percentile. **(b)** Since we are assuming that we don't know anything about the distribution, we will use Chebyshev's Rule. **(i)** From Exercises 24(c) and (e), $k = 2$. Therefore, at least $\left[1 - \left(\frac{1}{k}\right)^2\right]100\% = \left[1 - \left(\frac{1}{2}\right)^2\right]100\% = 75\%$ of American citizens spend between \$0 and \$616 annually on tobacco products. **(ii)** From Exercises 24(b) and (d), $k = 1$. Since $k = 1$ we can't use Chebyshev's Rule, so we can't do the problem.

27. Since we are assuming the distribution is bell-shaped, we will use the Empirical Rule. **(a)** Since we are assuming the distributions is bell-shaped, the 50th percentile is the mean. Therefore, the 50th percentile is 60 gallons. **(b)** First we need to find k.

$$k = \left|\frac{\text{data value} - \text{mean}}{\text{standard deviation}}\right| = \left|\frac{29 - 60}{40}\right| = 0.78 \qquad \text{and}$$

$$k = \left|\frac{\text{data value} - \text{mean}}{\text{standard deviation}}\right| = \left|\frac{109 - 60}{40}\right| = 1.23.$$

Since $k = 0.78$ is between 0 and 1, between 0% and (1/2) 68% = 34% of Americans drink between 29 and 60 gallons of carbonated beverages per year. Since $k = 1.23$ is between 1 and 2, between (1/2) 68% = 34% and (1/2) 95% = 47.5% of Americans drink between 60 and 109 gallons of carbonated beverages per year. Therefore, adding these results gives us that between 0% + 34% = 34% and 34% + 47.5% = 81.5% of Americans drink between 29 and 109 gallons of carbonated beverages per year. **(c)** We could not find the estimate in **(b)** without assuming the distribution is bell-shaped and one of the values of k is less than 1.
(d) Since $k = 1.23$ is between 1 and 2, between 2.5% and 16% of Americans drink more than 109 gallons of carbonated beverages per year.

28. Data set arranged in ascending order: 499, 501, 502, 510, 514, 515, 521, 523

(a) $n = 8$. Here, Q1 is the 25th percentile, so $p = 25$. Therefore,

$$i = \left(\frac{p}{100}\right)n = \left(\frac{25}{100}\right)8 = 2.$$

Since i is an integer, we know that Q1 is the mean of the numbers in the 2nd and 3rd positions. Thus

$$Q1 = \frac{501 + 502}{2} = 501.5.$$

(b) Here, Q2 is the 50th percentile, so $p = 50$. Therefore,

$$i = \left(\frac{p}{100}\right)n = \left(\frac{50}{100}\right)8 = 4.$$

Since i is an integer, we know that Q2 is the mean of the numbers in the 4th and 5th positions. Thus

$$Q2 = \frac{510 + 514}{2} = 512.$$

(c) Here, Q3 is the 75th percentile, so $p = 75$. Therefore,

$$i = \left(\frac{p}{100}\right)n = \left(\frac{75}{100}\right)8 = 6.$$

Since i is an integer, we know that Q3 is the mean of the numbers in the 6th and 7th positions. Thus

$$Q3 = \frac{515 + 521}{2} = 518.$$

29. IQR = Q3 − Q1 = 518 − 501.5 = 16.5

30. Minimum = 499, Q1 = 501.5, Median = 512, Q3 = 518, Maximum = 523

31. Lower fence = Q1 − 1.5 * IQR = 501.5 − 1.5(16.5) = 476.75

Upper fence = Q3 + 1.5 * IQR = 518 + 1.5(16.5) = 542.75

All of the SAT scores lie between 476.75 and 542.75, so there are no outliers.

32.

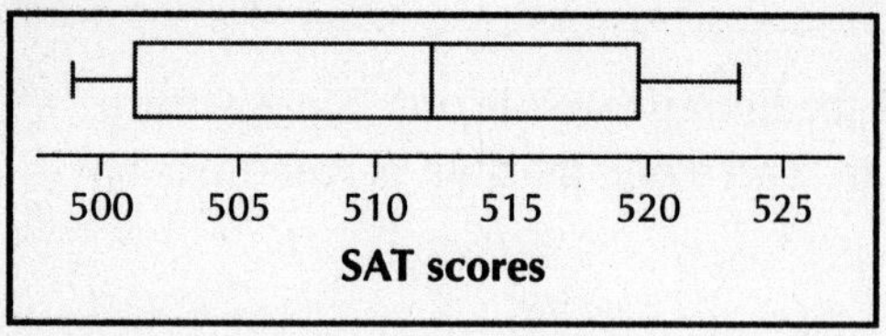

500 505 510 515 520 525

SAT scores

Chapter 4

Describing the Relationship between Two Variables

Section 4.1

1. The researchers may want to know if there are differences in the frequencies of the categories of one variable that depend on the category of the other variable. Some examples given in the text are differences in emotions felt after September 11, 2001 based on gender and differences in which careers are considered prestigious based on gender. Also, the researchers might want to predict the value of one variable using the other.

3. The headings of the columns and the rows of a crosstabulation are categories and not numbers. We could re-code the variables in order to use crosstabulation by dividing the numbers up into classes like we did for grouped frequency distributions. Then we could code each class as a category.

5.

	Beverages	Food	Household/ personal	Tobacco	All
Glass	2	0	0	0	2
Metal	0	0	2	0	2
Paper	0	0	1	3	4
Plastic	1	2	1	0	4
All	3	2	4	3	12

7.

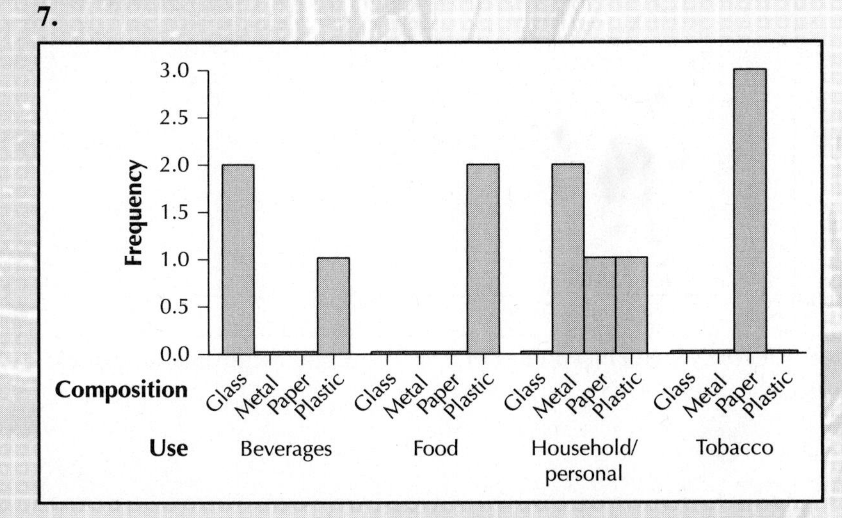

9.

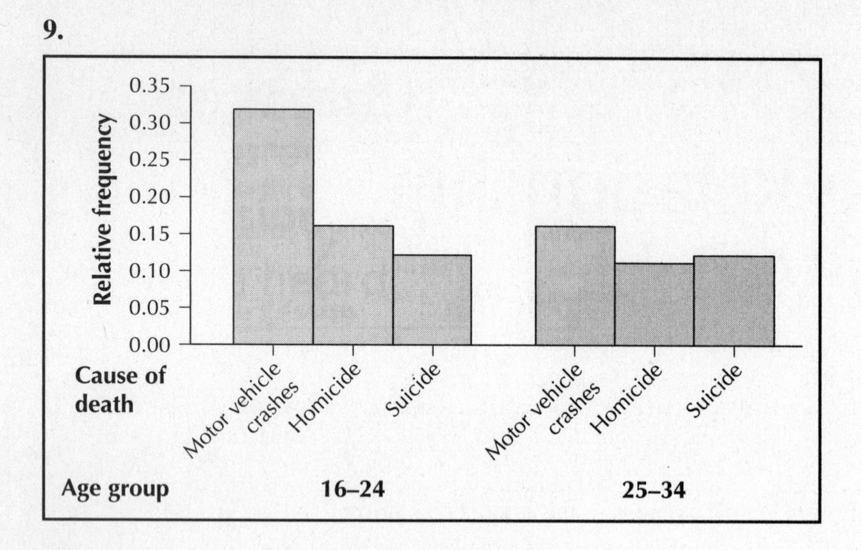

11. (a)

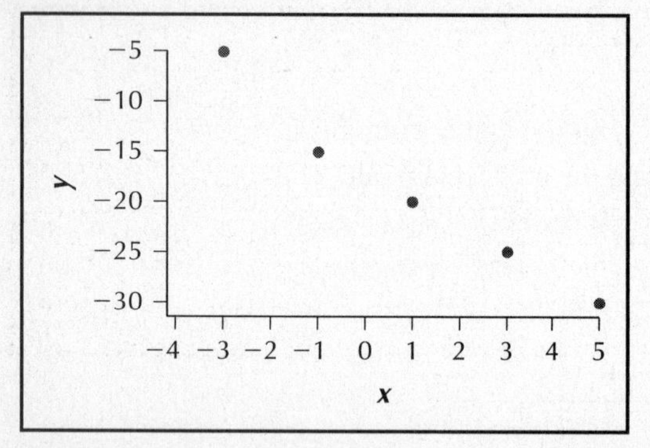

(b) Negative relationship. **(c)** As x increases, y decreases. Smaller values of x are associated with larger values of y; larger values of x are associated with smaller values of y.

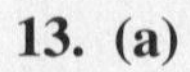

13. (a)

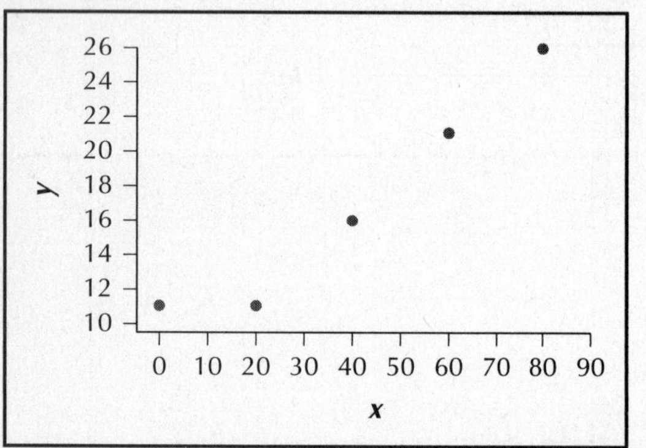

(b) Positive relationship. **(c)** As x increases, y increases. Smaller values of x are associated with smaller values of y; larger values of x are associated with larger values of y.

15.

	Freshman	Junior	Senior	Sophomore	All
Art	0	1	0	1	2
Business	0	1	0	2	3
Communication	1	0	0	2	3
Economics	0	0	2	0	2
English	4	0	0	1	5
History	0	3	0	0	3
Philosophy	0	0	1	0	1
Political Science	0	0	1	0	1
Psychology	0	2	0	3	5
All	5	7	4	9	25

(a) No apparent relationship: No patterns.

(b)

Tally for Discrete Variables: Major	
Major	**Count**
Art	2
Business	3
Communication	3
Economics	2
English	5
History	3
Philosophy	1
Political Science	1
Psychology	5
N =	25

Tally for Discrete Variables: Class	
Class	**Count**
Freshman	5
Junior	7
Senior	4
Sophomore	9
N =	25

17.

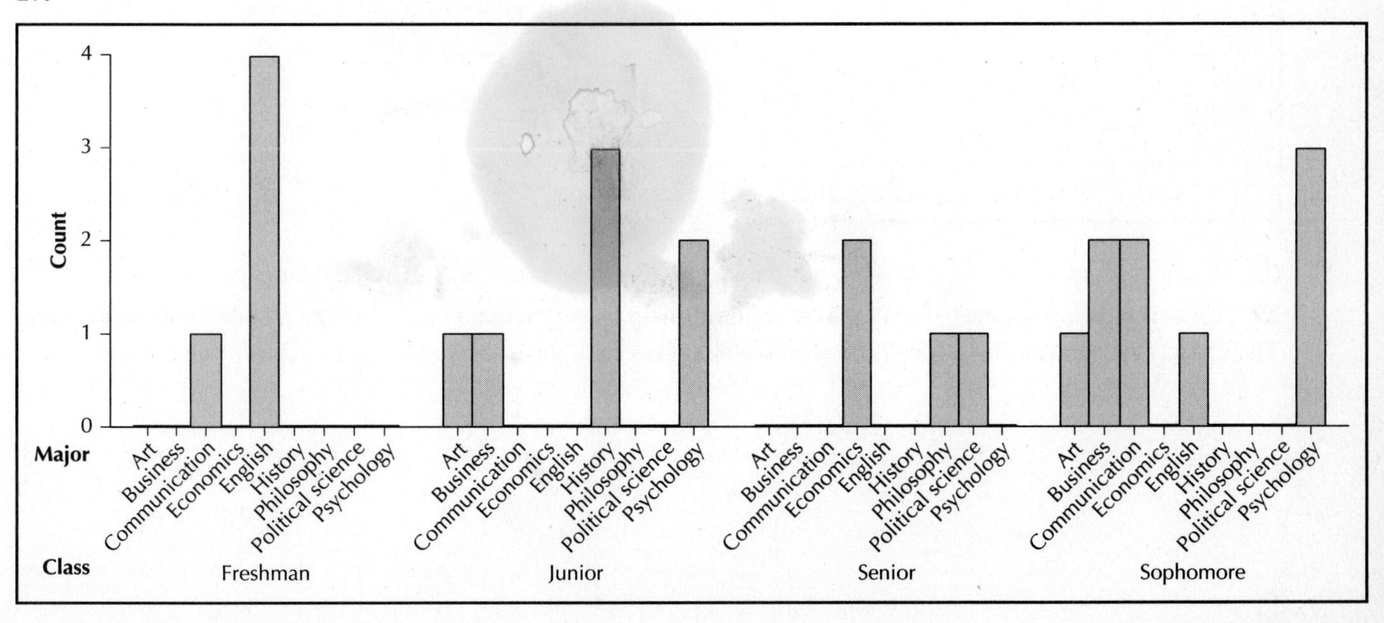

19. Missing values are bold.

	Gender		
"How much do you enjoy shopping?"	**Male**	**Female**	**Total**
A lot	**1338 − 950 = 388**	950	1338
Some	582	**1255 − 582 = 673**	1255
Only a little	662	497	**662 + 497 = 1159**
Not at all	497	**717 − 497 = 220**	717
Don't know/refused	**45 − 25 = 20**	25	45
Total	2149	**4514 − 2149 = 2365**	4514

21. **(a)** Women; **(b)** Women; **(c)** Men; **(d)** Men

23. **(a)** 0.7100; **(b)** 0.5363; **(c)** 0.4288; **(d)** 0.3068

25. **(a)**

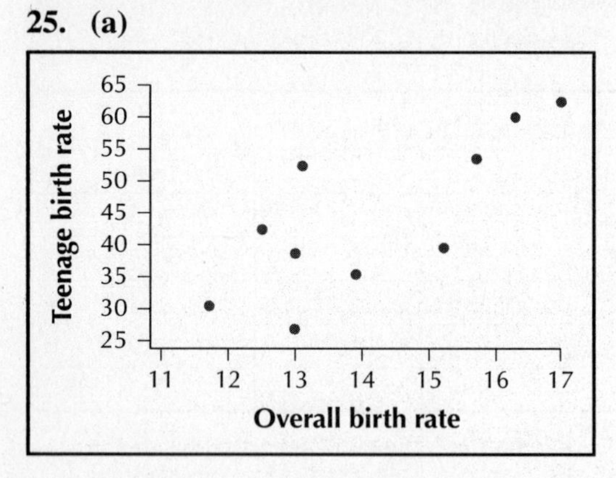

(b) The relationship between overall birthrate and teenage birthrate appears to be positive.

27. **(a)**

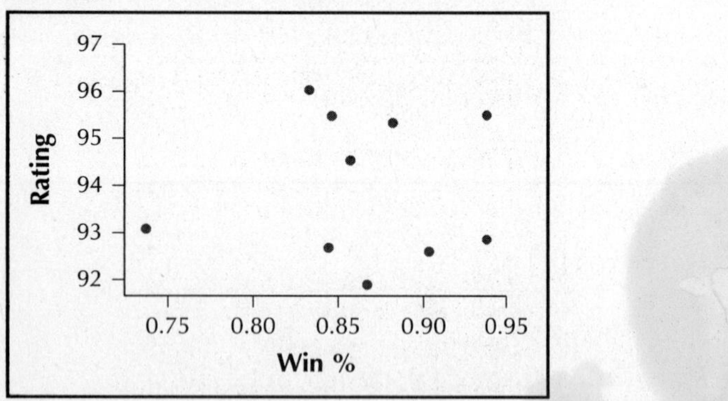

(b) There appears to be no relationship between a team's winning percentage and a team's power rating.

29. Knowing the age of the baseball player wouldn't help us very much to estimate the player's batting average. The scatterplot indicates that there is no relationship between a player's age and a player's batting average.

31.

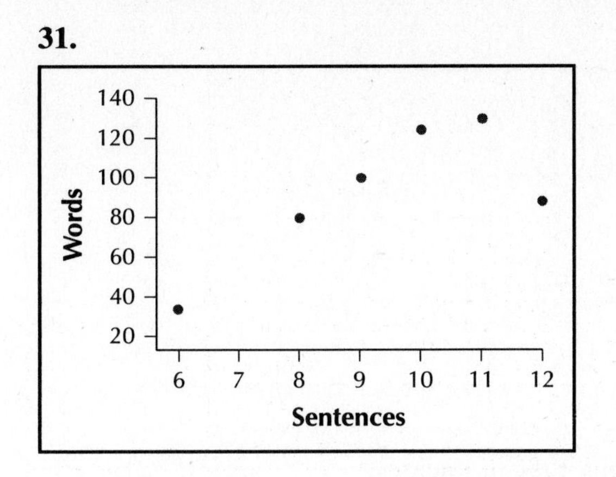

There appears to be a positive relationship between the number of words and the number of sentences in the *New Yorker* ads.

33. 0.09

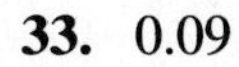

35.

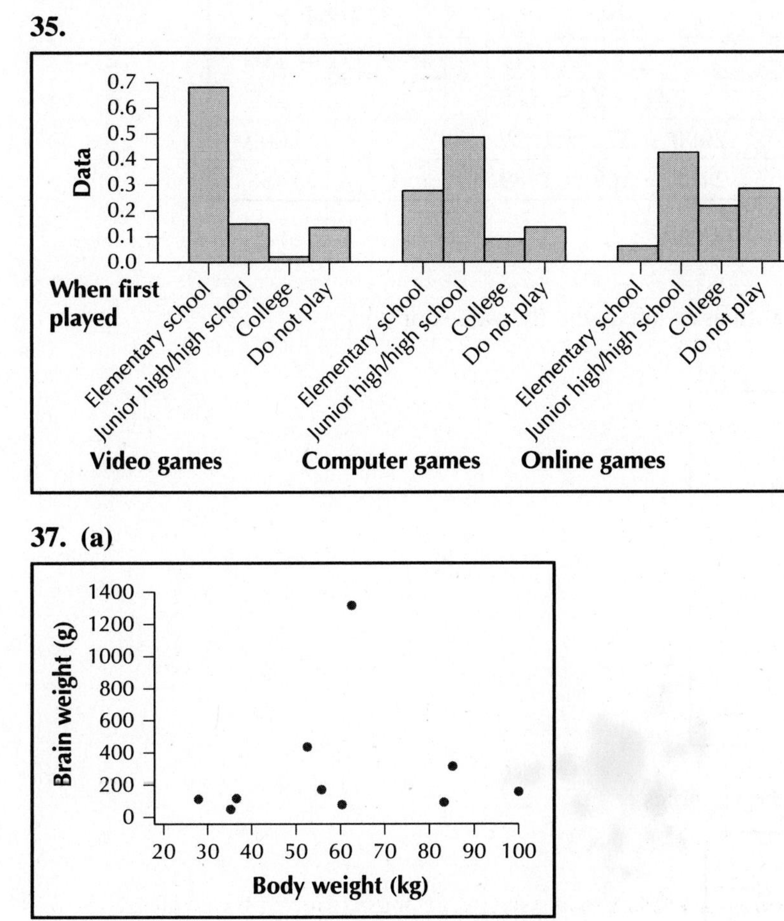

37. (a)

(b) There appears to be no relationship between body weight and brain weight. **(c)** Yes. **(d)** Humans.

39.

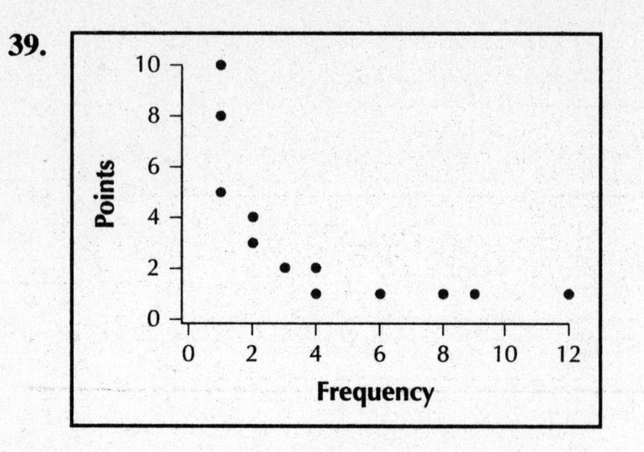

There appears to be a negative relationship between the points, but it is not linear.

41. Missing values are bold.

Visited Memorial Web Site

Student Status	Yes	No	Total
Full-time	**469 − 53 − 371 = 45**	171	**45 + 171 = 216**
Part-time	53	**242 − 53 = 189**	242
No	371	**2000 − 371 = 1629**	2000
Total	469	**2458 − 469 = 1989**	2458

43. The people who weren't students; full-time students.

45. The relationship is not apparent.

47. Low values of x and high values of x are both associated with the values of y.

49. Answers will vary.

51. (a)

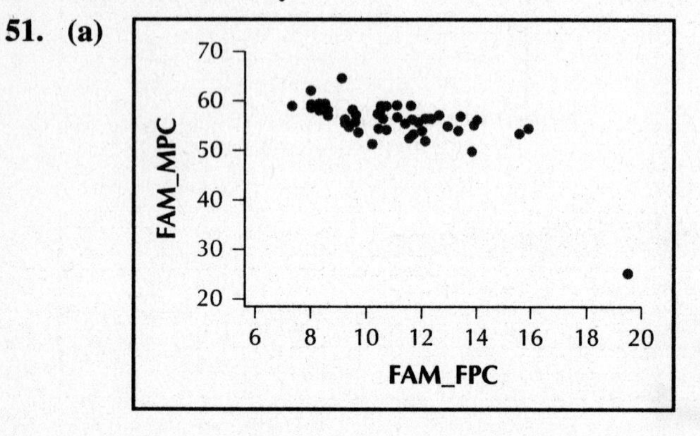

(b) There is a negative relationship between the variables.

53. (a)

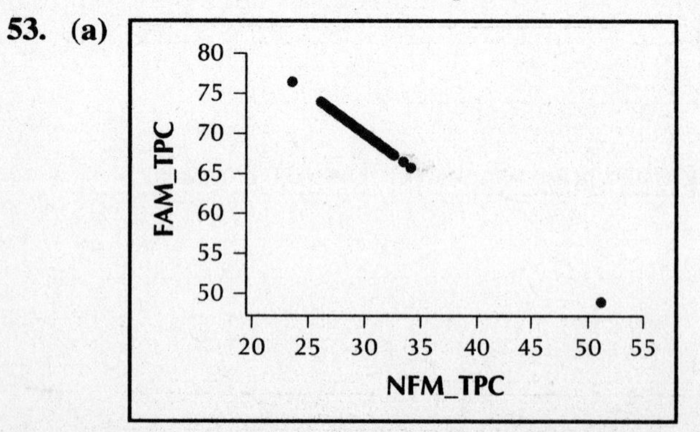

(b) Negative. The higher the percentage of nonfamily households, the lower the percentage of family households. **(c)** It is almost a perfect straight line. Two variables usually don't have this perfect of a linear relationship.

Section 4.2

1. Scatterplot.

3. Between -1 and 1, inclusive. Thus $-1 \leq r \leq 1$.

5. A value of r near 1 means that the points in the scatterplot lie close to a straight line with a positive slope. Thus the answer is graph **i.**

7. A value of r near 0 means that the points in the scatterplot don't lie in any apparent linear pattern. Thus the answer is graph **ii.**

9. A value of r near -1 means that the points in the scatterplot lie close to a straight line with a negative slope. Thus the answer is graph **iv.**

11. **(a)** (6,8), (7,7), (8,7), (9,8), (10,8), (11,8), (12,7), (13,8), (14,8), (15,7)

(b) From Minitab:

Correlations: x, y

```
Pearson correlation of x and y = 0.000
```

From the TI-83/84: $r = 0$.

Using the formula:

$$\bar{x} = \frac{\sum x}{n} = \frac{6+7+8+9+10+11+12+13+14+15}{10} = \frac{105}{10} = 10.5$$

$$\bar{y} = \frac{\sum y}{n} = \frac{8+7+7+8+8+8+7+8+8+7}{10} = \frac{76}{10} = 7.6$$

x	y	$x-\bar{x}$	$(x-\bar{x})^2$	$y-\bar{y}$	$(y-\bar{y})^2$	$(x-\bar{x})(y-\bar{y})$
6	8	−4.5	20.25	0.4	0.16	−1.8
7	7	−3.5	12.25	−0.6	0.36	2.1
8	7	−2.5	6.25	−0.6	0.36	1.5
9	8	−1.5	2.25	0.4	0.16	−0.6
10	8	−0.5	0.25	0.4	0.16	−0.2
11	8	0.5	0.25	0.4	0.16	0.2
12	7	1.5	2.25	−0.6	0.36	−0.9
13	8	2.5	6.25	0.4	0.16	1.0
14	8	3.5	12.25	0.4	0.16	1.4
15	7	4.5	20.25	−0.6	0.36	−2.7
			$\sum(x-\bar{x})^2 = 82.5$		$\sum(y-\bar{y})^2 = 2.4$	$\sum(x-\bar{x})(y-\bar{y}) = 0$

$$s_x = \sqrt{\frac{\sum(x-\bar{x})^2}{n-1}} = \sqrt{\frac{82.5}{10-1}} = 3.027650354$$

$$s_y = \sqrt{\frac{\sum(y-\bar{y})^2}{n-1}} = \sqrt{\frac{82.1}{10-1}} = 0.5163977795$$

$$r = \sqrt{\frac{\sum(x-\bar{x})(y-\bar{y})}{(n-1)s_x s_y}} = \frac{0}{(10-1)(3.027650354)(0.5163977795)} = 0$$

13. **(a)** (1,11), (2,9), (3,7), (4,7), (5,6), (6,6), (7,4), (8,3), (9,3), (10,1)

(b) From Minitab:

```
Correlations: x, y

        Pearson correlation of x and y = -0.978
```

From TI-83/84: $r = -0.9781316853$.

Using the formula:

$$\bar{x} = \frac{\sum x}{n} = \frac{1+2+3+4+5+6+7+8+9+10}{10} = \frac{55}{10} = 5.5$$

$$\bar{y} = \frac{\sum y}{n} = \frac{11+9+7+7+6+6+4+3+3+1}{10} = \frac{57}{10} = 5.7$$

x	y	$x-\bar{x}$	$(x-\bar{x})^2$	$y-\bar{y}$	$(y-\bar{y})^2$	$(x-\bar{x})(y-\bar{y})$
1	11	−4.5	20.25	5.3	28.09	−23.85
2	9	−3.5	12.25	3.3	10.89	−11.55
3	7	−2.5	6.25	1.3	1.69	−3.25
4	7	−1.5	2.25	1.3	1.69	−1.95
5	6	−0.5	0.25	0.3	0.09	−0.15
6	6	0.5	0.25	0.3	0.09	0.15
7	4	1.5	2.25	−1.7	2.89	−2.55
8	3	2.5	6.25	−2.7	7.29	−6.75
9	3	3.5	12.25	−2.7	7.29	−9.45
10	1	4.5	20.25	−4.7	22.09	−21.15
			$\sum(x-\bar{x})^2 = 82.5$		$\sum(y-\bar{y})^2 = 82.1$	$\sum(x-\bar{x})(y-\bar{y}) = -80.5$

$$s_x = \sqrt{\frac{\sum(x-\bar{x})^2}{n-1}} = \sqrt{\frac{82.5}{10-1}} = 3.027650354$$

$$s_y = \sqrt{\frac{\sum(y-\bar{y})^2}{n-1}} = \sqrt{\frac{82.1}{10-1}} = 3.020301677$$

$$r = \frac{\sum(x-\bar{x})(y-\bar{y})}{(n-1)s_x s_y} = \frac{-80.5}{(10-1)(3.027650354)(3.020301677)} = -0.9781316854$$

15.

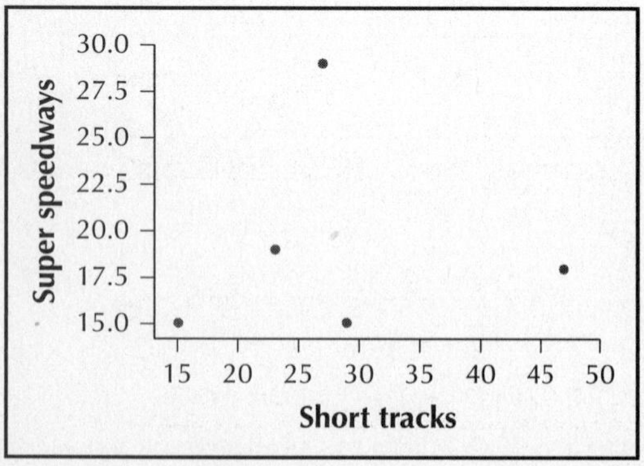

17. **(a)** $\bar{x} = \frac{\sum x}{n} = \frac{47+27+15+29+23}{5} = \frac{141}{5} = 28.2$

$$\bar{y} = \frac{\sum y}{n} = \frac{18+29+15+15+19}{5} = \frac{96}{5} = 19.2$$

(b)

x	y	$x - \bar{x}$	$(x - \bar{x})^2$	$y - \bar{y}$	$(y - \bar{y})^2$	$(x - \bar{x})(y - \bar{y})$
47	18	18.8	353.44	−1.2	1.44	−22.56
27	29	−1.2	1.44	9.8	96.04	−11.76
15	15	−13.2	174.24	−4.2	17.64	55.44
29	15	0.8	0.64	−4.2	17.64	−3.36
23	19	−5.2	27.04	−.2	.04	1.04
			$\Sigma(x - \bar{x})^2 = 556.8$		$\Sigma(y - \bar{y})^2 = 132.8$	$\Sigma(x - \bar{x})(y - \bar{y}) = 18.8$

(c) $s_x = \sqrt{\dfrac{\sum(x - \bar{x})^2}{n - 1}} = \sqrt{\dfrac{556.8}{5 - 1}} = 11.79830496$

$s_y = \sqrt{\dfrac{\sum(y - \bar{y})^2}{n - 1}} = \sqrt{\dfrac{132.8}{5 - 1}} = 5.761944116$

(d) $r = \dfrac{\sum(x - \bar{x})(y - \bar{y})}{(n - 1)s_x s_y} = \dfrac{18.8}{(5 - 1)(11.79830496)(5.761944116)} = 0.0691367879$

(e) From Minitab:

Correlations: Short Tracks, Super Speedways

```
Pearson correlation of Short Tracks and Super Speedways = 0.069
```

From the TI-83/84: $r = 0.0691367879$.

19.

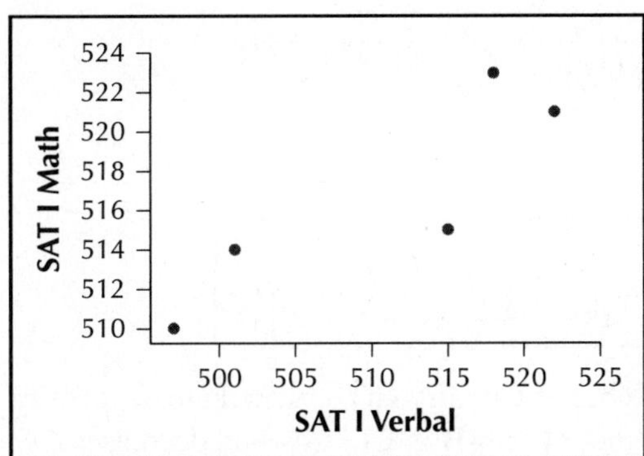

Positively correlated.

21. SAT I Verbal Scores and SAT I Math Scores are positively correlated. As the SAT I Verbal Score increases the SAT I Math Score increases. Yes.

23. From Minitab:

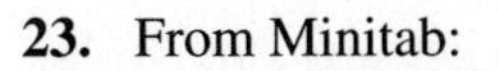

Correlations: Body Weight (kg), Brain Weight (g)

```
Pearson correlation of Body Weight (kg) and Brain Weight (g) = 0.099= 0.879
```

From the TI-83/84: $r = 0.0992032294$.

25. Positive.

27. The number of calories in a serving of breakfast cereal and the amount of sugar in a serving of breakfast cereal are positively correlated. As the amount of sugar in a serving of breakfast cereal increases, the number of calories in a serving of the breakfast cereal tends to increase.

29. A player's age and a player's batting average are not correlated. As a player's age increases, the player's batting average tends to remain unchanged.

31. Positive; positive

33. **(a)** 18 – 65 group. **(b)** No.

35. **(a)** $\bar{x} = \frac{\sum x}{n} = \frac{1 + 3 + 11 + 8 + 5 + 27}{6} = \frac{55}{6} = 9.16666667$

$\bar{y} = \frac{\sum y}{n} = \frac{10 + 12 + 1 + 3 + 6 + 1}{6} = \frac{33}{6} = 5.5$

x	y	$x - \bar{x}$	$(x - \bar{x})^2$	$y - \bar{y}$	$(y - \bar{y})^2$	$(x - \bar{x})(y - \bar{y})$
1	10	−8.16666667	66.69444449	4.5	20.25	−36.75
3	12	−6.16666667	38.0277778	6.5	42.25	−40.08333333
11	1	1.83333333	3.36111109	−4.5	20.25	−8.249999985
8	3	−1.16666667	1.36111111	−2.5	6.25	2.916666675
5	6	−4.16666667	17.3611111	.5	.25	−2.0833333335
27	1	17.83333333	318.027777	−4.5	20.25	−80.2499999
			$\sum(x - \bar{x})^2 = 444.8333333$		$\sum(y - \bar{y})^2 = 109.5$	$\sum(x - \bar{x})(y - \bar{y}) = -164.5$

$s_x = \sqrt{\frac{\sum(x - \bar{x})^2}{n - 1}} = \sqrt{\frac{444.8333333}{6 - 1}} = 9.432214303 \qquad s_y = \sqrt{\frac{\sum(y - \bar{y})^2}{n - 1}} = \sqrt{\frac{109.5}{6 - 1}} = 4.679743583$

$$r = \frac{\sum(x - \bar{x})(y - \bar{y})}{(n - 1)s_x s_y} = \frac{-164.5}{(6 - 1)(9.432214303)(4.679743583)} = -0.7453498716$$

From the TI-83/84: $r = -0.7453498716$.

From Minitab:

```
Correlations: Hip-Hop CDs owned, Country CDs owned

     Pearson correlation of Hip-Hop CDs owned and Country CDs owned = -0.745
```

(b) Yes. **(c)** The variables number of Hip-Hop CD's owned and number of Country CD's owned are negatively correlated. As the number of Country CD's owned increases, the number of Hip-Hop CD's owned decreases.

37. **(a)** Yes. **(b)** The variables birth weight and time born are not correlated. As the time born increases the birth weight tends to remain the same.

39. Answers will vary.

41. Answers will vary.

43. **(a)** Since r is positive, the relationship between the variables is positive. **(b)** Since r is negative, the relationship between the variables is negative. **(c)** Since r is zero, there is no apparent relationship between the variables.

45. **(a)** Answers will vary. For the particular example given, three of the four points fall in Regions 1 and 3, so it makes sense that r is positive. **(b)** Answers will vary. For the particular example given, all of the points fall in Regions 2 and 4, so it makes sense that r is negative. **(c)** Answers will vary. For the particular example given, there is one point in each Region. Therefore, it makes sense that r is near zero.

Section 4.3

1. To approximate the relationship between two numerical variables using the regression line and the regression equation.

3. We can find the predicted value of y by plugging a given value of x into the regression equation and simplifying.

5. Extrapolation is the process of making predictions based on x-values that are beyond the range of the x-values in our data set.

7. **(a)** $\bar{x} = \frac{\sum x}{n} = \frac{47 + 27 + 15 + 29 + 23}{5} = \frac{141}{5} = 28.2$

$\bar{y} = \frac{\sum y}{n} = \frac{18 + 29 + 15 + 15 + 19}{5} = \frac{96}{5} = 19.2$

x	y	$x - \bar{x}$	$(x - \bar{x})^2$	$y - \bar{y}$	$(y - \bar{y})^2$	$(x - \bar{x})(y - \bar{y})$
47	18	18.8	353.44	−1.2	1.44	−22.56
27	29	−1.2	1.44	9.8	96.04	−11.76
15	15	−13.2	174.24	−4.2	17.64	55.44
29	15	0.8	0.64	−4.2	17.64	−3.36
23	19	−5.2	27.04	−.2	.04	1.04
			$\Sigma(x - \bar{x})^2 = 556.8$		$\Sigma(y - \bar{y})^2 = 132.8$	$\Sigma(x - \bar{x})(y - \bar{y}) = 18.8$

$$b_1 = \frac{\sum(x - \bar{x})(y - \bar{y})}{\sum(x - \bar{x})^2} = \frac{18.8}{556.8} = 0.0337643678$$

(b) $b_0 = \bar{y} - (b_1 \cdot \bar{x}) = 19.2 - (0.0337643678)(28.2) = 18.24784483$

(c) From Minitab:

```
The regression equation is
          Super Speedways = 18.2 + 0.034 Short Tracks

          Predictor          Coef  SE Coef      T      P
          Constant         18.248    8.469   2.15  0.120
          Short Tracks     0.0338   0.2813   0.12  0.912
```

From the TI-83/84:

$$a = b_0 = 18.24784483 \text{ and } b = b_1 = .0337643678$$

$$\hat{y} = b_1 + b_1 x = 18.24784483 + 0.0337643678\, x$$

The estimated value of the number of Super Speedways won is 18.24784483 plus 0.0337643678 times the number of Short Tracks won.

9. **(a)** When $x = 30$, $\hat{y} = 18.24784483 + 0.0337643678(30) = 19.26077586 \cong 19$ is the estimate of the number of Super Speedways won. **(b)** Since the range of the x values of the data set is from 15 to 47, inclusive, $x = 50$ is not in the range of the x values of the data set. Therefore it is not appropriate to use the regression equation to estimate the number of Super Speedways won when $x = 50$. **(c)** When $x = 20$, $\hat{y} = 18.24784483 + 0.0337643678\,(20) = 18.92313219 \cong 19$ is the estimate of the number of Super Speedways won.

11. **(a)** The slope $b_1 = 0.4264339152$ means that the estimated SAT I Math Score increases by 0.4264339152 points for every increase of one point in the SAT I Verbal Score. **(b)** The y-intercept $b_0 = 298.8628429$ means that the estimated SAT I Math Score is 298.8628429 when the SAT I Verbal Score is 0. **(c)** The SAT I Verbal Score can't be 0, so this situation will never happen.

13. From Minitab:

```
The regression equation is
      Brain Weight (g) = 194 + 1.59 Body Weight (kg)

      Predictor             Coef  SE Coef     T      P
      Constant             193.5    360.7  0.54  0.606
      Body Weight (kg)     1.595    5.656  0.28  0.785
```

So $b_0 = 193.5$ and $b_1 = 1.595$.
(a) Therefore $\hat{y} = 193.5 + 1.595\,x$ **(b)** The estimated Brain Weight of a mammal is 193.5 grams plus 1.595 times the Body Weight.

15. **(a)** When $x = 50$ kg, the estimated brain weight is $193.5 + 1.595(50) = 273.25$ grams. **(b)** When $x = 100$ kg, the estimated brain weight is $193.5 + 1.595(100) = 353$ grams. **(c)** The value $x = 200$ kg is outside of the range of the x-values of the data set, which is between 35 kg and 100 kg. Thus it is not appropriate to use the regression equation to estimate the brain weight of a mammal with a body weight of 200 kg.

17. Curved.

19. **(a)** 10.5 **(b)** In a state with 0 households 10.5% of the households are headed by women. This does not make sense. Since all states have households, the value $x = 0$ would not occur. Since any percent of 0 is 0 there is no reason why the percent of 0 households that are headed by woman would have to be 10.5%. **(c)** This estimate would be considered extrapolation because the value of $x = 0$ is outside the range of x-values in the data set. **(d)** 0.000000282 **(e)** For each increase of one household the percentage of households headed by women increases by 0.000000282. **(f)** Percentage of households headed by women = 10.5 + 0.000000282 (Total number of households). The estimated percentage of households headed by women equals 10.5 plus 0.000000282 times the total number of households. **(g)** Positive, because the slope is positive.

21. **(a)** When $x = 7{,}000{,}000$ households, then the estimated percent of households headed by women is $10.5 + 0.000000282\,(7{,}000{,}000) = 12.474\%$. **(b)** Since $x = 100{,}000$ is not in the range of the x-values in the data set, it is not appropriate to use the regression equation to estimate the percentage of households headed by women.

23. **(a)** 1.032 **(b)** The estimated percent change in the stock prices selected by the darts increases by 1.032% for each unit percent increase in the Dow Jones Industrial Average. **(c)** Darts = −2.49 + 1.032 DJIA The estimated percent change in the stock prices selected by the darts is equal to −2.49 plus 1.032 times the percent change in the Dow Jones Industrial Average. **(d)** Positive, because the slope of the regression line is positive.

25. **(a)** When $x = 22$, the estimated percent change in the stocks selected by the darts is $-2.49 + 1.032\,(22) = 20.214\%$. **(b)** When $x = -10$, the estimated percent change in the stocks selected by the darts is $-2.49 + 1.032\,(-10) = -12.81\%$. **(c)** Since $x = -22$ is not in the range of the x-values of the data set, it is not appropriate to use the regression equation to estimate the net price change for the darts' portfolio.

27. For the original 10 towns:
From Minitab:

Regression Analysis: High temperature versus Low temperature

```
The regression equation is
      High temperature = 10.1 + 0.987 Low temperature

      Predictor            Coef  SE Coef      T      P
      Constant           10.053    3.386   2.97  0.018
      Low temperature   0.98654  0.07765  12.70  0.000


      S = 3.57275   R-Sq = 95.3%   R-Sq(adj) = 94.7%
```

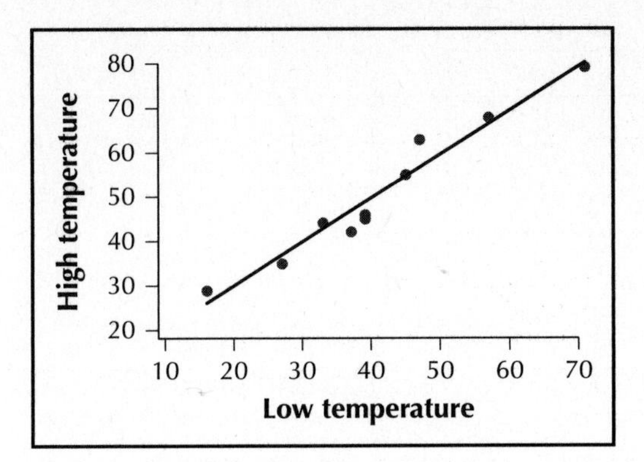

After a town with a low temperature of 10 degrees and a high temperature of 50 degrees is added to the data set.

Regression Analysis: High temperature versus Low temperature

```
The regression equation is
        High temperature = 23.7 + 0.703 Low temperature

        Predictor            Coef   SE Coef     T      P
        Constant           23.652     6.635  3.56  0.006
        Low temperature    0.7027    0.1592  4.41  0.002


        S = 8.71350    R-Sq = 68.4%    R-Sq(adj) = 64.9%
```

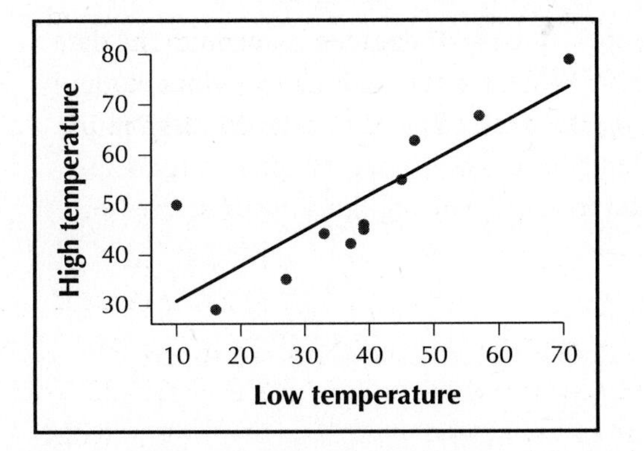

After a town with a low temperature of 10 degrees and a high temperature of 60 degrees is added to the data set.

Regression Analysis: High temperature versus Low temperature

```
The regression equation is
        High temperature = 28.2 + 0.608 Low temperature

        Predictor            Coef   SE Coef     T      P
        Constant           28.172     8.547  3.30  0.009
        Low temperature    0.6083    0.2051  2.97  0.016


        S = 11.2249    R-Sq = 49.4%    R-Sq(adj) = 43.8%
```

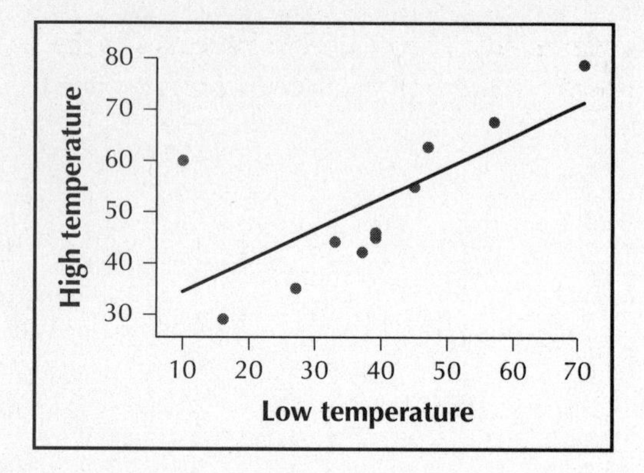

All three regression lines and scatterplots on the same graph.

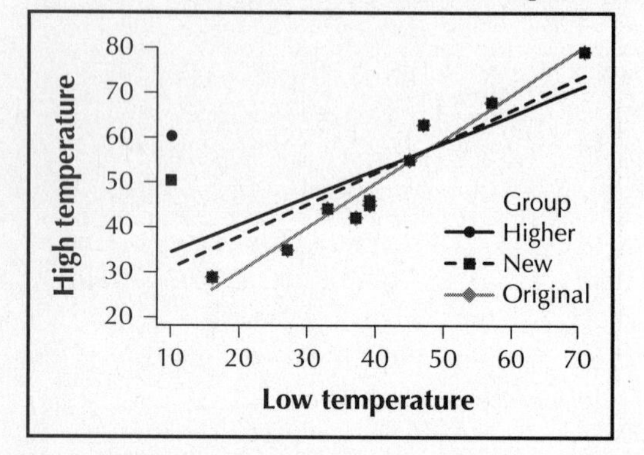

When a town with a low temperature of 10 degrees and a high temperature of 50 degrees is added to the data set, the slope of the regression line decreases from 0.9865 to 0.7027. When a town with a low temperature of 10 degrees and a temperature greater than 50 degrees is added to the data set, the slope of the regression line decreases to less than 0.7027. This is because a town with a low temperature of 10 degrees and a high temperature of at least 50 degrees will pull the regression line up at the left end, causing the slope to decrease.

29. Answers will vary.

31. Answers will vary.

33. **(a)** Positive relationship between x and y. **(b)** Negative relationship between x and y. **(c)** No apparent relationship between x and y.

Chapter 4 Review Exercises

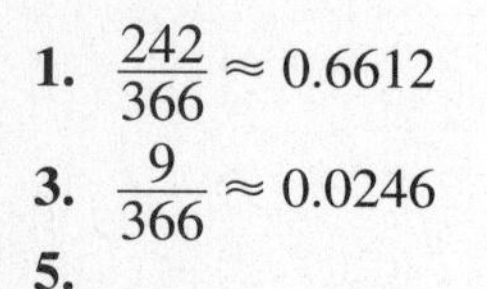

1. $\frac{242}{366} \approx 0.6612$

3. $\frac{9}{366} \approx 0.0246$

5.

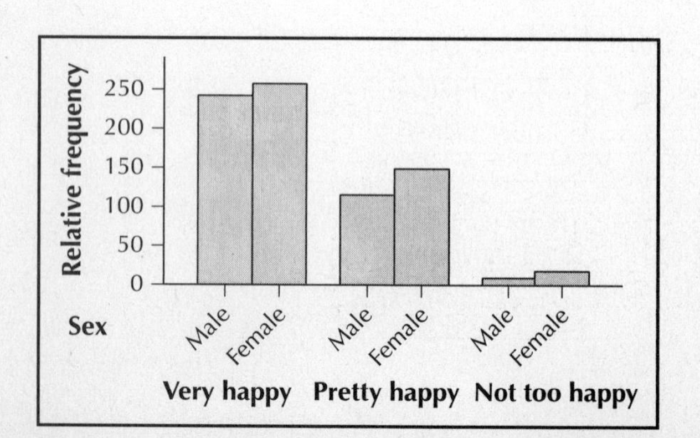

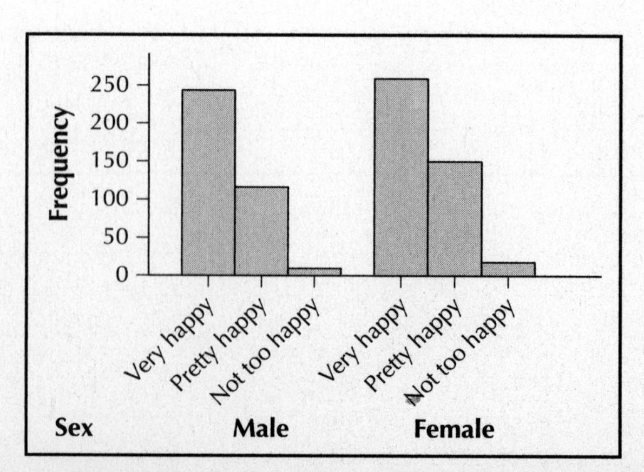

7. **(a)** The relationship in each of the scatterplots in Exercise 6 is positive. **(b)** The *y* variable increases as the *x* variable increases. **(c)** Low values of *x* are associated with low values of *y* and high values of *x* are associated with high values of *y*.

9. Near zero.

11. Positive.

13.

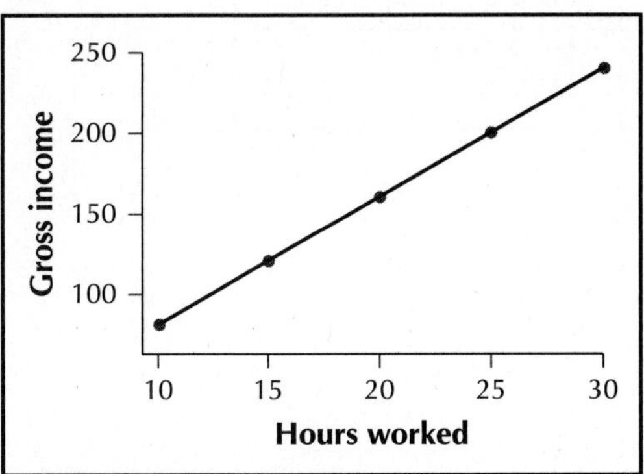

(a) 216; **(b)** 192 **(c)** Since all of the points lie on the regression line, this *x* variable is extremely useful in predicting this *y* variable.

15. $b_0 \approx 0$

Chapter 4 Quiz

1. True.

2. False. Scatterplots are constructed with the *x* variable on the horizontal axis and the *y* variable on the vertical axis.

3. False. The correlation coefficient *r* measures the strength of the linear relationship between two numerical variables.

4. Two-way tables; contingency tables

5. Estimate.

6. 0

7. Unit

8. Clustered bar graphs.

9. Extrapolation.

10. Negative.

11. Missing values are bold.

	Exciting	**Routine**	**Dull**	**Total**
Protestant	264	**623 − 264 − 33 = 326**	33	623
Catholic	107	128	6	**241**
Jewish	12	8	**20 − 12 − 8 = 0**	20
None	38	29	2	69
Other	9	5	**14 − 9 − 5 = 0**	14
Total	**430**	**496**	41	967

12.

	Exciting	Routine	Dull
Protestant	264/430 × 100% ≈ 61.40%	326/496 × 100% ≈ 65.73%	33/41 × 100% ≈ 80.49%
Catholic	107/430 × 100% ≈ 24.88%	128/496 × 100% ≈ 25.81%	6/41 × 100% ≈ 14.63%
Jewish	12/430 × 100% ≈ 2.79%	8/496 × 100% ≈ 1.61%	0/41 × 100% ≈ 0%
None	38/430 × 100% ≈ 8.84%	29/496 × 100% ≈ 5.85%	2/41 × 100% ≈ 4.88%
Other	9/430 × 100% ≈ 2.09%	5/496 × 100% ≈ 1.01%	0/41 × 100% ≈ 0%
Total	100.00%	100.01%	100.0%

13. Protestant.

14. Protestant.

15. Protestant.

16.

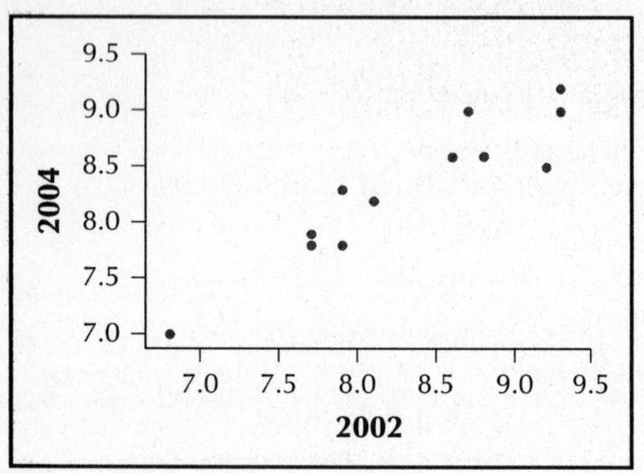

17. Positive.

18. As the percent of violent crimes committed in 2002 increases, the percent of violent crimes committed in 2004 increases.

19. **(a)**

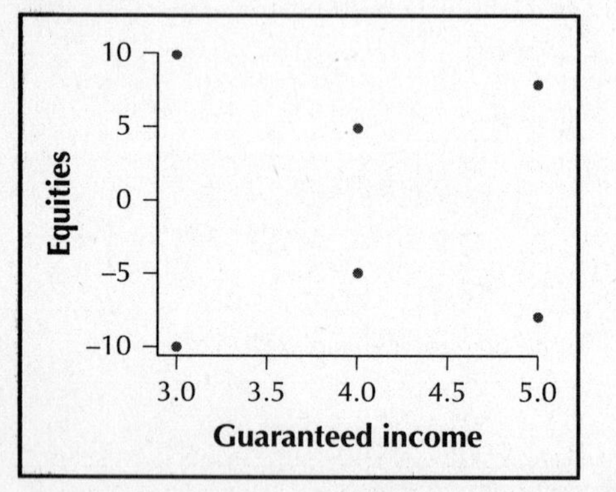

(b) No apparent relationship. **(c)** Near zero; near zero.

20. **(a)** $\bar{x} = \frac{\sum x}{n} = \frac{3 + 3 + 4 + 4 + 5 + 5}{6} = \frac{24}{6} = 4$

$$\bar{y} = \frac{\sum y}{n} = \frac{10 + (-10) + (-5) + 5 + 8 + (-8)}{6} = \frac{0}{6} = 0$$

x	y	$x - \bar{x}$	$(x - \bar{x})^2$	$y - \bar{y}$	$(y - \bar{y})^2$	$(x - \bar{x})(y - \bar{y})$
3	10	−1	1	10	100	−10
3	−10	−1	1	−10	100	10
4	−5	0	0	−5	25	0
4	5	0	0	5	25	0
5	8	1	1	8	64	8
5	−8	1	1	−8	64	−8
			$\Sigma(x - \bar{x})^2 = 4$		$\Sigma(y - \bar{y})^2 = 378$	$\Sigma(x - \bar{x})(y - \bar{y}) = 0$

$$s_x = \sqrt{\frac{\sum(x - \bar{x})^2}{n - 1}} = \sqrt{\frac{4}{6 - 1}} = 0.894427191$$

$$s_y = \sqrt{\frac{\sum(y - \bar{y})^2}{n - 1}} = \sqrt{\frac{378}{6 - 1}} = 8.694826048$$

$$r = \frac{\sum(x - \bar{x})(y - \bar{y})}{(n - 1)s_x s_y} = \frac{0}{(6 - 1)(0.894427191)(8.694826048)} = 0$$

$r = 0$ from the TI-83/84.

From Minitab:

Correlations: Equities, Guaranteed Income

```
Pearson correlation of Equities and Guaranteed Income = 0.000
```

(b) Yes. **(c)** There is no relationship between the net price change for equities and the net price change for guaranteed income investments. When the net price changes for guaranteed income increases, the net price changes for equities remain unchanged.

21.

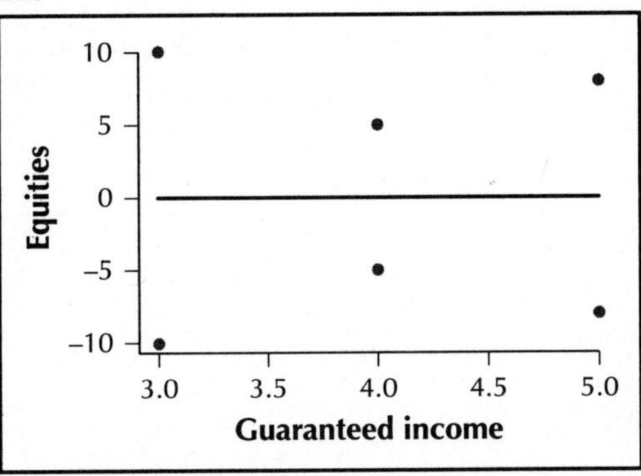

(a) 0 **(b)** 0; All of the estimated equities net price changes are 0. **(c)** Not very useful because all of the estimates are 0.

22. **(a)** y-intercept $b_0 = 97.1$ **(b)** A breakfast cereal with 0 grams of fat will have 97.1 calories per serving. This makes sense because there are other ingredients in cereal besides fat and they have calories. **(c)** No, because $x = 0$ is an x-value in the data set. **(d)** The number of calories in a serving of breakfast cereal increases by 9.65 calories for every increase of one gram of fat. **(e)** Calories = 97.1 + 9.65 Fat. The estimated number of calories in a serving of breakfast cereal equals 97.1 plus 9.65 times the number of grams of fat. **(f)** Positive because the slope of the regression line is positive.

23. **(a)** Cereal A has 2 (9.65) = 19.3 more calories per serving than does Cereal B. **(b)** Cereal C has 3 (9.65) = 28.95 fewer calories per serving than does Cereal D.

24. **(a)** When $x = 4$, the estimated number of calories per serving of breakfast cereal is 97.1 + (9.65) (4) = 135.7. **(b)** The value of $x = 10$ is outside the range of x-values in our data set, so it is not appropriate to use the regression equation to estimate the number of calories per serving of breakfast cereal with 10 grams of fat.

Chapter 5

Probability

Section 5.1

1. Answers will vary; chance, likelihood.

3. **(a), (b), (c)** Answers will vary.

5. The experiment has equally likely outcomes.

7. We consider all available information, tempered by our experience and intuition, and then assign a probability value that expresses our estimate of the likelihood that the outcome will occur.

9. First find out how many students are at your college and find out how many of them like Hip-Hop music. Then calculate the relative frequency of students who like Hip-Hop music. Use the relative frequency method.

11. It is not a probability model because the probability for females is greater than 1.

13. It is not a probability model because the sum of probabilities is greater than 1.

15. It is not a probability model because the sum of probabilities is less than 1.

For Exercises 17, 19, 21, the sample space is $\{1, 2, 3, 4, 5, 6\}$, so $N(S) = 6$.

17. Let E be the event roll an even number. Event E consists of $\{2, 4, 6\}$, so $N(E) = 3$. Therefore,

$$P(E) = \frac{N(E)}{N(S)} = \frac{3}{6} = \frac{1}{2}.$$

19. Let E be the event roll a number less than 3. Event E consists of $\{1, 2\}$, so $N(E) = 2$. Therefore,

$$P(E) = \frac{N(E)}{N(S)} = \frac{2}{6} = \frac{1}{3}.$$

21. Let E be the event roll a 3 and a 5. Event E consists of $\{\ \}$, so $N(E) = 0$. Therefore,

$$P(E) = \frac{N(E)}{N(S)} = \frac{0}{6} = 0.$$

For Exercises 23, 25, $N(S) = 52$.

23. Let E be the event draw a heart. Event E consists of the 13 hearts, $\{A\heartsuit, 2\heartsuit, 3\heartsuit, 4\heartsuit, 5\heartsuit, 6\heartsuit, 7\heartsuit, 8\heartsuit, 9\heartsuit, 10\heartsuit, J\heartsuit, Q\heartsuit, K\heartsuit\}$, so $N(E) = 13$. Therefore,

$$P(E) = \frac{N(E)}{N(S)} = \frac{13}{52} = \frac{1}{4}.$$

25. Let E be the event draw a black card. Event E consists of the 26 black cards, $\{A\spadesuit, 2\spadesuit, 3\spadesuit, 4\spadesuit, 5\spadesuit, 6\spadesuit, 7\spadesuit, 8\spadesuit, 9\spadesuit, 10\spadesuit, J\spadesuit, Q\spadesuit, K\spadesuit, A\clubsuit, 2\clubsuit, 3\clubsuit, 4\clubsuit, 5\clubsuit, 6\clubsuit, 7\clubsuit, 8\clubsuit, 9\clubsuit, 10\clubsuit, J\clubsuit, Q\clubsuit, K\clubsuit\}$, so $N(E) = 26$. Therefore,

$$P(E) = \frac{N(E)}{N(S)} = \frac{26}{52} = \frac{1}{2}.$$

27. {HHH, HHT, HTH, THH, TTH, THT, HTT, TTT}

29. Choose a branch for the first toss, then choose a branch for the second toss, then choose a branch for the third toss. This will give us one of the possible outcomes. If we follow all possible choices of branches for each toss, we get all of the outcomes in the sample space.

31. The probability we found in Exercise 16 was for both an event and an outcome. All of the other probabilities were for events.

33. **(a)** A sum of 7. **(b)** Let E be the event roll a sum of 7. Then the outcomes that belong to event E are: {(1,6), (2,5), (3,4), (4,3), (5,2), (6,1)}, so $N(E) = 6$.

From Example 5.3 $N(S) = 36$. Therefore,

$$P(E) = \frac{N(E)}{N(S)} = \frac{6}{36} = \frac{1}{6}.$$

35. **(a)** Let E be the event it rained that day. Then

$$P(E) \approx \frac{\text{frequency of } E}{\text{number of trials of experiment}} = \frac{33}{100} = 0.33.$$

(b) If it rained on 33 out of 100 days then it did not rain on $100 - 33 = 67$ days. Let F be the event that it did not rain that day. Then

$$P(F) \approx \frac{\text{frequency of } F}{\text{number of trials of experiment}} = \frac{67}{100} = 0.67.$$

(c) Relative frequency method.

37. **(a)** Let E be the event that a student cuts a particular class. Then

$$P(E) \approx \frac{\text{frequency of } E}{\text{number of trials of experiment}} = \frac{25}{1000} = 0.025.$$

(b) Relative frequency method. **(c)** No, the events cutting class and not cutting class are not equally likely.

39. **(a)**

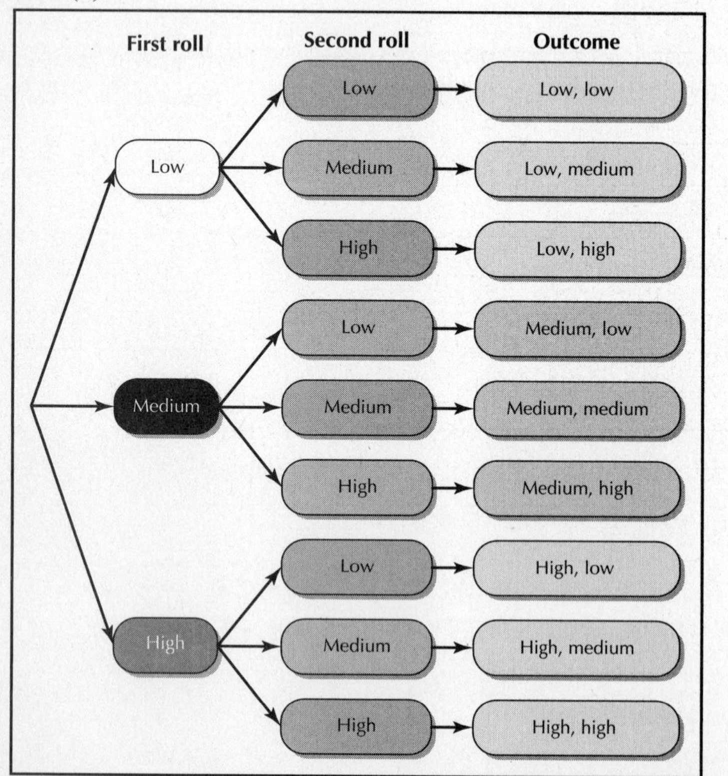

(b) {Low and Low, Low and Medium, Low and High, Medium and Low, Medium and Medium, Medium and High, High and Low, High and Medium, High and High}; **(c)** Since there are nine possible outcomes in the sample space and only one of them consists of two low die results, $P(\text{low, low}) = \frac{N(\text{low, low})}{N(S)} = \frac{1}{9}$. Since each outcome in the sample space is equally likely, the classical method of assigning probability was used.

41. (a)

	Frequency	Relative frequency
Girls	18	18/44 = 0.4091
Boys	26	26/44 = 0.5909
Total	44	44/44 = 1.0000

(b)

Outcome	Probability
Girl	18/44 = 0.4091
Boy	26/44 = 0.5909

Both $P(\text{Girl}) = 18/44 = 0.4091$ and $P(\text{Boy}) = 26/44 = 0.5909$ are between 0 and 1. $P(\text{Girl}) + P(\text{Boy}) = 18/44 + 26/44 = 0.4091 + 0.5909 = 44/44 = 1.0000$.

43. Answers will vary.

45. (a) E, Z, Yes; **(b)** a consonant; **(c)** 5/26; **(d)** 1/2; **(e)** One-half of the most popular letters are vowels but only 5/26 of all of the letters are vowels; **(f)** T, N, R, S, D.

47. (a) Let E be the event roll a sum of at least 9. Event E consists of rolling a sum of 9, 10, 11, or 12. Thus event E consists of the outcomes: {(3,6), (4,5), (5,4), (6,3), (4,6), (5,5), (6,4), (5,6), (6,5), (6,6)}, so $N(E) = 10$. From Example 5.3, $N(S) = 36$. Therefore,

$$P(E) = \frac{N(E)}{N(S)} = \frac{10}{36} = \frac{5}{18}.$$

(b) Since rolling a sum of at least 9 results in winning \$5, a sum of at most 8 will result in winning nothing. From (a), there are 10 ways to win \$5, there are $36 - 10 = 26$ ways to win nothing. Let F be the event wins nothing. Then

$$P(F) = \frac{N(F)}{N(S)} = \frac{26}{36} = \frac{13}{18}.$$

(c) $5/18 \cdot \$5 \approx \$1.39$

49. (a)

Type of music	Probability
Hip-hop/rap	0.27
Pop	0.23
Rock/punk	0.17
Alternative	0.07
Christian/gospel	0.06
R&B	0.06
Country	0.05
Techno/house	0.04
Jazz	0.01
Other	0.04

(b) All of the probabilities are between 0 and 1 and the sum of the probabilities is 1. **(c)** Yes. **(d)** Answers will vary. One possible answer:

Tally for Discrete Variables: C4

C4	Count	Percent
1	4	40.00
2	2	20.00
5	2	20.00
6	2	20.00
N=	10	

Tally for Discrete Variables: C4

C4	Count	Percent
1	26	26.00
2	23	23.00
3	19	19.00
4	8	8.00
5	7	7.00
6	3	3.00
7	5	5.00
8	6	6.00
9	2	2.00
10	1	1.00
N=	100	

Tally for Discrete Variables: C4

C4	Count	Percent
1	270	27.00
2	222	22.20
3	173	17.30
4	60	6.00
5	58	5.80
6	60	6.00
7	51	5.10
8	44	4.40
9	13	1.30
10	49	4.90
N=	1000	

Tally for Discrete Variables: C4

C4	Count	Percent
1	2601	26.01
2	2329	23.29
3	1695	16.95
4	729	7.29
5	637	6.37
6	595	5.95
7	512	5.12
8	386	3.86
9	113	1.13
10	403	4.03
N=	10000	

Here Hip-hop/rap = 1, Pop = 2, Rock/punk = 3, Alternative = 4, Christian/gospel = 5, R&B = 6, Country = 7, Techno/house = 8, Jazz = 9, Other = 10. **(e)** As the sample size increases the relative frequencies approach the probabilities.

51. (a) Answers will vary. One possible answer:

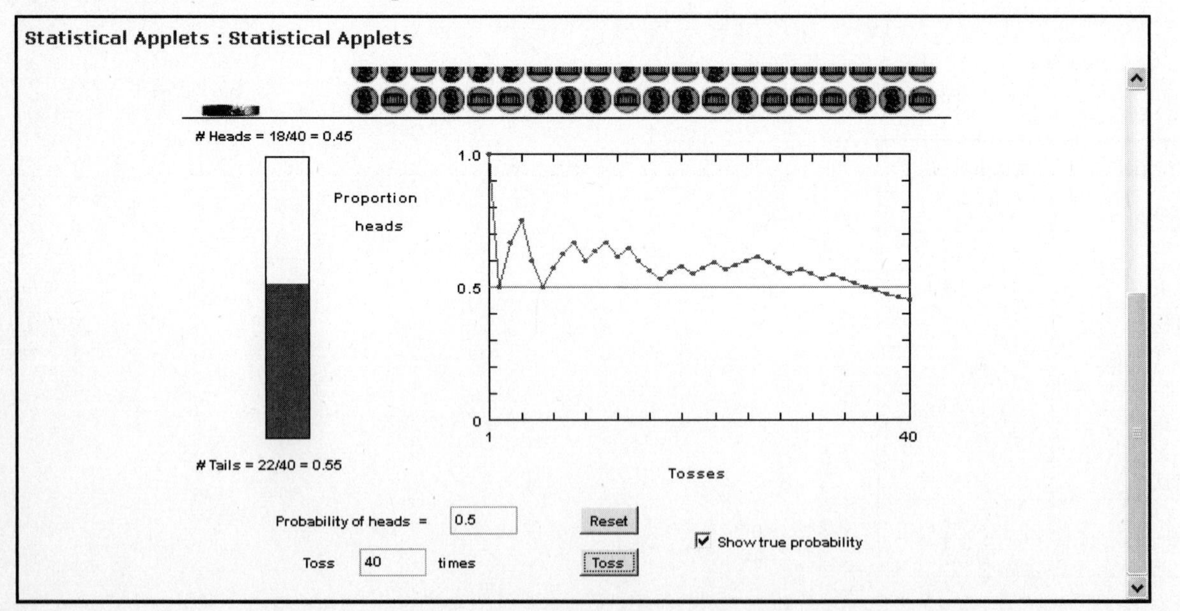

(b) Answers will vary. One possible answer:

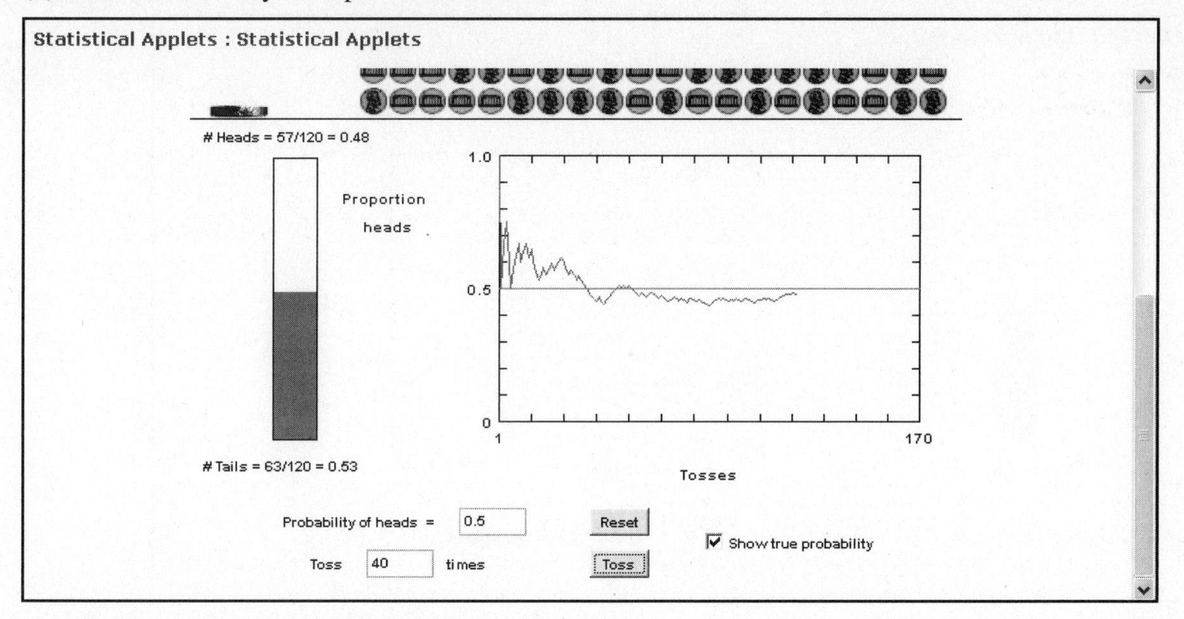

(c) Answers will vary. One possible answer:

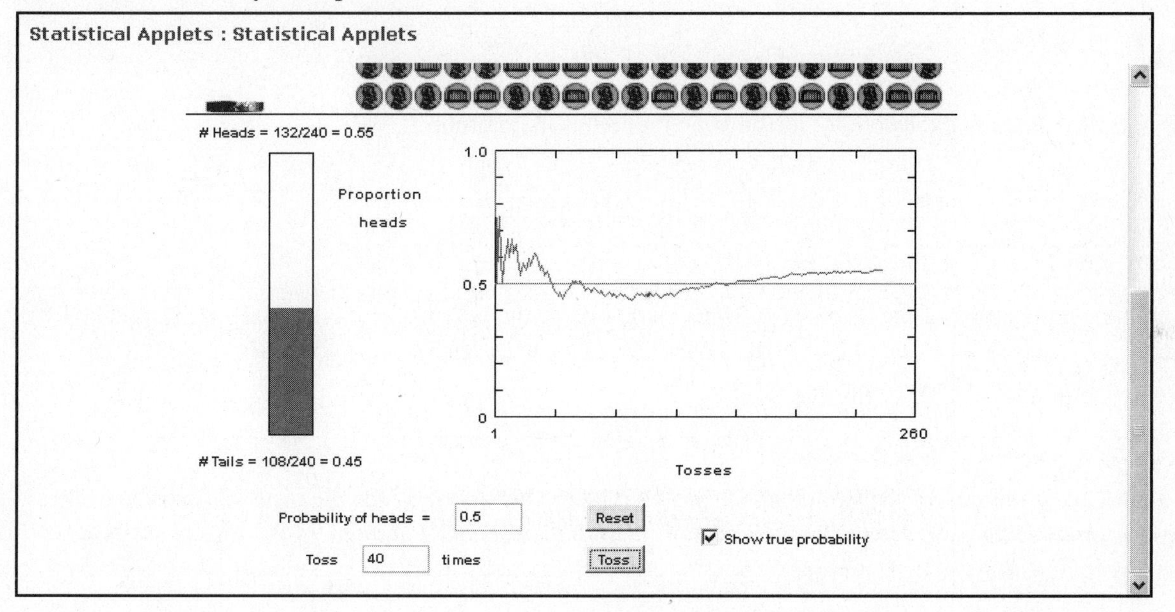

(d) Answers will vary. One possible answer:

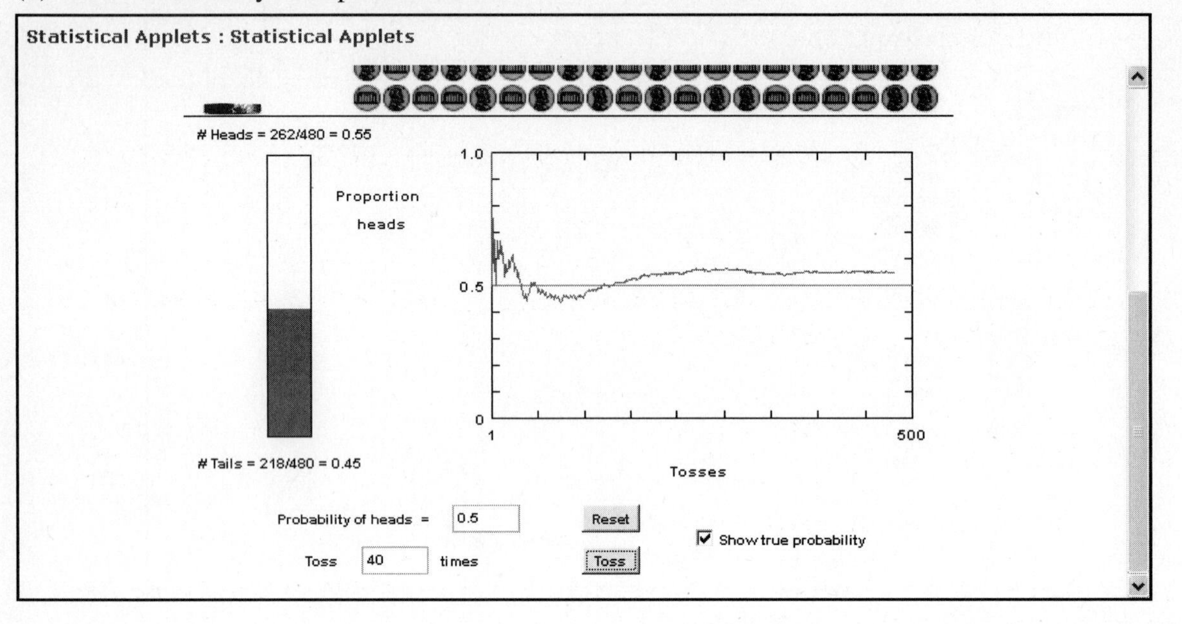

Section 5.2

1. Two events are mutually exclusive if they have no outcomes in common.

3. Yes.

5. Answers will vary; $1 - P(\text{will rain})$; they are complementary events.

For Exercises 7, 9, 11, $N(S) = 36$.

7. Let E be the event neither of the dice is a 4. Then E consists of the outcomes: {(1,1), (1,2), (1,3), (1,5), (1,6), (2,1), (2,2), (2,3), (2,5), (2,6), (3,1), (3,2), (3,3), (3,5), (3,6), (5,1), (5,2), (5,3), (5,5), (5,6), (6,1), (6,2), (6,3), (6,5), (6,6)}, so $N(E) = 25$. Therefore,

$$P(E) = \frac{N(E)}{N(S)} = \frac{25}{36}.$$

9. Let E be the event the sum of the two dice is 3 and let F be the event one of the dice is a 4. Then the event $E \cap F$ consists of all of the outcomes where the sum of the two dice equals 3 and one of the dice is a 4. Since this can't happen, $N(E \cap F) = 0$. Therefore,

$$P(E \cap F) = \frac{N(E \cap F)}{N(S)} = \frac{0}{36} = 0.$$

11. Let E be the event the sum of the two dice is 3. Then E consists of the outcomes: {(1,2), (2,1)}, so $N(E) = 2$. Therefore,

$$P(E) = \frac{N(E)}{N(S)} = \frac{2}{36} = \frac{1}{18}.$$

Let F be the event neither of the dice is a 4. Then F consists of the outcomes: {(1,1), (1,2), (1,3), (1,5), (1,6), (2,1), (2,2), (2,3), (2,5), (2,6), (3,1), (3,2), (3,3), (3,5), (3,6), (5,1), (5,2), (5,3), (5,5), (5,6), (6,1), (6,2), (6,3), (6,5), (6,6)}, so $N(F) = 25$. Therefore,

$$P(F) = \frac{N(F)}{N(S)} = \frac{25}{36}.$$

The event $E \cap F$ consists of all of the outcomes where the sum of the two dice equals 3 and neither of the dice is a 4. Then $E \cap F$ consists of the outcomes {(1,2), (2,1)}, so $N(E \cap F) = 2$. Therefore,

$$P(E \cap F) = \frac{N(E \cap F)}{N(S)} = \frac{2}{36} = \frac{1}{18}.$$

Then

$$P(E \cup F) = P(E) + P(F) - P(E \cap F) = \frac{2}{36} + \frac{25}{36} - \frac{2}{36} = \frac{25}{36}.$$

For Exercises 13, 15, 17, 19, 21, 23, $N(S) = 52$.

13. Let K be the event draw a king. Then event K consists of the 4 kings, {K♦, K♥, K♣, K♠}, so $N(K) = 4$. Therefore,

$$P(K) = \frac{N(K)}{N(S)} = \frac{4}{52} = \frac{1}{13}.$$

Let H be the event draw a heart. Then event H consists of the 13 hearts, {A♥, 2♥, 3♥, 4♥, 5♥, 6♥, 7♥, 8♥, 9♥, 10♥, J♥, Q♥, K♥}, so $N(H) = 13$. Therefore,

$$P(H) = \frac{N(H)}{N(S)} = \frac{13}{52} = \frac{1}{4}.$$

From Exercise 12, $P(K \cap H) = \frac{1}{52}$. Then

$$P(K \cup H) = P(K) + P(H) - P(K \cap H) = \frac{4}{52} + \frac{13}{52} - \frac{1}{52} = \frac{16}{52} = \frac{4}{13}.$$

15. Let H be the event draw a heart. Then event H consists of the 13 hearts, {A♥, 2♥, 3♥, 4♥, 5♥, 6♥, 7♥, 8♥, 9♥, 10♥, J♥, Q♥, K♥}, so $N(H) = 13$. Therefore,

$$P(H) = \frac{N(H)}{N(S)} = \frac{13}{52} = \frac{1}{4}.$$

Let D be the event draw a spade. Then event D consists of the 13 spades, {A♠, 2♠, 3♠, 4♠, 5♠, 6♠, 7♥, 8♠, 9♠, 10♠, J♠, Q♠, K♠}, so $N(D) = 13$. Therefore,

$$P(D) = \frac{N(D)}{N(S)} = \frac{13}{52} = \frac{1}{4}.$$

The events draw a heart and draw a spade are mutually exclusive, so

$$P(H \cup D) = P(H) + P(D) = \frac{13}{52} + \frac{13}{52} = \frac{26}{52} = \frac{1}{2}.$$

17. The event drawing a card that is not the king of hearts is the complement of the event drawing a card that is the king of hearts. From Exercise 12,

$$P(K \cap H) = \frac{1}{52},$$

so

$$P((K \cap H)^C) = 1 - P(K \cap H) = 1 - \frac{1}{52} = \frac{51}{52}.$$

19. Let R be the event drawing a red card. Then event R consists of the outcomes: {A♥, 2♥, 3♥, 4♥, 5♥, 6♥, 7♥, 8♥, 9♥, 10♥, J♥, Q♥, K♥, A♦, 2♦, 3♦, 4♦, 5♦, 6♦, 7♦, 8♦, 9♦, 10♦, J♦, Q♦, K♦}, so $N(R) = 26$. Therefore,

$$P(R) = \frac{N(R)}{N(S)} = \frac{26}{52} = \frac{1}{2}.$$

Since the event not draw a red card is the complement of the event draw a red card,

$$P(R^C) = 1 - P(R) = 1 - \frac{1}{2} = \frac{1}{2}.$$

Or you could do the problem using the fact that the event not drawing a red card is the same event as drawing a black card.

Let B be the event draw a black card. Event B consists of the 26 black cards:{A♠, 2♠, 3♠, 4♠, 5♠, 6♠, 7♠, 8♠, 9♠, 10♠, J♠, Q♠, K♠, A♣, 2♣, 3♣, 4♣, 5♣, 6♣, 7♣, 8♣, 9♣, 10♣, J♣, Q♣, K♣}, so $N(B) = 26$. Therefore,

$$P(R^C) = P(B) = \frac{N(B)}{N(S)} = \frac{26}{52} = \frac{1}{2}.$$

21. Let A be the event draw a face card that is not a diamond. Then A consists of the outcomes: {J♥, Q♥, K♥, J♠, Q♠, K♠, J♣, Q♣, K♣}, so $N(A) = 9$. Therefore,

$$P(F) = \frac{9}{52}.$$

23. Let F be the event drawing a face card and D be the event drawing a diamond. The event $F \cap D$ consists of the outcomes: {J♦, Q♦, K♦}, so $N(F \cap D) = 3$. Therefore,

$$P(F \cap D) = \frac{N(F \cap D)}{N(S)} = \frac{3}{52}.$$

For Exercises 25, 27, the sample space S = {HHH, HHT, HTH, THH, TTH, THT, HTT, TTT} and $N(S) = 8$. These were found in Exercise 27 in Section 5.1.

25. Let E be the event observing 3 heads. Then from Exercise 24, $P(E) = \frac{1}{8}$. Since the event not observing 3 heads is the complement of the event observing 3 heads,

$$P(E^C) = 1 - \frac{1}{8} = \frac{7}{8}.$$

27. Let E be the event observing 2 heads. Then from Exercise 26, $P(E) = \frac{3}{8}$. Since the event not observing 2 heads is the complement of the event observing 2 heads,

$$P(E^C) = 1 - \frac{3}{8} = \frac{5}{8}.$$

From Exercise 28, $N(S) = 8$ for Exercises 29, 31, 33.

29. Let E be the event 1 girl and 2 boys. Then event E consists of the outcomes: {GBB, BGB, BBG}, so $N(E) = 3$. Therefore,

$$P(E) = \frac{N(E)}{N(S)} = \frac{3}{8}.$$

31. Let E be the event 1 girl and 2 boys and let F be the event 1 boy and 2 girls. From Exercise 29, $P(E) = \frac{3}{8}$ and from Exercise 30, $P(F) = \frac{3}{8}$. Since the events E and F are mutually exclusive,

$$P(\text{2 of one gender and 1 of the other gender}) = P(E \cup F) = P(E) + P(F) = \frac{3}{8} + \frac{3}{8} = \frac{6}{8} = \frac{3}{4}.$$

33. Let E be the event 3 girls. Then event E consists of the outcome: {GGG}, so $N(E) = 1$. Therefore,

$$P(E) = \frac{N(E)}{N(S)} = \frac{1}{8}.$$

35. **(a)** Since frequencies are given, use the relative frequency method. Let A be the event that Jessica gets an A in a randomly chosen college course. Then

$P(A) \approx$ relative frequency of

$$A = \frac{\text{number of college courses in which Jessica has gotten an A}}{\text{number of college courses Jessica has taken}} = \frac{10}{30} = \frac{1}{3}.$$

(b) The event Jessica gets a grade other than an A in a randomly chosen college course is the complement of the event Jessica gets an A in a randomly chosen college course, so

$$P(A^C) = 1 - P(A) = 1 - \frac{1}{3} = \frac{2}{3}.$$

37. **(a)** Let K be the event that the author is Katherine Paterson. Then $P(K \cap F) \approx$ relative frequency of

$$K \cap F = \frac{\text{number of books whose author is Katherine Paterson and that are friction}}{\text{total number of books in the list}} = \frac{1}{10}.$$

(b) Let K be the event that the author is Katherine Paterson. Then

$P(K) \approx$ relative frequency of

$$K = \frac{\text{number of books whose author is Katherine Paterson}}{\text{total number of books in the list}} = \frac{1}{10}.$$

From Exercise 36 (a), $P(F) = \frac{5}{10} = \frac{1}{2}$. Therefore

$$P(K \cup F) = P(K) + P(F) - P(K \cap F) = \frac{1}{10} + \frac{5}{10} - \frac{1}{10} = \frac{5}{10} = \frac{1}{2}.$$

For Exercise 39, $N(S) = 36$.

39. **(a)** Let E be the event you roll a 9 and therefore land on Park Place. Then event E consists of the outcomes: $\{(3,6), (4,5), (5,4), (6,3)\}$, so $N(E) = 4$. Therefore,

$$P(E) = \frac{N(E)}{N(S)} = \frac{4}{36} = \frac{1}{9}.$$

(b) Let F be the event you roll an 11 and therefore land on Boardwalk. Then event F consists of the outcomes: $\{(5,6), (6,5)\}$, so $N(F) = 2$. Therefore,

$$P(F) = \frac{N(F)}{N(S)} = \frac{2}{36} = \frac{1}{18}.$$

(c) You will lose the game on this roll if either event E or event F occurs. Since events E and F are mutually exclusive,

$$P(\text{you lose the game on this roll}) = P(E \cup F) = P(E) + P(F) = \frac{4}{36} + \frac{2}{36} = \frac{6}{36} = \frac{1}{6}.$$

41. **(a)** Let F be the event choose a female and S be the event choose a sophomore. Then

$$P(F) \approx \text{relative frequency of choosing a female} = \frac{\text{number of females in the high school}}{\text{total number of students in the high school}} = \frac{200}{500} = \frac{2}{5},$$

$$P(S) \approx \text{relative frequency of choosing a sophomore} = \frac{\text{number of sophomores in the high school}}{\text{total number of students in the high school}}$$

$$= \frac{100}{500} = \frac{1}{5},$$

$P(F \cap S) \approx$ relative frequency of choosing a female and a sophomore

$$= \frac{\text{number of female sophomores in the high school}}{\text{total number of students in the high school}} = \frac{50}{500} = \frac{1}{10}$$

Therefore,

$$P(F \cup S) = P(F) + P(S) - P(F \cap S) = \frac{200}{500} + \frac{100}{500} - \frac{50}{500} = \frac{250}{500} = \frac{1}{2}.$$

(b) Let M be the event choose a male. Since there are 200 females in the high school there are $500 - 200 = 300$ males in the high school. Then

$$P(M) \approx \text{relative frequency of choosing a male} = \frac{\text{number of males in the high school}}{\text{total number of students in the high school}} = \frac{300}{500} = \frac{3}{5}.$$

Since there are 50 female sophomores in the high school there are $100 - 50 = 50$ male sophomores in the high school. Then

$P(M \cap S) \approx$ relative frequency of choosing a male and a sophomore

$$= \frac{\text{number of male sophomores in the high school}}{\text{total number of students in the high school}} = \frac{50}{500} = \frac{1}{10}.$$

Therefore,

$$P(M \cup S) = P(M) + P(S) - P(M \cap S) = \frac{300}{500} + \frac{100}{500} - \frac{50}{500} = \frac{350}{500} = \frac{7}{10}.$$

(c) Since there are 100 sophomores in the high school there are $500 - 100 = 400$ students who are not sophomores in the high school. Then

$P(S^C) \approx$ relative frequency of choosing a student who is not a sophomore

$$= \frac{\text{number of students who are not sophomores in the high school}}{\text{total number of students in the high school}} = \frac{400}{500} = \frac{4}{5}.$$

Since there are 50 female sophomores there are $200 - 50 = 150$ females who are not sophomores. Then $P(F \cap S^C) \approx$ relative frequency of choosing a female who is not a sophomore

$$= \frac{\text{number of females who are not sophomores in the high school}}{\text{total number of students in the high school}} = \frac{150}{500} = \frac{3}{10}.$$

Therefore,

$$P(F \cup S^C) = P(F) + P(S^C) - P(F \cap S^C) = \frac{200}{500} + \frac{400}{500} - \frac{150}{500} = \frac{450}{500} = \frac{9}{10}.$$

43. $N(S) = 52$.

(a) Let H be the event draw a heart. Then event H consists of the 13 hearts, {A♥, 2♥, 3♥, 4♥, 5♥, 6♥, 7♥, 8♥, 9♥, 10♥, J♥, Q♥, K♥}, so $N(H) = 13$. Therefore,

$$P(H) = \frac{N(H)}{N(S)} = \frac{13}{52} = \frac{1}{4}.$$

Let D be the event drawing a diamond. Then event D consists of the outcomes: {A♦, 2♦, 3♦, 4♦, 5♦, 6♦, 7♦, 8♦, 9♦, 10♦, J♦, Q♦, K♦}, so $N(D) = 13$. Therefore,

$$P(D) = \frac{N(D)}{N(S)} = \frac{13}{52} = \frac{1}{4}.$$

Since the events H and D are mutually exclusive,

$$P(H \cup D) = P(H) + P(D) = \frac{13}{52} + \frac{13}{52} = \frac{26}{52} = \frac{1}{2}.$$

(b) Let R be the event drawing a red card. Then event R consists of the outcomes: {A♥, 2♥, 3♥, 4♥, 5♥, 6♥, 7♥, 8♥, 9♥, 10♥, J♥, Q♥, K♥, A♦, 2♦, 3♦, 4♦, 5♦, 6♦, 7♦, 8♦, 9♦, 10♦, J♦, Q♦, K♦}, so $N(R) = 26$. Therefore,

$$P(R) = \frac{N(R)}{N(S)} = \frac{26}{52} = \frac{1}{2}.$$

Let J be the event draw a jack. Then event J consists of the 4 jacks, {J♦, J♥, J♣, J♠}, so $N(J) = 4$. Therefore,

$$P(J) = \frac{N(J)}{N(S)} = \frac{4}{52} = \frac{1}{13}.$$

Event $R \cap J$ consists of the outcomes: {J♦, J♥}, so $N(R \cap J) = 2$. Therefore,

$$P(R \cap J) = \frac{N(R \cap J)}{N(S)} = \frac{2}{52} = \frac{1}{26}.$$

Then

$$P(R \cup J) = P(R) + P(J) - P(R \cap J) = \frac{26}{52} + \frac{4}{52} - \frac{2}{52} = \frac{28}{52} = \frac{7}{13}.$$

(c) Let C be the event draw a club. Then event C consists of the 13 clubs: {A♣, 2♣, 3♣, 4♣, 5♣, 6♣, 7♣, 8♣, 9♣, 10♣, J♣, Q♣, K♣}, so $N(C) = 13$. Therefore,

$$(C) = \frac{N(C)}{N(S)} = \frac{13}{52} = \frac{1}{4}.$$

Let F be the event drawing a face card. Then F consists of the outcomes: {J♥, Q♥, K♥, J♦, Q♦, K♦, J♠, Q♠, K♠, J♣, Q♣, K♣}, so $N(F) = 12$. Therefore,

$$P(F) = \frac{12}{52} = \frac{3}{13}.$$

Then event $C \cap F$ consists of the outcomes: {J♣, Q♣, K♣}, so $N(C \cap F) = 3$. Therefore,

$$P(C \cap F) = \frac{N(C \cap F)}{N(S)} = \frac{3}{52}.$$

Then

$$P(C \cup F) = P(C) + P(F) - P(C \cap F) = \frac{13}{52} + \frac{12}{52} - \frac{3}{52} = \frac{22}{52} = \frac{11}{26}.$$

(d) Since the events H and D are mutually exclusive, $P(H \cap D) = 0$.

(e) Let E be the event draw a spade. Then event E consists of the 13 spades, {A♠, 2♠, 3♠, 4♠, 5♠, 6♠, 7♠, 8♠, 9♠, 10♠, J♠, Q♠, K♠}, so $N(E) = 13$. Therefore,

$$P(E) = \frac{N(E)}{N(S)} = \frac{13}{52} = \frac{1}{4},$$

so

$$P(E^C) = 1 - P(E) = 1 - \frac{1}{4} = \frac{3}{4}.$$

45. **(a)** Let T be the event the use of the litter item is tobacco. Then

$P(T) \approx$ relative frequency of litter items with a use of tobacco

$$= \frac{\text{number of litter items in the sample with a use of tobacco}}{\text{total number of litter items in the sample}} = \frac{3}{12} = \frac{1}{4}.$$

(b) Let NT be the event the use of the litter item is not tobacco. Then

$P(NT) \approx$ relative frequency of litter items with a use that is not tobacco

$$= \frac{\text{number of litter items in the sample with a use that is not tobacco}}{\text{total number of litter items in the sample}} = \frac{9}{12} = \frac{3}{4}.$$

$$P(T^C) = 1 - P(T) = 1 - \frac{1}{4} = \frac{3}{4}.$$

47. $P(\text{Mozart}) = 0.35$, $P(\text{Haydn}) = 0.15$, and $P(\text{Beethoven}) = 0.15$. All 3 events are mutually exclusive. **(a)** $P(\text{Mozart or Haydn}) = P(\text{Mozart}) + P(\text{Haydn}) = 0.35 + 0.50 = 0.85$; **(b)** $P(\text{Mozart or Beethoven}) = P(\text{Mozart}) + P(\text{Beethoven}) = 0.35 + 0.15 = 0.50$; **(c)** $P(\text{Haydn or Beethoven}) = P(\text{Haydn}) + P(\text{Beethoven}) = 0.50 + 0.15 = 0.65$; **(d)** $P(\text{Mozart and Haydn}) = 0$.

49. **(a)** Let ME be the event that a randomly selected anti-American terrorist incident in 2003 occurred in the Middle East. Then

$P(ME) \approx$ relative frequency of anti-American terrorist incidents in 2003 that took place in the Middle East

$$= \frac{\text{number of anti-American terrorist incidents in 2003 that took place in the Middle East}}{\text{total number of anti-American terrorist incidents in 2003}} = \frac{20}{60} = \frac{1}{3}.$$

(b) Relative frequency method; **(c)** $P(ME^C) = 1 - P(ME) = 1 - \frac{1}{3} = \frac{2}{3}.$

51. **(a)** P(not scoring with a runner on first with no outs) $= 1 - P$(not scoring with a runner on first with no outs) $= 1 - 0.43 = 0.57$; **(b)** P(not scoring with a runner on second with one out) $= 1 - P$(scoring with a runner on second with one out) $= 1 - 0.45 = 0.55$.

53. **(a)** Let M be the event male. Then

$$P(M) \approx \text{relative frequency of males} = \frac{\text{number of online dates that are male}}{\text{total number of online dates}} = \frac{23{,}952}{52{,}817} \approx 0.4535.$$

(b) Let ND be the event prefers not to describe their physical appearance. Then

$P(ND) \approx$ relative frequency of online dates who prefer not to describe their physical appearance

$$= \frac{\text{Number of online dates who prefer not to describe their physical appearance}}{\text{total number of online dates}} = \frac{6{,}287}{52{,}817} \approx 0.1190.$$

(c) $P(M \cap ND) \approx$ relative frequency of males who prefer not to describe their physical appearance

$$= \frac{\text{number of online dates that are male and prefer not to describe their physical appearance}}{\text{total number of online dates}}$$

$$= \frac{2{,}809}{52{,}817} \approx 0.0532.$$

(d) $P(M \cup ND) = P(M) + P(ND) - P(M \cap ND)$

$$= \frac{23{,}952}{52{,}817} + \frac{6{,}287}{52{,}817} - \frac{2{,}809}{52{,}817} \approx 0.5193.$$

55. **(a)** Let A be the event the cause of death was accidents and let K be the event the cause of death Parkinson's disease. Then $P(A \cap K) = 0$, so events A and K are mutually exclusive.

(b) $P(A \cup K) = P(A) + P(K) = \frac{108{,}694}{2{,}398{,}365} + \frac{18{,}018}{2{,}398{,}365} \approx 0.0528.$

Section 5.3

1. **(a)** Yes. **(b)** The probability of winning the football game depends on whether the star quarterback can play in the game.

3. For P(A | B), we assume that the event B has occurred, and now need to find the probability of event A, given event B. On the other hand, P(A∩B), we do not assume that event B has occurred, and instead need to determine the probability that both events occurred.

5. The Gambler's Fallacy is the mistaken belief that each time an event doesn't happen, the probability that it will happen increases. For example, if a coin is flipped several times and lands on tails every time, then one might think that the coin is bound to land on heads soon. But if the coin is fair, then the probability of it landing on heads is 0.5 on each flip no matter what happened on the previous flips.

7. **(a)** Independent; sampling with replacement. **(b)** Dependent; sampling without replacement.

9. A and B are independent events, so $P(A \cap B) = P(A) \cdot P(B) = 0.6 \cdot 0.4 = 0.24$.

11. A and B are independent events, so $P(B \mid A) = P(B) = 0.4$.

13. A and B are independent events, so $P(A \cap B) = P(A) \cdot P(B) = 0.5 \cdot 0.2 = 0.1$.

15. A and B are independent events, so $P(B \mid A) = P(B) = 0.2$.

17. $P(B \mid A) = \dfrac{P(A \cap B)}{P(A)} = \dfrac{0.05}{0.3} = 0.1667$.

For Exercise 19, 21, 23, $P(T) = \frac{1}{2}$.

19. $\frac{1}{2} \cdot \frac{1}{2} \cdot \frac{1}{2} = \frac{1}{8}$

21. $\frac{1}{2} \cdot \frac{1}{2} \cdot \frac{1}{2} \cdot \frac{1}{2} \cdot \frac{1}{2} = \frac{1}{32}$

23. A and B are independent events, so $P(A \text{ and } B) = P(A) \cdot P(B) = 0.4 \cdot 0.5 = 0.2$.

25. A and B are independent events, so $P(B \mid A) = P(B) = 0.5$.

27. P(A and B) ↑C= 1 − P(A and B) = 1 − 0.2 = 0.8

29. $P(C \mid D) = \dfrac{P(C \text{ and } D)}{P(D)} = \dfrac{0.21}{0.3} = 0.7$

31. Yes. P(C and D) = 0.21 = (0.7) (0.3) = P(C) P(D), P(C | D) = 0.7 = P(C), and P(D | C) = 0.3 = P(D).

33. P(E and F) or P(E | F) and P(F | E).

For Exercises 35, 37, $N(S) = 36$.

35. Let E be the event observe a 1 on the second roll. Then event E consists of the outcomes: {(1,1), (2,1), (3,1), (4,1), (5,1), (6,1)}, so $N(E) = 6$. Therefore,

$$P(E) = \frac{N(E)}{N(S)} = \frac{6}{36} = \frac{1}{6}.$$

37. Let E be the event observe an even number on the second roll. Then event E consists of the outcomes: {(1,2), (1,4), (1,6), (2,2), (2,4), (2,6), (3,2), (3,4), (3,6), (4,2), (4,4), (4,6), (5,2), (5,4), (5,6), (6,2), (6,4), (6,6)}, so $N(E) = 18$. Therefore,

$$P(E) = \frac{N(E)}{N(S)} = \frac{18}{36} = \frac{1}{2}.$$

39. Since P(observing a 1 on the second roll | observing a 1 on the first roll) $= \frac{1}{6} = P$(observing a 1 on the second roll) and P(observing an even number on the second roll | observing an even number on the first roll) $= \frac{1}{2} = P$(observing an even number on the second roll), they are independent.

41. **(a)** Let E be the event roll a 6. Then event E consists of the outcome:

{6}, so $N(E) = 1$. Since $N(S) = 6$, $P(E) = \dfrac{N(E)}{N(S)} = \dfrac{1}{6}$.

(b) Let F be the event roll an even number. Then event F consists of the outcomes: {2,4,6} and event $E \cap F$ consists of the outcome: {6}, so $N(F) = 3$ and $N(E \cap F) = 1$. Therefore,

$$P(E \mid F) = \frac{N(E \cap F)}{N(F)} = \frac{1}{3}.$$

Hence, this knowledge increases the probability of rolling a 6. **(c)** No, $P(\text{rolling a 6}) = \frac{1}{6}$; $P(\text{rolling a 6} \mid \text{odd number}) = 0$.

43. **(a)** $P(F \cap C) = \frac{N(F \cap C)}{N(S)} = \frac{100}{300} = \frac{1}{3}$

(b) $P(F \cap O) = \frac{N(F \cap O)}{N(S)} = \frac{30}{300} = \frac{1}{10}$

(c) $P(M \cap C) = \frac{N(M \cap C)}{N(S)} = \frac{50}{300} = \frac{1}{6}$

(d) $P(M \cap O) = \frac{N(M \cap O)}{N(S)} = \frac{20}{300} = \frac{1}{15}$

45. No; $P(C) = 1/2 \neq 5/9 = P(C \mid F)$, $P(C) = 1/2 \neq 5/12 = P(C \mid M)$, $P(F \text{ and } C) = 1/3 \neq 3/10 = (3/5)(1/2) = P(F)\,P(C)$, and $P(M \text{ and } C) = 1/6 \neq 1/5 = (2/5)(1/2) = P(M)\,P(C)$.

47. There are a total of $4 + 3 = 7$ balls in the urn, so $N(S) = 7$. **(a)** If we sample with replacement, then the events the first ball is blue and the second ball is blue are independent, so the probability that both balls are blue is $\frac{4}{7} \cdot \frac{4}{7} = \frac{16}{49}$. **(b)** If we sample without replacement, then the events the first ball is blue and the second ball is blue are dependent, so the probability that both balls are blue is $\frac{4}{7} \cdot \frac{3}{6} = \frac{2}{7}$.

49. Assume independent events. **(a)** $(0.019)^2 = 0.000361$; **(b)** $(1 - 0.019)^2 = 0.9624$; **(c)** $(1 - 0.019)^5 = 0.9085$; **(d)** P(at least 1 of 3 randomly selected persons received most of their information about September 11 from the Internet) $= 1 - P$(none of the 3 randomly selected persons received most of their information about September 11 from the Internet) $= 1 - (1 - 0.019)^3 = 0.0559$.

51. For this problem, $P(\text{Female}) = P(F) = 0.55$ and $P(\text{Business major} \mid \text{Female}) = P(B \mid F) = 0.10$. **(a)** $P(F \text{ and } B) = P(F) \cdot P(B \mid F) = 0.55 \cdot 0.10 = 0.055$; **(b)** $(0.55)^2 = 0.3025$; **(c)** 0.55; **(d)** 0.55; **(e)** $(0.055)^2 = 0.003025$; **(f)** It is not possible to find the probability that two randomly selected students are business majors because we are only given the percent of female students that are business majors and not the percent of all students that are business majors.

53. **(a)** $6/10 = 3/5$; **(b)** $1 - 3/5 = 2/5$; **(c)** $6/10 \cdot 5/9 = 1/3$; **(d)** $5/9$; **(e)** $1/3$.

55. **(a)** $P(F \cap \text{More}) = \frac{N(F \cap \text{More})}{N(S)} = \frac{212}{618} \approx 0.3430$; **(b)** $P(M \cap \text{More}) = \frac{N(M \cap \text{More})}{N(S)} = \frac{200}{618} \approx 0.3236$.

57. No. P(more serious than Pearl Harbor | female) $= 0.6752 \neq 0.6667 = P$(more serious than Pearl Harbor) P(more serious than Pearl Harbor | male) $= 0.6579 \neq 0.6667 = P$(more serious than Pearl Harbor).

Section 5.4

1. Tree diagram.

3. In a permutation, order is important. In a combination, order is not important.

5. Acceptance sampling refers to the process of (1) selecting a random sample from a batch of items, (2) evaluating the sample for defectives, and (3) either accepting or rejecting the entire batch based on the evaluation of the sample.

7. $6! = 6 \cdot 5 \cdot 4 \cdot 3 \cdot 2 \cdot 1 = 720$

9. $0! = 1$

11. $1! = 1$

13. ${}_7P_3 = \frac{7!}{(7-3)!} = \frac{7 \cdot 6 \cdot 5 \cdot 4!}{4!} = 7 \cdot 6 \cdot 5 = 210$

15. ${}_8P_5 = \frac{8!}{(8-5)!} = \frac{8 \cdot 7 \cdot 6 \cdot 5 \cdot 4 \cdot 3!}{3!} = 8 \cdot 7 \cdot 6 \cdot 5 \cdot 4 = 6720$

17. ${}_{100}P_1 = \frac{100!}{(100-1)!} = \frac{100 \cdot 99!}{99!} = 100$

19. Done on TI-Inspire calculator 93, 326, 215, 443, 944, 152, 681, 699, 238, 856, 266, 700, 490, 715, 968, 264, 381, 621, 468, 592, 963, 895, 217, 599, 993, 229, 915, 608, 941, 463, 976, 156, 518, 286, 253, 697, 920, 827, 223, 758, 251, 185, 210, 916, 864, 000, 000, 000, 000, 000, 000, 000, 000

21. ${}_7C_3 = \frac{7!}{3!(7-3)!} = \frac{7 \cdot 6 \cdot 5 \cdot 4!}{3 \cdot 2 \cdot 1 \cdot 4!} = \frac{210}{6} = 35$

23. ${}_{11}C_8 = \frac{11!}{8!(11-8)!} = \frac{11 \cdot 10 \cdot 9 \cdot 8!}{8! \cdot 3!} = \frac{990}{6} = 165$

25. ${}_{11}C_{10} = \frac{11!}{10!(11-10)!} = \frac{11 \cdot 10!}{10! \cdot 1!} = \frac{11}{1} = 11$

27. ${}_{100}C_0 = \frac{100!}{0!(100-0)!} = \frac{100!}{0! \cdot 100!} = 1$

29. ${}_7C_3 = \frac{7!}{3! \cdot 4!} = \frac{7!}{4! \cdot 3!} = {}_7C_4$

31. {Amy, Bob, Chris}, {Amy, Chris, Bob}, {Bob, Chris, Amy}, {Bob, Amy, Chris}, {Chris, Amy, Bob}, {Chris, Bob, Amy}, {Amy, Bob, Danielle}, {Amy, Danielle, Bob}, {Bob, Amy, Danielle}, {Bob, Danielle, Amy}, {Danielle, Amy, Bob}, {Danielle, Bob, Amy}, {Amy, Chris, Danielle}, {Amy, Danielle, Chris}, {Chris, Amy, Danielle}, {Chris, Danielle, Amy}, {Danielle, Amy, Chris}, {Danielle, Chris, Amy}, {Bob, Chris, Danielle}, {Bob, Danielle, Chris}, {Chris, Bob, Danielle}, {Chris, Danielle, Bob}, {Danielle, Bob, Chris}, {Danielle, Chris, Bob} ${}_4P_3 = 24$.

33. Order is important in a permutation, but not in a combination. Thus {Amy, Bob, Chris}, {Amy, Chris, Bob}, {Chris, Amy, Bob},{Chris, Bob, Amy}, {Bob, Amy, Chris}, and {Bob, Chris, Amy} are all different permutations but the same combination.

35. r!

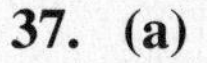

37. (a)

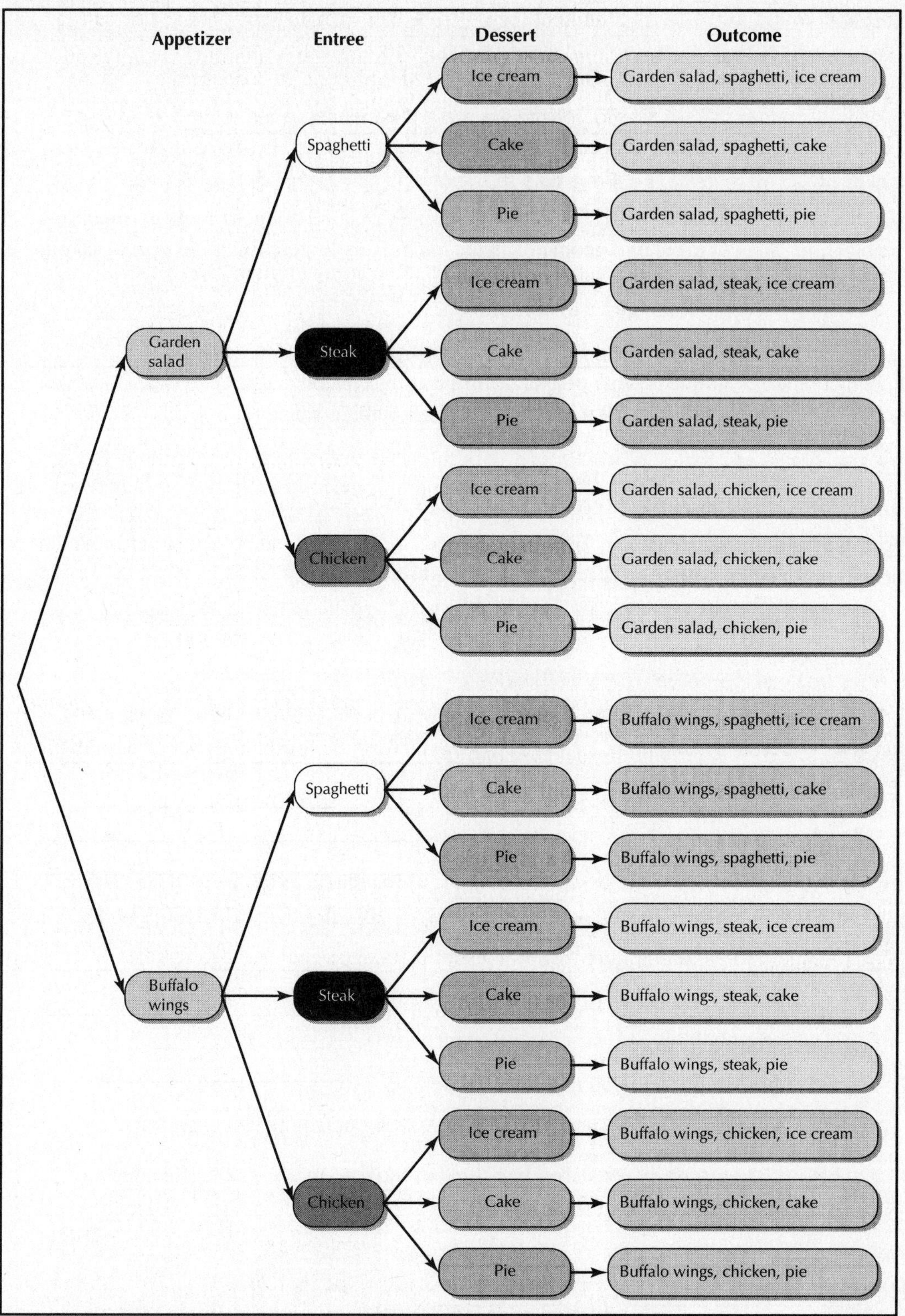

(b) 18

39. The key word is "ordering." This means that order is important, so this is a permutation of all $n = r = 9$ players. Therefore, there are ${}_9P_9 = 9! = 362{,}880$ ways the manager can arrange the 9 players.

41. What makes the routes different is the order in which the document delivery person visits the five different destinations, so this is a permutation of all $n = r = 5$ different destinations. Therefore, there are ${}_5P_5 = 5! = 120$ ways for the document delivery person to deliver the documents to the five different destinations.

43. What makes the routes different is the order in which the different countries are visited, so this is a permutation of all $n = r = 6$ countries. Therefore, there are ${}_6P_6 = 6! = 720$ ways to visit the 6 countries.

45. Since one child is throwing the ball and one child is catching the ball, order is important. Thus, this is a permutation with $n = 5$ and $r = 2$. Therefore, there are

$$ {}_5P_2 = \frac{5!}{(5-2)!} = \frac{5 \cdot 4 \cdot 3!}{3!} = 5 \cdot 4 = 20 $$

ways that one child can throw a ball to another child once.

47. Since both people that are shaking hands with each other are doing the same thing, order is unimportant. Therefore this is a combination. Since it takes two people to shake hands, this is a combination of $n = 25$ people taken $r = 2$ at a time. Therefore, there are a total of

$$ {}_{25}C_2 = \frac{25!}{2!(25-2)!} = \frac{25 \cdot 24 \cdot 23!}{2! \cdot 23!} = \frac{600}{2} = 300 \text{ handshakes.} $$

49. Since what matters in a random sample is what people or objects are selected and not the order in which they are selected, this is a combination with $n = 20$ and $r = 1$. Therefore, there are

$$ {}_{20}C_1 = \frac{20!}{1!(20-1)!} = \frac{20 \cdot 19!}{1! \cdot 19!} = \frac{20}{1} = 20 $$

random samples of size 1 that can be chosen from a population of size 20.

51. Since what matters in a random sample is what people or objects are selected and not the order in which they are selected, this is a combination with $n = 20$ and $r = 10$. Therefore, there are

$$ {}_{20}C_{10} = \frac{20!}{10!(20-10)!} = \frac{20 \cdot 19 \cdot 18 \cdot 17 \cdot 16 \cdot 15 \cdot 14 \cdot 13 \cdot 12 \cdot 11 \cdot 10!}{10 \cdot 9 \cdot 8 \cdot 7 \cdot 6 \cdot 5 \cdot 4 \cdot 3 \cdot 2 \cdot 1 \cdot 10!} = \frac{670{,}442{,}572{,}800}{3{,}628{,}800} = 184{,}756 $$

random samples of size 10 that can be chosen from a population of size 20.

53. We are seeking the number of permutations of $n = 8$ items, of which $n_1 = 3$ letter S's, $n_2 = 1$ letter B, $n_3 = 1$ letter U, $n_4 = 1$ letter I, $n_5 = 1$ letter N, and $n_6 = 1$ letter E. Using the formula for the number of permutations of nondistinct items,

$$ \frac{n!}{n_1! \cdot n_2! \cdot n_3! \cdot n_4! \cdot n_5! \cdot n_6!} = \frac{8!}{3! \cdot 1! \cdot 1! \cdot 1! \cdot 1! \cdot 1!} = \frac{40{,}320}{6 \cdot 1 \cdot 1 \cdot 1 \cdot 1 \cdot 1} = 6720. $$

There are 6720 distinct strings of letters that can be made using all of the letters in the word BUSINESS.

55. **(a)** Combination; order is unimportant. **(b)** ${}_5C_2 \cdot {}_{25}C_1 + {}_5C_3 \cdot {}_{25}C_0 = 10 \cdot 25 + 10 \cdot 1 = 260$; **(c)** The total number of samples is ${}_{30}C_3 = 4060$, so the probability that the bushel will be rejected is $\frac{260}{4060} = 0.0640$.

57. The total number of ways to select five songs is ${}_{12}C_5 = 792$.

(a) ${}_7C_5 \cdot {}_5C_0 = 21 \cdot 1 = 21$, so the probability that you like all five songs is $\frac{21}{792} = 0.0265$.

(b) ${}_7C_0 \cdot {}_5C_5 = 1 \cdot 1 = 1$, so the probability that you like none of the 5 songs is $\frac{1}{792} = 0.0013$.

(c) ${}_7C_2 \cdot {}_5C_3 = 21 \cdot 10 = 210$, so the probability that you like two of the 5 songs is $\frac{210}{792} = 0.2652$.

(d) ${}_7C_3 \cdot {}_5C_2 = 35 \cdot 10 = 350$, so the probability that you like three of the five songs is $\frac{350}{792} = 0.4419$.

Chapter 5 Review

For Exercises 1, 3, 5, the sample space is {HHH, HHT, HTH, THH, TTH, THT, HTT, TTT} and $N(S) = 8$.

1. Let E be the event 2 heads. Then event E consists of the outcomes: {HHT, HTH, THH}, so $N(E) = 3$. Therefore,

$$ P(E) = \frac{N(E)}{N(S)} = \frac{3}{8}. $$

3. Let E be the event 4 heads. Since the coin is only tossed 3 times, there can't be 4 heads. Therefore there are no outcomes in event E, so $N(E) = 0$. Therefore,

$$P(E) = \frac{N(E)}{N(S)} = \frac{0}{8} = 0.$$

5. Let E be the event at most 1 tail. Then event E consists of the outcomes with 0 or 1 tail: {HHH, HHT, HTH, THH}, so $N(E) = 4$. Therefore,

$$P(E) = \frac{N(E)}{N(S)} = \frac{4}{8} = \frac{1}{2}.$$

7. **(a)** Let E be the event that a noncitizen farm worker is a high school graduate and let F be the event that a noncitizen farm worker has some college. Then since the educational levels attained are mutually exclusive,

$$P(E \cup F) = P(E) + P(F) = \frac{59{,}784}{376{,}000} + \frac{20{,}304}{376{,}000} = \frac{80{,}088}{376{,}000} = 0.213.$$

(b) Let E be the event that a citizen farm worker is a high school graduate and let F be the event that a citizen farm worker has some college. Then, because the educational levels attained are mutually exclusive,

$$P(E \cup F) = P(E) + P(F) = \frac{222{,}144}{624{,}000} + \frac{187{,}200}{624{,}000} = \frac{409{,}344}{624{,}000} = 0.656.$$

(c) Since the education levels are mutually exclusive, the probability that a noncitizen farm worker has less than a 9th grade education and some college is 0.

9. **(a)** Yes, since no student in the sample repeated 12th grade. **(b)** Since events A and B are mutually exclusive, $P(A \cup B) = P(A) + P(B) = 0.013028169 + 0.016 = 0.029028169$. **(c)** Since events E and F are mutually exclusive, $P(E \cup F) = P(E) + P(F) = 0.0090140845 + 0.006 = 0.0150140845$.

11. **(a)** P(a randomly selected research study had an outcome that favored the drug) $= \frac{39}{40} = 0.975$. If we assume that the outcomes of the different research studies are independent, then the probability that three randomly selected research studies all favor this drug is $(0.975)^3 \approx 0.9269$. **(b)** P(a randomly selected research study had an outcome that did not favor the drug) $= 1 - P$(a randomly selected research study had an outcome that favored the drug) $= 1 - 0.975 = 0.025$. If we assume that the outcomes of the different research studies are independent, then the probability that none of the three randomly selected research studies favor this drug is $(0.025)^3 \approx 0$. **(c)** P(at least one of three randomly selected research studies favors the drug) $= 1 - P$(none of the research studies favors the drug) $= 1 - 0 = 1$.

13. **(a)** P(a randomly selected person would not vote for a political candidate who disagreed with his or her views on gay marriage) $= \frac{322}{1149}$. Therefore, P(a randomly selected person would vote for a political candidate who disagreed with his or her views on gay marriage) $= 1 - P$(a randomly selected person would not vote for a political candidate who disagreed with his or her views on gay marriage) $= 1 - \frac{322}{1149} = \frac{827}{1149}$.

If we assume that the responses to the survey are independent, then P(all of the respondents would vote for a political candidate who disagreed with his or her views on gay marriage)

$$= \left(\frac{827}{1149}\right)^2 \approx 0.3729.$$

(b) P(at least one of three randomly selected respondents would not vote for a political candidate who disagreed with his or her views on gay marriage) $= 1 - P$(all of the respondents would vote for a political candidate who disagreed with his or her views on gay marriage) $= 1 - 0.3729 = 0.6271$.

15. **(a)** $P(C) = \frac{N(C)}{N(S)} = \frac{150}{300} = \frac{1}{2}$

(b) $P(D) = \frac{N(D)}{N(S)} = \frac{100}{300} = \frac{1}{3}$

17. **(a)** $P(D \mid F) = \frac{N(F \cap D)}{N(F)} = \frac{50}{180} = \frac{5}{18}$

(b) $P(D \mid M) = \frac{N(M \cap D)}{N(M)} = \frac{50}{120} = \frac{5}{12}$

19. We are seeking the number of permutations of $n = 11$ items, of which $n_1 = 4$ letter S's, $n_2 = 4$ letter I's, $n_3 = 2$ letter P's, and $n_4 = 1$ letter M. Using the formula for the number of permutations of nondistinct items,

$$\frac{n!}{n_1! \cdot n_2! \cdot n_3! \cdot n_4!} = \frac{11!}{4! \cdot 4! \cdot 2! \cdot 1!} = \frac{39{,}916{,}800}{24 \cdot 24 \cdot 2 \cdot 1} = 34{,}650.$$

There are 34,650 distinct strings of letters that can be made using all of the letters in the word MISSISSIPPI.

21. **(a)** Combination, order is not important. **(b)** ${}_3C_2 = 3$; **(c)** There are ${}_{18}C_2 = 153$ ways to select 2 soldiers from the 18 soldiers. Therefore the probability that the entire squad will have to run a five-mile course in full gear is $\frac{3}{153} = 0.0196$.

Chapter 5 Quiz

1. False. An *event* is a collection of a series of events from the sample space of an experiment. An *outcome* is the result of a single trial of an experiment.

2. True.

3. True.

4. 0, 1

5. or, and

6. 0.5

7. permutation

8. 1

9. with replacement

10. the intersection of A and B

11. multiplication rule for counting

For Exercises 12–20, $N(S) = 36$.

12. Let E be the event roll a pair of 1s. Then event E consists of the outcome: {(1,1)}, so $N(E) = 1$. Therefore,

$$P(E) = \frac{N(E)}{N(S)} = \frac{1}{36}.$$

13. Let E be the event roll a pair of 2s. Then event E consists of the outcome: {(2,2)}, so $N(E) = 1$. Therefore,

$$P(E) = \frac{N(E)}{N(S)} = \frac{1}{36}.$$

14. Let E be the event one of the dice show 5. Then event E consists of the outcomes: {(1,5), (2,5), (3,5), (4,5), (6,5), (5,1), (5,2), (5,3), (5,4), (5,6)}, so $N(E) = 10$. Therefore,

$$P(E) = \frac{N(E)}{N(S)} = \frac{10}{36} = \frac{5}{18}.$$

15. Let E be the event sum of the two dice equals 7. Then event E consists of the outcomes: {(1,6), (2,5), (3,4), (4,3), (5,2), (6,1)}, so $N(E) = 6$. Therefore,

$$P(E) = \frac{N(E)}{N(S)} = \frac{6}{36} = \frac{1}{6}.$$

16. Let E be the event sum of the two dice equals 5. Then event E consists of the outcomes: {(1,4), (2,3), (3,2), (4,1)}, so $N(E) = 4$. Therefore,

$$P(E) = \frac{N(E)}{N(S)} = \frac{4}{36} = \frac{1}{9}.$$

17. Let E be the event sum of the two dice equals 5. Then the event the sum of the two dice does not equal 5 is the complement of the event the sum of the two dice equals 5, so $P(E^C) = 1 - P(E) = 1 - \frac{1}{9} = \frac{8}{9}$.

18. Let F be the event one of the two dice shows 2. Then event F consists of the outcomes: {(1,2), (3,2), (4,2), (5,2), (6,2), (2,1), (2,3), (2,4), (2,5), (2,6)}, so $N(F) = 10$. Therefore,

$$P(F) = \frac{N(F)}{N(S)} = \frac{10}{36} = \frac{5}{18}.$$

19. Let E be the event sum of the two dice equals 5 and let F be the event one of the two dice shows 2. Then event $E \cap F$ consists of the outcomes: {(2,3), (3,2)}, so $N(E \cap F) = 2$. Therefore,

$$P(E \cap F) = \frac{N(E \cap F)}{N(S)} = \frac{2}{36} = \frac{1}{18}.$$

20. Let E be the event sum of the two dice equals 5 and let F be the event one of the two dice shows 2. Then

$$P(E \cup F) = P(E) + P(F) - P(E \cap F) = \frac{4}{36} + \frac{10}{36} - \frac{2}{36} = \frac{12}{36} = \frac{1}{3}.$$

21. $P(A \mid B) = \frac{P(A \cap B)}{P(B)} = \frac{0.15}{0.75} = 0.2$

22. $P(A \cap B) = P(B) \cdot P(A \mid B) = 0.85 - 0.25 = 0.2125$.

23. $N(S) = 52$.

(a) Let E be the event draw a heart. Event E consists of the 13 hearts, {A♥, 2♥, 3♥, 4♥, 5♥, 6♥, 7♥, 8♥, 9♥, 10♥, J♥, Q♥, K♥}, so $N(E) = 13$. Therefore,

$$P(E) = \frac{N(E)}{N(S)} = \frac{13}{52} = \frac{1}{4}.$$

(b) Let F be the event drawing a face card. Then F consists of the outcomes: {J♥, Q♥, K♥, J♦, Q♦, K♦, J♠, Q♠, K♠, J♣, Q♣, K♣}, so $N(F) = 12$. Therefore,

$$P(F) = \frac{12}{52} = \frac{3}{13}.$$

(c) Let E be the event draw a 7. Event E consists of the four 7s, {7♦, 7♥, 7♣, 7♠}, so $N(E) = 4$. Therefore,

$$P(E) = \frac{N(E)}{N(S)} = \frac{4}{52} = \frac{1}{13}.$$

(d) Let R be the event drawing a red card. Then event R consists of the outcomes: {A♥, 2♥, 3♥, 4♥, 5♥, 6♥, 7♥, 8♥, 9♥, 10♥, J♥, Q♥, K♥, A♦,2♦, 3♦, 4♦, 5♦, 6♦, 7♦, 8♦, 9♦, 10♦, J♦, Q♦, K♦}, so $N(R) = 26$. Therefore,

$$P(R) = \frac{N(R)}{N(S)} = \frac{26}{52} = \frac{1}{2}.$$

(e) Let E be the event draw the 7 of hearts. Event E consists of the outcome: {7♥}, so $N(E) = 1$. Therefore,

$$P(E) = \frac{N(E)}{N(S)} = \frac{1}{52}.$$

(f) Let E be the event draw a red queen. Event E consists of the outcomes: {Q♦, Q♥}, so $N(E) = 2$. Therefore,

$$P(E) = \frac{N(E)}{N(S)} = \frac{2}{52} = \frac{1}{26}.$$

24. **(a)** $P(B) = \frac{N(B)}{N(S)} = \frac{11{,}031}{28{,}799} \approx 0.3830$.

(b) $P(A) = \frac{N(A)}{N(S)} = \frac{4{,}762}{28{,}799} \approx 0.1654$.

(c) $P(A \cap B) = \frac{N(A \cap B)}{N(S)} = \frac{3{,}096}{28{,}799} \approx 0.1075$.

(d) $P(A \cup B) = P(A) + P(B) - P(A \cap B) = 0.3830 + 0.1654 - 0.1075 = 0.4409$.

25. **(a)** $P(F) = \dfrac{N(F)}{N(S)} = \dfrac{423}{789} \approx 0.5361.$

(b) $P(M) = \dfrac{N(M)}{N(S)} = \dfrac{366}{789} \approx 0.4639.$

(c) $P(Not) = \dfrac{N(Not)}{N(S)} = \dfrac{26}{789} \approx 0.0330.$

26. **(a)** $P(F \cap \text{Not}) = \dfrac{N(F \cap \text{Not})}{N(S)} = \dfrac{17}{789} \approx 0\ .0215.$

(b) $P(M \cap \text{Not}) = \dfrac{N(M \cap \text{Not})}{N(S)} = \dfrac{9}{789} \approx 0\ .0114.$

27. **(a)** $P(\text{Not} \mid F) = (N(F \cap \text{Not}))/(N(F)) = 17/423 \approx 0.0402.$

(b) $P(\text{Not} \mid M) = (N(M \cap \text{Not}))/(N(M)) = 9/366 \approx 0.0246.$

28. No, because P(Not too happily married) = 0.0330 ≠ 0.0402 = P(Not too happily married | Female) and P(Not too happily married) = 0.0330 ≠ 0.0246 = P(Not too happily married | Male).

29. Since we are only interested in which three teams make the playoffs and not the order in which they are selected, this is a combination with $n = 4$ and $r = 3$. There are

$$_4C_3 = \frac{4!}{3! \cdot 1!} = \frac{4 \cdot 3!}{3! \cdot 1!} = \frac{4}{1} = 4$$

different sets of teams making the playoffs.

30. **(a)** Permutation; the order in which the numbers are selected is important. **(b)** $n = 20$ and $r = 3$, so there are $_{20}P_3 = 20 \cdot 19 \cdot 18 = 6840$ possible outcomes. **(c)** 1/6840.

Chapter 6

Random Variables and the Normal Distribution

Section 6.1

1. A random variable is a variable whose values are determined by chance. Answers will vary.

3. A discrete random variable can take either a finite or a countable number of values. Each value can be graphed as a separate point on the number line with space between each point. A continuous random variable can take infinitely many values. The values of a continuous random variable form an interval on the number line.

5. Since the number of siblings a person has is finite and may be written as a list of numbers, it represents a discrete random variable.

7. Volume is something that must be measured, not counted. Volume can take infinitely many different possible values, with these values forming an interval on the number line. Thus, how much coffee there is in your next cup of coffee is a continuous random variable.

9. Since the number of correct answers a person has on a multiple choice quiz is finite and may be written as a list of numbers, it represents a discrete random variable.

11. $\{0, 1, 2, 3, 4, 5, 6, 7, 8, 9, 10, 11, 12, 13, 14, 15\}$

13. $\{0, 1, 2, 3, 4\}$

15. No, the probabilities don't add up to 1.

17. No, $P(X = 1)$ is negative.

19. We will define our random variable X to be

$$X = \text{the number of CDs that Shirelle will listen to tonight.}$$

The probability distribution for X is:

X = the number of CDs that Shirelle will listen to tonight	0	1	2	3	4
$P(X)$	0.06	0.24	0.38	0.22	0.10

(a) $P(X = 0) + P(X = 1) = 0.06 + 0.24 = 0.30$; **(b)** $P(X = 0) = 0.06$; **(c)** $P(X = 5) = 0$.

21. We will define our random variable X to be X = financial gain in dollars. The probability distribution for X is:

X = financial gain in dollars	−10,000	10,000	50,000
$P(X)$	$\frac{1}{3}$	$\frac{1}{2}$	$\frac{1}{6}$

(a) $P(X = 10{,}000) + P(X = 50{,}000) = \frac{1}{2} + \frac{1}{6} = \frac{2}{3}$; **(b)** gaining \$10,000;
(c) $P(X = -10{,}000) = \frac{1}{3}$.

23. Mean:

$$\mu = \sum[X \cdot P(X)] = 2\left(\frac{1}{4}\right) + 3\left(\frac{1}{2}\right) + 4\left(\frac{1}{4}\right) = 3 \text{ guests}$$

Variance using the definition formula:

X	$P(X)$	$X - \mu$	$(X - \mu)^2$	$(X - \mu)^2 \cdot P(X)$
2	1/4	−1	1	1/4
3	1/2	0	0	0
4	1/4	1	1	1/4

$$\sigma^2 = \sum[(X - \mu)^2 \cdot P(X)] = \frac{1}{2} = 0.5$$

$$\sigma = \sqrt{\sigma^2} = \sqrt{0.5} \approx 0.7071 \text{ guests}$$

Variance using the computational formula:

X	$P(X)$	X^2	$X^2 \cdot P(X)$
2	1/4	4	1
3	1/2	9	4.5
4	1/4	16	4

$$\sum[X^2 \cdot P(X)] = 9.5$$

$$\sigma^2 = \sum[X^2 \cdot P(X)] - \mu^2 = 9.5 - 3^2 = 0.5$$

$$\sigma = \sqrt{\sigma^2} = \sqrt{0.5} \approx 0.7071 \text{ guests}$$

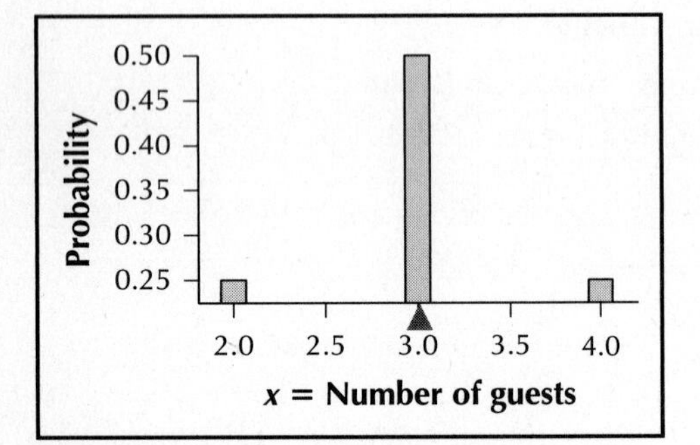

Yes.

25. Mean:

$$\mu = \sum [X \cdot P(X)] = 1(0.1) + 2(0.2) + 3(0.3) + 4(0.3) + 5(0.1) = 3.1 \text{ games}$$

Variance using the definition formula:

X	$P(X)$	$X - \mu$	$(X - \mu)^2$	$(X - \mu)^2 \cdot P(X)$
1	0.1	−2.1	4.41	0.441
2	0.2	−1.1	1.21	0.242
3	0.3	−0.1	0.01	0.003
4	0.3	0.9	0.81	0.243
5	0.1	1.9	3.61	0.361

$$\sigma^2 = \sum[(X - \mu)^2 \cdot P(X)] = 1.29$$

$$\sigma = \sqrt{\sigma^2} = \sqrt{1.29} \approx 1.1358 \text{ games}$$

Variance using the computational formula:

X	$P(X)$	X^2	$X^2 \cdot P(X)$
1	0.1	1	0.1
2	0.2	4	0.8
3	0.3	9	2.7
4	0.3	16	4.8
5	0.1	25	2.5

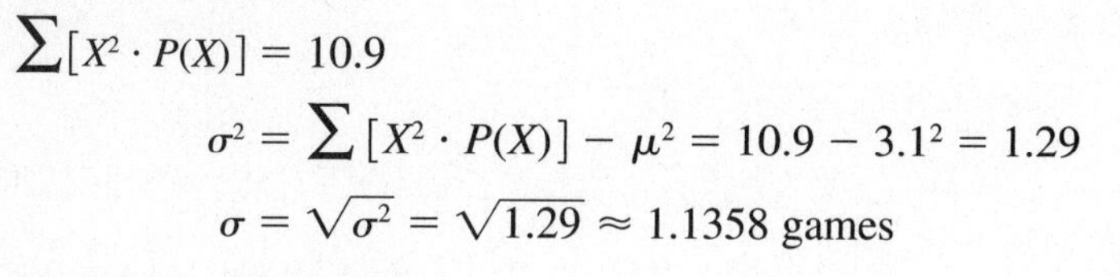

$$\sum[X^2 \cdot P(X)] = 10.9$$

$$\sigma^2 = \sum[X^2 \cdot P(X)] - \mu^2 = 10.9 - 3.1^2 = 1.29$$

$$\sigma = \sqrt{\sigma^2} = \sqrt{1.29} \approx 1.1358 \text{ games}$$

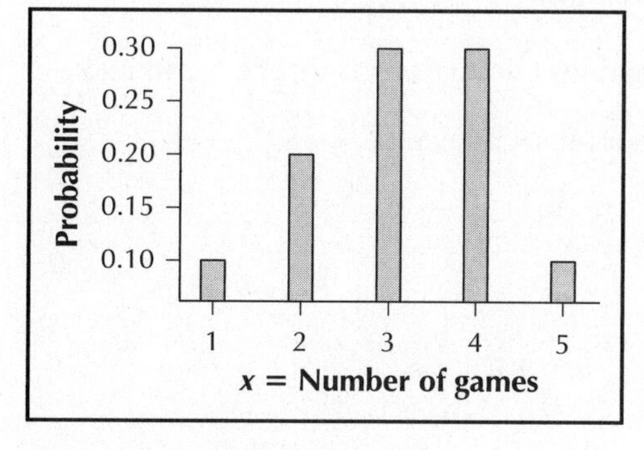

Yes.

27. About 5.4 games.

29. **(a)** Mean: $\mu = \sum[X \cdot P(X)] = 4(0.28) + 5(0.24) + 6(0.30) + 7(0.18) = 5.38$ games. **(b)** The estimate of the mean of 5.4 games is 0.02 games more than the actual mean of 5.38 games.

31. **(a)**

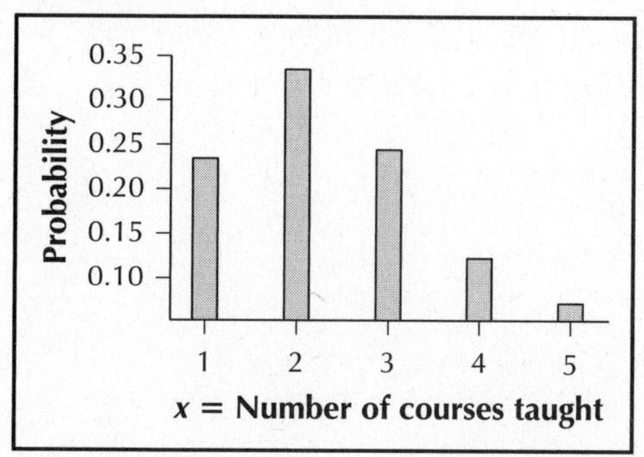

(b) 2 classes; 5 classes. **(c)** $\mu = E(X) = \sum[X \cdot P(X)] = 1(0.233) + 2(0.334) + 3(0.243) + 4(0.122) + 5(0.068) = 2.458$ classes. **(d)** The "typical" number of courses taught by faculty at all degree-granting institutions of higher learning in the United States is 2.458.

33. (a) From Exercise 31, $\mu = 2.458$ classes.
Variance using the computational formula:

X	$P(X)$	X^2	$X^2 \cdot P(X)$
1	0.233	1	0.233
2	0.334	4	1.336
3	0.243	9	2.187
4	0.122	16	1.952
5	0.068	25	1.700

$$\sum[X^2 \cdot P(X)] = 7.408$$

$$\sigma^2 = \sum[X^2 \cdot P(X)] - \mu^2 = 7.408 - 2.458^2 = 1.366236$$

$$\sigma = \sqrt{\sigma^2} = \sqrt{1.366236} \approx 1.1689 \text{ classes}$$

(b) Since the value $X = 5$ classes is $\left|\frac{X - \mu}{\sigma}\right| = \left|\frac{5 - 2.458}{1.1689}\right| \approx 2.17$ standard deviations from the mean and because $2 \leq 2.17 < 3$, teaching five courses would be considered moderately unusual.

35. (a)

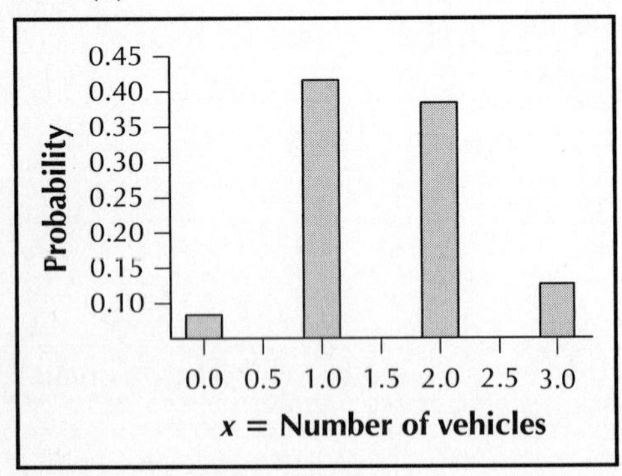

(b) 1 vehicle; 0 vehicles; **(c)** $\mu = \sum[X \cdot P(X)] = 0(0.081) + 1(0.414) + 2(0.382) + 3(0.123) = 1.547$ vehicles; **(d)** The "typical" number of vehicles owned by residents of Florida is 1.547.

37. (a) From Exercise 35, $\mu = 1.547$ vehicle.
Variance using the computational formula:

X	$P(X)$	X^2	$X^2 \cdot P(X)$
0	0.081	0	0.000
1	0.414	1	0.414
2	0.382	4	1.528
3	0.123	9	1.107

$$\sum[X^2 \cdot P(X)] = 3.049$$

$$\sigma^2 = \sum[X^2 \cdot P(X)] - \mu^2 = 3.049 - 1.547^2 = 0.655791$$

$$\sigma = \sqrt{\sigma^2} = \sqrt{0.655791} \approx 0.8098 \text{ vehicle}$$

(b) Since the value $X = 0$ vehicles is $\left|\frac{X - \mu}{\sigma}\right| = \left|\frac{0 - 1.547}{0.8098}\right| \approx 1.91$ standard deviations from the mean and because $0 \leq 1.91 < 2$, owning no vehicles would not be considered unusual.

39. The mean is about 7.

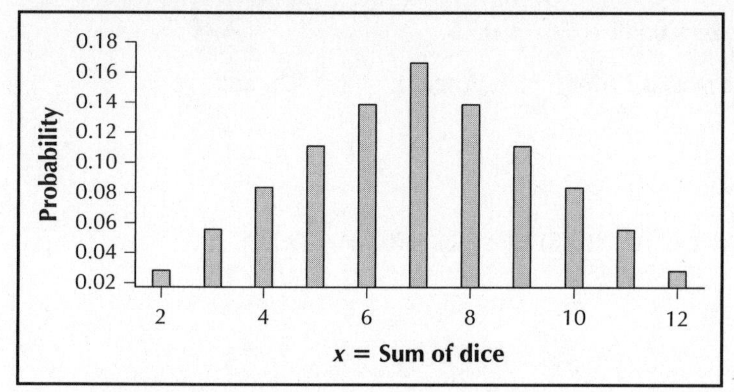

41. From Exercise 40, $\mu = 7$.
Variance using the computational formula:

X	$P(X)$	X^2	$X^2 \cdot P(X)$	X	$P(X)$	X^2	$X^2 \cdot P(X)$
2	1/36	4	4/36	8	5/36	64	320/36
3	2/36	9	18/36	9	4/36	81	324/36
4	3/36	16	48/36	10	3/36	100	300/36
5	4/36	25	100/36	11	2/36	121	242/36
6	5/36	36	180/36	12	1/36	144	144/36
7	6/36	49	294/36				

$$\sum[X^2 \cdot P(X)] = \frac{1974}{36}$$

$$\sigma^2 = \sum[X^2 \cdot P(X)] - \mu^2 = \frac{1974}{36} - 7^2 \approx 5.8333333333$$

$$\sigma = \sqrt{\sigma^2} = \sqrt{5.8333333333} \approx 2.4152$$

43. Since the value $X = 2$ is $\left|\frac{X - \mu}{\sigma}\right| = \left|\frac{2 - 7}{2.4152}\right| \approx 2.07$ standard deviations from the mean and because $2 \le 2.07 < 3$, snake eyes is considered moderately unusual. By symmetry, so is 12.

45. (a) Let X_i be the original data values and let $Y_i = X_i + k$ be the data values after some unknown amount k is added to each value of X. Let μ_X be the mean of the X's and μ_Y be the mean of the Y's. Since only the X's have been changed and not the probabilities, $P(X_i) = P(Y_i)$ for all i. Therefore,

$$\mu_Y = \sum[Y \cdot P(Y)] = \sum[(X + k) \cdot P(X)] = \sum[X \cdot P(X)] + k \cdot \sum[P(X)]$$
$$= \mu_X + k \cdot 1 = \mu_X + k, \text{ so the mean increases by } k.$$

(b) From (a), $\mu_Y = \mu_X + k$ and $P(X_i) = P(Y_i)$ for all i. Then $Y_i - \mu_Y = (X_i + k) - (\mu_X + k) = X_i - \mu_X$ for all i. Therefore,

$$\sigma_Y = \sum[(Y - \mu_Y)^2 \cdot P(Y)] = \sum[(X - \mu_X)^2 \cdot P(Y)] = \sigma_X,$$

so the standard deviation remains unchanged.

Section 6.2

1. Not binomial; the events "Person A comes to party" and "Person B comes to party" may not be independent.

3. Not binomial; more than two different ethnicities.

5. Not binomial; more than two possible ages.

7. Not binomial; more than two possible total number of spots.

9. Binomial; $n = 3$, X = number of even numbers, $p = 1/2$, $1 - p = 1/2$.

11. Binomial; $n = 2$, X = number of games won, $p = 0.25$, $1 - p = 0.75$.

For Exercises 13, 15, 17, $n = 3$ because the coin is tossed 3 times, $p = P(\text{head}) = \frac{1}{2} = 0.5$, and $1 - p = 1 - 0.5 = 0.5$.

13. No heads mean $X = 0$.
Therefore,

$$P(X = 0) = ({}_nC_X)p^X(1 - p)^{n-X} = ({}_3C_0)0.5^0(0.5)^{3-0} = (1)(1)(0.5)^3 = 0.125.$$

15. Two heads mean $X = 2$.
Therefore,

$$P(X = 2) = ({}_nC_X)p^X(1 - p)^{n-X} = ({}_3C_2)0.5^2(0.5)^{3-2} = (3)(0.5)^2(0.5)^1 = 0.375.$$

17. At most, two heads means either zero, one, or two heads. Therefore, find the probabilities for $X = 0$, $X = 1$, or $X = 2$ and add them up.

$$P(X = 0) + P(X = 1) + P(X = 2) = ({}_3C_0)0.5^0(0.5)^{3-0} + ({}_3C_1)0.5^1(0.5)^{3-1} + ({}_3C_2)0.5^2(0.5)^{3-2}$$
$$= (1)(1)(0.5)^3 + (3)(0.5)^1(0.5)^2 + (3)(0.5)^2(0.5)^1 = 0.125 + 0.375 + 0.375 = 0.875.$$

19. We have $n = 10$ because the coin is tossed 10 times,

$$p = P(\text{head}) = \frac{1}{2} = 0.5,\ 1 - p = 1 - 0.5 = 0.5,$$

and $X = 5$. Therefore,

$$P(X = 5) = ({}_nC_X)p^X(1 - p)^{n-X} = ({}_{10}C_5)0.5^5(0.5)^{10-5} = (252)(0.5)^5(0.5)^5 = 0.24609375 \approx 0.2461.$$

21. We have $n = 20$ because we are considering the next 20 people that you meet, $p = P(\text{left-handed}) = 0.1$, $1 - p = 1 - 0.1 = 0.9$, and $X = 0$. Therefore,

$$P(X = 0) = ({}_nC_X)p^X(1 - p)^{n-X} = ({}_{20}C_0)0.1^0(0.9)^{20-0} = (1)(1)(0.9)^{20} = 0.1215766546 \approx 0.1216.$$

23. We have $n = 10$ because the coin is tossed 10 times,

$$p = P(\text{head}) = \frac{1}{2} = 0.5, \text{ and } 1 - p = 1 - 0.5 = 0.5.$$

Thus,

$$\mu = n \cdot p = (10)(0.5) = 5,\ \sigma^2 = n \cdot p \cdot (1 - p) = (10)(0.5)(0.5) = 2.5, \text{ and } \sigma = \sqrt{\sigma^2} = \sqrt{2.5} \approx 1.5811.$$

The expected number of heads in 10 tosses of a fair coin is 5.

25. We have $n = 20$ because we are considering the next 20 people that you meet,

$$p = P(\text{left-handed}) = 0.1, \text{ and } 1 - p = 1 - 0.1 = 0.9.$$

Thus,

$$\mu = n \cdot p = (20)(0.1) = 2,\ \sigma^2 = n \cdot p \cdot (1 - p) = (20)(0.1)(0.9) = 1.8, \text{ and } \sigma = \sqrt{\sigma^2} = \sqrt{1.8} \approx 1.3416.$$

The expected number of people who are left-handed in a random sample of 20 people is 2.

27. Since you are tossing the dice three times, $n = 3$. We have $p = P(\text{doubles}) = \frac{1}{6}$ and $1 - p = 1 - \frac{1}{6} = \frac{5}{6}$.

Rolling doubles on at least two of the next three tosses of the dice means rolling doubles on either two or three of the next three tosses, so find the probabilities for $X = 2$ and $X = 3$ and add them up. Therefore,

$$P(X = 2) + P(X = 3) = ({}_3C_2)\left(\frac{1}{6}\right)^2\left(\frac{5}{6}\right)^{3-2} + ({}_3C_3)\left(\frac{1}{6}\right)^3\left(\frac{5}{6}\right)^{3-3}$$
$$= (3)\left(\frac{1}{6}\right)^2\left(\frac{5}{6}\right)^1 + (1)\left(\frac{1}{6}\right)^3\left(\frac{5}{6}\right)^0 = 0.0694444444 + 0.0046296296$$
$$= 0.0740740741 \approx 0.0741.$$

29. Since you have 17 classmates, $n = 17$. We have

$$p = P(\text{Canada}) = 0.045 \text{ and } 1 - p = 1 - 0.045 = 0.955.$$

At most, two classmates from Canada means either zero, one, or two classmates from Canada, so find the probabilities for $X = 0$, $X = 1$, and $X = 2$ and add them up. Therefore,

$$P(X=0)+P(X=1)+P(X=2)=({}_{17}C_0)(0.045)^0(0.955)^{17-0}+({}_{17}C_1)(0.045)^1(0.955)^{17-1}$$
$$+({}_{17}C_2)(0.045)^2(0.955)^{17-2}=(1)(0.045)^0(0.955)^{17}+(17)(0.045)^1(0.955)^{16}$$
$$+(136)(0.045)^2(0.955)^{15}=0.4571485188+0.3661975046+0.1380430364$$
$$=0.9613890598\approx 0.9614.$$

31. Since you have 17 classmates, $n = 17$. We have $p = P(\text{Canada}) = 0.045$ and $1 - p = 1 - 0.045 = 0.955$. Thus,

$$\mu = n \cdot p = (17)(0.045) = 0.765,$$
$$\sigma^2 = n \cdot p \cdot (1-p) = (17)(0.045)(0.955) = 0.730575 \approx 0.7306, \text{ and } \sigma = \sqrt{\sigma^2} = \sqrt{0.730575} \approx 0.8547.$$

The expected number of students that are from Canada in a random sample of 17 students is 0.765.

33. We have $n = 5$ because we are considering the next five cars on the interstate,

$$p = P(\text{obeys the speed limit}) = 0.5, \text{ and } 1 - p = 1 - 0.5 = 0.5.$$

Thus,

$$\mu = n \cdot p = (5)(0.5) = 2.5, \sigma^2 = n \cdot p \cdot (1-p) = (5)(0.5)(0.5) = 1.25, \text{ and } \sigma = \sqrt{\sigma^2} = \sqrt{1.25} \approx 1.1180.$$

The expected number of cars obeying the speed limit in a random sample of five cars on the Interstate is 2.5.

35. We have $n = 10$ because the coin is tossed 10 times and $p = P(\text{head}) = \frac{1}{2} = 0.5$. Tossing at least five heads means 5, 6, 7, 8, 9, or 10 heads, so find the probabilities for $X = 5$, $X = 6$, $X = 7$, $X = 8$, $X = 9$, and $X = 10$ and add them up. Using Table B in the Appendix,

$$P(X=5)+P(X=6)+P(X=7)+P(X=8)+P(X=9)+P(X=10)$$
$$=0.2461+0.2051+0.1172+0.0439+0.0098+0.0010=0.6231.$$

37. We have $n = 20$ because we are considering the 20 people that you survey and $p = P(\text{support an Independent candidate for president}) = 0.15$. At least five people would support an Independent for president means that 5, 6, 7, 8, 9, 10, 11, 12, 13, 14, 15, 16, 17, 18, 19, or 20 people would support an Independent for president, so find the probabilities for $X = 5$, $X = 6$, $X = 7$, $X = 8$, $X = 9$, $X = 10$, $X = 11$, $X = 12$, $X = 13$, $X = 14$, $X = 15$, $X = 16$, $X = 17$, $X = 18$, $X = 19$, and $X = 20$ and add them up. Using Table B in the Appendix,

$$P(X=5)+P(X=6)+P(X=7)+P(X=8)+P(X=9)+P(X=10)+P(X=11)+P(X=12)$$
$$+P(X=13)+P(X=14)+P(X=15)+P(X=16)+P(X=17)+P(X=18)$$
$$+P(X=19)+P(X=20)$$
$$=0.1028+0.0454+0.0160+0.0046+0.0011+0.0002+0+0+0+0+0+0$$
$$+0+0+0+0=0.1701.$$

39. We have $n = 40$ because the coin is tossed 40 times,

$$p = P(\text{head}) = \frac{1}{2} = 0.5, \text{ and } 1 - p = 1 - 0.5 = 0.5.$$

Thus,

$$\mu = n \cdot p = (40)(0.5) = 20, \sigma^2 = n \cdot p \cdot (1-p) = (40)(0.5)(0.5) = 10, \text{ and } \sigma = \sqrt{\sigma^2} = \sqrt{10} \approx 3.1623.$$

The expected number of heads observed in 40 tosses of a fair coin is 20.

41. We have $n = 20$ because we are considering the 20 people that you survey, $p = P(\text{support an Independent candidate for president}) = 0.15$, and $1 - p = 1 - 0.15 = 0.85$.

Thus,

$$\mu = n \cdot p = (20)(0.15) = 3, \sigma^2 = n \cdot p \cdot (1-p) = (20)(0.15)(0.85) = 2.55, \text{ and } \sigma = \sqrt{\sigma^2} = \sqrt{2.55} \approx 1.5969.$$

The expected number of people who will support an Independent candidate for president in the next election in a random sample of 20 people is three.

43. $n = 12$ and $p = 0.25$. Use Table B in the Appendix. **(a)** Exactly three sophomores means $X = 3$, so $P(X = 3) = 0.2581$. **(b)** At most, three sophomores means zero, one, two, or three sophomores, so find the probabilities for $X = 0$, $X = 1$, $X = 2$, and $X = 3$ and add them up.
Thus,

$$P(X = 0) + P(X = 1) + P(X = 2) + P(X = 3) = 0.0317 + 0.2062 + 0.2835 + 0.2362 = 0.6489.$$

(c) We can see from Table B in the Appendix that $P(X = 3)$ is the largest probability, so three sophomores is the most likely number of sophomores in the sample.

45. **(a)** $n = 12$, $p = 0.25$, and $1 - p = 1 - 0.25 = 0.75$. Thus,

$$\mu = n \cdot p = (12)(0.25) = 3, \sigma^2 = n \cdot p \cdot (1 - p) = (12)(0.25)(0.75) = 2.25, \text{ and } \sigma = \sqrt{\sigma^2} = \sqrt{2.25} = 1.5.$$

The expected number of sophomores in a random sample of 12 statistics students is 3.

(b) $X = 0$ sophomores is

$$\left|\frac{X - \mu}{\sigma}\right| = \left|\frac{0 - 3}{1.5}\right| = 2$$

standard deviations from the mean. Since $2 \le 2 < 3$, a random sample of 12 statistics students that contained no sophomores would be considered moderately unusual.

47. $n = 20$ and $p = P(\text{woman}) = 0.4$. Use Table B in the Appendix. **(a)** Exactly 10 women means $X = 10$, so $P(X = 10) = 0.1171$. **(b)** At least 10 women means $X = 10$, $X = 11$, $X = 12$, $X = 13$, $X = 14$, $X = 15$, $X = 16$, $X = 17$, $X = 18$, $X = 19$, and $X = 20$, so find the probabilities and add them up. Thus,

$$\begin{aligned} &P(X = 10) + P(X = 11) + P(X = 12) + P(X = 13) + P(X = 14) + P(X = 15) \\ &\quad + P(X = 16) + P(X = 17) + P(X = 18) + P(X = 19) + P(X = 20) \\ &\quad = 0.1171 + 0.0710 + 0.0355 + 0.0146 + 0.0049 + 0.0013 + 0.0003 \\ &\quad + 0 + 0 + 0 + 0 = 0.2447. \end{aligned}$$

(c) From Table B in the Appendix we can see that $P(X = 8)$ is the largest probability, so eight is the most likely number of women.

49. **(a)** $n = 20$, $p = P(\text{woman}) = 0.4$, and $1 - p = 1 - 0.4 = 0.6$. Thus, $\mu = n \cdot p = (20)(0.4) = 8$, $\sigma^2 = n \cdot p \cdot (1 - p) = (20)(0.4)(0.6) = 4.8$, and $\sigma = \sqrt{\sigma^2} = \sqrt{4.8} \approx 2.1909$. The expected number of women in a random sample of 20 students from management courses is eight.

(b) $X = 12$ women is

$$\left|\frac{X - \mu}{\sigma}\right| = \left|\frac{12 - 8}{2.1909}\right| = 1.83$$

standard deviations from the mean. Since $0 < 1.83 < 2$, a random sample of 20 students from management courses that contains 12 women is not considered unusual.

51. **(a)** $n = 10$ and $p = P(\text{"Music I Like"}) = 0.87$. Since $p = 0.87$ is not in Table B in the Appendix, we need to use the formula or technology. Minitab generated the following table for $n = 10$ and $p = 0.87$.

X	Probability	X	Probability
0	0.0000	6	0.0260
1	0.0000	7	0.0995
2	0.0000	8	0.2496
3	0.0000	9	0.3712
4	0.0006	10	0.2484
5	0.0047		

$X = 8$, so from the table, $P(X = 8) = 0.2496$.

(b) From the table in (a), we can see that $P(X = 9)$ is the biggest probability, so nine women is the most likely number of women who report that "Music I Like" is the biggest factor in deciding which radio station to tune to.

(c) $\mu = n \cdot p = (10)(0.87) = 8.7$ women. The expected number of women who reported that "Music I Like" is the biggest factor in deciding which radio station to listen to in a random sample of 10 women is 8.7.

53. (a) $n = 10, p = P(\text{"Music I Like"}) = 0.87$, and $1 - p = 1 - 0.87 = 0.13$. Thus,

$$\sigma^2 = n \cdot p \cdot (1 - p) = (10)(0.87)(0.13) = 1.131, \text{ and } \sigma = \sqrt{\sigma^2} = \sqrt{1.131} \approx 1.0635.$$

(b) From Exercise 51(c), $\mu = 8.7$ women. Thus, $X = 2$ women is

$$\left|\frac{X - \mu}{\sigma}\right| = \left|\frac{2 - 8.7}{1.0635}\right| = 6.30$$

standard deviations from the mean. Since $3 \leq 6.30$, a random sample of 10 women that contains 2 women who report "Music I Like" as the biggest factor in deciding which radio station to tune to is unusual.

55. (a) $n = 100$ and $p = P(\text{access to Internet}) = 0.61$. Since $n = 100$ and $p = 0.61$ are not in Table B in the Appendix, we need to use the formula or technology. Minitab generated the following table for $n = 100$ and $p = 0.61$.

X	Probability	X	Probability	X	Probability
0	0.0000	27	0.0000	54	0.0290
1	0.0000	28	0.0000	55	0.0380
2	0.0000	29	0.0000	56	0.0477
3	0.0000	30	0.0000	57	0.0576
4	0.0000	31	0.0000	58	0.0668
5	0.0000	32	0.0000	59	0.0744
6	0.0000	33	0.0000	60	0.0795
7	0.0000	34	0.0000	61	0.0816
8	0.0000	35	0.0000	62	0.0803
9	0.0000	36	0.0000	63	0.0757
10	0.0000	37	0.0000	64	0.0685
11	0.0000	38	0.0000	65	0.0593
12	0.0000	39	0.0000	66	0.0492
13	0.0000	40	0.0000	67	0.0390
14	0.0000	41	0.0000	68	0.0296
15	0.0000	42	0.0001	69	0.0215
16	0.0000	43	0.0001	70	0.0149
17	0.0000	44	0.0002	71	0.0098
18	0.0000	45	0.0004	72	0.0062
19	0.0000	46	0.0008	73	0.0037
20	0.0000	47	0.0015	74	0.0021
21	0.0000	48	0.0025	75	0.0012
22	0.0000	49	0.0042	76	0.0006
23	0.0000	50	0.0067	77	0.0003
24	0.0000	51	0.0102	78	0.0001
25	0.0000	52	0.0151	79	0.0001
26	0.0000	53	0.0213		

$P(X) \approx 0$ for $X = 80$ through 100. $X = 60$, so from the table, $P(X = 60) = 0.0795$. **(b)** $P(X = 60) + P(X = 61) + P(X = 62) = 0.0795 + 0.0816 + 0.0803 = 0.2414$ **(c)** From the table in (a), $P(X = 61)$ is the highest probability, so 61 is the most likely number of women who had access to the Internet. **(d)** $\mu = n \cdot p = (100)(0.61) = 61$ women. The expected number of women with access to the Internet in a random sample of 100 women is 61.

57. **(a)** $n = 100$, $p = P(\text{access to Internet}) = 0.61$, and $1 - p = 1 - 0.61 = 0.39$. Thus,

$$\sigma^2 = n \cdot p \cdot (1 - p) = (100)(0.61)(0.39) = 23.79 \text{ and } \sigma = \sqrt{\sigma^2} = \sqrt{23.79} \approx 4.8775.$$

(b) From Exercise 55(d), $\mu = 61$ women. Thus, $X = 49$ women is

$$\left|\frac{X - \mu}{\sigma}\right| = \left|\frac{49 - 61}{4.8775}\right| = 2.46$$

standard deviations from the mean. Since $2 \leq 2.46 < 3$, a random sample of 100 women that contains 49 women that had access to the Internet is moderately unusual.

59. **(a)** $n = 100$ and $p = P(\text{women is depressed}) = 0.12$. Since $n = 100$ and $p = 0.12$ are not in Table B in the Appendix, we need to use the formula or technology. Minitab generated the following table for $n = 100$ and $p = 0.12$.

X	Probability	*X*	Probability	*X*	Probability
0	0.0000	9	0.0871	18	0.0229
1	0.0000	10	0.1080	19	0.0135
2	0.0003	11	0.1205	20	0.0074
3	0.0012	12	0.1219	21	0.0039
4	0.0038	13	0.1125	22	0.0019
5	0.0100	14	0.0954	23	0.0009
6	0.0215	15	0.0745	24	0.0004
7	0.0394	16	0.0540	25	0.0002
8	0.0625	17	0.0364		

$P(X) \approx 0$ for $X = 26$ through 100.

At most, 10 women means 0, 1, 2, 3, 4, 5, 6, 7, 8, 9, or 10 women. Therefore, find the probabilities for $X = 0$, $X = 1$, $X = 2$, $X = 3$, $X = 4$, $X = 5$, $X = 6$, $X = 7$, $X = 8$, $X = 9$, or $X = 10$ and add them up. From the table above,

$$\begin{aligned} &P(X = 0) + P(X = 1) + P(X = 2) + P(X = 3) + P(X = 4) + P(X = 5) + P(X = 6) \\ &\quad + P(X = 7) + P(X = 8) + P(X = 9) + P(X = 10) \\ &\quad = 0 + 0 + 0.0003 + 0.0012 + 0.0038 + 0.0100 + 0.0215 + 0.0394 + 0.0625 \\ &\qquad + 0.0871 + 0.1080 = 0.3338 \end{aligned}$$

(b) From the table above, we can see that $P(X = 12)$ is the highest probability, so the most likely number of women who are affected by a depressive disorder is 12 women. **(c)** $\mu = n \cdot p = (100)(0.12) = 12$ women. The expected number of women who are affected by a depressive disorder in a random sample of 100 women is 12.

61. **(a)** $n = 100$, $p = P(\text{women is depressed}) = 0.12$, and $1 - p = 1 - 0.12 = 0.88$. Thus,

$$\sigma^2 = n \cdot p \cdot (1 - p) = (100)(0.12)(0.88) = 10.56 \text{ and } \sigma = \sqrt{\sigma^2} = \sqrt{10.56} \approx 3.2496.$$

(b) From Exercise 59(c), $\mu = 12$ women. Thus, $X = 10$ women is

$$\left|\frac{X - \mu}{\sigma}\right| = \left|\frac{10 - 12}{3.2496}\right| = 0.62$$

standard deviations from the mean. Since $0 \leq 0.62 < 2$, a random sample of 100 women that contains 10 women who are affected by a depressive disorder is not unusual.

63. **(a)** $n = 400$ and $p = P(\text{using paid services}) = 0.17$. Since $n = 400$ and $p = 0.17$ are not in Table B in the Appendix, we need to use the formula or technology. Minitab generated the following table for $n = 400$ and $p = 0.17$.

X	Probability	X	Probability	X	Probability
0	0.0000	33	0.0000	66	0.0518
1	0.0000	34	0.0000	67	0.0529
2	0.0000	35	0.0000	68	0.0530
3	0.0000	36	0.0000	69	0.0523
4	0.0000	37	0.0000	70	0.0506
5	0.0000	38	0.0000	71	0.0482
6	0.0000	39	0.0000	72	0.0451
7	0.0000	40	0.0000	73	0.0415
8	0.0000	41	0.0000	74	0.0376
9	0.0000	42	0.0001	75	0.0334
10	0.0000	43	0.0001	76	0.0293
11	0.0000	44	0.0002	77	0.0252
12	0.0000	45	0.0003	78	0.0214
13	0.0000	46	0.0005	79	0.0179
14	0.0000	47	0.0008	80	0.0147
15	0.0000	48	0.0013	81	0.0119
16	0.0000	49	0.0019	82	0.0095
17	0.0000	50	0.0027	83	0.0074
18	0.0000	51	0.0038	84	0.0057
19	0.0000	52	0.0052	85	0.0044
20	0.0000	53	0.0070	86	0.0033
21	0.0000	54	0.0092	87	0.0024
22	0.0000	55	0.0119	88	0.0018
23	0.0000	56	0.0150	89	0.0013
24	0.0000	57	0.0185	90	0.0009
25	0.0000	58	0.0225	91	0.0006
26	0.0000	59	0.0267	92	0.0004
27	0.0000	60	0.0311	93	0.0003
28	0.0000	61	0.0354	94	0.0002
29	0.0000	62	0.0397	95	0.0001
30	0.0000	63	0.0436	96	0.0001
31	0.0000	64	0.0470	97	0.0001
32	0.0000	65	0.0498		

$P(X) \approx 0$ for $X = 98$ through 400.

$X = 60$, so from the table above, $P(X = 60) = 0.0311$

(b) From the table above, $P(X = 68)$ is the largest probability, so the most likely number of music downloaders who are using a paid service is 68 music downloaders. **(c)** $\mu = n \cdot p = (400)(0.17) = 68$ music downloaders. The expected number of music downloaders who say that they are using paid services in a sample of 400 music downloaders is 68.

65. **(a)** $n = 400$, $p = P(\text{using paid services}) = 0.17$, and $1 - p = 1 - 0.17 = 0.83$.

Thus, $\sigma^2 = n \cdot p \cdot (1 - p) = (400)(0.17)(0.83) = 56.44$ and $\sigma = \sqrt{\sigma^2} = \sqrt{56.44} \approx 7.5127$.

(b) From Exercise 63(c), $\mu = 68$ music downloaders. Thus, $X = 90$ music downloaders is

$$\left|\frac{X - \mu}{\sigma}\right| = \left|\frac{90 - 68}{7.5127}\right| = 2.93$$

standard deviations from the mean. Since $2 \leq 2.93 < 3$, a random sample of 400 music downloaders that contains 90 music downloaders who are using paid services is moderately unusual.

Section 6.3

1. For a continuous random variable X, the probability that X equals some particular value is always zero.

3. Area under the normal distribution curve above an interval.

5. $P(X = 3285) = 0$

7. $P(X \geq 3285) = 0.5$

9. Greater than 0.5. Since $X = 4285$ is greater than the mean of 3285 and the area to the left of $\mu = 3285$ is 0.5, the area to the left of $X = 4285$ is greater than the area to the left of $X = 3285$.

Use the following graph for Exercise 11.

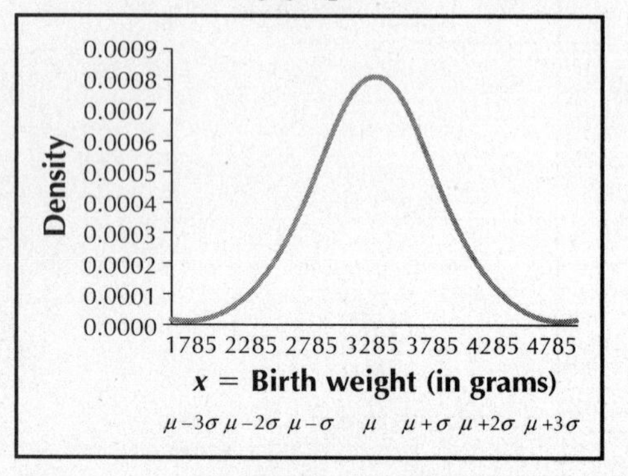

11. The graph shows that the area under the curve between 1785 grams and 4785 grams is the area between $\mu - 3\sigma$ and $\mu + 3\sigma$. Courtesy of the Empirical Rule, the area between $\mu - 3\sigma$ and $\mu + 3\sigma$ is about 0.997. Therefore, the probability of a birth weight between 1785 and 4785 grams is about about 0.997.

13. Normal distribution A has mean 10 and normal distribution B has mean 25. We know this because the peak of a normal distribution curve is located at the mean and the peak of curve A is left of the peak of curve B, so the mean of normal distribution A is less than the mean of normal distribution B.

15. $\mu = 0, \sigma = 1$

17. $\mu = -10, \sigma = 10$

19. $\mu = 100$, 125 is 2 standard deviations from the mean of $\mu = 100$, so $\sigma = \frac{1}{2}(125 - 100) = 12.5$.

21. $\mu = 0$. 6 is 3 standard deviations above the mean of $\mu = 0$, so $\sigma = \frac{1}{3}(6 - 0) = 2$.

23.

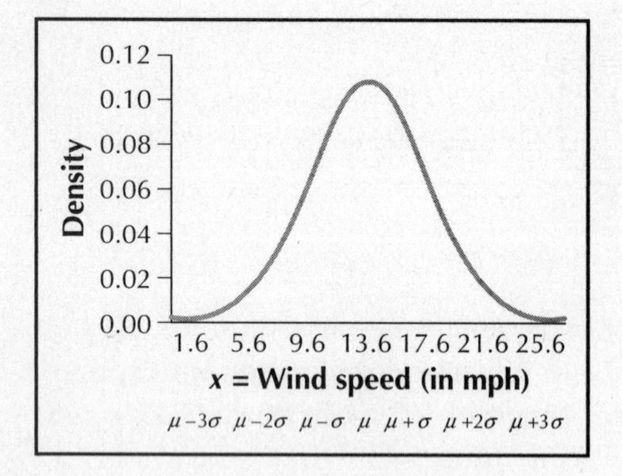

(a)

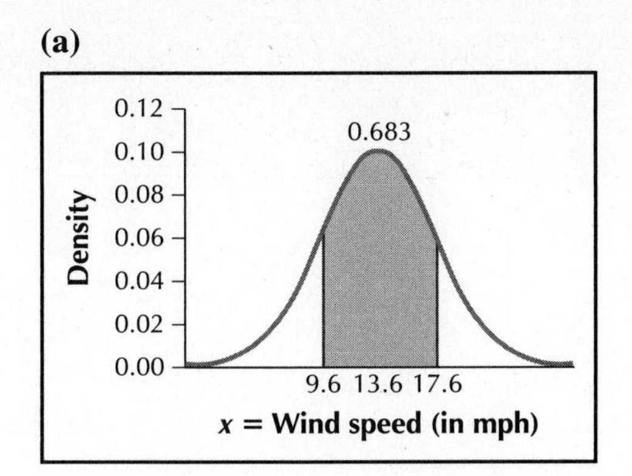

(b) From the first graph, the area between 9.6 mph and 17.6 mph is the area between $\mu - \sigma$ and $\mu + \sigma$. Courtesy of the Empirical Rule, the area between $\mu - \sigma$ and $\mu + \sigma$ is about 0.68. Therefore, about 0.68 of wind speeds in San Francisco in July are between 9.6 mph and 17.6 mph.

25.

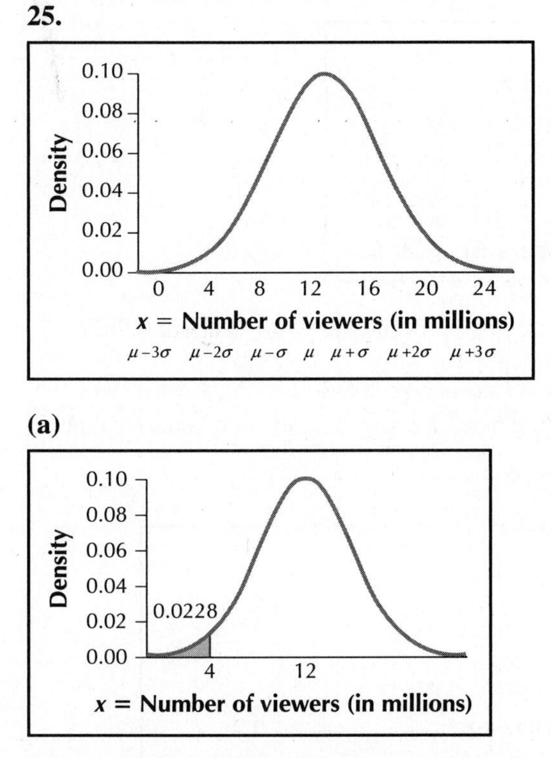

(b) From the first graph, the area to the left of 4 million viewers is the area to the left of $\mu - 2\sigma$. Courtesy of the Empirical Rule, the area to the left of $\mu - 2\sigma$ is about 0.025. Therefore, the probability that fewer than 4 million viewers will watch *60 Minutes* is about 0.025.

27.

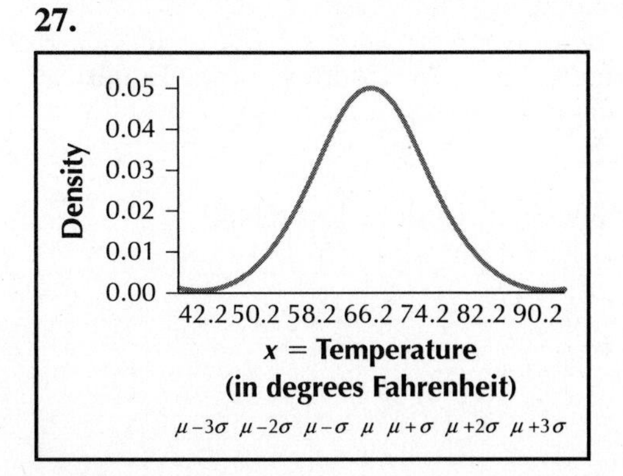

(a)

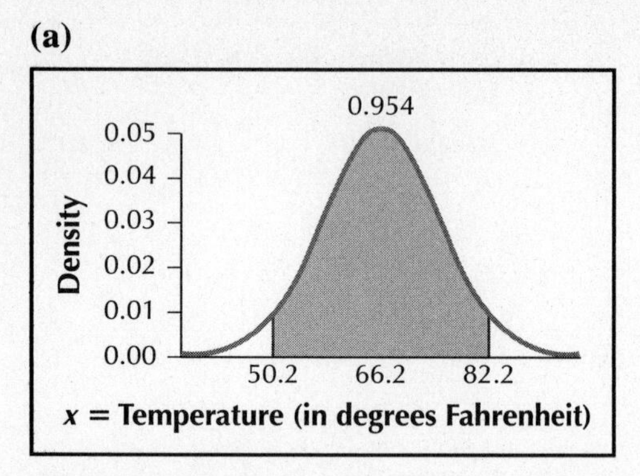

(b) From the first graph, the area between 50.2°F and 82.2°F is the area between $\mu - 2\sigma$ and $\mu + 2\sigma$. Courtesy of the Empirical Rule, the area between $\mu - 2\sigma$ and $\mu + 2\sigma$ is about 0.95. Therefore, the probability that the temperature in Los Angeles is between 50.2°F and 82.2°F is about 0.95.

29.

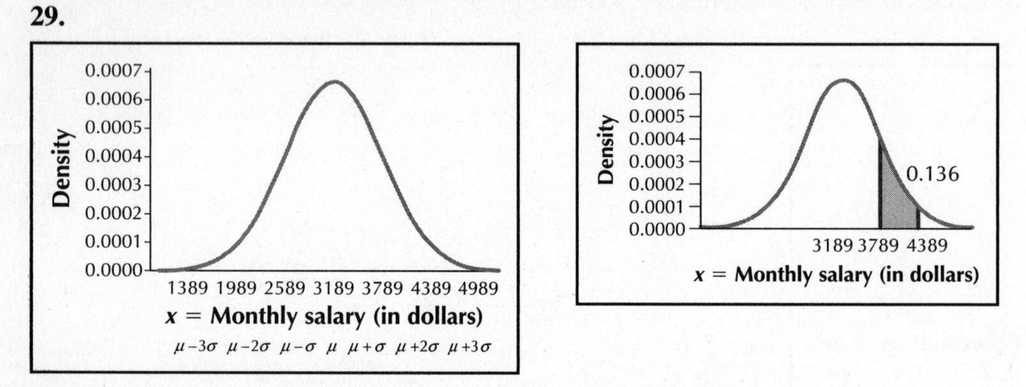

From the first graph, the area between \$3789 and \$4389 is the area between $\mu + \sigma$ and $\mu + 2\sigma$. Courtesy of the Empirical Rule, the area between $\mu + \sigma$ and $\mu + 2\sigma$ is about 0.135. Therefore, the proportion of salaries that lies between \$3789 and \$4389 is about 0.135.

31.

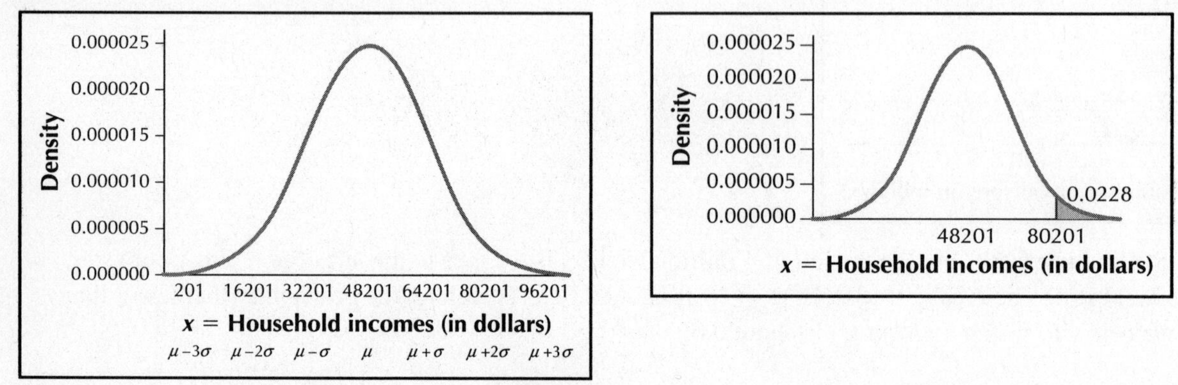

From the first graph, the area to the right of \$80,201 is the area to the right of $\mu + 2\sigma$. Courtesy of the Empirical Rule, the area to the right of $\mu + 2\sigma$ is about 0.025. Therefore, the probability that a randomly selected household has an income of greater than \$80,201 is about 0.025.

33. There is no minimum value of X for a normal distribution, but birth weights can't be less than 0.

Section 6.4

1. $\mu = 0$

3. True.

5. **(a)** Greater than 0.5. **(b)** From Table C, the area to the left of $Z = 1.96$ is 0.9750. **(c)** 0.9750 is greater than 0.5.

7. **(a)** Less than 0.5. **(b)** From Table C, the area between $Z = 0$ and $Z = 2.10$ is $0.9821 - 0.5 = 0.4821$. **(c)** 0.4821 is less than 0.5.

9. **(a)** Less than 0.5. **(b)** From Table C, the area to the right of $Z = 2.10$ is $1 - 0.9821 = 0.0179$. **(c)** 0.0179 is less than 0.5.

11. **(a)** Greater than 0.5. **(b)** From Table C, the area between $Z = -2.20$ and $Z = 0.90$ is $0.8159 - 0.0139 = 0.8020$. **(c)** 0.8020 is greater than 0.5.

13. **(a)** Less than 0.5. **(b)** From Table C, the area between $Z = 0.80$ and $Z = 1.90$ is $0.9713 - 0.7881 = 0.1832$. **(c)** 0.1832 is less than 0.5.

15. **(a)** Greater than 0.5. **(b)** From Table C, the area to the left of $Z = 1.80$ is 0.9641. **(c)** 0.9641 is greater than 0.5.

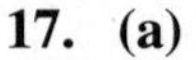

17. **(a)**

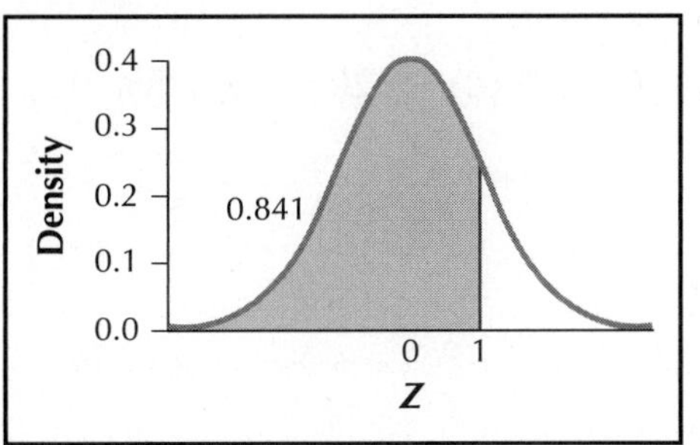

(b) Greater than 0.5. **(c)** From Table C, the area to the left of $Z = 1.00$ is 0.8413. **(d)** 0.8413 is greater than 0.5.

19. **(a)**

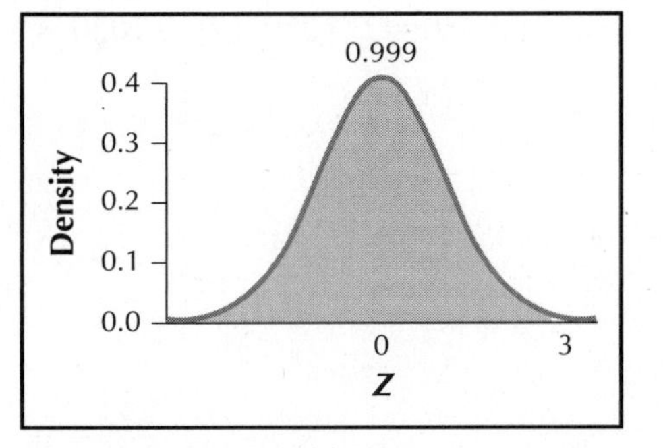

(b) Greater than 0.5. **(c)** From Table C, the area to the left of $Z = 3.00$ is 0.9987. **(d)** 0.9987 is greater than 0.5.

21. (a)

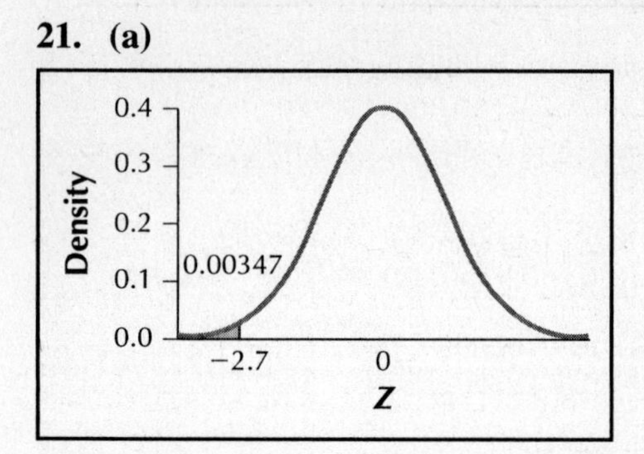

(b) Less than 0.5. **(c)** From Table C, the area to the left of $Z = -2.70$ is 0.0035. **(d)** 0.0035 is less than 0.5.

23. (a)

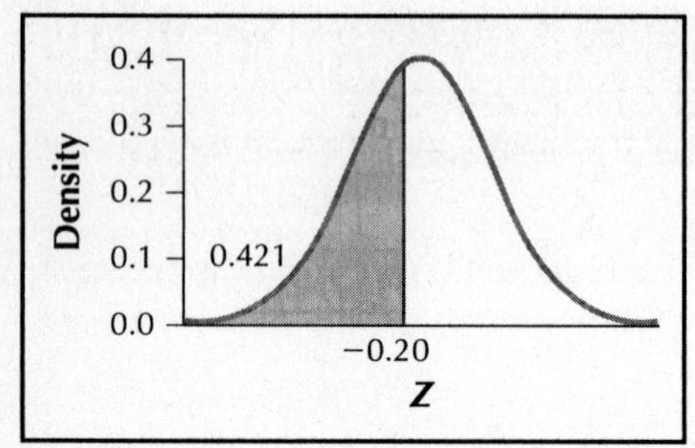

(b) Less than 0.5. **(c)** From Table C, the area to the left of $Z = -0.20$ is 0.4207. **(d)** 0.4207 is less than 0.5.

25. (a)

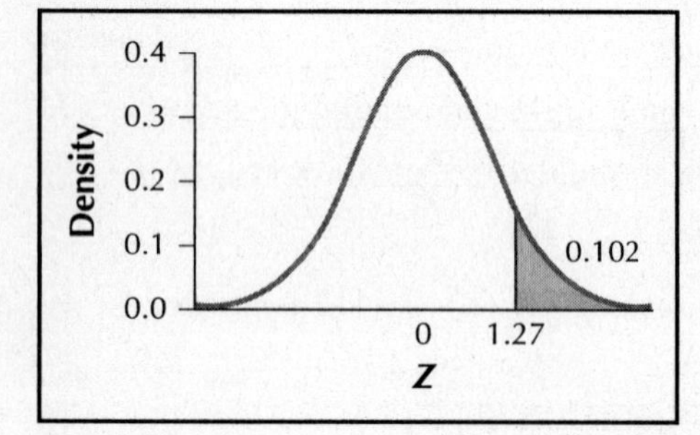

(b) Less than 0.5. **(c)** From Table C, the area to the right of $Z = 1.27$ is $1 - 0.8980 = 0.1020$. **(d)** 0.1020 is less than 0.5.

27. (a)

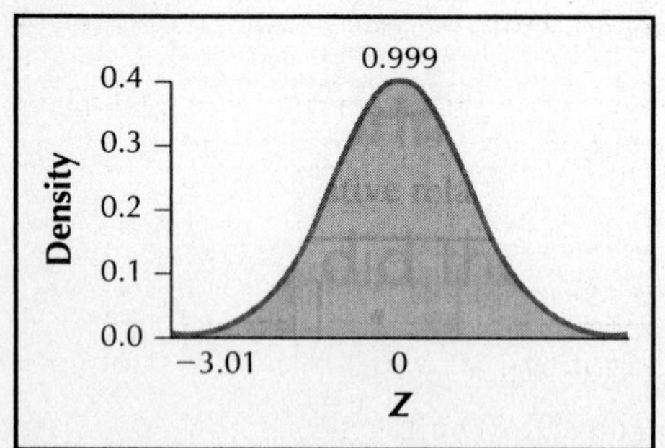

(b) Greater than 0.5. **(c)** From Table C, the area to the right of $Z = -3.01$ is $1 - 0.0013 = 0.9987$.
(d) 0.9987 is greater than 0.5.

29. (a)

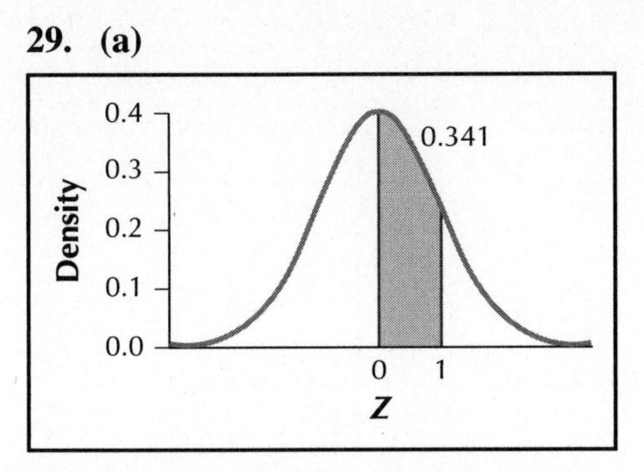

(b) Less than 0.5. **(c)** From Table C, the area between Z = 0 and Z = 1.00 is 0.8413 − 0.5 = 0.3413.
(d) 0.3143 is less than 0.5.

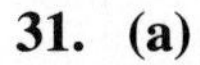
31. (a)

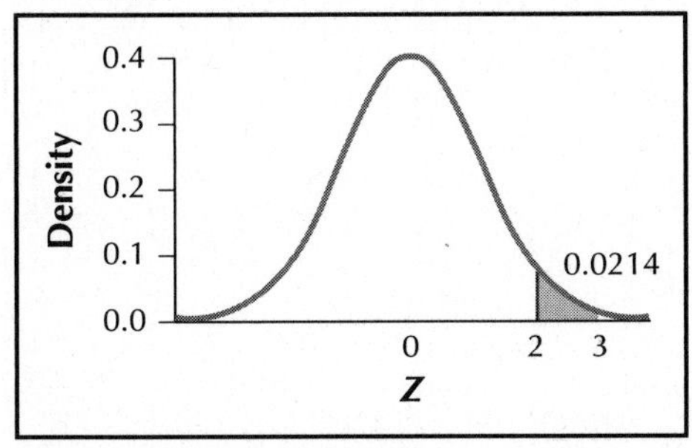

(b) Less than 0.5. **(c)** From Table C, the area between Z = 2.00 and Z = 3.00 is 0.9987 − 0.9772 = 0.0215.
(d) 0.0215 is less than 0.5.

33. (a)

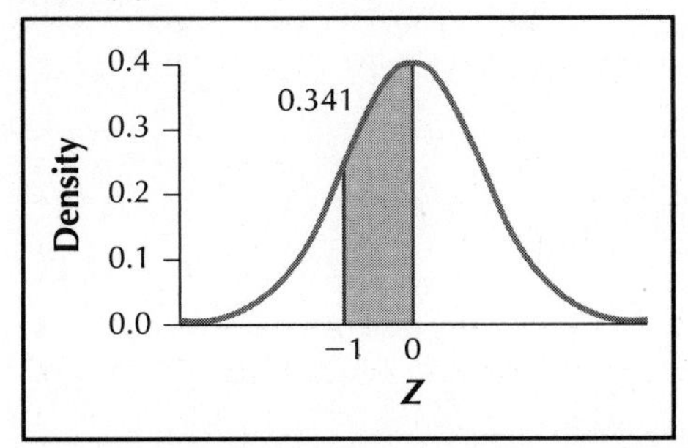

(b) Less than 0.5. **(c)** From Table C, the area between Z = −1.00 and Z = 0 is 0.5000 − 0.1587 = 0.3413.
(d) 0.3413 is less than 0.5.

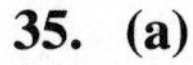
35. (a)

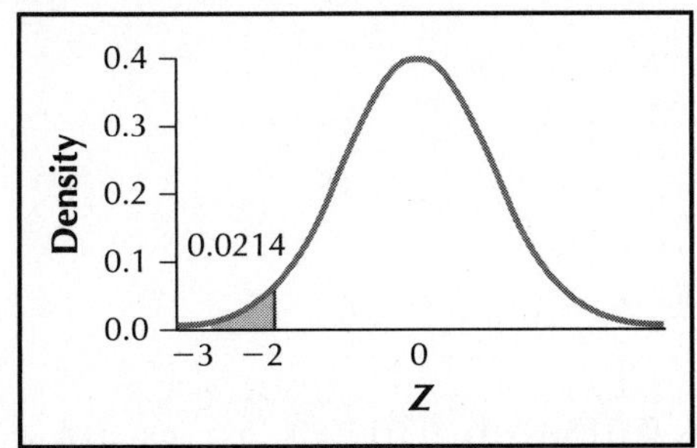

(b) Less than 0.5. **(c)** From Table C, the area between Z = −3.00 and Z = −2.00 is 0.0228 − 0.0013 = 0.0215. From Minitab, the area is 0.0214. **(d)** Both 0.0215 and 0.0214 are less than 0.5.

37. (a)

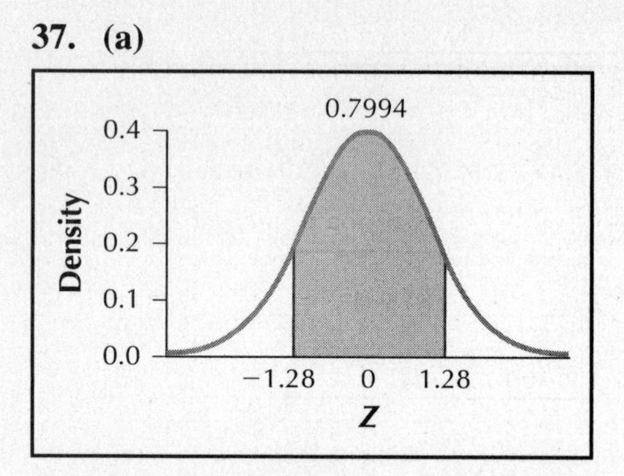

(b) Greater than 0.5. **(c)** From Table C, the area between $Z = -1.28$ and $Z = 1.28$ is $0.8997 - 0.1003 = 0.7994$. **(d)** 0.7994 is greater than 0.5.

39. (a)

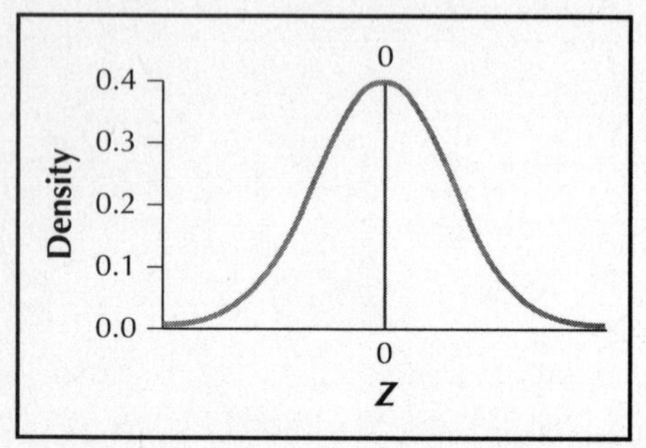

(b) 0 **(c)** 0 **(d)** The estimate equals the actual value.

41. (a)

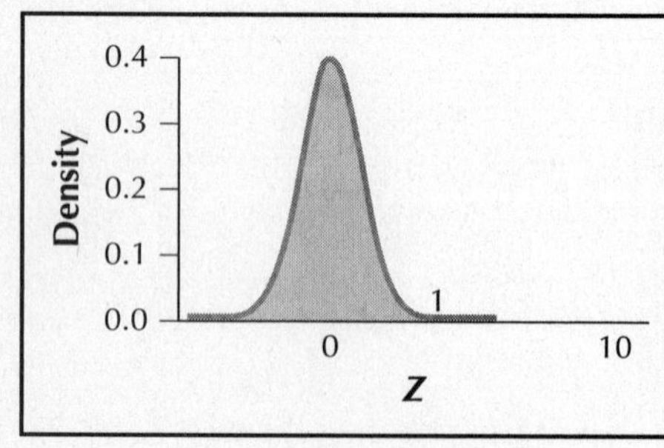

(b) 1 **(c)** 1 **(d)** The estimate equals the actual value.

43. (a)

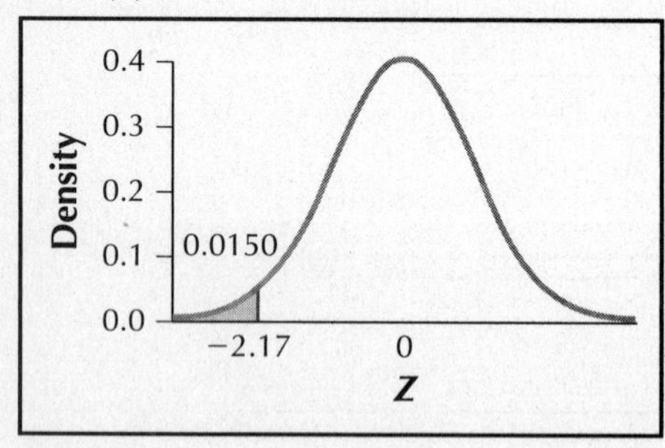

(b) Less than 0.5. **(c)** From Table C, the area to the left of $Z = -2.17$ is 0.0150. **(d)** 0.0150 is less than 0.5.

45. (a)

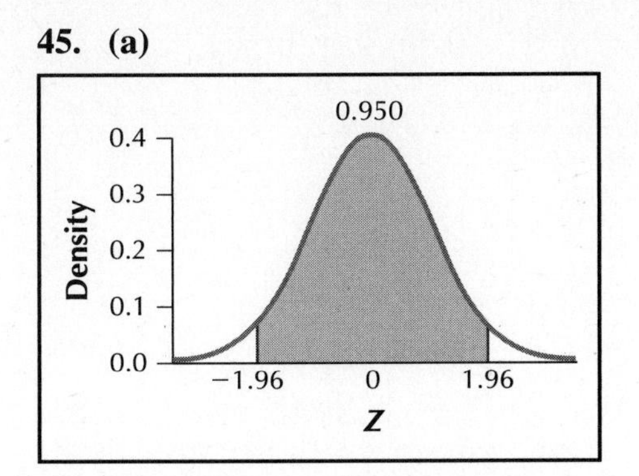

(b) Greater than 0.5. **(c)** From Table C, the area between $Z = -1.96$ and $Z = 1.96$ is $0.9750 - 0.0250 = 0.9500$. **(d)** 0.9500 is greater than 0.5.

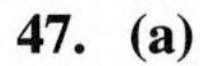
47. (a)

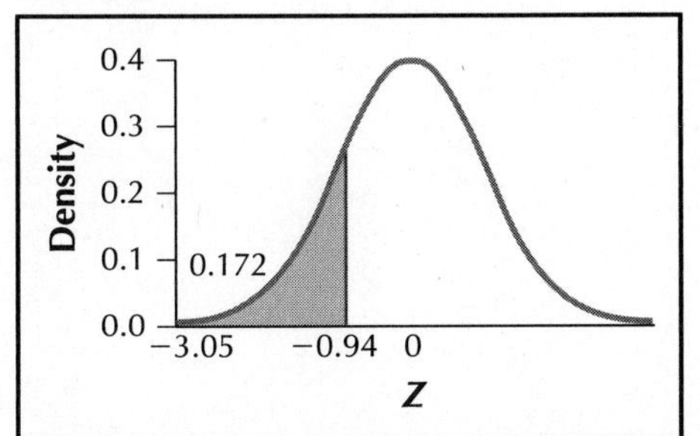

(b) Less than 0.5. **(c)** From Table C, the area between $Z = -3.05$ and $Z = -0.94$ is $0.1736 - 0.0011 = 0.1725$. **(d)** 0.1725 is less than 0.5.

49. (a)

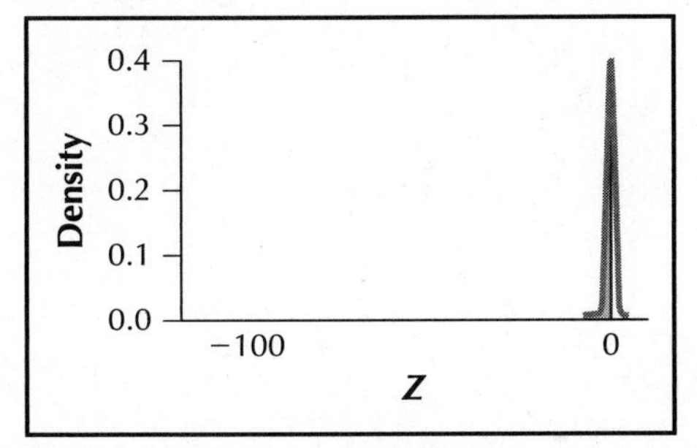

(b) About 0.5. **(c)** 0.5000 **(d)** The estimate equals the mean.

51.

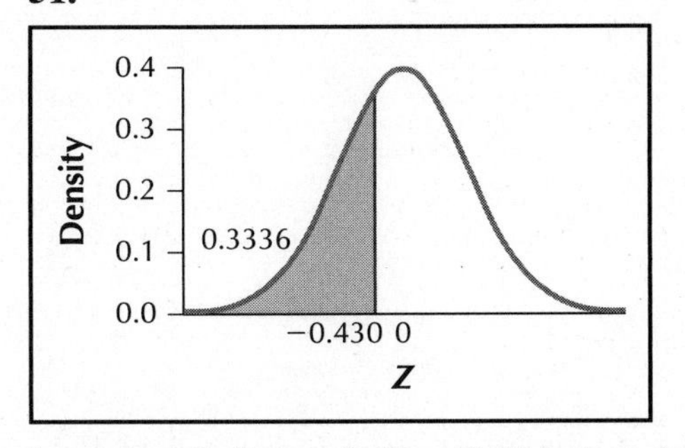

Less than 0, $Z = -0.43$, -0.43 is less than 0.

53.

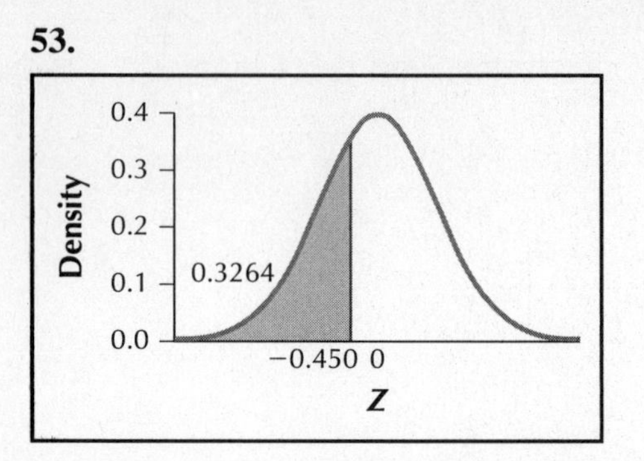

Less than 0, $Z = -0.45$, -0.45 is less than 0.

55.

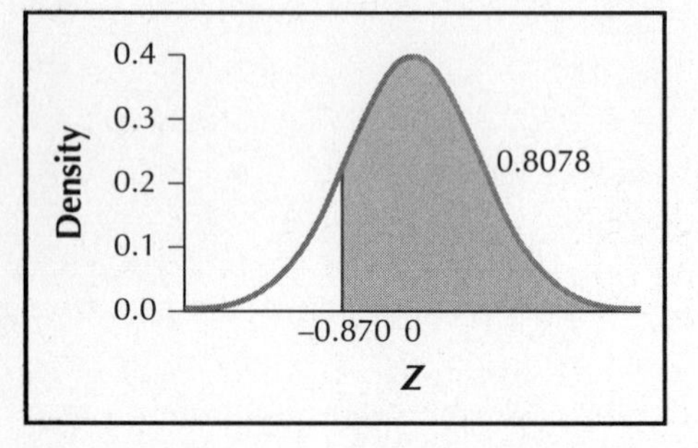

Less than 0, $Z = -0.87$, -0.87 is less than 0.

57.

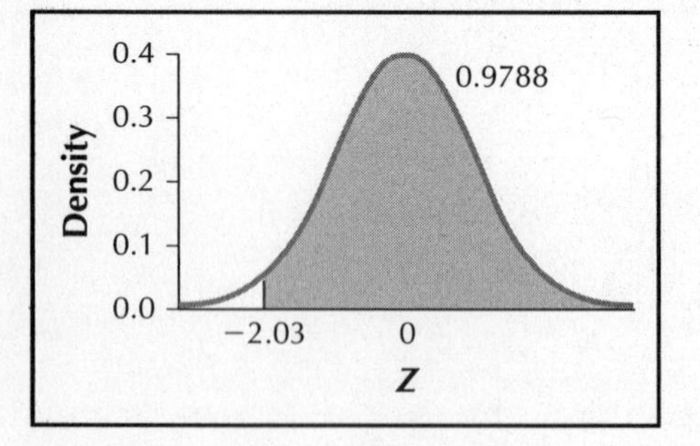

Less than 0, $Z = -2.03$, -2.03 is less than 0.

59. $Z = 0$

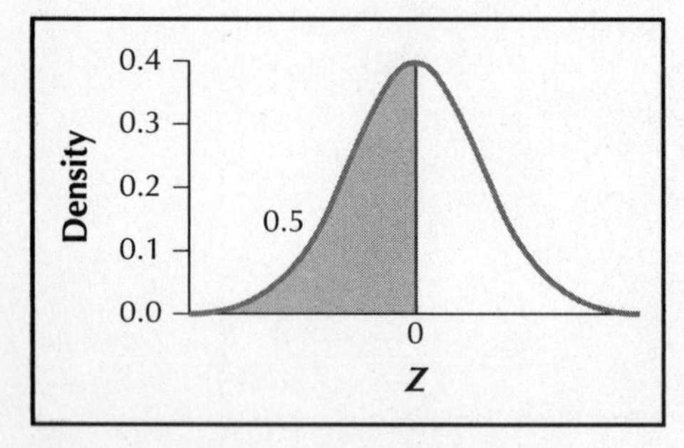

61. $Z = 2.58$

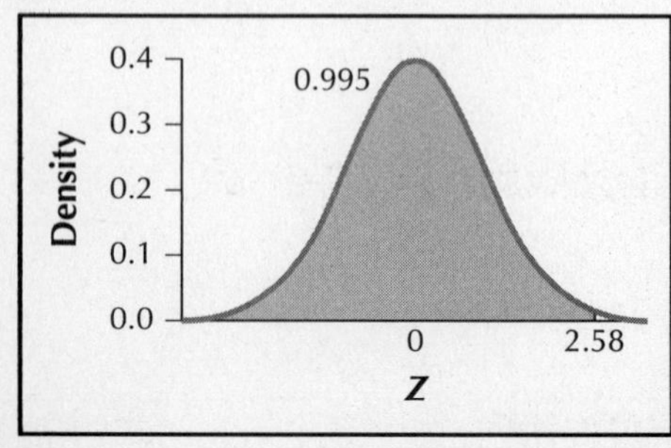

63.

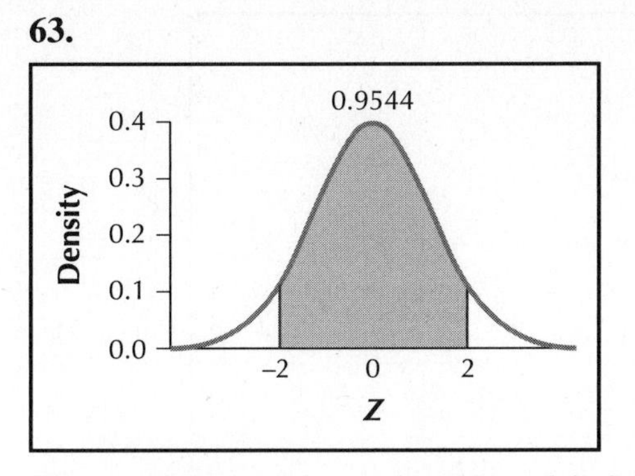

The area between $Z = -2$ and $Z = 2$ is 0.9544.

From Table C, the area between $Z = -2.00$ and $Z = 2.00$ is $0.9772 - 0.0228 = 0.9544$. By the Empirical Rule, the area between $Z = -2$ and $Z = 2$ is about 0.95.

65. **(a)** By symmetry, the area to the right of $Z = 1.5$ equals the area to the left of $Z = -1.5$. Thus, the area to the right of $Z = 1.5$ is 0.0668. **(b)** The area to the right of $Z = -1.5$ is 1 – the area to the left of $Z = -1.5 = 1 - 0.0668 = 0.9332$. **(c)** By symmetry, the area to the left of $Z = 1.5$ equals the area to the right of $Z = -1.5$. Thus, from (b), the area to the left of $Z = 1.5$ equals 0.9332. Therefore, the area between $Z = -1.5$ and $Z = 1.5$ is the area to the left of $Z = 1.5$ − the area to the left of $Z = -1.5 = 0.9332 - 0.0668 = 0.8664$.

67. $Z = -2.58$ and $Z = 2.58$

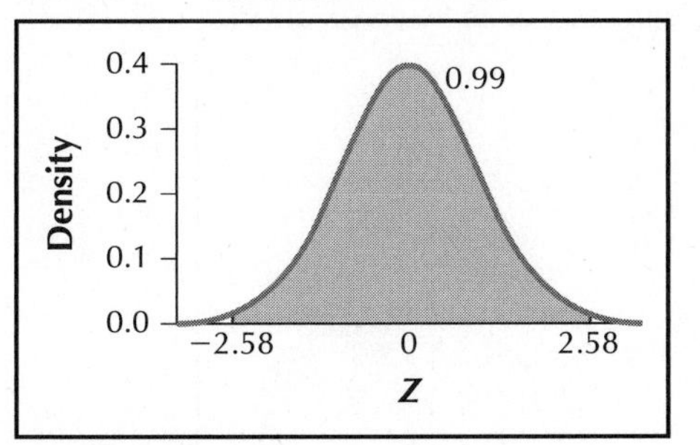

69. For the 25th percentile:

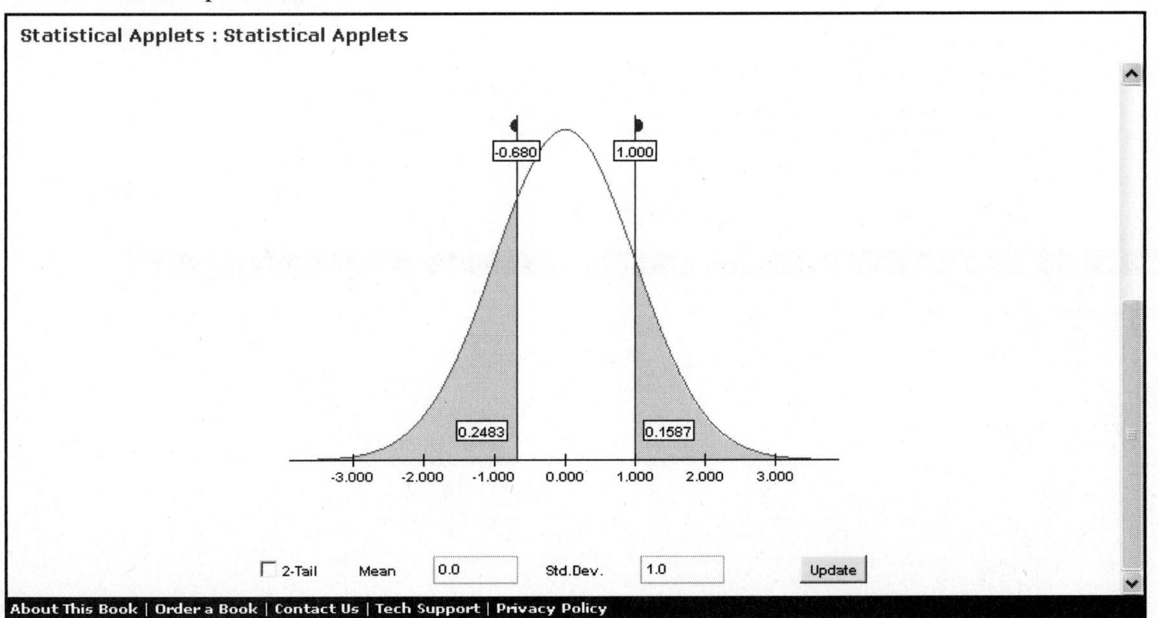

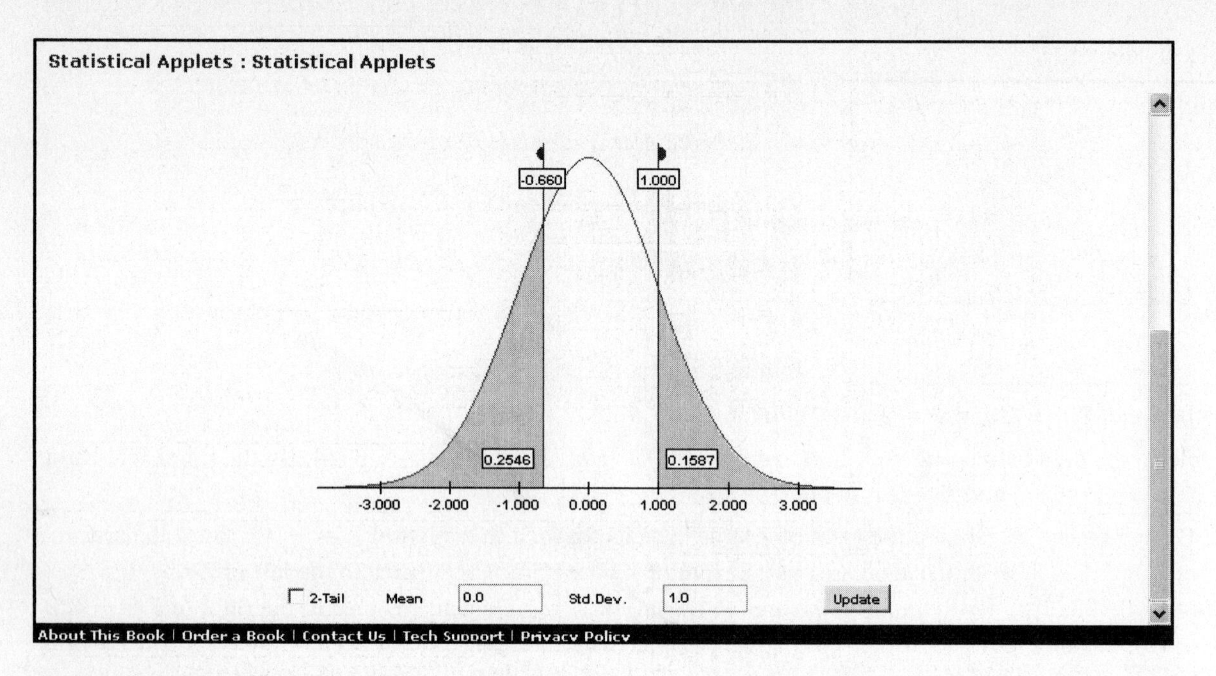

So the 25th percentile is about $Z = -0.67$.
For the 50th percentile:

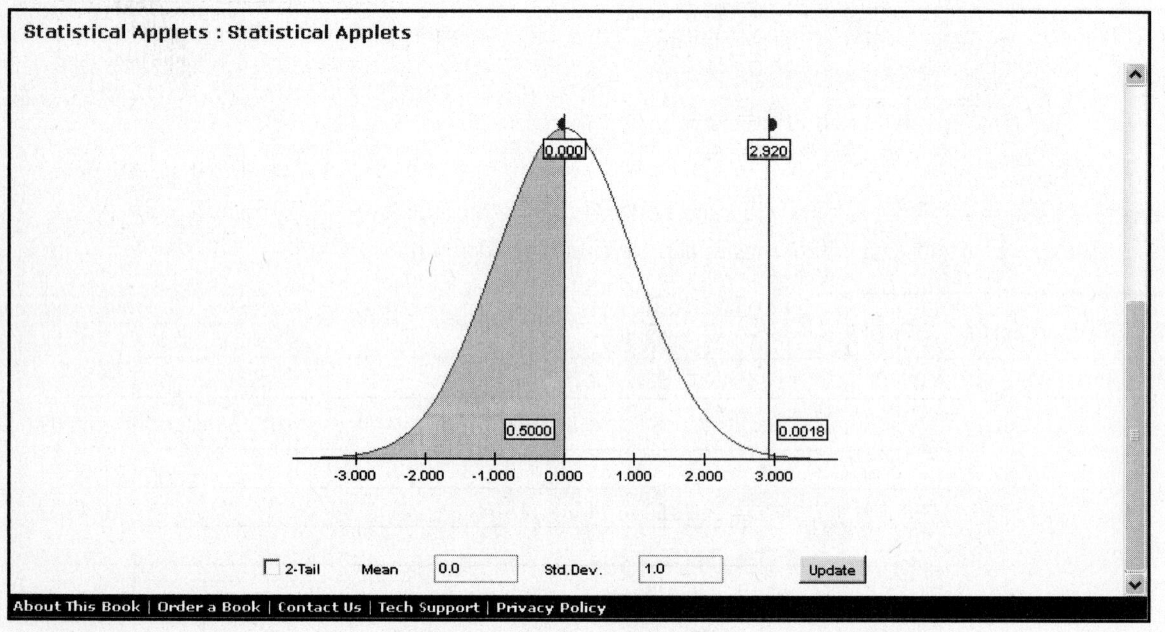

So the 50th percentile is $Z = 0$.

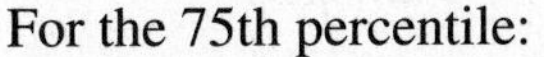

For the 75th percentile:

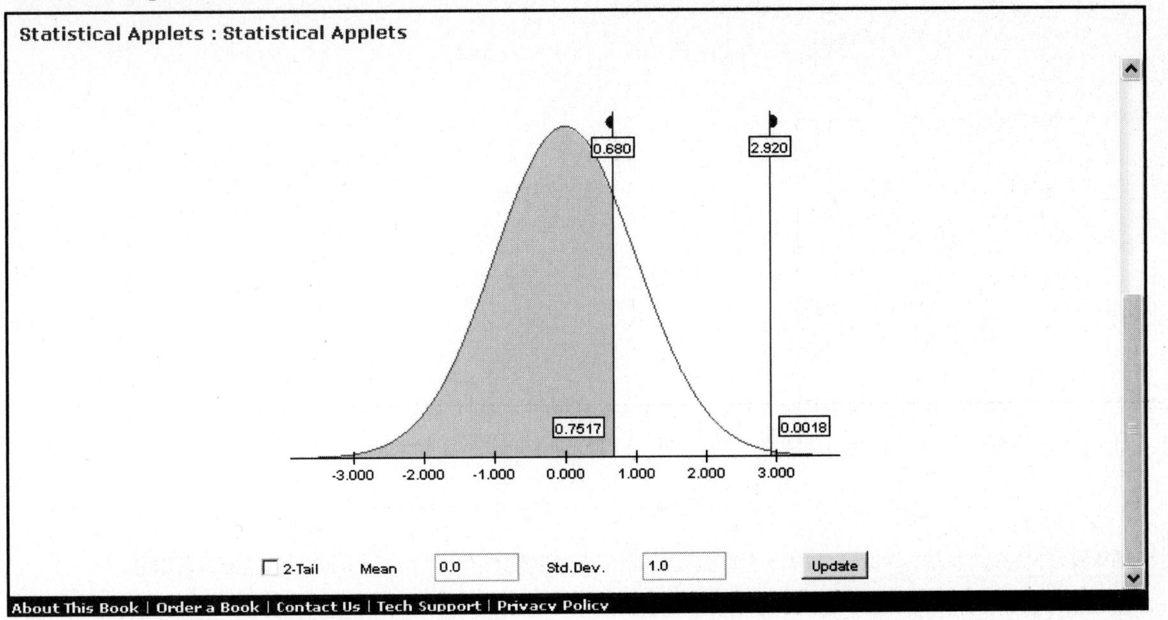

So the 75th percentile is about $Z = 0.67$.

Section 6.5

1. To standardize things means to make them all the same, uniform, or equivalent. To standardize a normal random variable X, we transform X into the standard normal random variable Z by the formula $Z = \dfrac{X - \mu}{\sigma}$. We do this so that we can use the standard normal table to find the probabilities for X.

3. You would get the wrong probability.

For Exercises 5, 7, 9, 11, 13, 15, $\mu = 70$ and $\sigma = 10$, so $Z = \dfrac{X - \mu}{\sigma} = \dfrac{X - 70}{10}$.

5.

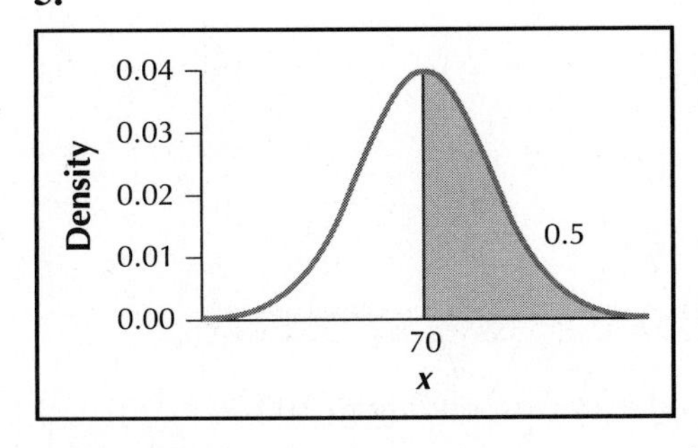

The Z-value corresponding to 70 is $Z = \dfrac{70 - 70}{10} = 0$.

Therefore,

$$P(X > 70) = P(Z > 0) = 0.5.$$

7.

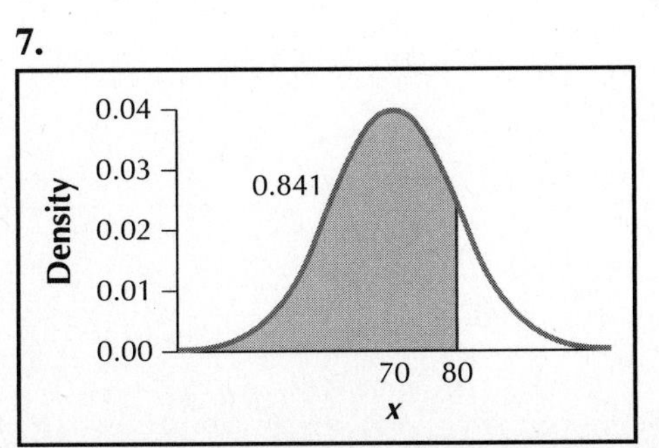

The Z-value corresponding to 80 is $Z = \dfrac{80 - 70}{10} = 1.00$. Therefore,

$$P(X < 80) = P(Z < 1.00) - 0.8413.$$

9.

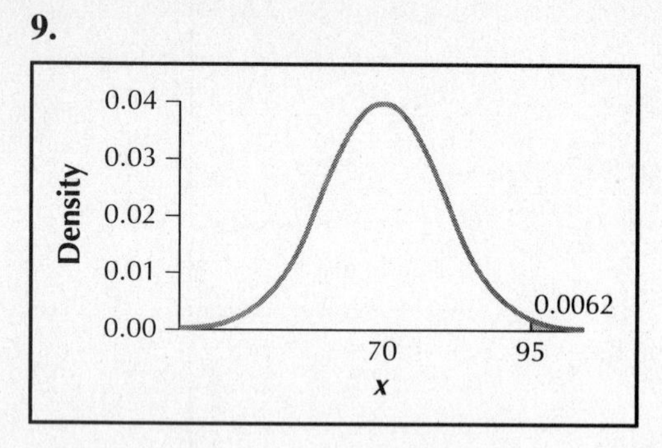

The Z-value corresponding to 95 is $Z = \dfrac{95 - 70}{10} = 2.50$. Therefore,

$$P(X \geq 95) = P(Z \geq 2.50) = 1 - 0.9938 = 0.0062.$$

11.

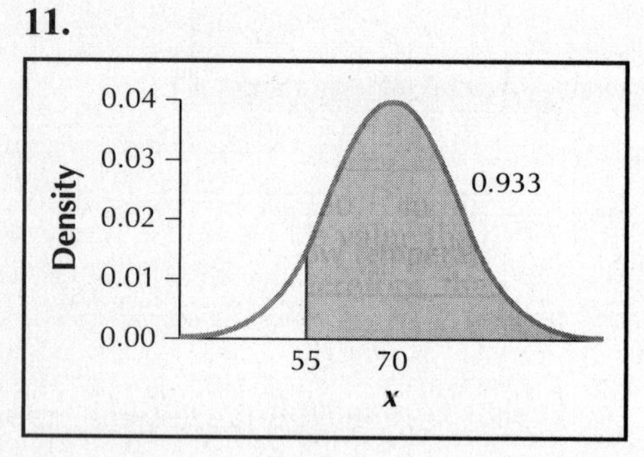

The Z-value corresponding to 55 is $Z = \dfrac{55 - 70}{10} = -1.50$. Therefore,

$$P(X \geq 55) = P(Z \geq -1.50) = 1 - 0.0668 = 0.9332.$$

13.

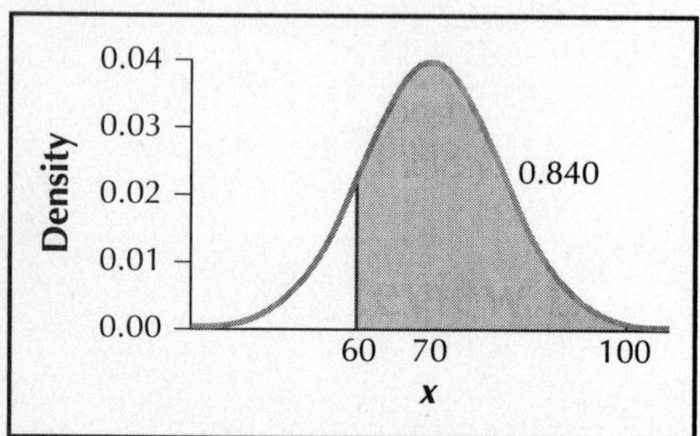

The Z-value corresponding to 60 is $Z = \dfrac{60 - 70}{10} = -1.00$ and the Z-value corresponding to 100 is $Z = \dfrac{100 - 70}{10} = 3.00$. Therefore,

$$P(60 \leq X \leq 100) = P(-1.00 \leq Z \leq 3.00) = 0.9987 - 0.1587 = 0.8400.$$

15.

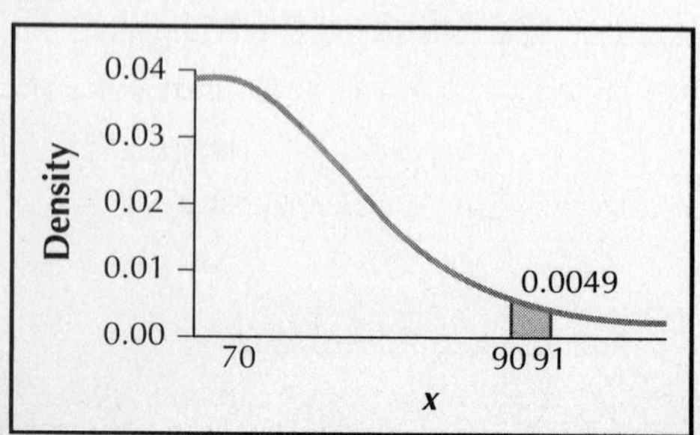

The Z-value corresponding to 90 is $Z = \frac{90 - 70}{10} = 2.00$ and the Z-value corresponding to 91 is $Z = \frac{91 - 70}{10} = 2.10$. Therefore, $P(90 \le X \le 91) = P(2.00 \le Z \le 2.10) = 0.9821 - 0.9772 = 0.0049$.

17.

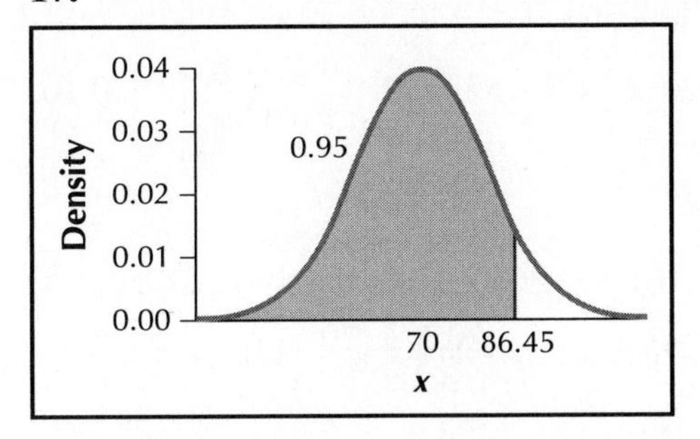

From Table C, the value of Z with an area of 0.9500 to the left of it is $Z = 1.645$. Therefore, the 95th percentile is $X = Z\sigma + \mu = (1.645)(10) + 70 = 86.45$.

19.

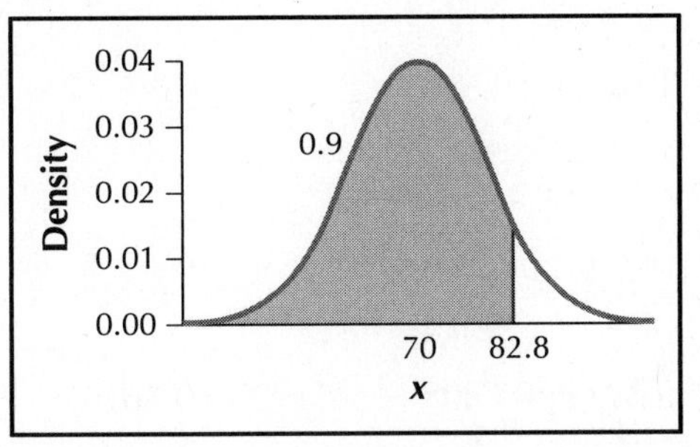

From Table C, the value of Z with an area of 0.9000 to the left of it is $Z = 1.28$. Therefore, the 90th percentile is $X = Z\sigma + \mu = (1.28)(10) + 70 = 82.8$.

21.

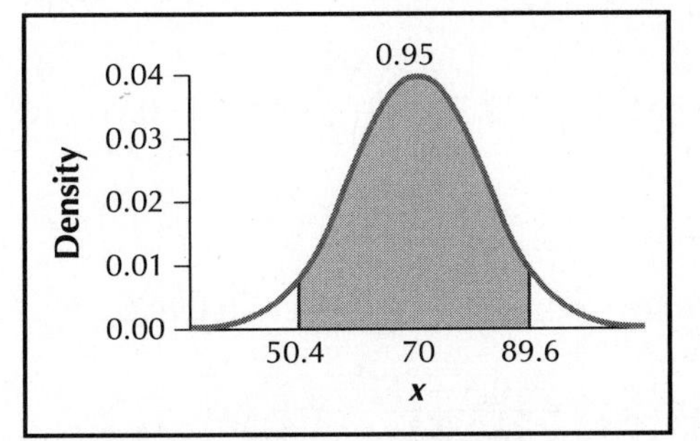

The two symmetric values of X that contain the central 95% of X between them are the 2.5th percentile and the 97.5th percentile. From Table C, the value of Z with an area of 0.0250 to the left of it is $Z = -1.96$. Therefore, the 2.5th percentile is $X = Z\sigma + \mu = (-1.96)(10) + 70 = 50.4$. From Table C, the value of Z with an area of 0.9750 to the left of it is $Z = 1.96$. Therefore, the 97.5th percentile is $X = Z\sigma + \mu = (1.96)(10) + 70 = 89.6$.

23. We have $n \cdot p = 10 \cdot 0.5 = 5 \ge 5$ and $n \cdot (1 - p) = 10(1 - 0.5) = 5 \ge 5$. Therefore, it is appropriate to use the normal approximation to the binomial probability distribution.

25. We have $n \cdot p = 10 \cdot 0.4 = 4 < 5$. Therefore, it is not appropriate to use the normal approximation to the binomial probability distribution.

27. We have $n \cdot p = 45 \cdot 0.1 = 4.5 < 5$. Therefore, it is not appropriate to use the normal approximation to the binomial probability distribution.

For Exercises 29, 31, 33, 35, $n = 40, p = 0.5$, and $1 - p = 1 - 0.5 = 0.5$. We have $n \cdot p = 40 \cdot 0.5 = 20 \geq 5$ and $n \cdot (1 - p) = 40\,(1 - 0.5) = 20 \geq 5$. Therefore, it is appropriate to use the normal approximation to the binomial probability distribution. $\mu = n \cdot p = 40 \cdot 0.5 = 20$ and

$$\sigma = \sqrt{n \cdot p \cdot (1-p)} = \sqrt{40 \cdot 0.5 \cdot (1 - 0.5)} \approx 3.1623$$

29. $P(X_{binomial} = 20) \approx P(19.5 \leq Y_{normal} \leq 20.5)$. The Z-value corresponding to 19.5 is $Z = \dfrac{19.5 - 20}{3.1623} = -0.16$, and the Z-value corresponding to 20.5 is $Z = \dfrac{20.5 - 20}{3.1623} = 0.16$.
Therefore,

$$P(X_{binomial} = 20) \approx P(19.5 \leq Y_{normal} \leq 20.5) = P(-0.16 \leq Z \leq 0.16) = 0.5636 - 0.4364 = 0.1272.$$

31. $P(X_{binomial} > 20) \approx P(Y_{normal} > 20.5)$. The Z-value corresponding to 20.5 is $Z = \dfrac{20.5 - 20}{3.1623} = 0.16$.
Therefore,

$$P(X_{binomial} > 20) \approx P(Y_{normal} > 20.5) = P(Z > 0.16) = 1 - 0.5636 = 0.4364.$$

33. $P(X_{binomial} < 20) \approx P(Y_{normal} < 19.5)$. The Z-value corresponding to 19.5 is $Z = \dfrac{19.5 - 20}{3.1623} = -0.16$.
Therefore,

$$P(X_{binomial} < 20) \approx P(Y_{normal} < 19.5) = P(Z < -0.16) = 0.4364.$$

35. $P(18 < X_{binomial} < 22) \approx P(18.5 < Y_{normal} < 21.5)$. The Z-value corresponding to 18.5 is $Z = \dfrac{18.5 - 20}{3.1623} = -0.47$, and the Z-value corresponding to 21.5 is $Z = \dfrac{21.5 - 20}{3.1623} = 0.47$.
Therefore,

$$P(18 < X_{binomial} < 22) \approx P(18.5 < Y_{normal} < 21.5) = P(-0.47 < Z < 0.47) = 0.6808 - 0.3192 = 0.3616.$$

For Exercises 37, 39, 41, 43, $n = 120, p = 0.1$, and $1 - p = 1 - 0.1 = 0.9$. We have $n \cdot p = 120 \cdot 0.1 = 12 \geq 5$ and $n \cdot (1 - p) = 120\,(1 - 0.1) = 108 \geq 5$. Therefore, it is appropriate to use the normal approximation to the binomial probability distribution. $\mu = n \cdot p = 120 \cdot 0.1 = 12$ and $\sigma = \sqrt{n \cdot p \cdot (1-p)} = \sqrt{120 \cdot 0.1 \cdot (1 - 0.1)} \approx 3.2863$

37. $P(X_{binomial} = 10) \approx P(9.5 \leq Y_{normal} \leq 10.5)$. The Z-value corresponding to 9.5 is $Z = \dfrac{9.5 - 12}{3.2863} = -0.76$, and the Z-value corresponding to 10.5 is $Z = \dfrac{10.5 - 12}{3.2863} = -0.46$.
Therefore,

$$P(X_{binomial} = 10) \approx P(9.5 \leq Y_{normal} \leq 10.5) = P(-0.76 \leq Z \leq -0.46) = 0.3228 - 0.2236 = 0.0992.$$

39. $P(X_{binomial} > 10) \approx P(Y_{normal} > 10.5)$. The Z-value corresponding to 10.5 is $Z = \dfrac{10.5 - 12}{3.2863} = -0.46$.
Therefore,

$$P(X_{binomial} > 10) \approx P(Y_{normal} > 10.5) = P(Z > -0.46) = 1 - 0.3228 = 0.6772.$$

41. $P(X_{binomial} < 8) \approx P(Y_{normal} < 7.5)$. The Z-value corresponding to 7.5 is $Z = \dfrac{7.5 - 12}{3.2863} = -1.37$.
Therefore,

$$P(X_{binomial} < 8) \approx P(Y_{normal} < 7.5) = P(Z < -1.37) = 0.0853.$$

43. $P(9 < X_{binomial} < 11) \approx P(9.5 < Y_{normal} < 10.5)$. The Z-value corresponding to 9.5 is $Z = \dfrac{9.5 - 12}{3.2863} = -0.76$, and the Z-value corresponding to 10.5 is $Z = \dfrac{10.5 - 12}{3.2863} = -0.46$.
Therefore,

$$P(9 < X_{binomial} < 11) \approx P(9.5 < Y_{normal} < 10.5) = P(-0.76 < Z < -0.46) = 0.3228 - 0.2236 = 0.0992.$$

45. $Z = \dfrac{X - \mu}{\sigma} = \dfrac{X - 15}{2}$

(a) The Z-value corresponding to 17 is $Z = \dfrac{17 - 15}{2} = 1.00$ and the Z-value corresponding to 19 is $Z = \dfrac{19 - 15}{2} = 2.00$.
Therefore,

$$P(17 \leq X \leq 19) = P(1.00 \leq Z \leq 2.00) = 0.9772 - 0.8413 = 0.1359.$$

(b) The Z-value corresponding to 11 is $Z = \dfrac{11 - 15}{2} = -2.00$. Therefore,

$$P(X < 11) = P(Z < -2.00) = 0.0228.$$

47. $Z = \dfrac{X - \mu}{\sigma} = \dfrac{X - 13.6}{6}$

(a) The Z-value corresponding to 7.2 mph is $Z = \dfrac{7.2 - 13.6}{6} = -1.07$.

Therefore,

$$P(X < 7.2) = P(Z < -1.07) = 0.8577.$$

(b) The Z-value corresponding to 20 mph is $Z = \dfrac{20 - 13.6}{6} = 1.07$.
Therefore,

$$P(X > 20) = P(Z > 1.07) = 1 - 0.8577 = 0.1423.$$

So 0.1423 of the days in July have a wind speed greater than 20 mph.

(c) The Z-value corresponding to 15 mph is $Z = \dfrac{15 - 13.6}{6} = 0.23$, and the Z-value corresponding to 20 mph is $Z = \dfrac{20 - 13.6}{6} = 1.07$.
Therefore,

$$P(15 \leq X \leq 20) = P(0.23 \leq Z \leq 1.07) = 0.8577 - 0.5910 = 0.2667.$$

Therefore, 26.67% of days in July have wind speeds between 15 and 20 mph.

(d) The value of X higher than 99% of all other wind speeds in July is the 99th percentile. From Table C, the value of Z with an area to the left of it equal to 0.9900 is $Z = 2.33$. Therefore, the 99th percentile is $X = Z\sigma + \mu = (2.33)(6) + 13.6 = 27.6$ mph.

(e) The Z-score for $X = 0$ mph is $Z = \dfrac{X - \mu}{\sigma} = \dfrac{0 - 13.6}{6} = -2.27$.
Since $-3 < -2.27 \leq -2$, a day in July with no wind at all should be considered moderately unusual.

49. $Z = \dfrac{X - \mu}{\sigma} = \dfrac{X - 12}{4}$

(a) The Z-value corresponding to 10 million people is $Z = \dfrac{10 - 12}{4} = -0.50$.
Thus,

$$P(X < 10) = P(Z < -0.50) = 0.3085.$$

(b) The Z-value corresponding to 18 million viewers is $Z = \dfrac{18 - 12}{4} = 1.50$.
Thus,

$$P(X > 18) = P(Z > 1.50) = 1 - 0.9332 = 0.0668.$$

Therefore, 6.68% of *60 Minutes* broadcasts are watched by more than 18 million viewers.

(c) The Z-value corresponding to 10 million viewers is $Z = \dfrac{10 - 12}{4} = -0.50$, and the Z-value corresponding to 11 million viewers is $Z = \dfrac{11 - 12}{4} = -0.25$.
Thus,

$$P(10 \leq X \leq 11) = P(-0.50 \leq Z \leq -0.25) = 0.4013 - 0.3085 = 0.0928.$$

Therefore, 0.0928 of *60 Minutes* broadcasts are watched by between 10 million and 11 million viewers.

(d) From Table C, the value of Z with an area to the left of it equal to 0.7500 is $Z = 0.67$. Therefore, the 75th percentile is $X = Z\sigma + \mu = (0.67)(4) + 12 = 14.68$ million viewers. Using technology, the 75th percentile is $X = 14.70$ million viewers.

(e) The Z-score for $X = 24$ million viewers is $Z = \dfrac{X - \mu}{\sigma} = \dfrac{24 - 12}{4} = 3$.
Since $3 \geq 3$, a night in which 24 million people watched *60 Minutes* is unusual.

51. $Z = \dfrac{X - \mu}{\sigma} = \dfrac{X - 100}{15}$

(a) The Z-value corresponding to 70 is $Z = \dfrac{70 - 100}{15} = -2.00$.
Thus,

$$P(X > 70) = P(Z > -2.00) = 1 - 0.0228 = 0.9772.$$

(b) The Z-value corresponding to 125 is $Z = \dfrac{125 - 100}{15} = 1.67$.
Thus,

$$P(X > 125) = P(Z > 1.67) = 1 - 0.9525 = 0.0475.$$

Therefore, 4.75% of tortoises live longer than 125 years.

(c) The Z-value corresponding to 115 is $Z = \dfrac{115 - 100}{15} = 1.00$,
and the Z-value corresponding to 145 million viewers is $Z = \dfrac{145 - 100}{15} = 3.00$.

Thus,

$$P(115 \leq X \leq 145) = P(1.00 \leq Z \leq 3.00) = 0.9987 - 0.8413 = 0.1574.$$

Therefore, 0.1574 of tortoises live for between 115 and 145 years.

(d) From Table C, the value of Z with an area to the left of it equal to 0.9900 is $Z = 2.33$. Therefore, the 99th percentile is $X = Z\sigma + \mu = (2.33)(15) + 100 = 134.95$ years.

(e) The Z-score for $X = 150$ years is

$$Z = \frac{X - \mu}{\sigma} = \frac{150 - 100}{15} = 3.33.$$

Since $3.33 \geq 3$, a 150-year old tortoise is unusual.

53. $Z = \dfrac{X - \mu}{\sigma} = \dfrac{X - 5}{2}$

(a) The Z-value corresponding to 4 million people is $Z = \dfrac{4 - 5}{2} = -0.50$.
Thus,

$$P(X > 4) = P(Z > -0.50) = 1 - 0.3085 = 0.6915.$$

(b) The Z-value corresponding to 1 million people is $Z = \dfrac{1 - 5}{2} = -2.00$.
Thus,

$$P(X < 1) = P(Z < -2.00) = 0.0228.$$

(c) The Z-value corresponding to 3 million people is $Z = \dfrac{3 - 5}{2} = -1.00$.
Thus,

$$P(X > 3) = P(Z > -1.00) = 1 - 0.1587 = 0.8413.$$

(d) From Table C, the value of Z with an area to the left of it equal to 0.7500 is $Z = 0.67$. Therefore, the 75th percentile is $X = Z\sigma + \mu = (0.67)(2) + 5 = 6.34$ million tobacco-related deaths.

(e) The Z-score for $X = 8$ million people is $Z = \dfrac{X - \mu}{\sigma} = \dfrac{8 - 5}{2} = 1.5$. Since $-2 < 1.5 < 2$, a year in which 8 million people died from tobacco-related causes is not unusual.

55. $Z = \dfrac{X - \mu}{\sigma} = \dfrac{X - 11.4}{6}$

(a) The Z-value corresponding to 0 yards is $Z = \dfrac{0 - 11.4}{6} = -1.90$.
Thus,

$$P(X < 0) = P(Z < -1.90) = 0.0287.$$

(b) The Z-value corresponding to 20 yards is $Z = \frac{20 - 11.4}{6} = 1.43$.
Thus,

$$P(X > 20) = P(Z > 1.43) = 1 - 0.9236 = 0.0764.$$

Therefore, 7.64% of passing plays gain more than 20 yards.

(c) The Z-value corresponding to 5 yards is $Z = \frac{5 - 11.4}{6} = -1.07$
and the Z-value corresponding to 10 yards is $Z = \frac{10 - 11.4}{6} = -0.23$.
Thus,

$$P(5 \leq X \leq 10) = P(-1.07 \leq Z \leq -0.23) = 0.4090 - 0.1423 = 0.2667.$$

(d) The two symmetric values of X that contain the central 95% of X between them are the 2.5th percentile and the 97.5th percentile. From Table C, the value of Z with an area of 0.0250 to the left of it is $Z = -1.96$. Therefore, the 2.5th percentile is $X = Z\sigma + \mu = (-1.96)(6) + 11.4 = -0.36$. From Table C, the value of Z with an area of 0.9750 to the left of it is $Z = 1.96$. Therefore, the 97.5th percentile is $X = Z\sigma + \mu = (1.96)(6) + 11.4 = 23.16$. So the two symmetric values of X that contain the central 95% of X between them are -0.36 yards and 23.16 yards.

(e) The Z-score for $X = 35$ yards is $Z = \frac{X - \mu}{\sigma} = \frac{35 - 11.4}{6} = 3.93$. Since $3.93 \geq 3$, a passing play that went for 35 yards is unusual.

57. $Z = \frac{X - \mu}{\sigma} = \frac{X - 30{,}000}{4{,}000}$

(a) The Z-value corresponding to \$40,000 is $Z = \frac{40{,}000 - 30{,}000}{4{,}000} = 2.50$.
Thus,

$$P(X > 40{,}000) = P(Z > 2.50) = 1 - 0.9938 = 0.0062.$$

(b) The Z-value corresponding to \$22,000 is $Z = \frac{22{,}000 - 30{,}000}{4{,}000} = -2.00$.
Thus,

$$P(X < 22{,}000) = P(Z < -2.00) = 0.0228.$$

(c) The Z-value corresponding to \$27,000 is $Z = \frac{27{,}000 - 30{,}000}{4{,}000} = -0.75$, and the Z-value corresponding to \$38,000 is $Z = \frac{38{,}000 - 30{,}000}{4{,}000} = 2.00$.
Thus,

$$P(27{,}000 \leq X \leq 38{,}000) = P(-0.75 \leq Z \leq 2.00) = 0.9772 - 0.2266 = 0.7506.$$

(d) The Z-value corresponding to \$32,000 is $Z = \frac{32{,}000 - 30{,}000}{4{,}000} = 0.50$, and the Z-value corresponding to \$39,000 is $Z = \frac{39{,}000 - 30{,}000}{4{,}000} = 2.25$.
Thus,

$$P(32{,}000 \leq X \leq 39{,}000) = P(0.50 \leq Z \leq 2.25) = 0.9878 - 0.6915 = 0.2963.$$

(e) The Z-value corresponding to \$27,000 is $Z = \frac{27{,}000 - 30{,}000}{4{,}000} = -0.75$.
Thus,

$$P(X > 27{,}000) = P(Z > -0.75) = 1 - 0.2266 = 0.7734.$$

(f) From Table C, the value of Z with an area to the left of it equal to 0.1000 is $Z = -1.28$. Therefore, the 10th percentile is $X = Z\sigma + \mu = (-1.28)(4{,}000) + 30{,}000 = \$24{,}880$.

(g) The two symmetric values of X that contain the middle 60% of X between them are the 20th percentile and the 80th percentile. From Table C, the value of Z with an area of 0.2000 to the left of it is $Z = -0.84$. Therefore, the 20th percentile is $X = Z\sigma + \mu = (-0.84)(4{,}000) + 30{,}000 = \$26{,}640$. From Table C, the value of Z with an area of 0.8000 to the left of it is $Z = 0.84$. Therefore, the 80th percentile is $X = Z\sigma + \mu = (0.84)(4{,}000) + 30{,}000 = \$33{,}360$. So the two symmetric values of X that contain the middle 60% of X between them are \$26,640 and \$33,360.

59. **(a)** From Table C, the value of Z with an area to the left of it equal to 0.0500 is $Z = -1.645$. Therefore, the 5th percentile is $X = Z\sigma + \mu = (-1.645)\,(2) + 2.25 = -1.04$ calories per gram. **(b)** The answer in (a) does not make sense. No food can have a negative number of calories per gram. **(c)** Yes. **(d)** The distribution of the number of calories per gram are right-skewed and therefore not normal.

61. **(a)** 0.0026

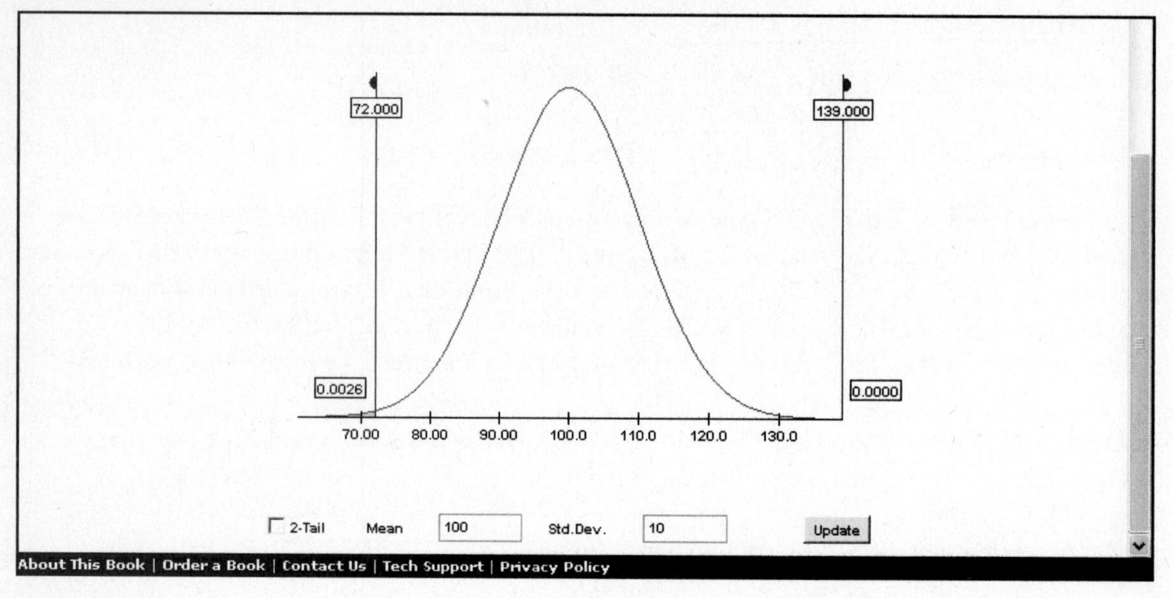

(b) 0.2313

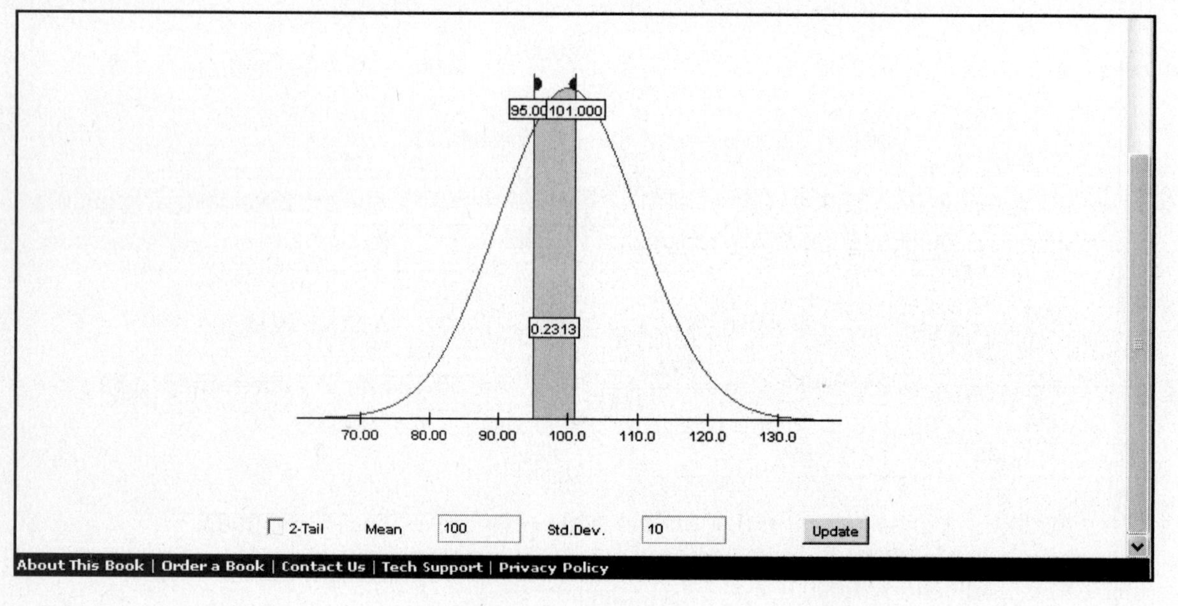

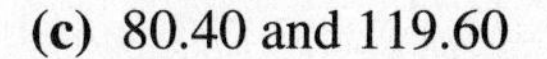

(c) 80.40 and 119.60

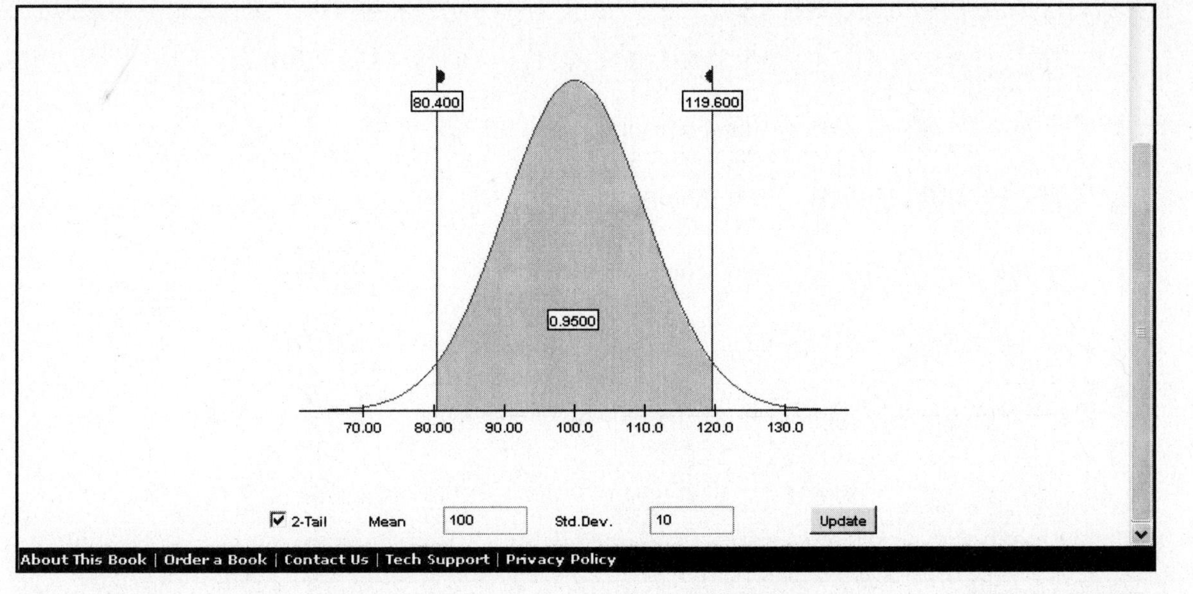

63. In 2002, $n = 50$, $p = 0.53$, and $1 - p = 1 - 0.53 = 0.47$. We have $n \cdot p = 50 \cdot 0.53 = 26.5 \geq 5$ and $n \cdot (1 - p) = 50\,(1 - 0.53) = 23.5 \geq 5$. Therefore, it is appropriate to use the normal approximation to the binomial probability distribution. $\mu = n \cdot p = 50 \cdot 0.53 = 26.5$ and $\sigma = \sqrt{n \cdot p \cdot (1 - p)}$ $= \sqrt{50 \cdot 0.53 \cdot (1 - 0.53)} \approx 3.5292$. In 2007, $n = 50$, $p = 0.38$, and $1 - p = 1 - 0.38 = 0.62$. We have

$$n \cdot p = 50 \cdot 0.38 = 19 \geq 5 \text{ and } n \cdot (1 - p) = 50\,(1 - 0.38) = 31 \geq 5.$$

Therefore, it is appropriate to use the normal approximation to the binomial probability distribution. $\mu = n \cdot p = 50 \cdot 0.38 = 19$ and $\sigma = \sqrt{n \cdot p \cdot (1 - p)} = \sqrt{50 \cdot 0.38 \cdot (1 - 0.38)} \approx 3.4322$.

(a) $P(X_{binomial} \geq 25) \approx P(Y_{normal} \geq 24.5)$. The Z-value corresponding to 24.5 is $Z = \dfrac{24.5 - 26.5}{3.5292} = -0.57$. Therefore,

$$P(X_{binomial} \geq 25) \approx P(Y_{normal} \geq 24.5) = P(Z \geq -0.57) = 1 - 0.2843 = 0.7157.$$

(b) $P(X_{binomial} \geq 25) \approx P(Y_{normal} \geq 24.5)$. The Z-value corresponding to 24.5 is $Z = \dfrac{24.5 - 19}{3.4322} = 1.60$. Therefore,

$$P(X_{binomial} \geq 25) \approx P(Y_{normal} \geq 24.5) = P(Z \geq 1.60) = 1 - 0.9452 = 0.0548.$$

(c) $P(X_{binomial} < 15) \approx P(Y_{normal} < 14.5)$. The Z-value corresponding to 14.5 is $Z = \dfrac{14.5 - 26.5}{3.5292} = -3.40$. Therefore,

$$P(X_{binomial} < 15) \approx P(Y_{normal} < 14.5) = P(Z < -3.40) = 0.0003.$$

(d) $P(X_{binomial} < 15) \approx P(Y_{normal} < 14.5)$. The Z-value corresponding to 14.5 is $Z = \dfrac{14.5 - 19}{3.4322} = -1.31$. Therefore,

$$P(X_{binomial} < 15) \approx P(Y_{normal} < 14.5) = P(Z < -1.31) = 0.0951.$$

65. $n = 100$, $p = 0.09$, and $1 - p = 1 - 0.09 = 0.91$. We have $n \cdot p = 100 \cdot 0.09 = 9 \geq 5$ and $n \cdot (1 - p) = 100\,(1 - 0.09) = 91 \geq 5$. Therefore, it is appropriate to use the normal approximation to the binomial probability distribution. $\mu = n \cdot p = 100 \cdot 0.09 = 9$ and

$$\sigma = \sqrt{n \cdot p \cdot (1 - p)} = \sqrt{100 \cdot 0.09 \cdot (1 - 0.09)} \approx 2.8618.$$

(a) $P(X_{binomial} = 9) \approx P(8.5 \leq Y_{normal} \leq 9.5)$. The Z-value corresponding to 8.5 is $Z = \dfrac{8.5 - 9}{2.8618} = -0.17$ and the Z-value corresponding to 9.5 is $Z = \dfrac{9.5 - 9}{2.8618} = 0.17$.

Therefore,

$$P(X_{binomial} = 9) \approx P(8.5 \leq Y_{normal} \leq 9.5) = P(-0.17 \leq Z \leq 0.17) = 0.5675 - 0.4325 = 0.1350.$$

(b) $P(X_{binomial} \geq 9) \approx P(Y_{normal} \geq 8.5)$. The Z-value corresponding to 8.5 is $Z = \dfrac{8.5 - 9}{2.8618} = -0.17$. Therefore,

$$P(X_{binomial} \geq 9) \approx P(Y_{normal} \geq 8.5) = P(Z \geq -0.17) = 1 - 0.4325 = 0.5675.$$

(c) $P(X_{binomial} > 9) \approx P(Y_{normal} > 9.5)$. The Z-value corresponding to 9.5 is $Z = \dfrac{9.5 - 9}{2.8618} = 0.17$. Therefore,

$$P(X_{binomial} > 9) \approx P(Y_{normal} > 9.5) = P(Z > 0.17) = 1 - 0.5675 = 0.4325.$$

(d) $P(X_{binomial} \leq 9) \approx P(Y_{normal} \leq 9.5)$. The Z-value corresponding to 9.5 is $Z = \dfrac{9.5 - 9}{2.8618} = 0.17$. Therefore,

$$P(X_{binomial} \leq 9) \approx P(Y_{normal} \leq 9.5) = P(Z \leq 0.17) = 0.5675.$$

(e) $P(X_{binomial} < 9) \approx P(Y_{normal} < 8.5)$. The Z-value corresponding to 8.5 is $Z = \dfrac{8.5 - 9}{2.8618} = -0.17$. Therefore,

$$P(X_{binomial} < 9) \approx P(Y_{normal} < 8.5) = P(Z < -0.17) = 0.4325.$$

67.

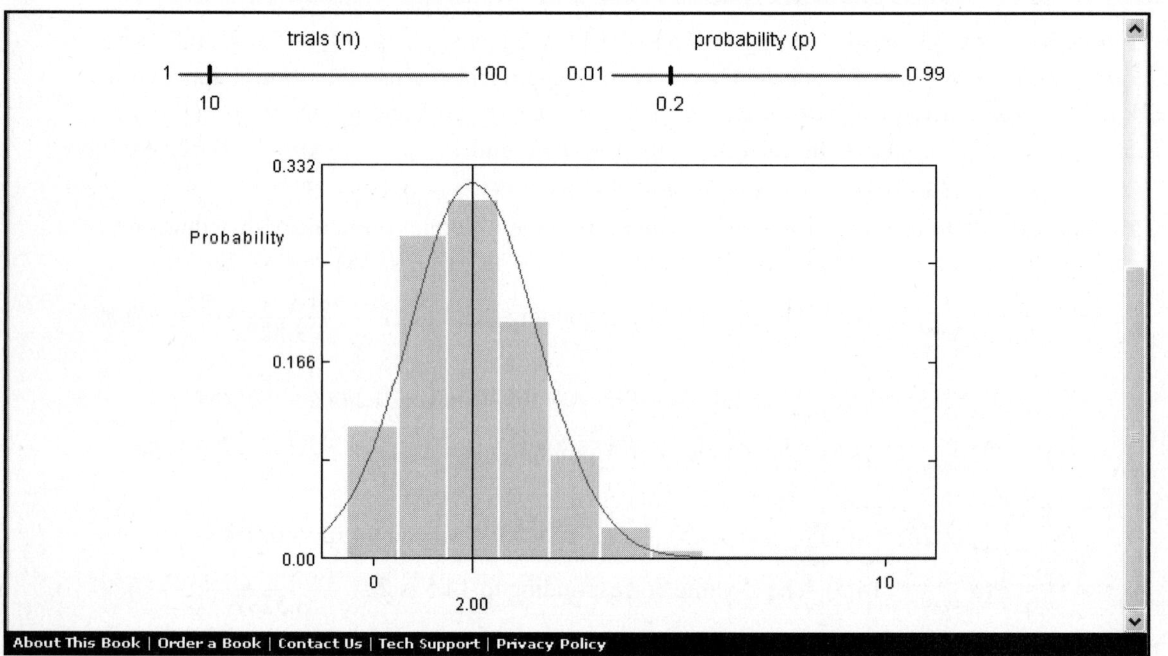

(a) No. **(b)** The normal distribution is not a good approximation to the binomial distribution. **(c)** We have $n \cdot p = 10 \cdot 0.2 = 2 < 5$. Therefore, it is not appropriate to use the normal approximation to the binomial probability distribution.

Chapter 6 Review

1. We will define our random variable X to be $X =$ the number of games attended. The probability distribution for X is:

X = Number of games attended	0	1	2	3	4
$P(X)$	0.05	0.20	0.35	0.25	0.15

(a) $P(X \geq 1) = P(X = 1) + P(X = 2) + P(X = 3) + P(X = 4) = 0.20 + 0.35 + 0.25 + 0.15 = 0.95$ Or, $P(X \geq 1) = 1 - P(X < 1) = 1 - P(X = 0) = 1 - 0.05 = 0.95$ **(b)** Two games. **(c)** $P(X \leq 2) = P(X = 0) + P(X = 1) + P(X = 2) = 0.05 + 0.20 + 0.35 = 0.60$

3. (a) $\mu = \sum[X \cdot P(X)] = -10{,}000(0.3) + 5{,}000(0.5) + 10{,}000(0.2) = \$1500$
Variance using the computational formula:

X	$P(X)$	X^2	$X^2 \cdot P(X)$
−10,000	0.3	100,000,000	30,000,000
5,000	0.5	25,000,000	12,500,000
10,000	0.2	100,000,000	20,000,000

$$\sum[X^2 \cdot P(X)] = 62{,}500{,}000$$

$$\sigma^2 = \sum[X^2 \cdot P(X)] - \mu^2 = 62{,}500{,}000 - 1500^2$$
$$= 60{,}250{,}000 \text{ dollars squared}$$
$$\sigma = \sqrt{\sigma^2} = \sqrt{60{,}250{,}000} \approx \$7762.09$$

(b) and **(c)**

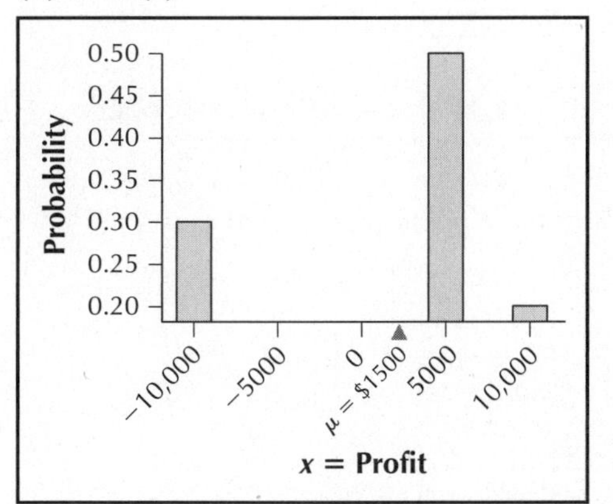

This value for the mean makes sense as the point where the distribution balances.

5. (a)

X	2	3	4	5	7	11
$P(X)$	7/13	2/13	1/13	1/13	1/13	1/13

(b)

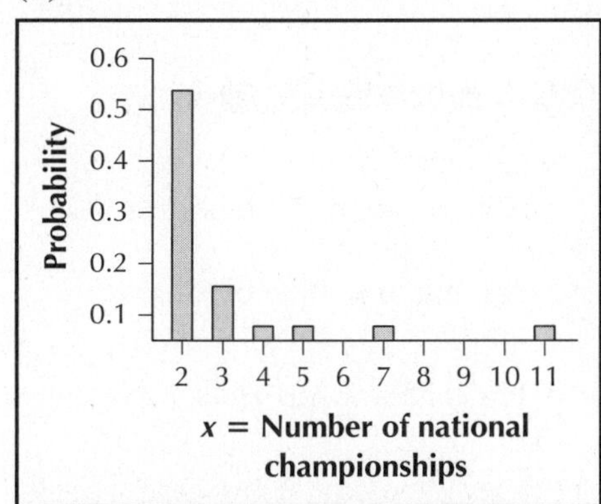

(c) Two national championships.

(d) $\mu = \sum[X \cdot P(X)] = 2(7/13) + 3(2/13) + 4(1/13) + 5(1/13) + 7(1/13) + 11(1/13) = 3.62$ national championships

7. **(a)** From Exercise 6, $\mu = 25.7748$ years.

X	$P(X)$	X^2	$X^2 \cdot P(X)$
15.5	0.0115	240.25	2.762875
18.5	0.2126	342.25	72.76235
20.5	0.2031	420.25	85.352775
23	0.1817	529	96.1193
27	0.1320	729	96.220
32	0.0764	1024	78.2336
40	0.1828	1600	292.48

$$\sum[X^2 \cdot P(X)] = 723.9309$$

$$\sigma^2 = \sum[X^2 \cdot P(X)] - \mu^2 = 723.9309 - 25.7748^2 = 59.59058496 \text{ years squared}$$

$$\sigma = \sqrt{\sigma^2} = \sqrt{59.59058496} \approx 7.7195 \text{ years}$$

(b) The Z-score for $X = 40$ years is $Z = \dfrac{40 - 25.7748}{7.7195} = 1.84$. Since $-2 < 1.84 < 2$, a 40-year-old college student would not be considered unusual.

9. $n = 10$ and $p = P(\text{female}) = 0.57$. Since $p = 0.57$ is not in Table B in the Appendix, we need to use the formula or technology.
Minitab generated the following table for $n = 10$ and $p = 0.57$.

X	**Probability**	X	**Probability**	X	**Probability**
0	0.0002	4	0.1401	8	0.0927
1	0.0029	5	0.2229	9	0.0273
2	0.0171	6	0.2462	10	0.0036
3	0.0604	7	0.1865		

(a) From the above table, $P(X = 10) = 0.0036$. Or you can use the formula with $n = 10$, $p = 0.57$, $1 - p = 1 - 0.57 = 0.43$, and $X = 10$. Therefore,

$$P(X = 10) = ({}_nC_X)p^X(1-p)^{n-X} = ({}_{10}C_{10})0.57^{10}(0.43)^{10-10} = (1)(0.57)^{10}(1) = 0.0036.$$

(b) $P(X \leq 9) = 1 - P(X = 10) = 1 - 0.0036 = 0.9964$

(c) From the above table,

$$P(X \geq 8) = P(X = 8) + P(X = 9) + P(X = 10) = 0.0927 + 0.0273 + 0.0036 = 0.1236.$$

11. $n = 20$, $P = P(\text{gestational diabetes}) = 0.08$, and $1 - p = 1 - 0.08 = 0.92$. Thus,
$\mu = n \cdot p = (20)(0.08) = 1.6$, $\sigma^2 = n \cdot p \cdot (1 - p)$
$= (20)(0.08)(0.92) = 1.472$, and $\sigma = \sqrt{\sigma^2} = \sqrt{1.472} \approx 1.2133$.
The expected number of pregnancies in which gestational diabetes occurs in a random sample of 20 pregnancies is 1.6.

13. $n = 15$, $p = P(\text{price of gas followed most closely}) = 0.35$, and $1 - p = 1 - 0.35 = 0.65$. Thus,
$\mu = n \cdot p = (15)(0.35) = 5.25$, $\sigma^2 = n \cdot p \cdot (1 - p) = (15)(0.35)(0.65)$
$= 3.4125$, and $\sigma = \sqrt{\sigma^2} = \sqrt{3.4125} \approx 1.8473$.
The expected number of Americans who said that the price of gasoline was the news story they followed more closely than any other news story in a random sample of 15 Americans is 5.25.

15. $n = 100$ and $p = P(\text{consumed alcohol}) = 0.17$. Since $n = 100$ and $p = 0.17$ are not in Table B in the Appendix, we need to use the formula or technology. Minitab generated the following table for $n = 100$ and $p = 0.17$.

X	Probability	X	Probability	X	Probability
0	0.0000	7	0.0020	14	0.0817
1	0.0000	8	0.0047	15	0.0960
2	0.0000	9	0.0098	16	0.1044
3	0.0000	10	0.0182	17	0.1057
4	0.0001	11	0.0305	18	0.0998
5	0.0002	12	0.0463	19	0.0882
6	0.0007	13	0.0642	20	0.0732

(a) From the above table, $P(X = 17) = 0.1057$. **(b)** From the above table, $P(X = 18) + P(X = 19) = 0.0998 + 0.0882 = 0.1880$.

17. **(a)** $n = 100$, $p = P(\text{consumed alcohol}) = 0.17$, and $1 - p = 1 - 0.17 = 0.83$. Thus, $\mu = n \cdot p = (100)(0.17) = 17$, $\sigma^2 = n \cdot p \cdot (1 - p) = (100)(0.17)(0.83) = 14.11$, and $\sigma = \sqrt{\sigma^2} = \sqrt{14.11} \approx 3.7563$.

The expected number of eighth graders who have consumed alcohol in the last month in a random sample of 100 eighth graders is 17.

(b) The Z-score for $X = 27$ eighth graders is

$$Z = \frac{X - \mu}{\sigma} = \frac{27 - 17}{3.7563} = 2.66.$$

Since $2 \leq 2.66 < 3$, a random sample of 100 eighth graders that contained 27 who had consumed alcohol during the past month is moderately unusual.

19. $n = 100$ and $p = P(\text{Baptist}) = 0.163$. Since $n = 100$ and $p = 0.163$ are not in Table B in the Appendix, we need to use the formula or technology. **(a)** From the TI-84, $P(X = 16) = 0.1079$. **(b)** From the TI-84, $P(X \geq 16) = 0.5742$.

21. **(a)** $n = 100$, $p = P(\text{Baptist}) = 0.163$, and $1 - p = 1 - 0.163 = 0.837$. Thus, $\mu = n \cdot p = (100)(0.163) = 16.3$, $\sigma^2 = n \cdot p \cdot (1 - p) = (100)(0.163)(0.837) = 13.6431$, and $\sigma = \sqrt{\sigma^2} = \sqrt{13.6431} \approx 3.6937$. The expected number of Baptists in a random sample of 100 church members is 16.3.

(b) The Z-score for $X = 25$ is $Z = \dfrac{X - \mu}{\sigma} = \dfrac{25 - 16.3}{3.6937} = 2.36$.

Since $2 \leq 2.36 < 3$, a random sample of 100 church members that contained 25 Baptists is moderately unusual.

23. $n = 100$ and $p = P(\text{death due to motor vehicle accident}) = 0.32$. Since $n = 100$ and $p = 0.32$ are not in Table B in the Appendix, we need to use the formula or technology. **(a)** From the TI-84, $P(X = 32) = 0.0853$. **(b)** From the TI-84, $P(X \leq 32) = 0.5477$.

25. **(a)** $n = 100$, $p = P(\text{death due to motor vehicle accident}) = 0.32$, and $1 - p = 1 - 0.32 = 0.68$. Thus,

$$\mu = n \cdot p = (100)(0.32) = 32, \sigma^2 = n \cdot p \cdot (1 - p) = (100)(0.32)(0.68) = 21.76,$$
$$\text{and } \sigma = \sqrt{\sigma^2} = \sqrt{21.76} \approx 4.6648.$$

The expected number of deaths due to motor vehicle accidents in a random sample of 100 deaths of people aged 16–24 is 32.

(b) The Z-score for $X = 25$ is $Z = \dfrac{X - \mu}{\sigma} = \dfrac{25 - 32}{4.6648} = -1.50$. Since $-2 < -1.50 < 2$, a random sample of 100 deaths of people aged 16–24 that contains 25 deaths due to motor vehicle accidents is not unusual.

Use the following graph for Exercises 27, 29, 31, 33

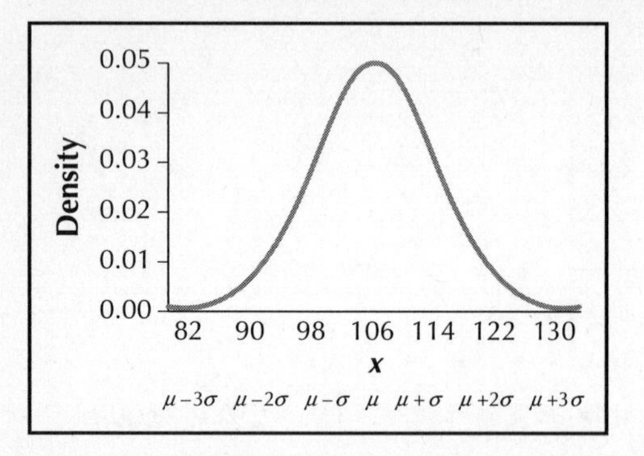

27. Since 106 mm Hg is the mean, $P(X > 106) = 0.5$.

29. Less than 0.5. Since the area to the right of the mean $\mu = 106$ mm is 0.5 and $X = 110$ mm is greater than the mean $\mu = 106$ mm, the area to the right of $X = 110$ mm is less than the area to the right of the mean $\mu = 106$.

31. From the graph, the area between 98 mm Hg and 114 mm Hg is the area between $\mu - \sigma$ and $\mu + \sigma$. Courtesy of the Empirical Rule, the area between $\mu - \sigma$ and $\mu + \sigma$ is about 0.68. Therefore, the probability that a randomly chosen systolic blood pressure is between 98 and 114 mm Hg is about 0.68.

33. From the graph, the area between 82 mm Hg and 130 mm Hg is the area between $\mu - 3\sigma$ and $\mu + 3\sigma$. Courtesy of the Empirical Rule, the area between $\mu - 3\sigma$ and $\mu + 3\sigma$ is about 0.997. Therefore, the probability that a randomly chosen systolic blood pressure is between 82 and 130 mm Hg is about 0.997.

35. **(a)**

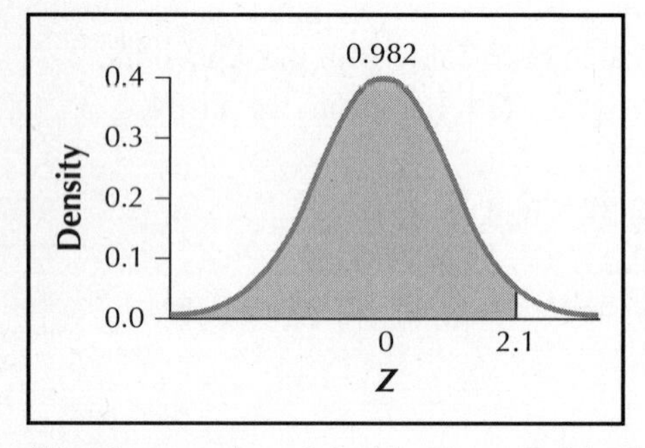

(b) Greater than 0.5. **(c)** From Table C, $P(Z < 2.1) = 0.9821$. **(d)** 0.9821 is greater than 0.5.

37. **(a)**

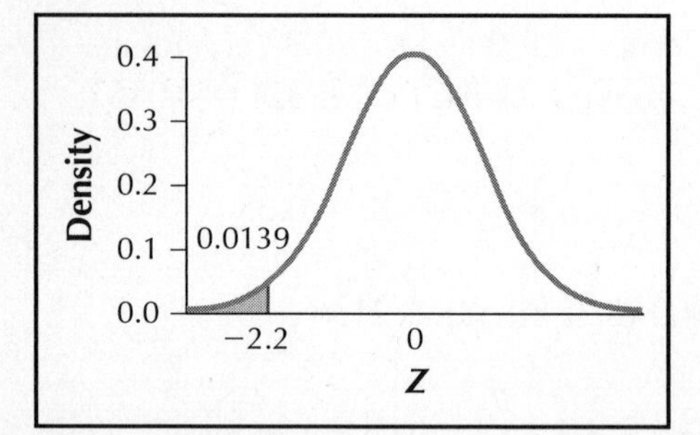

(b) Less than 0.5. **(c)** From Table C, $P(Z < -2.2) = 0.0139$. **(d)** 0.0139 is less than 0.5.

39. (a)

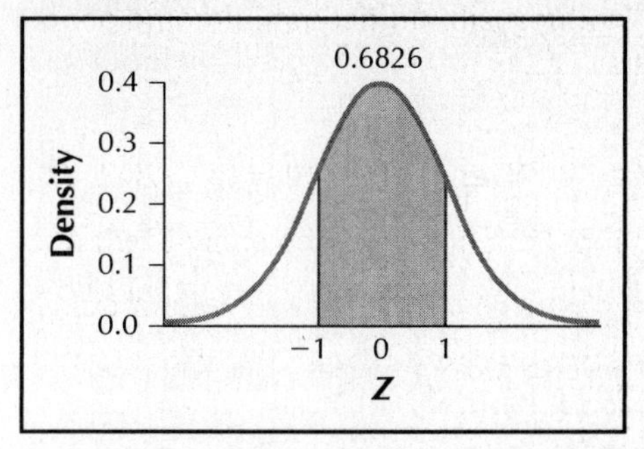

(b) Greater than 0.5. **(c)** From Table C, $P(-1 < Z < 1) = P(Z < 1) - P(Z < -1) = 0.8413 - 0.1587 = 0.6826$. From Minitab, $P(-1 < Z < 1) = 0.6827$. **(d)** 0.6826 and 0.6827 are both greater than 0.5.

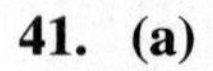

41. (a)

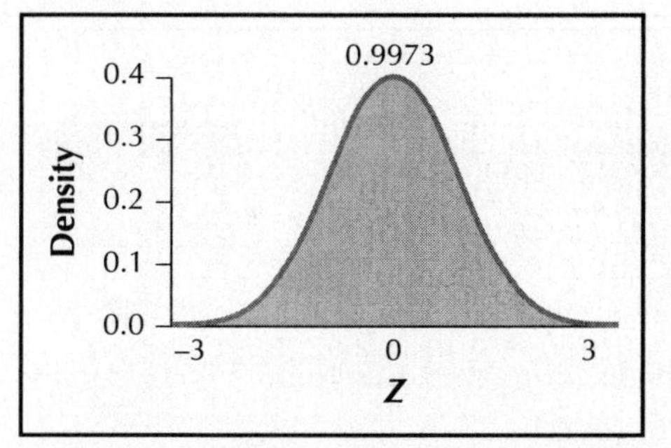

(b) Greater than 0.5. **(c)** From Table C, $P(-3 < Z < 3) = P(Z < 3) - P(Z < -3) = 0.9987 - 0.0013 = 0.9974$. From Minitab, $P(-3 < Z < 3) = 0.9973$. **(d)** 0.9974 and 0.9973 are both greater than 0.5.

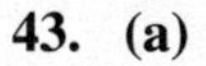

43. (a)

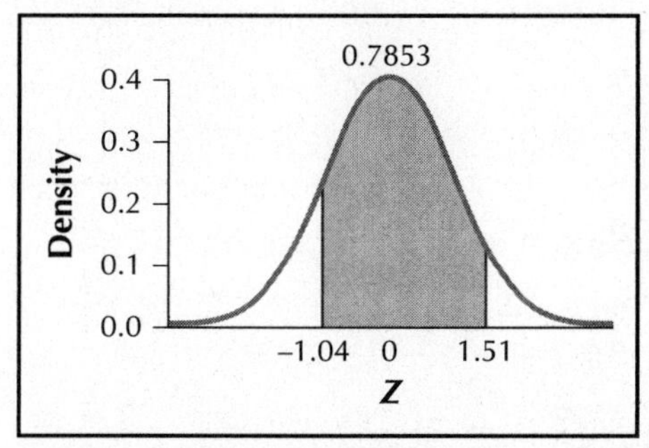

(b) Greater than 0.5. **(c)** From Table C, $P(-1.04 < Z < 1.51) = P(Z < 1.51) - P(Z < -1.04) = 0.9345 - 0.1492 = 0.7853$. **(d)** 0.7853 is greater than 0.5.

45. (a)

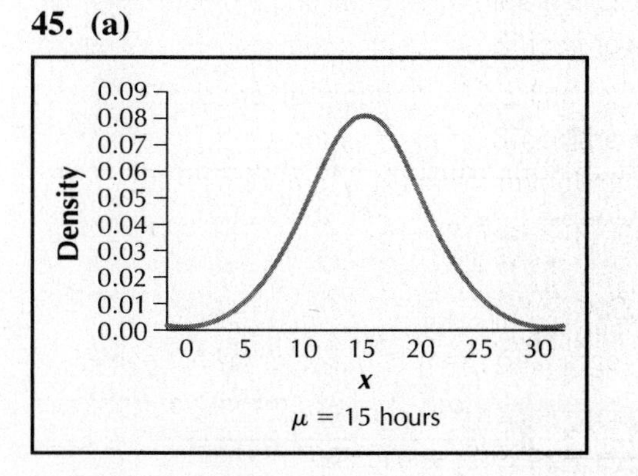

(b)

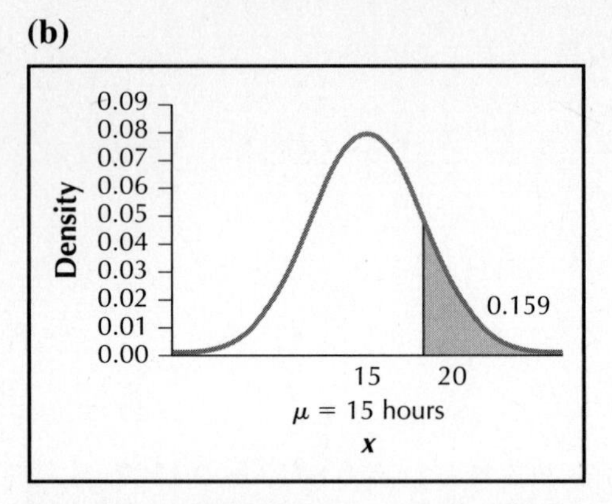

(c) The probability that a randomly selected reader of the magazine *Computer Gaming World* plays computer games for more than 20 hours a week is 0.1587. **(d)** $P(X < 20) = 1 - P(X > 20) = 1 - 0.1587 = 0.8413$
(e) Since the mean is 15 hours, $P(X < 15) = 0.5$, and $P(X > 15) = 0.5$.

47. $n = 100, p = 0.48$, and $1 - p = 1 - 0.48 = 0.52$.
We have $n \cdot p = 100 \cdot 0.48 = 48 \geq 5$ and $n \cdot (1 - p) = 100\,(1 - 0.48) = 52 \geq 5$.
Therefore, it is appropriate to use the normal approximation to the binomial probability distribution.

$$\mu = n \cdot p = 100 \cdot 0.48 = 48 \text{ and } \sigma = \sqrt{n \cdot p \cdot (1 - p)} = \sqrt{100 \cdot 0.48 \cdot (1 - 0.48)} \approx 4.9960.$$

(a) $P(X_{binomial} = 50) \approx P(49.5 \leq Y_{normal} \leq 50.5)$.

The Z-value corresponding to 49.5 is $Z = \dfrac{49.5 - 48}{4.9960} = 0.30$, and the Z-value corresponding to 50.5 is $Z = \dfrac{50.5 - 48}{4.9960} = 0.50$.

Therefore,

$$P(X_{binomial} = 50) \approx P(49.5 \leq Y_{normal} \leq 50.5) = P(0.30 \leq Z \leq 0.50) = 0.6915 - 0.6179 = 0.0736.$$

(b) $P(X_{binomial} \geq 50) \approx P(Y_{normal} \geq 49.5)$.

The Z-value corresponding to 49.5 is $Z = \dfrac{49.5 - 48}{4.9960} = 0.30$.
Therefore,

$$P(X_{binomial} \geq 50) \approx P(Y_{normal} \geq 49.5) = P(Z \geq 0.30) = 1 - P(Z \leq 0.30) = 1 - 0.6179 = 0.3821.$$

(c) $P(X_{binomial} > 50) \approx P(Y_{normal} > 50.5)$.

The Z-value corresponding to 50.5 is $Z = \dfrac{50.5 - 48}{4.9960} = 0.50$.
Therefore,

$$P(X_{binomial} > 50) \approx P(Y_{normal} > 50.5) = P(Z > 0.50) = 1 - P(Z < 0.50) = 1 - 0.6915 = 0.3085.$$

(d) $P(X_{binomial} \leq 50) \approx P(Y_{normal} \leq 50.5)$.

The Z-value corresponding to 50.5 is $Z = \dfrac{50.5 - 48}{4.9960} = 0.50$.
Therefore,

$$P(X_{binomial} \leq 50) \approx P(Y_{normal} \leq 50.5) = P(Z \leq 0.50) = 0.6915.$$

(e) $P(X_{binomial} < 50) \approx P(Y_{normal} < 49.5)$.

The Z-value corresponding to 49.5 is $Z = \dfrac{49.5 - 48}{4.9960} = 0.30$.
Therefore,

$$P(X_{binomial} < 50) \approx P(Y_{normal} < 49.5) = P(Z < 0.30) = 0.6179.$$

(f) $P(48 \leq X_{binomial} \leq 52) \approx P(47.5 \leq Y_{normal} \leq 52.5)$. The Z-value corresponding to 47.5 is

$$Z = \frac{47.5 - 48}{4.9960} = -0.10,$$

and the Z-value corresponding to 52.5 is $Z = \frac{52.5 - 48}{4.9960} = 0.90$.
Therefore,

$$P(48 \leq X_{binomial} \leq 52) \approx P(47.5 \leq Y_{normal} \leq 52.5) = P(-0.10 \leq Z \leq 0.90) = 0.8159 - 0.4602 = 0.3557.$$

49. $Z = \frac{X - \mu}{\sigma} = \frac{X - 5}{1}$

(a) The Z-value corresponding to \$3 is $Z = \frac{3 - 5}{1} = -2.00$.
From Table C,

$$P(X < 3) = P(Z < -2.00) = 0.0228.$$

(b) The Z-value corresponding to \$7 is $Z = \frac{7 - 5}{1} = 2.00$.
From Table C,

$$P(X > 7) = P(Z > 2.00) = 1 - P(Z \leq 2.00) = 1 - 0.9772 = 0.0228.$$

Therefore, 2.28% of 18-year-old babysitters earn more than \$7 per hour.

(c) The Z-value corresponding to \$5 is $Z = \frac{5 - 5}{1} = 0.00$, and the Z-value corresponding to \$6 is $Z = \frac{6 - 5}{1} = 1.00$.

From Table C,

$$P(5 \leq X \leq 6) = P(0.00 \leq Z \leq 1.00) = 0.8413 - 0.5000 = 0.3413.$$

Therefore, 0.3413 of 18-year-old babysitters earn between \$5 and \$6 per hour.

(d) From Table C, the value of Z with an area of 0.0100 to the right of it is the value of Z with an area of $1 - 0.0100 = 0.9900$ to the left of it, which is $Z = 2.33$. Therefore, the 99th percentile is $X = Z\sigma + \mu = (2.33)(1) + 5 = \$7.33$ per hour.

(e) The Z-score for $X = \$6$ per hour is $Z = \frac{X - \mu}{\sigma} = \frac{6 - 5}{1} = 1$.
Since $-2 < 1 < 2$, an18-year-old who babysits for \$6 per hour is not unusual.

51. $Z = \frac{X - \mu}{\sigma} = \frac{X - 32{,}257}{8{,}000}$

(a) The Z-value corresponding to \$40,000 is $Z = \frac{40{,}000 - 32{,}257}{8{,}000} = 0.97$.
From Table C,

$$P(X > \$40{,}000) = P(Z > 0.97) = 1 - P(Z \leq 0.97) = 1 - 0.8340 = 0.1660.$$

(b) The Z-value corresponding to \$30,000 is $Z = \frac{30{,}000 - 32{,}257}{8{,}000} = -0.28$.
From Table C,

$$P(X < \$30{,}000) = P(Z < -0.28) = 0.3897.$$

Thus, 38.97% of white Americans living alone earn less than \$30,000.

(c) The Z-value corresponding to \$30,000 is $Z = \frac{30{,}000 - 32{,}257}{8{,}000}$ $= -0.28$ and the Z-value corresponding to \$35,000 is $Z = \frac{35{,}000 - 32{,}257}{8{,}000} = 0.34$.

From Table C,

$$P(30{,}000 \leq X \leq 35{,}000) = P(-0.28 \leq Z \leq 0.34) = 0.6331 - 0.3897 = 0.2434.$$

Thus, 0.2434 of white Americans living alone earned between \$30,000 and \$35,000.

(d) From Table C, the value of Z with an area of 0.0200 to the left of it is $Z = -2.05$.
Therefore, the 2nd percentile is $X = Z\sigma + \mu = (-2.05)(\$8{,}000) + \$32{,}257 = \$15{,}857$.

(e) The Z-score for $X = \$55{,}000$ is $Z = \frac{X - \mu}{\sigma} = \frac{55{,}000 - 32{,}257}{8{,}000} = 2.84$.

Since $2 \leq 2.84 < 3$, a white American living alone who makes \$55,000 is moderately unusual.

53. **(a)** For a normal distribution, the mean equals the median. Since the mean and median of the population of incomes of white Americans living alone are not equal, then the assumption that the distribution of incomes of white Americans living alone is normal is incorrect. **(b)** All of our results in Exercise 51 are incorrect.

55. $Z = \frac{X - \mu}{\sigma} = \frac{X - 23.4}{5}$

(a) The Z-value corresponding to 18.4 is $Z = \frac{18.4 - 23.4}{5} = -1.00$.

From Table C,

$$P(X < 18.4) = P(Z < -1.00) = 0.1587.$$

(b) The Z-value corresponding to 15 is $Z = \frac{15 - 23.4}{5} = -1.68$, and the Z-value corresponding to 20 is $Z = \frac{20 - 23.4}{5} = -0.68$.

From Table C,

$$P(15 \leq X \leq 20) = P(-1.68 \leq Z \leq -0.68) = 0.2483 - 0.0465 = 0.2018.$$

(c) From Table C, the value of Z with an area of 0.1000 to the left of it is $Z = -1.28$. Therefore, the 10th percentile is $X = Z\sigma + \mu = (-1.28)(5) + 23.4 = 17$.

(d) The Z-score for $X = 8.4$ is $Z = \frac{X - \mu}{\sigma} = \frac{8.4 - 23.4}{5} = -3$. Since $-3 \leq -3$, a rating below 8.4 is unusual.

57. $Z = \frac{X - \mu}{\sigma} = \frac{X - 5.8}{1.8}$

(a) The Z-value corresponding to 6 hours is $Z = \frac{6 - 5.8}{1.8} = 0.11$.

From Table C,

$$P(X > 6) = P(Z > 0.11) = 1 - P(Z \leq 0.11) = 1 - 0.5438 = 0.4562.$$

(b) The Z-value corresponding to 3 hours is $Z = \frac{3 - 5.8}{1.8} = -1.56$

and the Z-value corresponding to 4 hours is $Z = \frac{4 - 5.8}{1.8} = -1.00$.

From Table C,

$$P(3 \leq X \leq 4) = P(-1.56 \leq Z \leq -1.00) = 0.1587 - 0.0594 = 0.0993.$$

Thus, 9.93% of British females spend between 3 and 4 hours shopping per week.

(c) From Table C, the value of Z with an area of 0.9800 to the left of it is $Z = 2.05$. Therefore, the 98th percentile is

$$X = Z\sigma + \mu = (2.05)(1.8) + 5.8 = 9.49 \text{ hours}.$$

(d) The Z-score for $X = 0.5$ hour is $Z = \frac{X - \mu}{\sigma} = \frac{0.5 - 5.8}{1.8} = -2.94$.

Since $-3 < -2.94 \leq -2$, a British female who only spends 0.5 hour a week shopping is moderately unusual.

Chapter 6 Quiz

1. True.

2. False. This is not binomial since there are more than two possible outcomes.

3. False.

4. 0.5

5. 0

6. 0

7. Discrete.

8. Binomial.

9. Mean $\mu = 0$, standard deviation $\sigma = 1$.

10. (a) We will define our random variable X to be X = amount won.
The outcomes for which $X = \$5$ are {(3,6), (4,5), (5,4), (6,3), (4,6), (5,5), (6,4), (5,6), (6,5), (6,6)}. There are 36 possible outcomes total, and $X = \$5$ for 10 of them. Therefore, $P(X = \$5) = 10/36 = 5/18$. Then, $P(X = \$0) = 1 - P(X = \$5) = 1 - 5/18 = 13/18$. The probability distribution for X is:

X = Amount won	0	5
$P(X)$	13/18	5/18

(b) $\mu = E(X) = \sum [X \cdot P(X)] = 0(13/18) + 5(5/18) \approx \$1.39$ **(c)** \$1.39

11. $n = 100$ and $p = P(\text{drives a luxury car}) = 0.19$. Since $n = 100$ and $p = 0.19$ are not in Table B in the Appendix, we need to use the formula or technology. **(a)** From the TI-84, $P(X = 20) = 0.0962$. Or you can use the formula with $n = 100$, $p = 0.19$, $1 - p = 1 - 0.19 = 0.81$, and $X = 20$.
Therefore,

$$P(X = 20) = ({}_nC_X)\, p^X (1 - p)^{n-X} = ({}_{100}C_{20})\, 0.19^{20} (0.81)^{100-20}$$
$$= (5.359833704 \times 10^{20})(0.19)^{20}(0.81)^{80} = 0.0962.$$

(b) Minitab generated the following table for $n = 100$ and $p = 0.19$.

X	Probability	X	Probability	X	Probability	X	Probability
0	0.0000	26	0.0209	52	0.0000	78	0.0000
1	0.0000	27	0.0134	53	0.0000	79	0.0000
2	0.0000	28	0.0082	54	0.0000	80	0.0000
3	0.0000	29	0.0048	55	0.0000	81	0.0000
4	0.0000	30	0.0027	56	0.0000	82	0.0000
5	0.0000	31	0.0014	57	0.0000	83	0.0000
6	0.0001	32	0.0007	58	0.0000	84	0.0000
7	0.0004	33	0.0003	59	0.0000	85	0.0000
8	0.0012	34	0.0002	60	0.0000	86	0.0000
9	0.0029	35	0.0001	61	0.0000	87	0.0000
10	0.0062	36	0.0000	62	0.0000	88	0.0000
11	0.0118	37	0.0000	63	0.0000	89	0.0000
12	0.0206	38	0.0000	64	0.0000	90	0.0000
13	0.0327	39	0.0000	65	0.0000	91	0.0000
14	0.0476	40	0.0000	66	0.0000	92	0.0000
15	0.0640	41	0.0000	67	0.0000	93	0.0000
16	0.0798	42	0.0000	68	0.0000	94	0.0000
17	0.0924	43	0.0000	69	0.0000	95	0.0000
18	0.1000	44	0.0000	70	0.0000	96	0.0000
19	0.1012	45	0.0000	71	0.0000	97	0.0000
20	0.0962	46	0.0000	72	0.0000	98	0.0000
21	0.0859	47	0.0000	73	0.0000	99	0.0000
22	0.0724	48	0.0000	74	0.0000	100	0.0000
23	0.0576	49	0.0000	75	0.0000		
24	0.0433	50	0.0000	76	0.0000		
25	0.0309	51	0.0000	77	0.0000		

From the above table, we see that $P(X = 19)$ is the largest probability, so 19 CEOs is the most likely number of CEOs who drive luxury cars.

(c) $\mu = n \cdot p = (100)(0.19) = 19$, $\sigma^2 = n \cdot p \cdot (1 - p) = (100)(0.19)(0.81) = 15.39$, and

$$\sigma = \sqrt{\sigma^2} = \sqrt{15.39} \approx 3.9230.$$

The expected number of CEOs who drive luxury cars in a random sample of 100 CEOs is 19.

(d) The Z-score for $X = 40$ CEOs is

$$Z = \frac{X - \mu}{\sigma} = \frac{40 - 19}{3.9230} = 5.35.$$

Since $5.35 \geq 3$, a random sample of 100 CEOs that contains 40 who drive luxury cars is unusual.

12. $Z = \frac{X - \mu}{\sigma} = \frac{X - 2849}{900}$

(a) The Z-value corresponding to \$4000 is $Z = \frac{4000 - 2849}{900} = 1.28$.
From Table C,

$$P(X > 4000) = P(Z > 1.28) = 1 - P(Z \leq 1.28) = 1 - 0.8997 = 0.1003.$$

(b) The Z-value corresponding to \$3000 is $Z = \frac{3000 - 2849}{900} = 0.17$,

and the Z-value corresponding to \$4000 is $Z = \frac{4000 - 2849}{900} = 1.28$.
From Table C,

$$P(3000 \leq X \leq 4000) = P(0.17 \leq Z \leq 1.28) = 0.8997 - 0.5675 = 0.3322.$$

Thus, 33.22% of males lost between \$3000 and \$4000.

(c) From Table C, the value of Z with an area of 0.9500 to the left of it is $Z = 1.645$. Therefore, the 95th percentile is $X = Z\sigma + \mu = (1.645)(\$900) + \$2849 = \$4329.50$.

(d) The Z-score for $X = \$1000$ is $Z = \frac{X - \mu}{\sigma} = \frac{1000 - 2849}{900} = -2.05$.

Since $-3 < -2.05 \leq -2$, a male gambler who lost \$1000 in four weeks and then approached a treatment provider is moderately unusual.

13. (a) For cats, $n = 1000$ and $p = P(\text{had a cat}) = 0.314$. Since $n = 1000$ and $p = 0.314$ are not in Table B in the Appendix, we need to use the formula or technology. From the TI-84,

$$P(X = 320) = 0.0249.$$

(b) For dogs, $n = 1000$ and $p = P(\text{had a dog}) = 0.343$. Since $n = 1000$ and $p = 0.343$ are not in Table B in the Appendix, we need to use the formula or technology. From the TI-84,

$$P(X = 320) = 0.0083.$$

(c) For cats, $n = 1000$, $p = P(\text{had a cat}) = 0.314$, and $1 - p = 1 - 0.314 = 0.686$.
$\mu = n \cdot p = (1000)(0.314) = 314$,
$\sigma^2 = n \cdot p \cdot (1 - p) = (1000)(0.314)(0.686) = 215.404$, and

$$\sigma = \sqrt{\sigma^2} = \sqrt{215.404} \approx 14.6766.$$

The expected number of American households that have a cat in a random sample of 1000 households is 314. For dogs: $n = 1000$, $p = P(\text{had a dog}) = 0.343$, and $1 - p = 1 - 0.343 = 0.657$. $\mu = n \cdot p = (1000)(0.343) = 343$, $\sigma^2 = n \cdot p \cdot (1 - p) = (1000)(0.343)(0.657) = 225.351$, and $\sigma = \sqrt{\sigma^2} = \sqrt{225.351} \approx 15.0117$.
The expected number of American households that have a dog in a random sample of 1000 households is 343.

(d) The Z-score for $X = 290$ households is

$$Z = \frac{X - \mu}{\sigma} = \frac{290 - 314}{14.6766} = -1.64.$$

Since $-2 < -1.64 < 2$, a random sample of 1000 household that contains 290 households with a cat is not unusual.

14. $Z = \frac{X - \mu}{\sigma} = \frac{X - 38{,}000}{10{,}000}$

(a) The Z-value corresponding to 40,000 people is $Z = \frac{40{,}000 - 38{,}000}{10{,}000} = 0.20$.
From Table C,

$$P(X > 40{,}000) = P(Z > 0.20) = 1 - P(Z \leq 0.20) = 1 - 0.5793 = 0.4207.$$

(b) The Z-value corresponding to 35,000 people is $Z = \frac{35{,}000 - 38{,}000}{10{,}000} = -0.30$, and the Z-value corresponding to 41,000 people is $Z = \frac{41000 - 38{,}000}{10{,}000} = 0.30$.

From Table C,

$$P(35{,}000 \leq X \leq 41{,}000) = P(-0.30 \leq Z \leq 0.30) = 0.6179 - 0.3821 = 0.2358.$$

(c) From Table C, the value of Z with an area of 0.1000 to the right of it is the value of Z with an area of $1 - 0.1000 = 0.9000$ to the left of it, which is $Z = 1.28$. Therefore, the 90th percentile is $X = Z\sigma + \mu = (1.28)\,(10{,}000) + 38{,}000 = 50{,}800$ visitors.

(d) The Z-score for $X = 15{,}000$ visitors is $Z = \dfrac{X - \mu}{\sigma} = \dfrac{15{,}000 - 38{,}000}{10{,}000} = -2.3$.

Since $-3 < -2.3 \leq -2$, a day at the Magic Kingdom with only 15,000 visitors is moderately unusual.

Chapter 7

Sampling Distributions

Section 7.1

1. *Sampling error* is the distance between the point estimate and its target parameter. It measures how far the point estimate misses the actual target parameter. In the real world, the only time we know the actual value of the sampling error is when we know the true value of the target parameter. Usually, the true value of the target parameter is unknown.

3. The larger the sample size, the more precise the analysis is. However, there comes a point somewhere around $n = 80$ when the trade-off between higher precision and higher cost becomes less beneficial. After that point, further increases in sample size lead to very small decreases in $\sigma_{\bar{x}}$. That is, increasing the sample size past a certain point yields diminishing returns.

5. We have $\sigma_{\bar{x}} = \sigma/\sqrt{n}$, so the standard deviation of the sampling distribution for $\bar{x}$ is smaller than the population standard deviation.

7. $|\bar{x} - \mu| = |95 - 100| = |-5| = 5$

9. $|s - \sigma| = |15 - 20| = |-5| = 5$

11. $|\hat{p} - p| = |0.30 - 0.25| = |0.05| = 0.05$

13. $\mu_{\bar{x}} = \mu = 100, \sigma_{\bar{x}} = \dfrac{\sigma}{\sqrt{n}} = \dfrac{20}{\sqrt{25}} = 4$

15. $\mu_{\bar{x}} = \mu = 0, \sigma_{\bar{x}} = \dfrac{\sigma}{\sqrt{n}} = \dfrac{10}{\sqrt{9}} = 3.3333$

17. $\mu_{\bar{x}} = \mu = -10, \sigma_{\bar{x}} = \dfrac{\sigma}{\sqrt{n}} = \dfrac{5}{\sqrt{100}} = 0.5$

19. The sampling distribution of $\bar{x}$ for $n = 16$ is normal with mean $\mu_{\bar{x}} = \mu = 10$ and standard deviation $\sigma_{\bar{x}} = \dfrac{\sigma}{\sqrt{n}} = \dfrac{4}{\sqrt{16}} = 1$. That is, $\bar{x}$ is Normal (10, 1).

21. Using symmetry, $P(\bar{x} < 9) = P\left(\dfrac{\bar{x} - 10}{1} < \dfrac{9 - 10}{1}\right) = P(Z < -1) = 0.1587$.

23. From Table C, $Z = 1.96$. Therefore, $\bar{x} = Z \cdot \sigma_{\bar{x}} + \mu = 1.96 \cdot 1 + 10 = 11.96$.

25. 0.95

27. **(a)** $\bar{x} = 3021$ pounds **(b)** Another sample of vehicles would probably not have the same sample average weight. Different samples may have different average weights. **(c)** It is unlikely that the population mean weight of all crash test vehicles equals the sample mean weight. Since samples are subsets of the population, they are not perfect representations of the population.

29. **(a)** $\mu = 2930.34$ pounds **(b)** $|\bar{x} - \mu| = |3021 - 2930.34| = |90.66| = 90.66$ pounds

31. **(a)** $\sigma = 627.13$ pounds **(b)** $s = 607$ pounds **(c)** $|s - \sigma| = |607 - 627.13| = |-20.13| =$ 20.13 pounds

33. **(a)** $\mu_{\bar{x}} = \mu = 1.7$ seconds, $\sigma_{\bar{x}} = \frac{\sigma}{\sqrt{n}} = \frac{0.2}{\sqrt{9}} = 0.0667$ seconds **(b)** $\mu_{\bar{x}} = \mu = 1.7$ seconds, $\sigma_{\bar{x}} = \frac{\sigma}{\sqrt{n}} = \frac{0.2}{\sqrt{16}} = 0.05$ seconds **(c)** $\mu_{\bar{x}} = \mu = 1.7$ seconds, $\sigma_{\bar{x}} = \frac{\sigma}{\sqrt{n}} = \frac{0.2}{\sqrt{25}} = 0.04$ seconds **(d)** $\mu_{\bar{x}} = \mu = 1.7$ seconds, $\sigma_{\bar{x}} = \frac{\sigma}{\sqrt{n}} = \frac{0.2}{\sqrt{36}} = 0.0333$ seconds

35. **(a)** The mean of the sampling distribution of the sample mean $\mu_{\bar{x}}$ remains the same as the sample size n increases. **(b)** Since $\sigma_{\bar{x}} = \sigma/\sqrt{n}$, the standard deviation of the sampling distribution of the sample means decreases as the sample size n increases.

37. $\mu = \$100{,}000{,}000$, $\sigma = \$40{,}000{,}000$, $n = 4$

(a) $P(\bar{x} > 125{,}000{,}000) = P\left(\frac{\bar{x} - 100{,}000{,}000}{40{,}000{,}000/\sqrt{4}} > \frac{125{,}000{,}000 - 100{,}000{,}000}{40{,}000{,}000/\sqrt{4}}\right)$
$= P(Z > 1.25) = 1 - 0.8944 = 0.1056$

(b) $P(120{,}000{,}000 < \bar{x} < 140{,}000{,}000) =$

$$P\left(\frac{120{,}000{,}000 - 100{,}000{,}000}{40{,}000{,}000/\sqrt{4}} < \frac{\bar{x} - 100{,}000{,}000}{40{,}000{,}000/\sqrt{4}} < \frac{140{,}000{,}000 - 100{,}000{,}000}{40{,}000{,}000/\sqrt{4}}\right)$$
$$= P(1.00 < Z < 2.00) = 0.9772 - 0.8413 = 0.1359$$

(c) $P(\bar{x} < 95{,}000{,}000) = P\left(\frac{\bar{x} - 100{,}000{,}000}{40{,}000{,}000/\sqrt{4}} < \frac{95{,}000{,}000 - 100{,}000{,}000}{40{,}000{,}000/\sqrt{4}}\right) = P(Z < -0.25) = 0.4013$

39. **(a)** \$100 million, the 50th percentile equals the mean for a normal distribution

(b) From Table C, $Z = 1.645$. Therefore,

$$\bar{x} = Z \cdot \sigma_{\bar{x}} + \mu = Z \cdot \frac{\sigma}{\sqrt{n}} + \mu = 1.645 \cdot \frac{\$40 \text{ million}}{\sqrt{4}} + \$100 \text{ million} = \$132.90 \text{ million}$$

(c) From Table C, $Z = -1.645$. Therefore,

$$\bar{x} = Z \cdot \sigma_{\bar{x}} + \mu = Z \cdot \frac{\sigma}{\sqrt{n}} + \mu = -1.645 \cdot \frac{\$40 \text{ million}}{\sqrt{4}} + \$100 \text{ million} = \$67.10 \text{ million.}$$

(d) The middle 90% of the sample mean IPO amounts

41. $\mu = 365$ ppm, $\sigma = 100$ ppm, $n = 25$

(a) $P(\bar{x} > 385) = P\left(\frac{\bar{x} - 365}{100/\sqrt{25}} > \frac{385 - 365}{100/\sqrt{25}}\right) = P(Z > 1.00) = 1 - 0.8413 = 0.1587$

(b) $P(305 < \bar{x} < 425) = P\left(\frac{305 - 365}{100/\sqrt{25}} < \frac{\bar{x} - 365}{100/\sqrt{25}} < \frac{425 - 365}{100/\sqrt{25}}\right)$

$$= P(-3.00 < Z < 3.00) = 0.9987 - 0.0013 = 0.9974$$

(c) $P(\bar{x} < 345) = P\left(\frac{\bar{x} - 365}{100/\sqrt{25}} < \frac{345 - 365}{100/\sqrt{25}}\right) = P(Z < -1.00) = 0.1587$

43. **(a)** From Table C, $Z = 1.96$. Therefore,

$$\bar{x} = Z \cdot \sigma_{\bar{x}} + \mu = Z \cdot \frac{\sigma}{\sqrt{n}} + \mu = 1.96 \cdot \frac{100}{\sqrt{25}} + 365 = 404.20 \text{ ppm.}$$

(b) From Table C, $Z = -1.96$. Therefore,

$$\bar{x} = Z \cdot \sigma_{\bar{x}} + \mu = Z \cdot \frac{\sigma}{\sqrt{n}} + \mu = -1.96 \cdot \frac{100}{\sqrt{25}} + 365 = 325.80 \text{ ppm.}$$

(c) The middle 95% of sample mean carbon dioxide concentrations of samples of size $n = 25$

45. **(a)** ${}_5C_2 = 10$ samples of size $n = 2$ **(b)** The population mean is $\mu = 64.12$. We usually do not know the value of the population mean in a typical real-world problem because either the population is too large to gather all of the data for the population, it is too hard, if not impossible, to get the data for all of it, it is too expensive to get the data for all of it, or getting the data for the entire population may destroy the entire population. **(c)** The population standard deviation is $\sigma = 2.9991$. We usually do not know the value of the population standard deviation in a real-world situation because either the population is too big to get the data for all of it, it is too hard, if not impossible, to get the data for all of it, it is too expensive to get the data for all of it, or getting the data for the entire population may destroy the entire population.

47. **(a)** About 64.1

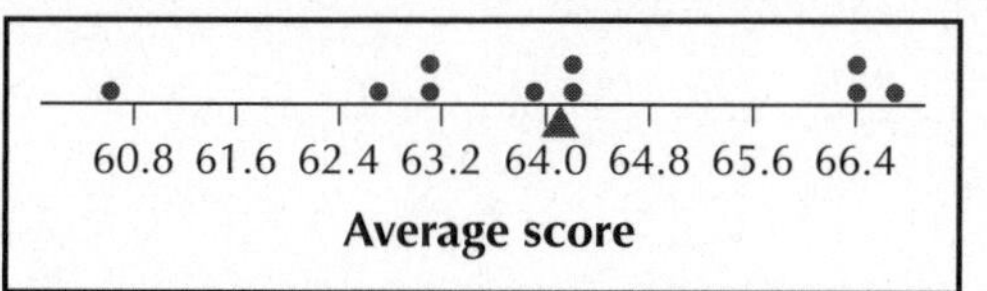

(b) About 64.1

49. **(a)** $\mu_{\bar{x}} = \mu = \sum[X \cdot P(X)] = 1 \cdot \frac{1}{6} + 2 \cdot \frac{1}{6} + 3 \cdot \frac{1}{6} + 4 \cdot \frac{1}{6} + 5 \cdot \frac{1}{6} + 6 \cdot \frac{1}{6} = \frac{21}{6} = 3.5$

(b) $\sigma^2 = \sum[X^2 \cdot P(X)] - \mu^2 = 1^2 \cdot \frac{1}{6} + 2^2 \cdot \frac{1}{6} + 3^2 \cdot \frac{1}{6} + 4^2 \cdot \frac{1}{6} + 5^2 \cdot \frac{1}{6} + 6^2 \cdot \frac{1}{6} - 3.5^2$
$= \frac{91}{6} - 32.5^2 = 2.916666667 \cdot \sigma = \sqrt{\sigma^2} = \sqrt{2.916666667}$
$= 1.707825128 \cdot \sigma_{\bar{x}} = \frac{\sigma}{\sqrt{n}} = \frac{1.707825128}{\sqrt{16}} = 0.4270$

51. $\mu = \$4, \sigma = \$6, n = 9$ **(a)** $\mu_{\bar{x}} = \mu = \$4$ **(b)** $\sigma_{\bar{x}} = \frac{\sigma}{\sqrt{n}} = \frac{\$6}{\sqrt{9}} = \$2$ **(c)** Since the distribution of the net gain in stock prices is normal with mean $\mu = \$4$ and standard deviation $\sigma = \$6$, the sampling distribution of the $\bar{x}$ is normal with a mean of $\mu_{\bar{x}} = \$4$ and a standard deviation of $\sigma_{\bar{x}} = \$2$ by Fact 4.

53. **(a)** Changes in the population standard deviation do not affect changes in the population mean. Thus, μ would remain the same. **(b)** Changes in the population standard deviation do not affect the sample size. Thus, n would remain the same. **(c)** Since $0 - \mu$ is negative, a decrease in σ will result in a decrease in $(0 - \mu)/\sigma$. Since $P(X < 0) = P\left(\frac{X-\mu}{\sigma} < \frac{0-\mu}{\sigma}\right)$, a decrease in σ will result in a decrease in $P(X < 0)$. **(d)** Since $\mu_{\bar{x}} = \mu$, changes in the population standard deviation will leave $\mu_{\bar{x}}$ the same. **(e)** Since $\sigma_{\bar{x}} = \sigma/\sqrt{n}$, a decrease in σ will result in a decrease in $\sigma_{\bar{x}}$.

55. $P(X < 500) = P\left(Z < \frac{500 - 515}{116}\right) = P(Z < -0.13) = 0.4483$

57. **(a)** $P(\bar{x} < 500) = P\left(Z < \frac{500 - 515}{29}\right) = P(Z < -0.52) = 0.3015$ **(b)** Sample means are less variable than individual observations, so 500 is $-0.13 \cdot \sigma$ below the mean of 515 but $-0.52 \cdot \sigma_{\bar{x}}$ below the mean of 500.

59. **(a)** From Exercise 58 (c), $\sigma_{\bar{x}}$ increases. Since $500 - \mu$ is negative, $\frac{500-\mu}{\sigma_{\bar{x}}}$ increases. Since $P(\bar{x} < 500) = P\left(Z < \frac{500-\mu}{\sigma_{\bar{x}}}\right)$, $P(\bar{x} < 500)$ increases. **(b)** From Exercise 58 (c), $\sigma_{\bar{x}}$ increases. Therefore, the 99.5th percentile $= 2.58 \cdot \sigma_{\bar{x}} + \mu$ increases. **(c)** From Exercise 58 (c), $\sigma_{\bar{x}}$ increases. Therefore, the 0.5th percentile $= -2.58 \cdot \sigma_{\bar{x}} + \mu$ decreases.

61. The sampling error decreases as the sample size increases.

63. The sampling error decreases as the sample size increases.

Section 7.2

1. The sampling distribution of the sample mean becomes approximately normal quicker for a population distribution that is symmetric than for a population distribution that is skewed. A population distribution that is symmetric will have as many values a given distance below the mean as the given distance above the mean, so most of the random samples taken from a symmetric distribution will have close to the same number of values below the mean as above the mean. Therefore, most of the sample means will be fairly close to the center. A population distribution that is skewed will have more values on one side of the mean than on the other side, so most random samples taken from a skewed distribution will contain more values on that side of the mean than on the other side of the mean. Therefore, most of the sample means will be on that side of the mean and it will take larger sample sizes to diminish the influence of the extreme values.

3. The variability of the sampling distribution gets smaller as the sample size gets larger.

5. Since the SAT scores are normally distributed, Case 1 applies.

7. Since accountant incomes are normally distributed, Case 1 applies.

9. Since the sample size is large, Case 2 applies.

11. Since body-mass indices of college students are not normally distributed and the sample size is small, Case 3 applies.

13. **(a)** $\mu_{\bar{x}} = \mu = 516$ **(b)** $\sigma_{\bar{x}} = \frac{\sigma}{\sqrt{n}} = \frac{116}{\sqrt{9}} \approx 38.6667$

(c) Since SAT scores are normally distributed, the sampling distribution of $\bar{x}$ is normal by Case 1.

15. **(a)** $\mu_{\bar{x}} = \mu = \$60{,}000$ **(b)** $\sigma_{\bar{x}} = \frac{\sigma}{\sqrt{n}} = \frac{\$10{,}000}{\sqrt{16}} = \$2{,}500$

(c) Since accountant incomes are normally distributed, the sampling distribution of $\bar{x}$ is normal by Case 1.

17. **(a)** $\mu_{\bar{x}} = \mu = 80$ **(b)** $\sigma_{\bar{x}} = \frac{\sigma}{\sqrt{n}} = \frac{8}{\sqrt{64}} = 1$

(c) Since the sample size of $n = 64$ is large ($n \geq 30$), the sampling distribution of $\bar{x}$ is approximately normal by Case 2.

19. **(a)** $\mu_{\bar{x}} = \mu = 50$ miles per gallon **(b)** $\sigma_{\bar{x}} = \frac{\sigma}{\sqrt{n}} = \frac{6}{\sqrt{16}} = 1.5$ miles per gallon **(c)** Since gas mileage for 2007 Toyota Prius hybrid automobiles is not normally distributed and the sample size of $n = 16$ is small ($n < 30$), the sampling distribution of $\bar{x}$ is unknown by Case 3.

21. Since the exchange rate for \$100 in U.S. dollars in Euros for 2007 is not normally distributed and the sample size of $n = 9$ is small ($n < 30$), the sampling distribution of $\bar{x}$ is unknown by Case 3. Therefore, it is not possible to find $P(\bar{x} > 75)$.

23. Since the 2007 prices for boned trout are normally distributed, the sampling distribution of $\bar{x}$ is normal by Case 1. Therefore,

$$P(\bar{x} < \$3) = P\left(Z < \frac{\$3 - \$3.10}{\$0.30/\sqrt{16}}\right) = P(Z < -1.33) = 0.0918.$$

25. Since the sample size of $n = 64$ is large ($n \geq 30$), the sampling distribution of $\bar{x}$ is approximately normal by Case 2. Therefore,

$$P(78 < \bar{x} < 82) = P\left(\frac{78 - 80}{8/\sqrt{64}} < Z < \frac{82 - 80}{8/\sqrt{64}}\right) = P(-2.00 < Z < 2.00) = 0.9772 - 0.0228 = 0.9544.$$

27. Since the pollen count distribution for Los Angeles in September is not normally distributed and the sample size of $n = 16$ is small ($n < 30$), the sampling distribution of $\bar{x}$ is unknown by Case 3. Therefore, it is not possible to find $P(\bar{x} > 9.0)$.

29. Since SAT scores are normally distributed, the sampling distribution of $\bar{x}$ is normal by Case 1. From Table C, $Z = 1.645$. Therefore, the sample mean SAT score that is larger than 95% of all sample means is

$$Z \cdot \frac{\sigma}{\sqrt{n}} + \mu = 1.645 \cdot \frac{116}{\sqrt{9}} + 516 = 579.61.$$

31. Since the 2007 prices for boned trout are normally distributed, the sampling distribution of $\bar{x}$ is normal by Case 1. By Table C, $Z = -1.28$. Thus, the sample mean price that is smaller than 90% of all sample means is

$$Z \cdot \frac{\sigma}{\sqrt{n}} + \mu = -1.28 \cdot \frac{\$0.30}{\sqrt{16}} + \$3.10 = \$3.00.$$

33. Since accountant incomes are normally distributed, the sampling distribution of $\bar{x}$ is normal by Case 1. By Table C, $Z = -1.28$. Thus, the sample mean price that is smaller than 90% of all sample means is

$$Z \cdot \frac{\sigma}{\sqrt{n}} + \mu = -1.28 \cdot \frac{\$10{,}000}{\sqrt{16}} + \$60{,}000 = \$56{,}800.$$

35. Since the sample size of $n = 64$ is large ($n \geq 30$), the sampling distribution of $\bar{x}$ is approximately normal by Case 2. From Table C, $Z = 1.645$. Thus, the 95th percentile is

$$Z \cdot \frac{\sigma}{\sqrt{n}} + \mu = 1.645 \cdot \frac{8}{\sqrt{64}} + 80 = 81.645.$$

37. Since the pollen count distribution for Los Angeles in September is not normally distributed and the sample size of $n = 16$ is small ($n < 30$), Case 3 applies and the sampling distribution of $\bar{x}$ is unknown. Thus, the 75th percentile cannot be found.

39. (a) Since the distribution of serum cholesterol levels in Americans in 2005 is unknown, Case 1 does not apply. **(b)** Since the sample size of $n = 36$ is large ($n \geq 30$), Case 2 applies and the sampling distribution of $\bar{x}$ is approximately normal. Thus,

$$P(\bar{x} > 212) = P\left(Z > \frac{212 - 202}{45/\sqrt{36}}\right) = P(Z > 1.33) = 1 - 0.9082 = 0.0918.$$

41. (a) Since the points per game for Shaquille O'Neal are normally distributed, Case 1 applies and the sampling distribution of $\bar{x}$ is normal. Thus,

$$P(\bar{x} > 25) = P\left(Z > \frac{25 - 21.5}{5/\sqrt{4}}\right) = P(Z > 1.40) = 1 - 0.9192 = 0.0808.$$

(b) Since the sample size of $n = 4$ is small ($n < 30$), Case 2 does not apply.

43. (a) Since the distribution of the salaries of all new sociology professors is right-skewed, it is not normal. Hence, Case 1 does not apply. **(b)** Since the sample size of $n = 16$ is small ($n < 30$), Case 2 does not apply.

45. (a) Since the distribution of the salaries of all new sociology professors is right-skewed, it is not normal. Hence, Case 1 does not apply. **(b)** Since the sample size of $n = 36$ is large ($n \geq 30$), Case 2 applies so the sampling distribution of $\bar{x}$ is approximately normal. Therefore,

$$P(\bar{x} > \$48{,}000) = P\left(Z > \frac{\$48{,}000 - 45{,}722}{\$6000/\sqrt{36}}\right) =$$

$$P(Z > 2.28) = 1 - 0.9887 = 0.0113.$$

47. Since the sample size of $n = 36$ is large ($n \geq 30$), Case 2 applies so the sampling distribution of $\bar{x}$ is approximately normal. Therefore,

$$P(\bar{x} > 50) = P\left(Z > \frac{50 - 42.7}{38.6/\sqrt{36}}\right) = P(Z > 1.13) = 1 - 0.8708 = 0.1292.$$

49. $n = 30$

51. (a) Since the sampling distribution of the changes in SAT scores are not normally distributed, Case 1 does not apply. **(b)** Since the sample size of $n = 40$ is large ($n \geq 30$), Case 2 applies, so the sampling distribution of $\bar{x}$ is approximately normal. Thus,

$$P(\bar{x} < 0) = P\left(\frac{\bar{x} - \mu}{\sigma/\sqrt{n}} < \frac{0 - \mu}{\sigma/\sqrt{n}}\right) = P\left(\frac{\bar{x} - 18}{12/\sqrt{40}} < \frac{0 - 18}{12/\sqrt{40}}\right) = P(Z < -9.49) \approx 0.$$

53. (a) Smaller. An increase in the sample size will result in a decrease in $\sigma_{\bar{x}} = \frac{\sigma}{\sqrt{n}}$. Since $0 - \mu$ is negative, $\frac{0 - \mu}{\sigma_{\bar{x}}}$ would decrease. Since

$$P(\bar{x} < 0) = P\left(Z < \frac{0 - \mu}{\sigma_{\bar{x}}}\right),$$

$P(\bar{x} < 0)$ would decrease.

(b) $P(\bar{x} < 0) = P\left(\frac{\bar{x} - \mu}{\sigma/\sqrt{n}} < \frac{0 - \mu}{\sigma/\sqrt{n}}\right) = P\left(\frac{\bar{x} - 18}{12/\sqrt{80}} < \frac{0 - 18}{12/\sqrt{80}}\right) = P(Z < -13.42) \approx 0$

55. (a) From Exercise 54(a) and (b), $\mu_{\bar{x}}$ remains unchanged and $\sigma_{\bar{x}}$ decreases. Since the sample mean increase in SAT math scores that is larger than 97.5% of all sample means is $1.96 \cdot \sigma_{\bar{x}} + \mu$, the sample mean increase in SAT math scores that is larger than 97.5% of all sample means would decrease. **(b)** From Exercise 54(a) and (b), $\mu_{\bar{x}}$ remains unchanged and $\sigma_{\bar{x}}$ decreases. Since the sample mean increase in SAT math scores that is smaller than 97.5% of all sample means is $-1.96 \cdot \sigma_{\bar{x}} + \mu$, the sample mean increase in SAT math scores that is smaller than 97.5% of all sample means would increase. **(c)** Since the two sample means that together span the middle 95% of all sample means are the sample mean increase in SAT math scores that is larger than 97.5% of all sample means and the sample mean increase in SAT math scores that is smaller than 97.5% of all sample means, the sample mean increase in SAT math scores that is larger than 97.5% of all sample means would decrease and the sample mean increase in SAT math scores that is smaller than 97.5% of all sample means would increase.

57. **(a)** Increase since n is in the denominator
(b) Smaller. From part (a), $\sigma_{\bar{x}}$ would decrease. Since $0 - \mu$ is negative, $\frac{0-\mu}{\sigma_{\bar{x}}}$ would decrease.
Since $P(\bar{x} > 0) = P\left(Z > \frac{0-\mu}{\sigma_{\bar{x}}}\right)$, $P(\bar{x} > 0)$ would increase.

59. **(a)** $3 \cdot \sigma_{\bar{x}} = 0.32$, so $\sigma_{\bar{x}} = 0.32/3 = 0.1067$ grams **(b)** Since $\sigma_{\bar{x}} = \frac{\sigma}{\sqrt{n}} = \frac{\sigma}{\sqrt{100}} = \frac{\sigma}{10} = 0.1067$ grams, $\sigma = 10 \cdot 0.1067$ grams $= 1.067$ grams. **(c)** Since $P(127.68 < \bar{x} < 128.32) = P(-3 < Z < 3)$, the Empirical Rule tells us that about 0.997 of the mean weights lie between 127.68 and 128.32 grams.

61. **(a)**
$$P(127.68 < \bar{x} < 128.32) = P\left(\frac{127.68-\mu}{\sigma_{\bar{x}}} < \frac{\bar{x}-\mu}{\sigma_{\bar{x}}} < \frac{128.32-\mu}{\sigma_{x}}\right)$$
$$= P\left(\frac{127.68-127.3}{0.1067} < \frac{\bar{x}-127.3}{0.1067} < \frac{128.32-127.3}{0.1067}\right)$$
$$= P(3.56 < Z < 9.56) = 0.0002$$

(b) 0.0002, 1 – 0.0002 = 0.9998 **(c)** The probability that he will get away with it found in part (b) is smaller than the probability that he will get away with it found in the original Case Study in the text, so the value found in the original Case Study in the text favors the Master of the Mint.

63. Solve $P\left(\frac{127.68-\mu}{0.16} < Z < \frac{128.32-\mu}{0.16}\right) = 1 - 0.25 = 0.75$ for μ to get $\mu = 127.788142$ grams.
(See chart.)

μ	$P\left(\frac{127.68-\mu}{0.16} < Z < \frac{128.32-\mu}{0.16}\right)$	μ	$P\left(\frac{127.68-\mu}{0.16} < Z < \frac{128.32-\mu}{0.16}\right)$
127.78814150	0.749999	127.78814166	0.75
127.78814151	0.749999	127.78814167	0.75
127.78814152	0.75	127.78814168	0.75
127.78814153	0.75	127.78814169	0.75
127.78814154	0.75	127.78814170	0.75
127.78814155	0.75	127.78814171	0.75
127.78814156	0.75	127.78814172	0.75
127.78814157	0.75	127.78814173	0.75
127.78814158	0.75	127.78814174	0.75
127.78814159	0.75	127.78814175	0.75
127.78814160	0.75	127.78814176	0.75
127.78814161	0.75	127.78814177	0.75
127.78814162	0.75	127.78814178	0.75
127.78814163	0.75	127.78814179	0.75
127.78814164	0.75	127.78814180	0.75
127.78814165	0.75	127.78814181	0.75

(Continued)

μ	$P\left(\frac{127.68-\mu}{0.16}<Z<\frac{128.32-\mu}{0.16}\right)$	μ	$P\left(\frac{127.68-\mu}{0.16}<Z<\frac{128.32-\mu}{0.16}\right)$
127.78814182	0.75	127.78814194	0.75
127.78814183	0.75	127.78814195	0.75
127.78814184	0.75	127.78814196	0.75
127.78814185	0.75	127.78814197	0.75
127.78814186	0.75	127.78814198	0.75
127.78814187	0.75	127.78814199	0.75
127.78814188	0.75	127.78814200	0.75
127.78814189	0.75	127.78814201	0.75
127.78814190	0.75	127.78814202	0.750001
127.78814191	0.75	127.78814203	0.750001
127.78814192	0.75	127.78814204	0.750001
127.78814193	0.75	127.78814205	0.750001

Section 7.3

1. Example 7.17 shows that a sample of size $n = 30$ may not be enough for the sampling distribution of $\hat{p}$ to be approximately normal.

3. The closer p is to 0, the larger n has to be in order for $np \geq 5$, and the closer p is to 1, the larger n has to be in order for $n(1-p) \geq 5$.

5. **(a)** As the sample size increases, the range of the horizontal scale decreases. **(b)** As n increases, $\sigma_{\hat{p}} = \sqrt{\frac{p(1-p)}{n}}$ decreases. Thus, the sampling distribution of $\hat{p}$ is less variable.

7. **(a)** $\mu_{\hat{p}} = p = 0.5$ **(b)** $\sigma_{\hat{p}} = \sqrt{\frac{p(1-p)}{n}} = \sqrt{\frac{0.5(1-0.5)}{100}} = 0.05$ **(c)** We have $np = (100)(0.5) = 50 \geq 5$ and $n(1-p) = (100)(1-0.5) = 50 \geq 5$, so the sampling distribution of $\hat{p}$ is approximately normal.

9. **(a)** $\mu_{\hat{p}} = p = 0.01$ **(b)** $\sigma_{\hat{p}} = \sqrt{\frac{p(1-p)}{n}} = \sqrt{\frac{0.01(1-0.01)}{100}} \approx 0.0099$ **(c)** We have $np = (100)(0.01) = 1 < 5$, so the sampling distribution of $\hat{p}$ is unknown.

11. **(a)** $\mu_{\hat{p}} = p = 0.9$ **(b)** $\sigma_{\hat{p}} = \sqrt{\frac{p(1-p)}{n}} = \sqrt{\frac{0.9(1-0.9)}{40}} \approx 0.0474$ **(c)** We have $np = (40)(0.9) = 36 \geq 5$ but $n(1-p) = 40(1-0.9) = 4 < 5$, so the sampling distribution of $\hat{p}$ is unknown.

13. $n_1 = \frac{5}{p} = \frac{5}{0.5} = 10$ and $n_2 = \frac{5}{1-p} = \frac{5}{1-0.5} = 10$. Therefore, 10 is the minimum sample size that produces a sampling distribution of $\hat{p}$ that is approximately normal.

15. $n_1 = \frac{5}{p} = \frac{5}{0.1} = 50$ and $n_2 = \frac{5}{1-p} = \frac{5}{1-0.1} \approx 5.5556$. Therefore, 50 is the minimum sample size that produces a sampling distribution of $\hat{p}$ that is approximately normal.

17. $n_1 = \frac{5}{p} = \frac{5}{0.01} = 500$ and $n_2 = \frac{5}{1-p} = \frac{5}{1-0.01} \approx 5.0505$. Therefore, 500 is the minimum sample size that produces a sampling distribution of $\hat{p}$ that is approximately normal.

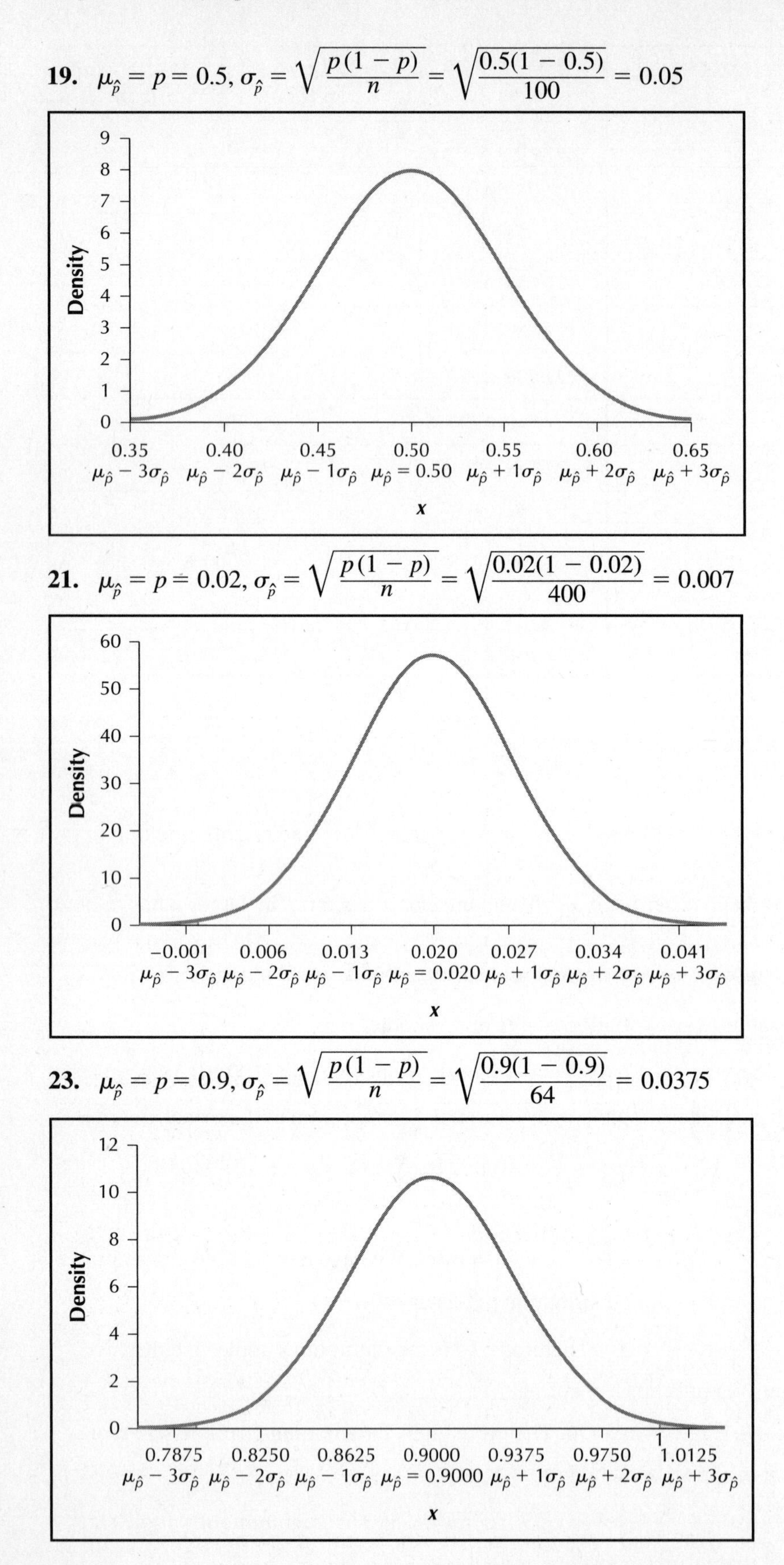

19. $\mu_{\hat{p}} = p = 0.5, \sigma_{\hat{p}} = \sqrt{\dfrac{p(1-p)}{n}} = \sqrt{\dfrac{0.5(1-0.5)}{100}} = 0.05$

21. $\mu_{\hat{p}} = p = 0.02, \sigma_{\hat{p}} = \sqrt{\dfrac{p(1-p)}{n}} = \sqrt{\dfrac{0.02(1-0.02)}{400}} = 0.007$

23. $\mu_{\hat{p}} = p = 0.9, \sigma_{\hat{p}} = \sqrt{\dfrac{p(1-p)}{n}} = \sqrt{\dfrac{0.9(1-0.9)}{64}} = 0.0375$

25. $\mu_{\hat{p}} = p = 0.1, \sigma_{\hat{p}} = \sqrt{\frac{p(1-p)}{n}} = \sqrt{\frac{0.1(1-0.1)}{64}} = 0.0375$

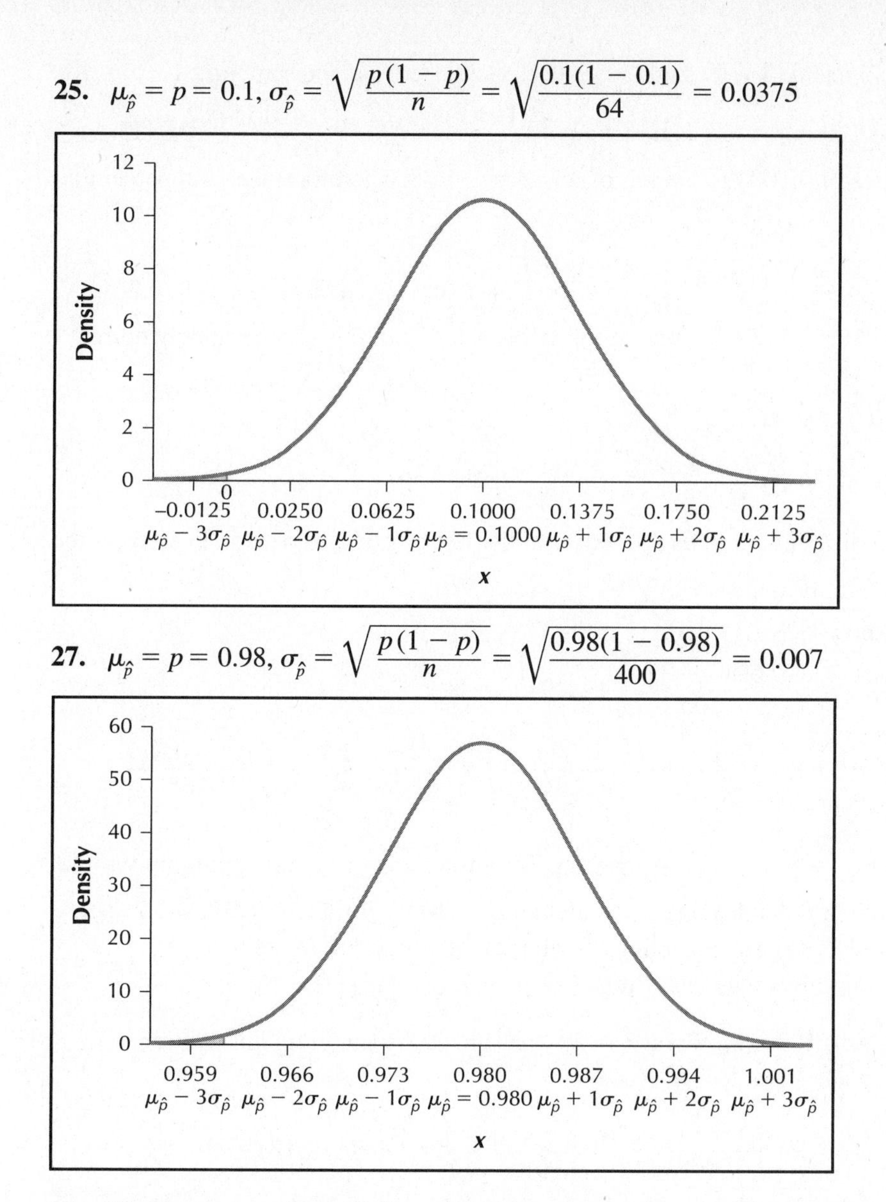

27. $\mu_{\hat{p}} = p = 0.98, \sigma_{\hat{p}} = \sqrt{\frac{p(1-p)}{n}} = \sqrt{\frac{0.98(1-0.98)}{400}} = 0.007$

29. We have $n\,p = (100)\,(0.5) = 50 \geq 5$ and $n\,(1-p) = 100\,(1-0.5) = 50 \geq 5$, so the sampling distribution of $\hat{p}$ is approximately normal. $\mu_{\hat{p}} = p = 0.5$ and $\sigma_{\hat{p}} = \sqrt{\frac{p(1-p)}{n}} = \sqrt{\frac{0.5(1-0.5)}{100}} = 0.05$. Therefore,

$$P(\hat{p} > 0.55) = P\left(\frac{\hat{p}-p}{\sigma_{\hat{p}}} > \frac{0.55-p}{\sigma_{\hat{p}}}\right) = P\left(\frac{\hat{p}-0.05}{0.05} > \frac{0.55-0.5}{0.05}\right) = P(Z > 1) = 1 - 0.8413 = 0.1587.$$

31. We have $n\,p = (100)\,(0.01) = 1 < 5$, so the sampling distribution of $\hat{p}$ is unknown. Therefore, $P(\hat{p} > 0.011)$ can't be found.

33. We have $n\,p = (40)\,(0.9) = 36 \geq 5$ but $n\,(1-p) = 40\,(1-0.9) = 4 < 5$, so the sampling distribution of $\hat{p}$ is unknown. Thus, $P(0.88 < \hat{p} < 0.91)$ can't be found.

35. We have $n\,p = (100)\,(0.5) = 50 \geq 5$ and $n\,(1-p) = 100\,(1-0.5) = 50 \geq 5$, so the sampling distribution of $\hat{p}$ is approximately normal. $\mu_{\hat{p}} = p = 0.5$ and $\sigma_{\hat{p}} = \sqrt{\frac{p(1-p)}{n}} = \sqrt{\frac{0.5(1-0.5)}{100}} = 0.05$
From Table C, the value of Z with an area of 0.9000 to the left of it is $Z = 1.28$. Therefore, the 90th percentile is $\hat{p} = Z \cdot \sigma + \mu = (1.28)\,(0.05) + 0.5 = 0.564$.

37. We have $n\,p = (64)\,(0.9) = 57.6 \geq 5$ and $n\,(1-p) = 64\,(1-0.9) = 6.4 \geq 5$, so the sampling distribution of $\hat{p}$ is approximately normal. $\mu_{\hat{p}} = p = 0.9$ and $\sigma_{\hat{p}} = \sqrt{\frac{p(1-p)}{n}} = \sqrt{\frac{0.9(1-0.9)}{64}} = 0.0375$.
From Table C, the value of Z with an area of 0.9500 to the left of it is $Z = 1.645$. Therefore, the 95th percentile is $\hat{p} = Z \cdot \sigma + \mu = (1.645)\,(0.0375) + 0.9 = 0.962$.

39. We have $n\,p = (64)\,(0.1) = 6.4 \geq 5$ and $n\,(1 - p) = 64\,(1 - 0.1) = 57.6 \geq 5$, so the sampling distribution of $\hat{p}$ is approximately normal. $\mu_{\hat{p}} = p = 0.9$ and $\sigma_{\hat{p}} = \sqrt{\frac{p(1-p)}{n}} = \sqrt{\frac{0.1(1-0.1)}{64}} = 0.0375$. From Table C, the value of Z with an area of 0.1000 to the left of it is $Z = -1.28$. Therefore, the 10th percentile is $\hat{p} = Z \cdot \sigma + \mu = (-1.28)\,(0.0375) + 0.1 = 0.052$.

41. **(a)** $\mu_{\hat{p}} = p = 0.25$ **(b)** $\sigma_{\hat{p}} = \sqrt{\frac{p(1-p)}{n}} = \sqrt{\frac{0.25(1-0.25)}{100}} \approx 0.0433$ **(c)** We have $n\,p = (100)\,(0.25) = 25 \geq 5$ and $n\,(1 - p) = 100\,(1 - 0.25) = 75 \geq 5$, so the sampling distribution of $\hat{p}$ is approximately normal with mean $\mu_{\hat{p}} = 0.25$ and $\sigma_{\hat{p}} \approx 0\ 0.043$.

43. From Exercise 41, $\mu_{\hat{p}} = 0.25$ and $\sigma_{\hat{p}} \approx 0.0433$.

(a) $P(\hat{p} > 0.30) = P\left(\frac{\hat{p} - p}{\sigma_{\hat{p}}} > \frac{0.30 - p}{\sigma_{\hat{p}}}\right) = P\left(\frac{\hat{p} - 0.25}{0.0433} > \frac{0.30 - 0.25}{0.0433}\right) = P(Z > 1.15) = 1 - 0.8749 = 0.1251$ **(b)** From Table C, the value of Z with an area of 0.9500 to the left of it is $Z = 1.645$. Therefore, the 95th percentile is $\hat{p} = Z \cdot \sigma + \mu = (1.645)\,(0.0433) + 0.25 \approx 0.321$.

45. **(a)** $n\,p = (100)\,(0.61) = 61 \geq 5$ and $n\,(1 - p) = 100\,(1 - 0.61) = 39 \geq 5$

(b) $\mu_{\hat{p}} = p = 0.61$ and $\sigma_{\hat{p}} = \sqrt{\frac{p(1-p)}{n}} = \sqrt{\frac{0.61(1-0.61)}{100}} \approx 0.0488$

(c) $P(X < 60) = P\left(\frac{X}{n} < \frac{60}{100}\right) = P(\hat{p} < 0.60) = P\left(\frac{\hat{p} - p}{\sigma_{\hat{p}}} < \frac{0.60 - p}{\sigma_{\hat{p}}}\right) = P\left(\frac{\hat{p} - 0.61}{0.0488} < \frac{0.60 - 0.61}{0.0488}\right)$
$= P(Z < -0.20) = 0.4207$

47. **(a)** $n_1 = \frac{5}{p} = \frac{5}{0.40} = 12.5$ and $n_2 = \frac{5}{1-p} = \frac{5}{1-0.40} \approx 8.33$. Therefore, $n^* = 13$ is the minimum sample size that produces a sampling distribution of $\hat{p}$ that is approximately normal. **(b)** $n^*\,p = (13)\,(0.40) = 5.2 \geq 5$ and $n^*\,(1 - p) = (13)\,(1 - 0.40) = 7.8 \geq 5$ **(c)** The sampling distribution of $\hat{p}$ is approximately normal. The Central Limit Theorem for Proportions allows us to say this. **(d)** $\mu_{\hat{p}} = p = 0.40$ and $\sigma_{\hat{p}} = \sqrt{\frac{p(1-p)}{n}} = \sqrt{\frac{0.40(1-0.40)}{20}} \approx 0.1095$ **(e)** Since $n = 20 \geq n^* = 13$, the sampling distribution of $\hat{p}$ is approximately normal for $n = 20$. Thus,

$$P(\hat{p} < 0.38) = P\left(\frac{\hat{p} - p}{\sigma_{\hat{p}}} < \frac{0.38 - p}{\sigma_{\hat{p}}}\right) = \left(\frac{\hat{p} - 0.40}{0.1095} < \frac{0.38 - 0.40}{0.1095}\right) = P(Z < -0.18) = 0.4286.$$

49. **(a)** Since $\sigma_{\hat{p}} = \sqrt{\frac{p(1-p)}{n}}$ an increase in the sample size n will result in a decrease of $\sigma_{\hat{p}}$.
(b) Any value of the sample size n larger than 20 will result in a decrease of $\sigma_{\hat{p}}$.
For example, if $n = 25$, then and $\sigma_{\hat{p}} = \sqrt{\frac{p(1-p)}{n}} = \sqrt{\frac{0.40(1-0.40)}{25}} \approx 0.0980$.

51.

Response	Frequency	Relative frequency (frequency/total)
Many people	2058	2058/4395 = 0.47
Just a few people	1806	1806/4395 = 0.41
Hardly any people	485	485/4395 = 0.11
No one/None	46	46/4395 = 0.01
Total	**4395**	

53. **(a)** $\mu_{\hat{p}} = p = 0.01$ and $\sigma_{\hat{p}} = \sqrt{\frac{p(1-p)}{n}} = \sqrt{\frac{0.01(1-0.01)}{500}} \approx 0.0044$ **(b)** This value of $\sigma_{\hat{p}}$ is smaller than most values of $\sigma_{\hat{p}}$ that we have dealt with so far because p is so small and n is so large. **(c)** Since a sample size of 200 respondents is smaller than the minimum sample size of 500 respondents required to produce a sampling distribution of $\hat{p}$ that is approximately normal, the sampling distribution of $\hat{p}$ is unknown. Thus, we can't find the probability that a sample of 200 respondents will have a proportion of people with no one to turn to that is greater than 2%. In order to find this probability, the sample size would need to be increased to at least 500 respondents.

55. **(a)** Since p is the population proportion of people with someone to turn to, p would increase. **(b)** Since p would increase, $(1 - p)$ would decrease. **(c)** Since $\mu_{\hat{p}} = p$, $\mu_{\hat{p}}$ would increase.

(d) Since $\sigma_{\hat{p}} = \sqrt{\frac{p(1+p)}{n}}$, an increase of p up to 0.99 will result in an increase in $\sigma_{\hat{p}}$, an increase of p to 0.99 will result in the same $\sigma_{\hat{p}}$, and an increase of p to a value greater than 0.99 but less than or equal to 1 will result in a decrease in $\sigma_{\hat{p}}$.

A graph of $\sigma_{\hat{p}} = \sqrt{\frac{p(1+p)}{n}}$ for $n = 500$ follows:

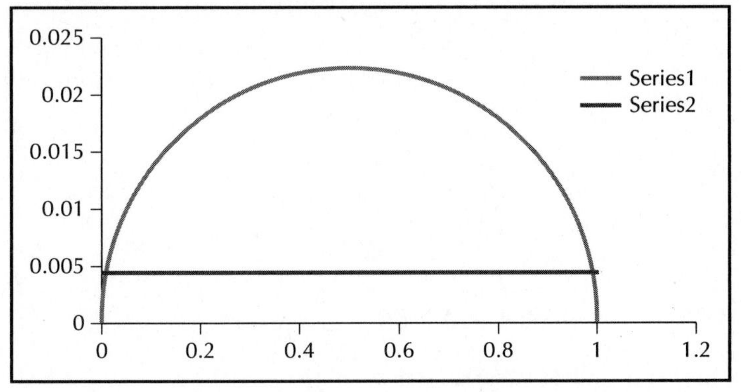

(e) Since n^* is the maximum of $n_1 = \frac{5}{p}$ and $n_2 = \frac{5}{1-p}$, the graph of both of these functions is given below. From the graph, we can see that $n_1 \geq n_2$ for $0 < p \leq 0.5$ and $n_2 \geq n_1$ for $0.5 \leq p < 1$. Thus, $n^* = n_1$ for $0 < p \leq 0.5$ and $n^* = n_2$ for $0.5 \leq p < 1$. Thus, n^* is less than 500 for $0.01 < p < 0.99$, equal to 500 when $p = 0.99$, and greater than 500 when $0.99 < p < 1$.

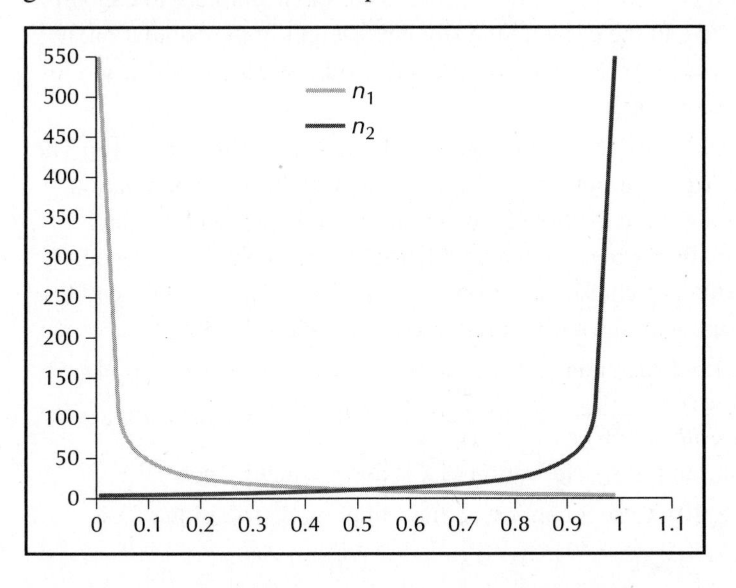

57. **(a)** The value of p will remain unchanged since p does not depend on n. **(b)** The value of $1 - p$ will remain unchanged since $1 - p$ does not depend on n. **(c)** The value of $\mu_{\hat{p}}$ will remain unchanged since $\mu_{\hat{p}} = p$ does not depend on n. **(d)** Since $\sigma_{\hat{p}} = \sqrt{\frac{p(1-p)}{n}}$ and n is in the denominator, an increase in n will result in a decrease in $\sigma_{\hat{p}}$. **(e)** The minimum sample size n^* required for approximate normality in the sampling distribution for $\hat{p}$ will remain unchanged since n^* does not depend on n.

59. **(a)** The standard deviation is

$$\sqrt{\frac{\left(\frac{2}{3}-0.6\right)^2+\left(\frac{1}{3}-0.6\right)^2+\left(\frac{2}{3}-0.6\right)^2+\left(\frac{2}{3}-0.6\right)^2+\left(\frac{3}{3}-0.6\right)^2+\left(\frac{2}{3}-0.6\right)^2+\left(\frac{1}{3}-0.6\right)^2+\left(\frac{2}{3}-0.6\right)^2+\left(\frac{1}{3}-0.6\right)^2+\left(\frac{2}{3}-0.6\right)^2}{10}}$$

$= 0.2$

(b) $\sigma_{\hat{p}} = \sqrt{\dfrac{0.6 \cdot (1-0.6)}{3}} = 0.2828$

$$\sqrt{\frac{N-n}{N-1}} \cdot \sigma_{\hat{p}} = \sqrt{\frac{5-3}{5-1}} \cdot \sqrt{\frac{0.6 \cdot (1-0.6)}{3}} = 0.2$$

Chapter 7 Review

1. $|\bar{x} - \mu| = |95 - 95| = 0$

3. $|\hat{p} - p| = |0 - 0.1| = 0.1$

5. $\mu_{\bar{x}} = \mu = 2,\ \sigma_{\bar{x}} = \dfrac{\sigma}{\sqrt{n}} = \dfrac{0.5}{\sqrt{36}} \approx 0.0833$

7. $\mu_{\bar{x}} = \mu = 50,\ \sigma_{\bar{x}} = \dfrac{\sigma}{\sqrt{n}} = \dfrac{40}{\sqrt{16}} = 10$

9. $P(\bar{x} > 11) = P\left(\dfrac{\bar{x} - 10}{0.8} > \dfrac{11 - 10}{0.8}\right) = P(Z > 1.25) = 1 - 0.8944 = 0.1056$

11. Using the results from Exercise 9, $P(9 \le \bar{x} \le 11) = 1 - (P(\bar{x} < 9) + P(\bar{x} > 11))$ $= 1 - (0.1056 + 0.1056) = 0.7888$.

13. **(a)** The point estimate for the population range of weights of all crash-test vehicles is the sample range of all crash-test vehicles, which is 2562 pounds. **(b)** The estimate in (a) is a statistic since it comes from a sample. The population range is a parameter since it comes from a population. **(c)** The population range is the largest weight in the population minus the smallest weight in the population. Since a sample is unlikely to contain both the car with the largest weight in the population and the car with the smallest weight in the population, it will most likely contain a car lighter than the heaviest car or a car heavier than the lightest car (or both), so the sample range will tend to underestimate the population range.

15. **(a)** $|\text{sample range} - \text{population range}| = |2562 - 4029| = 1467$ pounds. **(b)** The sampling error. **(c)** The sample range underestimated the population range. **(d)** The true range is the largest weight in the population minus the smallest weight in the population. Since a sample is unlikely to contain both the car with the largest weight in the population and the car with the smallest weight in the population, it will most likely contain a car lighter than the heaviest car or a car heavier than the lightest car (or both), so the sample range will tend to underestimate the population range. **(e)** This concurs with the earlier prediction in Problem 13 (c).

17. Since hummingbird wing beats are not normally distributed and the sample size is large, Case 2 applies.

19. **(a)** $\mu_{\bar{x}} = \mu = 50$ beats per second **(b)** $\sigma_{\bar{x}} = \dfrac{\sigma}{\sqrt{n}} = \dfrac{10}{\sqrt{100}} = 1$ beat per second **(c)** Since the sample size of $n = 100$ is large ($n \ge 30$), Case 2 applies, so the sampling distribution of $\bar{x}$ is approximately normal.

21. Since the sample size of $n = 100$ is large ($n \ge 30$), Case 2 applies, so the sampling distribution of $\bar{x}$ is approximately normal.
Therefore,

$$P(\bar{x} > 53) = P\left(\frac{\bar{x} - 50}{10/\sqrt{100}} > \frac{53 - 50}{10/\sqrt{100}}\right) = P(Z > 3.00) = 1 - 0.9987 = 0.0013.$$

23. Since the sample size of $n = 100$ is large ($n \ge 30$), Case 2 applies, so the sampling distribution of $\bar{x}$ is approximately normal. From Table C, the value of Z with an area to the left of it equal to 0.1000 is $Z = -1.28$. Thus, the 10th percentile is $Z \cdot \dfrac{\sigma}{\sqrt{n}} + \mu = -1.28 \cdot \dfrac{10}{\sqrt{100}} + 50 = 48.72$ beats per minute.

25. **(a)** From Table C, the value of Z with an area of 0.9750 to the left of it is $Z = 1.96$. Thus, the 97.5th percentile is $Z \cdot \frac{\sigma}{\sqrt{n}} + \mu = 1.96 \cdot \frac{10}{\sqrt{38}} + 44 = 47.18$ years. **(b)** From Table C, the value of Z with an area of 0.0250 to the left of it is $Z = -1.96$. Thus, the 2.5th percentile is $Z \cdot \frac{\sigma}{\sqrt{n}} + \mu = -1.96 \cdot \frac{10}{\sqrt{38}} + 44 = 40.82$ years. **(c)** 40.82 years and 47.18 years

(d)

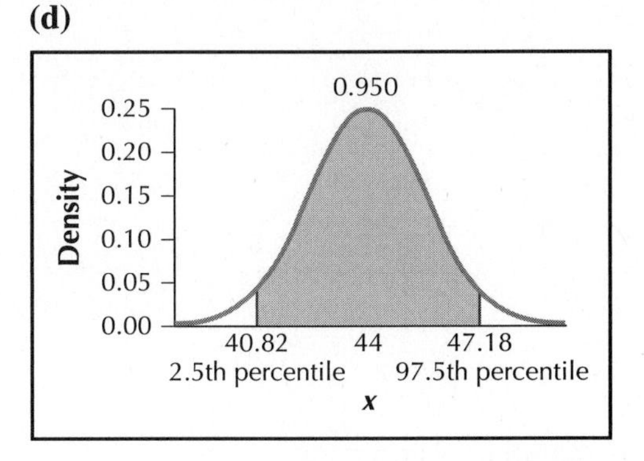

27. **(a)** $\mu_{\hat{p}} = p = 0.1$ **(b)** $\sigma_{\hat{p}} = \sqrt{\frac{p(1-p)}{n}} = \sqrt{\frac{0.1(1-0.1)}{50}} \approx 0.0424$ **(c)** We have $np = (50)(0.1) = 5 \geq 5$ and $n(1-p) = (50)(1-0.1) = 45 \geq 5$, so the sampling distribution of $\hat{p}$ is approximately normal.

29. $n_1 = \frac{5}{p} = \frac{5}{0.9} \approx 5.56$ and $n_2 = \frac{5}{1-p} = \frac{5}{1-0.9} = 50$. Therefore, $n^* = 50$ is the minimum sample size that produces a sampling distribution of $\hat{p}$ that is approximately normal.

31. $\mu_{\hat{p}} = p = 0.1$ and $\sigma_{\hat{p}} = \sqrt{\frac{p(1-p)}{n}} = \sqrt{\frac{0.1(1-0.1)}{144}} = 0.025$

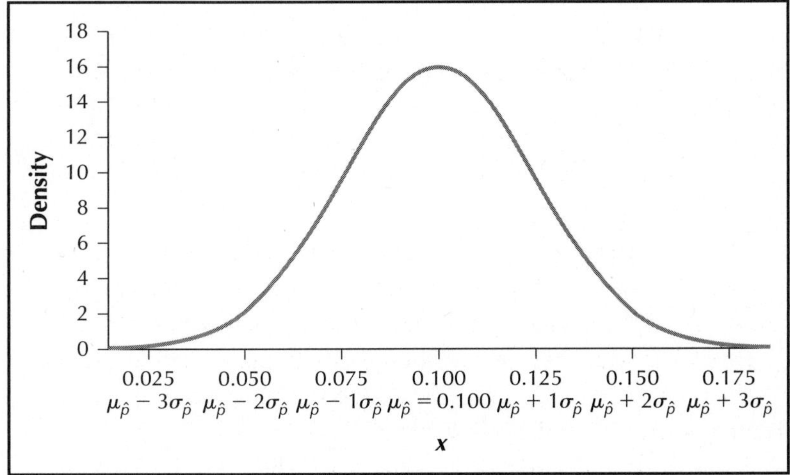

33. We have $np = (50)(0.1) = 5 \geq 5$ and $n(1-p) = (50)(1-0.1) = 45 \geq 5$, so the sampling distribution of $\hat{p}$ is approximately normal.

$$\mu_{\hat{p}} = p = 0.1 \text{ and } \sigma_{\hat{p}} = \sqrt{\frac{p(1-p)}{n}} = \sqrt{\frac{0.1(1-0.1)}{50}} \approx 0.0424.$$

Thus,

$$P(\hat{p} < 0.12) = P\left(\frac{\hat{p}-p}{\sigma_{\hat{p}}} < \frac{0.12-p}{\sigma_{\hat{p}}}\right) = P\left(\frac{\hat{p}-0.1}{0.0424} < \frac{0.12-0.1}{0.0424}\right) = P(Z < 0.47) = 0.6808.$$

35. We have $np = (625)(0.02) = 12.5 \geq 5$ and $n(1-p) = (625)(1-0.02) = 612.5 \geq 5$, so the sampling distribution of $\hat{p}$ is approximately normal. $\mu_{\hat{p}} = p = 0.02$ and $\sigma_{\hat{p}} = \sqrt{\frac{p(1-p)}{n}} = \sqrt{\frac{0.02(1-0.02)}{625}} = 0.0056$. The value of $\hat{p}$ that is smaller than 75% of all values of $\hat{p}$ is the value of $\hat{p}$ that is larger than $100 - 75 = 25\%$ of all values of $\hat{p}$.

From Table C, the value of Z that has an area of 0.2500 to the left of it is $Z = -0.67$. Thus, the 25th percentile is $Z \cdot \sigma_{\hat{p}} + \mu_{\hat{p}} = (-0.67)(0.0056) + 0.02 = 0.0162$.

37. **(a)** We have $np = (120)(0.11) = 13.2 \geq 5$ and $n(1-p) = (120)(1-0.11) = 106.8 \geq 5$, so the sampling distribution of $\hat{p}$ is approximately normal.

(b) $\mu_{\hat{p}} = p = 0.11$ and $\sigma_{\hat{p}} = \sqrt{\frac{p(1-p)}{n}} = \sqrt{\frac{0.11(1-0.11)}{120}} \approx 0.0286$

(c) $P(X < 10) = P\left(\frac{X}{n} < \frac{10}{120}\right) = P(\hat{p} < 0.0833) = P\left(\frac{\hat{p}-p}{\sigma_{\hat{p}}} < \frac{0.0833-p}{\sigma_{\hat{p}}}\right) = P\left(\frac{\hat{p}-0.11}{0.0286} < \frac{0.0833-0.11}{0.0286}\right)$
$= P(Z < -0.93) = 0.1762$

Chapter 7 Quiz

1. True.

2. False. Increasing the sample size will decrease the variability of the sampling distribution.

3. False. Compared to a symmetric distribution, a skewed distribution would generate means that behave *less* nicely, so the approximation to normality takes effect for *larger* samples.

4. Sampling error.

5. 30

6. Approximately normal.

7. Normal probability plot.

8. No.

9. $np \geq 5$ and $n(1-p) \geq 5$

10. Since the sample size of $n = 100$ is large ($n \geq 30$), Case 2 applies, so the sampling distribution of $\bar{x}$ is approximately normal.

(a) $P(\bar{x} < 38) = P\left(\frac{\bar{x}-40}{20/\sqrt{100}} < \frac{38-40}{20/\sqrt{100}}\right) = P(Z < -1.00) = 0.1587$

(b) $P(36.08 < \bar{x} < 43.92) = P\left(\frac{36.08-40}{20/\sqrt{100}} < Z < \frac{43.92-40}{20/\sqrt{100}}\right) = P(-1.96 < Z < 1.96)$

$$= 0.9750 - 0.0250 = 0.9500$$

(c) $P(\bar{x} > 42.5) = P\left(\frac{\bar{x}-40}{20/\sqrt{100}} > \frac{42.5-40}{20/\sqrt{100}}\right) = P(Z > 1.25) = 1 - 0.8944 = 0.1056$

11. **(a)** From Table C, the value of Z with an area of 0.9950 to the left of it is $Z = 2.58$. Thus, the 99.5th percentile is $Z \cdot \frac{\sigma}{\sqrt{n}} + \mu = 2.58 \cdot \frac{20}{\sqrt{100}} + 40 = 45.16$ grams.

(b) From Table C, the value of Z with an area of 0.0050 to the left of it is $Z = -2.58$. Thus, the 0.5th percentile is $Z \cdot \frac{\sigma}{\sqrt{n}} + \mu = -2.58 \cdot \frac{20}{\sqrt{100}} + 40 = 34.84$ grams. **(c)** 34.84 grams and 45.16 grams

12. **(a)** $P(\bar{x} > 68.6) = P\left(\frac{\bar{x}-68}{3/\sqrt{100}} > \frac{68.6-68}{3/\sqrt{100}}\right) = P(Z > 2.00) = 1 - 0.9772 = 0.0228$

(b) $P(\bar{x} < 67.4) = P\left(\frac{\bar{x}-68}{3/\sqrt{100}} < \frac{67.4-68}{3/\sqrt{100}}\right) = P(Z < -2.00) = 0.0228$

(c) $P(67.4 < \bar{x} < 68.6) = P\left(\frac{67.4-68}{3/\sqrt{100}} < Z < \frac{68.6-68}{3/\sqrt{100}}\right) =$
$P(-2.00 < Z < 2.00) = 0.9772 - 0.0228 = 0.9544$

13. **(a)** From Table C, the value of Z with an area of 0.9950 to the left of it is $Z = 2.58$.
Thus, the 99.5th percentile is $Z \cdot \frac{\sigma}{\sqrt{n}} + \mu = 2.58 \cdot \frac{3}{\sqrt{100}} + 68 = 68.77$ inches.
(b) From Table C, the value of Z with an area of 0.0050 to the left of it is $Z = -2.58$.
Thus, the 0.5th percentile is $Z \cdot \frac{\sigma}{\sqrt{n}} + \mu = -2.58 \cdot \frac{20}{\sqrt{100}} + 68 = 67.23$ inches.
(c) 67.23 inches and 68.77 inches

14. Since scores on a psychological test are not normally distributed and the sample size is small, Case 3 applies, so the sampling distribution of $\bar{x}$ is unknown.

15. Since scores on a psychological test are normally distributed, Case 1 applies, so the sampling distribution of $\bar{x}$ is normal.

16. **(a)** $\mu_{\bar{x}} = \mu = 100$ **(b)** $\sigma_{\bar{x}} = \frac{\sigma}{\sqrt{n}} = \frac{15}{\sqrt{25}} = 3$ **(c)** Since scores on a psychological test are not normally distributed and the sample size of $n = 25$ is small ($n < 30$), Case 3 applies, so the sampling distribution of $\bar{x}$ is unknown.

17. **(a)** $\mu_{\bar{x}} = \mu = 100$ **(b)** $\sigma_{\bar{x}} = \frac{\sigma}{\sqrt{n}} = \frac{15}{\sqrt{25}} = 3$ **(c)** Since scores on a psychological test are normally distributed, Case 1 applies, so the sampling distribution of $\bar{x}$ is normal.

18. Since scores on a psychological test are not normally distributed and the sample size of $n = 25$ is small ($n < 30$), Case 3 applies, so the sampling distribution of $\bar{x}$ is unknown. Therefore, $P(94 < \bar{x} < 103)$ can't be found.

19. Since scores on a psychological test are normally distributed, Case 1 applies, so the sampling distribution of $\bar{x}$ is normal. Thus,

$$P(94 < \bar{x} < 103) = P\left(\frac{94 - 100}{15/\sqrt{25}} < Z < \frac{103 - 100}{15/\sqrt{25}}\right) = P(-2.00 < Z < 1.00) = 0.8413 - 0.0228 = 0.8185.$$

20. Since scores on a psychological test are not normally distributed and the sample size of $n = 25$ is small ($n < 30$), Case 3 applies, so the sampling distribution of $\bar{x}$ is unknown. Therefore, the 50th percentile can't be found.

21. Since scores on a psychological test are normally distributed, Case 1 applies, so the sampling distribution of $\bar{x}$ is normal. Thus, the 50th percentile is the mean, which is 100.

22. **(a)** Since the distribution is unknown, Case 1 does not apply. **(b)** Since the sample size of $n = 15$ is small ($n < 30$), Case 2 does not apply.

23. **(a)** Since the distribution is unknown, Case 1 does not apply. **(b)** Since the sample size of $n = 30$ is large ($n \geq 30$), Case 2 applies. Thus,

$$P(\bar{x} < 100) = P\left(\frac{\bar{x} - \mu}{\sigma/\sqrt{n}} < \frac{100 - \mu}{\sigma/\sqrt{n}}\right) = P\left(\frac{\bar{x} - 124}{49/\sqrt{36}} < \frac{100 - 124}{49/\sqrt{36}}\right) = P(Z < -2.94) = 0.0016.$$

24. **(a)** $\mu_{\hat{p}} = p = 0.99$ **(b)** $\sigma_{\hat{p}} = \sqrt{\frac{p(1-p)}{n}} = \sqrt{\frac{0.99(1 - 0.99)}{100}} \approx 0.0099$

(c) We have $np = (100)(0.99) = 99 \geq 5$ but $n(1-p) = (100)(1 - 0.99) = 1 < 5$, so the sampling distribution of $\hat{p}$ is unknown.

25. **(a)** $\mu_{\hat{p}} = p = 0.99$ **(b)** $\sigma_{\hat{p}} = \sqrt{\frac{p(1-p)}{n}} = \sqrt{\frac{0.99(1 - 0.99)}{500}} \approx 0.0044$ **(c)** We have $np = (500)(0.99) = 495 \geq 5$ and $n(1-p) = (500)(1 - 0.99) = 5 \geq 5$, so the sampling distribution of $\hat{p}$ is approximately normal.

26. $n_1 = \frac{5}{p} = \frac{5}{0.95} \approx 5.26$ and $n_2 = \frac{5}{1-p} = \frac{5}{1 - 0.95} = 100$. Therefore, $n^* = 100$ is the minimum sample size that produces a sampling distribution of $\hat{p}$ that is approximately normal.

27. $n_1 = \frac{5}{p} = \frac{5}{0.99} \approx 5.05$ and $n_2 = \frac{5}{1-p} = \frac{5}{1 - 0.99} = 500$. Therefore, $n^* = 500$ is the minimum sample size that produces a sampling distribution of $\hat{p}$ that is approximately normal.

28. $\mu_{\hat{p}} = p = 0.98$ and $\sigma_{\hat{p}} = \sqrt{\dfrac{p(1-p)}{n}} = \sqrt{\dfrac{0.98(1-0.98)}{400}} = 0.007$

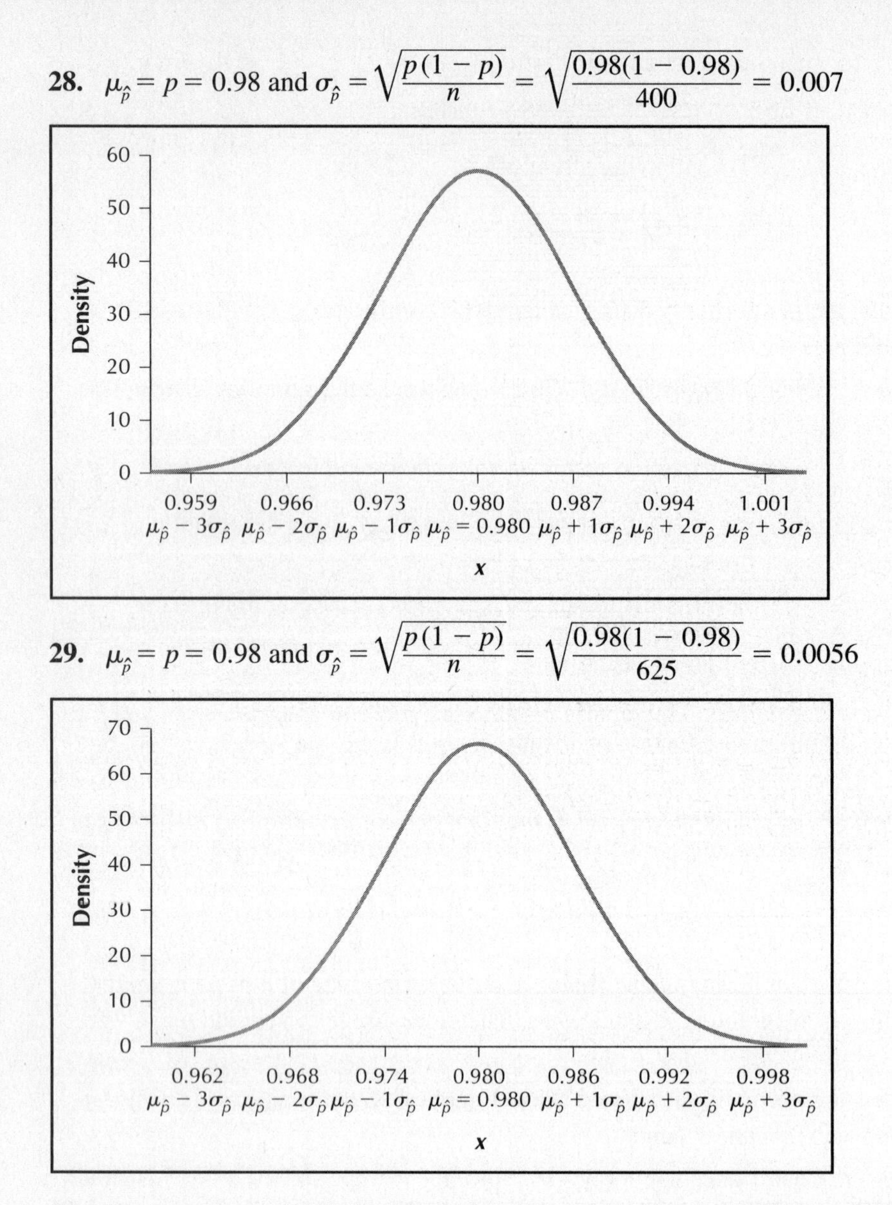

29. $\mu_{\hat{p}} = p = 0.98$ and $\sigma_{\hat{p}} = \sqrt{\dfrac{p(1-p)}{n}} = \sqrt{\dfrac{0.98(1-0.98)}{625}} = 0.0056$

30. We have $np = (100)(0.99) = 99 \geq 5$ but $n(1-p) = (100)(1-0.99) = 1 < 5$, so the sampling distribution of $\hat{p}$ is unknown. Thus, $P(0.985 < \hat{p} < 0.994)$ can't be found.

31. We have $np = (500)(0.99) = 495 \geq 5$ and $n(1-p) = (500)(1-0.99) = 5 \geq 5$, so the sampling distribution of $\hat{p}$ is approximately normal.

$$\mu_{\hat{p}} = p = 0.99 \text{ and } \sigma_{\hat{p}} = \sqrt{\frac{p(1-p)}{n}} = \sqrt{\frac{0.99(1-0.99)}{500}} \approx 0.0044.$$

Thus,

$$\begin{aligned} P(0.985 < \hat{p} < 0.994) &= P\left(\frac{0.985-p}{\sigma_{\hat{p}}} < \frac{\hat{p}-p}{\sigma_{\hat{p}}} < \frac{0.994-p}{\sigma_{\hat{p}}}\right) \\ &= P\left(\frac{0.985-0.99}{0.0044} < \frac{\hat{p}-0.99}{0.0044} < \frac{0.994-0.99}{0.0044}\right) \\ &= P(-1.14 < Z < 0.91) = 0.8186 - 0.1271 = 0.6915. \end{aligned}$$

32. We have $np = (400)(0.98) = 392 \geq 5$ and $n(1-p) = (400)(1-0.98) = 8 \geq 5$, so the sampling distribution of $\hat{p}$ is approximately normal. Thus, the 50th percentile is the mean, which is $\mu_{\hat{p}} = p = 0.98$.

33. We have $n\,p = (625)\,(0.98) = 612.5 \geq 5$ and $n\,(1 - p) = (625)\,(1 - 0.98) = 12.5 \geq 5$, so the sampling distribution of $\hat{p}$ is approximately normal. Thus, the 50th percentile is the mean, which is $\mu_{\hat{p}} = p = 0.98$.

34. **(a)** We have $n\,p = (100)\,(0.12) = 12 \geq 5$ and $n\,(1 - p) = (100)\,(1 - 0.12) = 88 \geq 5$, so the sampling distribution of $\hat{p}$ is approximately normal.

(b) $\mu_{\hat{p}} = p = 0.12$ and $\sigma_{\hat{p}} = \sqrt{\dfrac{p(1-p)}{n}} = \sqrt{\dfrac{0.12(1-0.12)}{100}} \approx 0.0325$

(c) $P(X > 10) = P\left(\dfrac{X}{n} > \dfrac{10}{100}\right) = P(\hat{p} > 0.1) = P\left(\dfrac{\hat{p}-p}{\sigma_{\hat{p}}} > \dfrac{0.1-p}{\sigma_{\hat{p}}}\right) = P\left(\dfrac{\hat{p}-0.12}{0.0325} > \dfrac{0.1-0.12}{0.0325}\right)$

$= P(Z > -0.62) = 1 - 0.2676 = 0.7324$

35. **(a)** We have $n\,p = (400)\,(0.17) = 68 \geq 5$ and $n\,(1 - p) = (400)\,(1 - 0.17) = 332 \geq 5$, so the sampling distribution of $\hat{p}$ is approximately normal.

(b) $\mu_{\hat{p}} = p = 0.17$ and $\sigma_{\hat{p}} = \sqrt{\dfrac{p(1-p)}{n}} = \sqrt{\dfrac{0.17(1-0.17)}{400}} \approx 0.0188$

(c) $P(X > 175) = P\left(\dfrac{X}{n} > \dfrac{175}{400}\right) = P(\hat{p} > 0.4375) = P\left(\dfrac{\hat{p}-p}{\sigma_{\hat{p}}} > \dfrac{0.4375-p}{\sigma_{\hat{p}}}\right)$

$= P\left(\dfrac{\hat{p}-0.17}{0.0188} > \dfrac{0.4375-0.17}{0.0188}\right) = P(Z > 14.23) \approx 0$

Chapter 8

Confidence Intervals

Section 8.1

1. A *confidence interval estimate* of a parameter consists of an interval of numbers generated by a point estimate, together with an associated *confidence level* specifying the probability that the interval contains the parameter.

3. **(a)** As the confidence level increases, $Z_{\alpha/2}$ increases.
(b) Since the confidence level is $(1 - \alpha) \times 100\%$, as the confidence level increases, $1 - \alpha$ increases. Thus, α and $\alpha/2$ will decrease. Since $\alpha/2$ is the area underneath the standard normal curve to the right of $Z_{\alpha/2}$, a decrease in $\alpha/2$ will result in an increase in $Z_{\alpha/2}$. For a 95% confidence interval, $Z_{\alpha/2} = 1.96$.

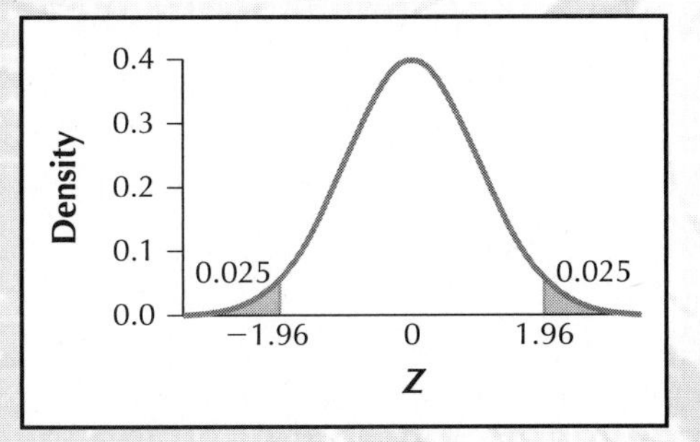

For a 99% confidence interval, $Z_{\alpha/2} = 2.576$.

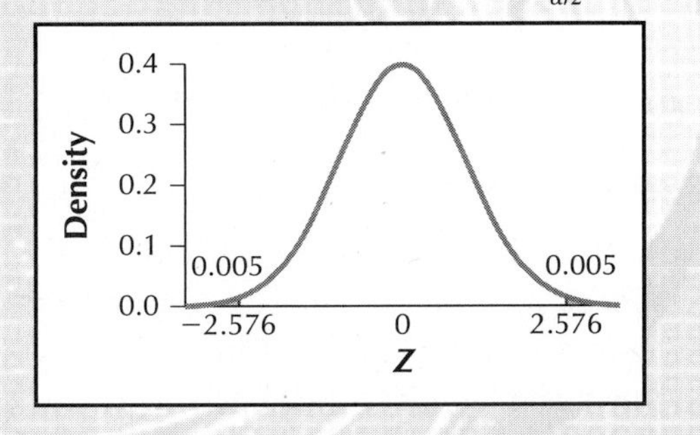

5. Shorter, tighter confidence intervals are better since the maximum difference between the sample mean and the true mean is reduced. Confidence intervals that are too wide are ineffectual and often useless.

7. A higher confidence level means that we will be more confident that our interval will contain the true value of the parameter. But a higher confidence level for the same sample data will result in a wider, less precise confidence interval.

9. **(a)** Since σ is unknown, we may not use the Z interval. **(b)** Since the original population is normal and σ is known, Case 1 applies, so we may use the Z interval. **(c)** Since the sample size is large ($n \geq 30$), Case 2 applies, so we may use the Z interval.

11. $Z_{\alpha/2} = 2.576$

13. $Z_{\alpha/2} = 2.576$

15. $Z_{\alpha/2} = 1.645$

17. $Z_{\alpha/2} = 1.645$

19. **(a)** Since the sample size of $n = 64$ is large ($n \geq 30$) and σ is known, Case 2 applies and we may use the Z interval. **(b)** $\sigma/\sqrt{n} = 4/\sqrt{64} = 0.5$ **(c)** $Z_{\alpha/2} = 1.96$ **(d)** The margin of error is $E = Z_{\alpha/2}(\sigma/\sqrt{n}) = (1.96)(0.5) = 0.98$. We may estimate μ to within 0.98 with 95% confidence. **(e)** $\bar{x} \pm Z_{\alpha/2}(\sigma/\sqrt{n}) = 10 \pm 0.98 = (9.02, 10.98)$. We are 95% confident that the true mean μ lies between 9.02 and 10.98.

21. **(a)** Since the original population is normal and σ is known, Case 1 applies and we may use the Z interval. **(b)** $\sigma/\sqrt{n} = 2/\sqrt{16} = 0.5$ **(c)** $Z_{\alpha/2} = 1.96$ **(d)** The margin of error is $E = Z_{\alpha/2}(\sigma/\sqrt{n}) = (1.96)(0.5) = 0.98$. We may estimate μ to within 0.98 with 95% confidence. **(e)** $\bar{x} \pm Z_{\alpha/2}(\sigma/\sqrt{n}) = 20 \pm 0.98 = (34.02, 35.98)$. We are 95% confident that the true mean μ lies between 34.02 and 35.98.

23. **(a)** Since the sample size of $n = 81$ is large ($n \geq 30$) and σ is known, Case 2 applies and we may use the Z interval. **(b)** $\sigma/\sqrt{n} = 18/\sqrt{81} = 2$ **(c)** $Z_{\alpha/2} = 1.96$ **(d)** The margin of error is $E = Z_{\alpha/2}(\sigma/\sqrt{n}) = (1.96)(2) = 3.92$. We may estimate μ to within 3.92 with 95% confidence. **(e)** $\bar{x} \pm Z_{\alpha/2}(\sigma/\sqrt{n}) = 100 \pm 3.92 = (96.08, 103.92)$. We are 95% confident that the true mean μ lies between 96.08 and 103.92.

25. **(a)** $\bar{x} \pm Z_{\alpha/2}(\sigma/\sqrt{n}) = 10 \pm 1.645\,(2/\sqrt{25}) = 10 \pm 0.658 = (9.342, 10.658)$. We are 90% confident that the true mean μ lies between 9.342 and 10.658. **(b)** $\bar{x} \pm Z_{\alpha/2}(\sigma/\sqrt{n}) = 10 \pm 1.96(2/\sqrt{25}) = 10 \pm 0.784 = (9.216, 10.784)$. We are 95% confident that the true mean μ lies between 9.216 and 10.784. **(c)** $\bar{x} \pm Z_{\alpha/2}(\sigma/\sqrt{n}) = 10 \pm 2.576(2/\sqrt{25}) = 10 \pm 1.0304 = (8.9696, 11.0304)$. We are 99% confident that the true mean μ lies between 8.9696 and 11.0304. **(d)** The confidence interval for a given sample size gets wider as the confidence level increases. **(e)** Since the original population is normal and σ is known, Case 1 applies and we may use the Z interval.

27. **(a)** $\bar{x} \pm Z_{\alpha/2}(\sigma/\sqrt{n}) = 90 \pm 1.96(10/\sqrt{25}) = 90 \pm 3.92 = (86.08, 93.92)$. We are 95% confident that the true mean μ lies between 86.08 and 93.92. **(b)** $\bar{x} \pm Z_{\alpha/2}(\sigma/\sqrt{n}) = 90 \pm 1.96(10/\sqrt{100}) = 90 \pm 1.96 = (88.04, 91.96)$. We are 95% confident that the true mean μ lies between 88.04 and 91.96. **(c)** $\bar{x} \pm Z_{\alpha/2}(\sigma/\sqrt{n}) = 90 \pm 1.96(10/\sqrt{400}) = 90 \pm 0.98 = (89.02, 90.98)$. We are 95% confident that the true mean μ lies between 89.02 and 90.98. **(d)** For a given confidence level, as the sample size increases, the width of the confidence interval decreases. **(e)** Since the original population is normal and σ is known, Case 1 applies and we may use the Z interval.

29. **(a)** $\bar{x} = 7$ ounces **(b)** $\sigma/\sqrt{n} = 1/\sqrt{36} \approx 0.17$ ounces **(c)** $Z_{\alpha/2} = 1.96$ **(d)** The margin of error is $E = Z_{\alpha/2}(\sigma/\sqrt{n}) = (1.96)(1/\sqrt{36}) \approx 0.33$ ounces. We may estimate μ, the mean amount of soda dispensed, to within 0.33 ounces with 95% confidence. **(e)** $\bar{x} \pm Z_{\alpha/2}(\sigma/\sqrt{n}) = 7 \pm 0.33 = (6.67, 7.33)$. We are 95% confident that the true mean amount of soda dispensed μ lies between 6.67 ounces and 7.33 ounces.

31. **(a)** $\bar{x} = 2600$ people **(b)** $\sigma/\sqrt{n} = 1000/\sqrt{50} \approx 141.42$ people **(c)** $Z_{\alpha/2} = 2.576$ **(d)** The margin of error is $E = Z_{\alpha/2}(\sigma/\sqrt{n}) = (2.576)(1000/\sqrt{50}) \approx 364.30$ people. We may estimate μ, the mean number of members in a Roman Catholic church parish, to within 364.30 people with 99% confidence. **(e)** $\bar{x} \pm Z_{\alpha/2}(\sigma/\sqrt{n}) = 2600 \pm 364.30 = (2235.70, 2964.30)$. We are 99% confident that the true mean number of members in a Roman Catholic church parish μ lies between 2235.70 people and 2964.30 people.

33. **(a)** Since the distribution of the lengths of time that boys remain engaged with a science exhibit at a museum is unknown, Case 1 doesn't apply. Since the sample size of $n = 10$ is small ($n < 30$), Case 2 doesn't apply. Thus, we should not use a Z interval. **(b)** $\sigma/\sqrt{n} = 117/\sqrt{36} = 19.5$ seconds **(c)** $Z_{\alpha/2} = 1.96$ **(d)** The margin of error is $E = Z_{\alpha/2}(\sigma/\sqrt{n}) = (1.96)(117/\sqrt{36}) = 38.22$ seconds. We may estimate μ, the mean length of time that boys remain engaged with a science exhibit at a museum, to within 38.22 seconds with 95% confidence. **(e)** $\bar{x} \pm Z_{\alpha/2}(\sigma/\sqrt{n}) = 107 \pm 38.22 = (68.78, 145.22)$. We are 95% confident that the true mean length of time that boys remain engaged with a science exhibit at a museum μ lies between 68.78 seconds and 145.22 seconds.

35. **(a)** The margin of error is $E = Z_{\alpha/2}(\sigma/\sqrt{n}) = 1.96(3/\sqrt{30}) \approx 1.07$ miles. We may estimate μ, the mean commuting distance, to within 1.07 miles with 95% confidence.
(b) $\bar{x} = \frac{\Sigma x}{n} = \frac{298}{30} \approx 9.93$, $\bar{x} \pm Z_{\alpha/2}(\sigma/\sqrt{n}) = 9.93 \pm 1.07 = (8.86, 11.00)$. We are 95% confident that the true mean commuting distance μ lies between 8.86 miles and 11.00 miles.

37. **(a)** The margin of error is $E = Z_{\alpha/2}(\sigma/\sqrt{n}) = 1.96(20/\sqrt{112}) \approx 3.70$ ng/g. We may estimate μ, the mean concentration of the herbicide *dicamba* in Iowa homes, to within 3.70 ng/g with 95% confidence.
(b) $\bar{x} \pm Z_{\alpha/2}(\sigma/\sqrt{n}) = 180 \pm 3.70 = (176.30, 183.70)$. We are 95% confident that the true mean concentration of the herbicide *dicamba* in Iowa homes μ lies between 176.30 ng/g and 183.70 ng/g.

39. **(a)** The margin of error is $E = Z_{\alpha/2}(\sigma/\sqrt{n}) = 1.645(3/\sqrt{45}) \approx 0.74$ IQ points. We may estimate μ, the mean increase in IQ points for all children after listening to a Mozart piano sonata for about 10 minutes, to within 0.74 IQ points with 90% confidence. **(b)** $\bar{x} \pm Z_{\alpha/2}(\sigma/\sqrt{n}) = 8.5 \pm 0.74 = (7.76, 9.24)$. We are 90% confident that the true mean increase in IQ points in all children after listening to a Mozart piano sonata for about 10 minutes μ lies between 7.76 IQ points and 9.24 IQ points.

41. **(a)** $Z_{\alpha/2} = 1.96$ **(b)** $\bar{x}$ is the midpoint of the confidence interval. Thus, $\bar{x} = (12 + 276)/2 = 144$ fewer steps. **(c)** Since the length of the confidence interval is 2 E, the margin of error is one-half of the length of the confidence interval. Thus, $E = (276 - 12)/2 = 132$ fewer steps. Or, since the confidence interval is $\bar{x} \pm E$, the distance between $\bar{x}$ and either end of the confidence interval is the margin of error. Thus, $E = 276 - 144 = 144 - 12 = 132$ fewer steps. We may estimate the true mean number of fewer steps for each additional hour of television viewing within 132 fewer steps with 95% confidence. **(d)** Since $E = Z_{\alpha/2}(\sigma/\sqrt{n})$, $132 = 1.96(\sigma/\sqrt{100}) = 0.196\,\sigma$, so $\sigma = 132/0.196 \approx 673$ fewer steps.

43. **(a)** Since σ is a population characteristic, it stays constant and is unaffected by a decrease in confidence level. **(b)** The quantity $\sigma/\sqrt{n}$ is unaffected by a decrease in confidence level. **(c)** A decrease in the confidence level will result in a decrease in $Z_{\alpha/2}$. The width of the confidence interval is $2\,Z_{\alpha/2}(\sigma/\sqrt{n})$. Thus, a decrease in $Z_{\alpha/2}$ will result in a decrease in the width of the confidence interval. **(d)** The quantity $\bar{x}$ only depends on the sample taken, so it will remain unaffected by a decrease in confidence level. **(e)** A decrease in the confidence level will result in a decrease in $Z_{\alpha/2}$. Since the margin of error is $E = Z_{\alpha/2}(\sigma/\sqrt{n})$, a decrease in $Z_{\alpha/2}$ will result in a decrease in the margin of error.

45. **(a)** $\bar{x} = 6199$ small firms
(b)

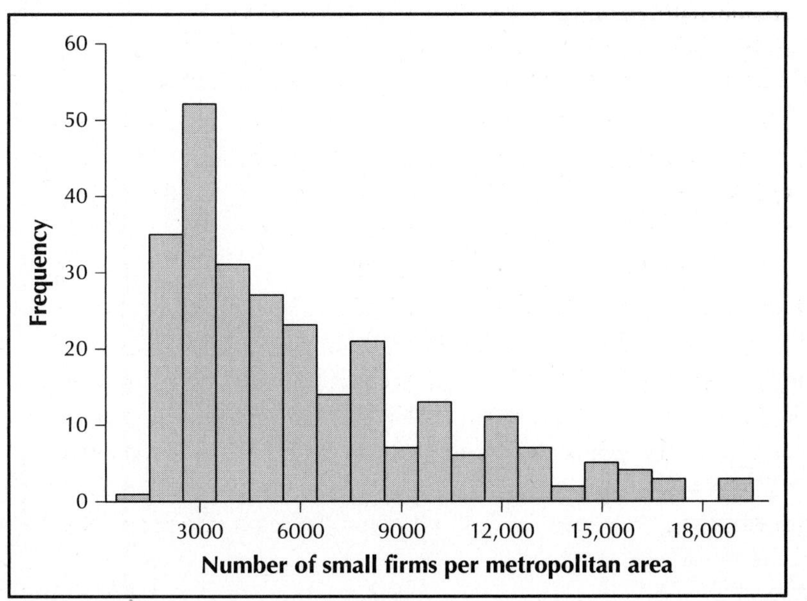

(c)

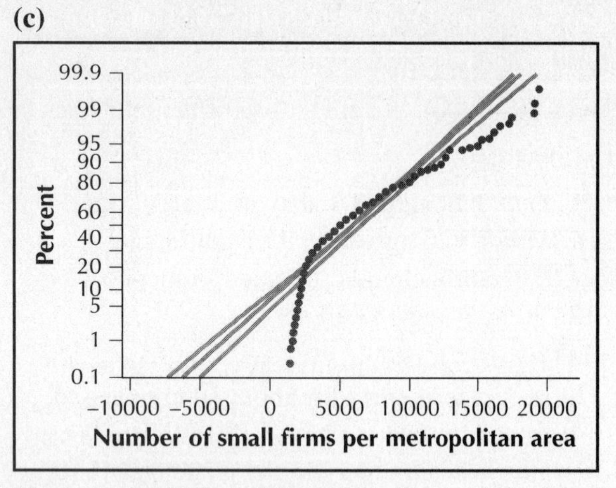

Since the majority of the points lie outside of the curved lines, the normality assumption is not valid.

(d) $\bar{x} \pm Z_{\alpha/2}(\sigma/\sqrt{n}) = 6199 \pm 1.96(25{,}000/\sqrt{265}) = 6199 \pm 3010 = (3189, 9209)$. We are 95% confident that the average number of small firms per metropolitan area lies between 3189 and 9209.

(e)

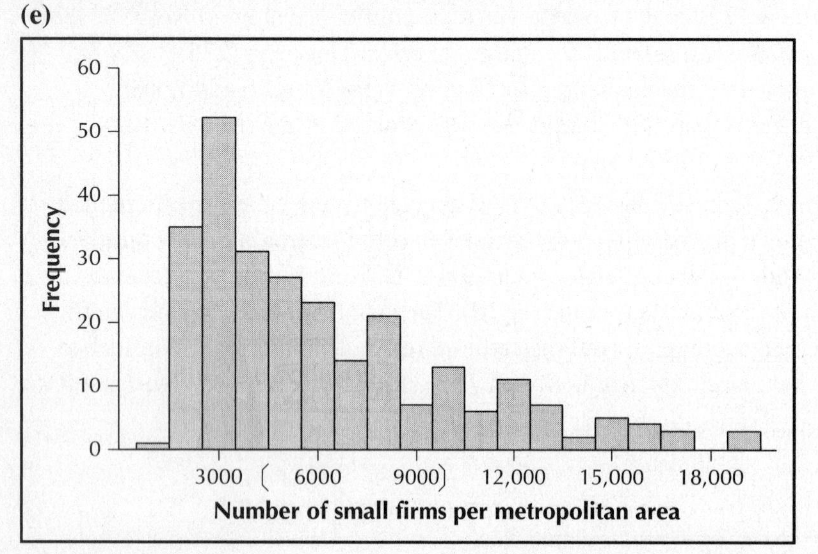

47. **(a)** Answers will vary but should be close to 90%. One possible answer:

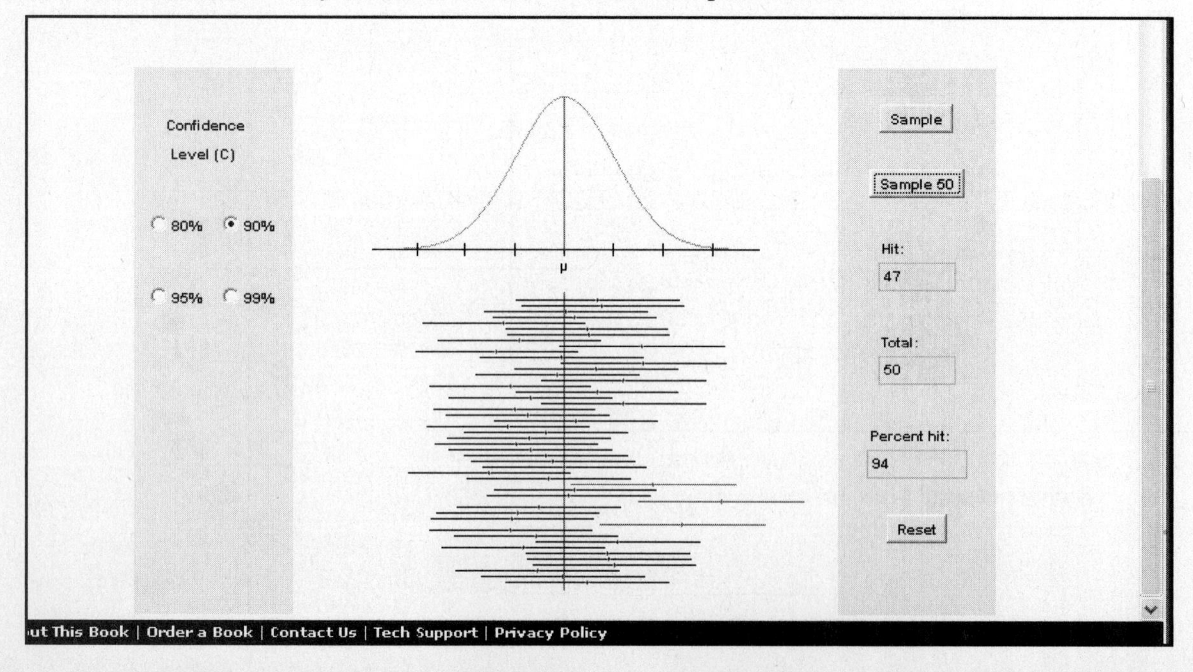

For this particular example, the percent hit is 94%.

(b) Answers will vary but should be close to 90%. One possible answer:

For this particular example, the percent hit is 90%.

(c) If we take sample after sample for a very long time, then in the long run, the proportion of intervals that will contain the parameter μ will equal 90%. One thousand confidence intervals is not enough to qualify for the long run.

49. The answers to the first part of this problem will vary.

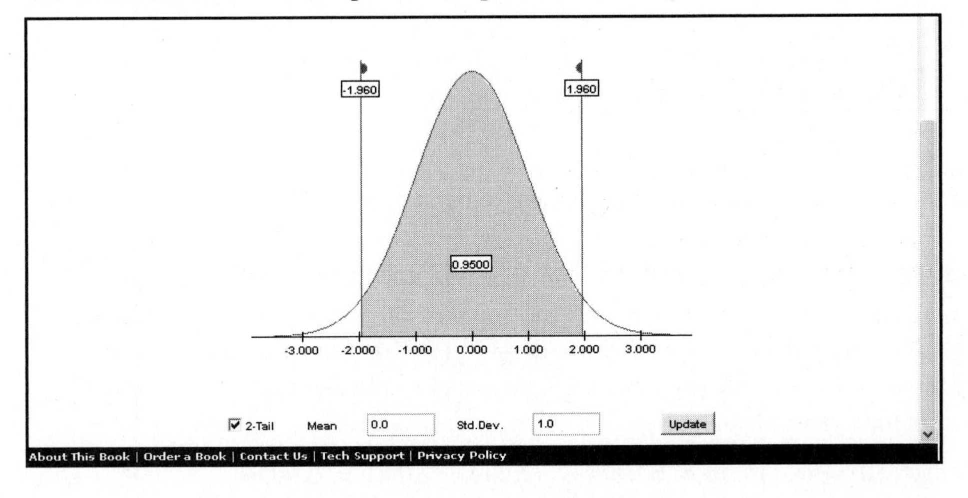

Section 8.2

1. In most real-world problems, the population standard deviation σ is unknown, so we may not use the Z interval.

3. As the sample size gets larger and larger, the t curve gets closer and closer to the Z curve.

5. **(a)** $\text{df} = n - 1 = 10 - 1 = 9$, $\alpha = 0.10$, $t_{\alpha/2} = 1.833$ **(b)** $\text{df} = n - 1 = 10 - 1 = 9$, $\alpha = 0.05$, $t_{\alpha/2} = 2.262$ **(c)** $\text{df} = n - 1 = 10 - 1 = 9$, $\alpha = 0.01$, $t_{\alpha/2} = 3.250$

7. **(a)** For a given sample size, the value of $t_{\alpha/2}$ increases as the confidence level increases. **(b)** $t_{\alpha/2} = 1.833$ for a 90% confidence interval with 9 degrees of freedom.

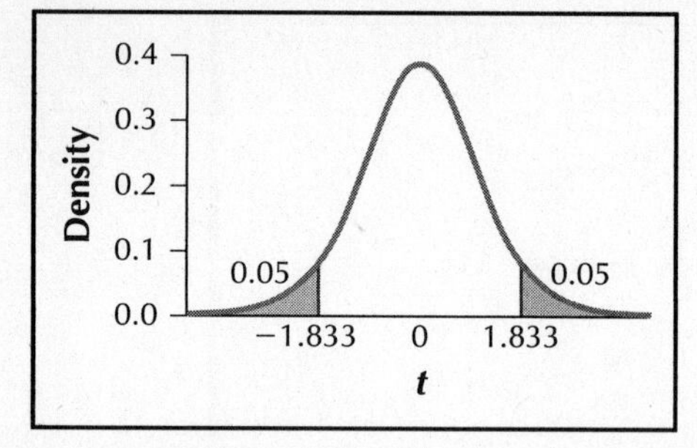

$t_{\alpha/2} = 2.262$ for a 95% confidence interval with 9 degrees of freedom.

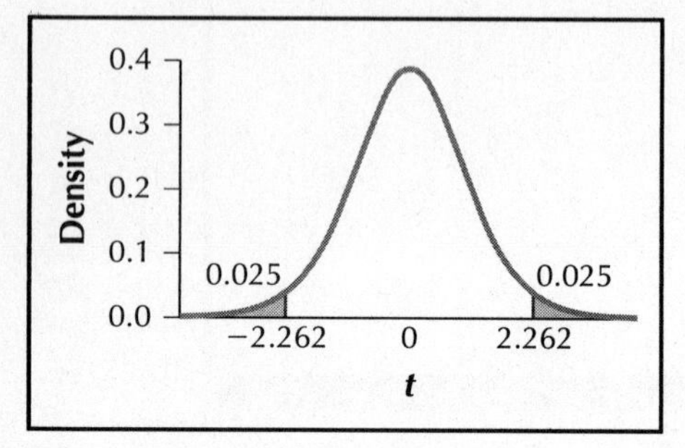

$t_{\alpha/2} = 3.250$ for a 99% confidence interval with 9 degrees of freedom.

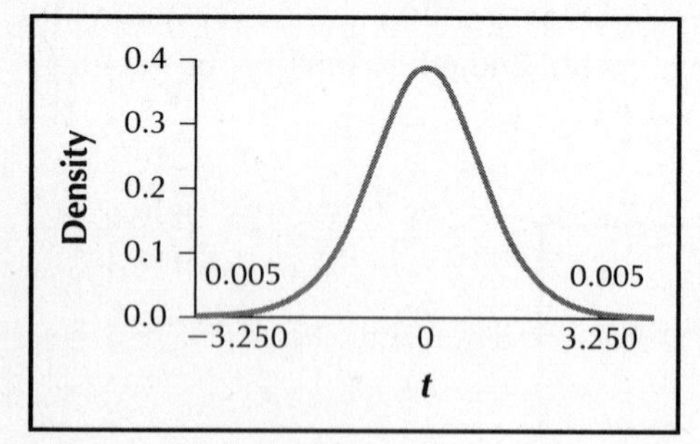

The larger the value of $1 - \alpha$ is, the larger the value of $t_{\alpha/2}$ will have to be in order to have an area of $1 - \alpha$ between $-t_{\alpha/2}$ and $t_{\alpha/2}$.

9. df $= n - 1 = 55 - 1 = 54$. The value of 54 for the df does not appear in the df column. The conservative approach would be to use the next row with df smaller than 54. That would be df = 50. Thus, the "conservative" $t_{\alpha/2}$ is 2.009. Alternatively, you could interpolate as follows. Since df = 54 is 4/10 of the distance between 50 and 60, we can estimate $t_{\alpha/2}$ by taking 4/10 of the distance between the associated t values for df = 50 and df = 60 and subtracting that result from the t value for df = 50:

$$\frac{4}{10}[(t_{\alpha/2} \text{ for df} = 50) - (t_{\alpha/2} \text{ for df} = 60)] = \frac{4}{10}(2.009 - 2.000) = 0.0036.$$

Thus, $t_{\alpha/2}$ for df = 54 would be $2.009 - 0.0036 = 2.0054$, using interpolation.

11. df $= n - 1 = 46 - 1 = 45$. The value of 45 for the df does not appear in the df column. The conservative approach would be to use the next row with df smaller than 45. That would be df = 40. Thus, the "conservative" $t_{\alpha/2}$ is 1.684. Alternatively, you could interpolate as follows. Since df = 40 is 5/10 of the distance between 40 and 50, we can estimate $t_{\alpha/2}$ by taking 5/10 of the distance between the associated t values for df = 40 and df = 50 and subtracting that result from the t value for df = 40:

$$\frac{5}{10}[(t_{\alpha/2} \text{ for df} = 40) - (t_{\alpha/2} \text{ for df} = 50)] = \frac{5}{10}(1.684 - 1.676) = 0.004.$$

Thus, $t_{\alpha/2}$ for df = 40 would be $1.684 - 0.004 = 1.680$, using interpolation.

13. **(a)** df $= n - 1 = 25 - 1 = 24$, $t_{\alpha/2} = 2.064$ **(b)** The margin of error $E = t_{\alpha/2} \cdot (s/\sqrt{n}) = 2.064(5/\sqrt{25}) = 2.064$. **(c)** $\bar{x} \pm t_{\alpha/2} \cdot (s/\sqrt{n}) = 10 \pm 2.064(5/\sqrt{25}) = 10 \pm 2.064 = (7.936, 12.064)$

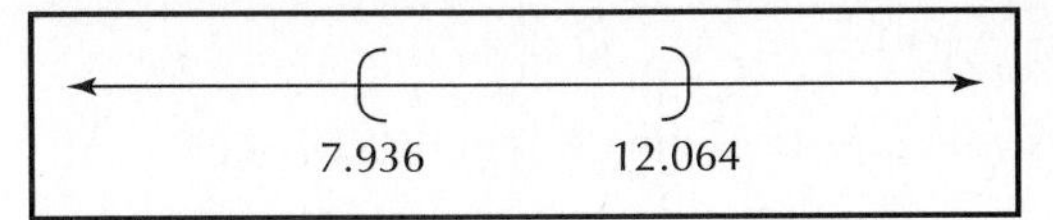

15. **(a)** df $= n - 1 = 4 - 1 = 3$, $t_{\alpha/2} = 3.182$ **(b)** The margin of error $E = t_{\alpha/2} \cdot (s/\sqrt{n}) = 3.182(6/\sqrt{4}) = 9.546$. **(c)** $\bar{x} \pm t_{\alpha/2} \cdot (s/\sqrt{n}) = 50 \pm 3.182(6/\sqrt{4}) = 50 \pm 9.546 = (40.454, 59.546)$

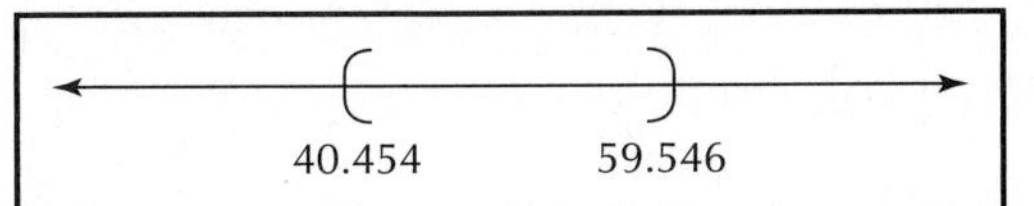

17. **(a)** df $= n - 1 = 9 - 1 = 8$, $t_{\alpha/2} = 1.860$ **(b)** The margin of error $E = t_{\alpha/2} \cdot (s/\sqrt{n}) = 1.860(6/\sqrt{9}) = 3.720$. **(c)** $\bar{x} \pm t_{\alpha/2} \cdot (s/\sqrt{n}) = -20 \pm 1.860(6/\sqrt{9}) = -20 \pm 3.720 = (-23.720, -16.280)$

−23.720 −16.280

19. **(a)** df $= n - 1 = 100 - 1 = 99$. The value of 99 for the df does not appear in the df column. The conservative approach would be to use the next row with df smaller than 99. That would be df $= 90$. Thus, the "conservative" $t_{\alpha/2}$ is 1.987. **(b)** The margin of error $E = t_{\alpha/2} \cdot (s/\sqrt{n}) = 1.987(10/\sqrt{100}) = 1.987$. **(c)** $\bar{x} \pm t_{\alpha/2} \cdot (s/\sqrt{n}) = 100 \pm 1.987(10/\sqrt{100}) = 100 \pm 1.987 = (98.013, 101.987)$

98.013 101.987

21. **(a)** df $= n - 1 = 64 - 1 = 63$. The value of 63 for the df does not appear in the df column. The conservative approach would be to use the next row with df smaller than 63. That would be df $= 60$. Thus, the "conservative" $t_{\alpha/2}$ is 2.660. **(b)** The margin of error $E = t_{\alpha/2} \cdot (s/\sqrt{n}) = 2.660(8/\sqrt{64}) = 2.660$. **(c)** $\bar{x} \pm t_{\alpha/2} \cdot (s/\sqrt{n}) = 35 \pm 2.660(8/\sqrt{64}) = 35 \pm 2.660 = (32.340, 37.660)$

32.340 37.660

23. **(a)** df $= n - 1 = 81 - 1 = 80$, $t_{\alpha/2} = 1.664$ **(b)** The margin of error $E = t_{\alpha/2} \cdot (s/\sqrt{n}) = 1.664(6/\sqrt{81}) \approx 1.1093$. **(c)** $\bar{x} \pm t_{\alpha/2} \cdot (s/\sqrt{n}) = -20 \pm 1.664(6/\sqrt{81}) = -20 \pm 1.1093 = (-21.1093, -18.8907)$

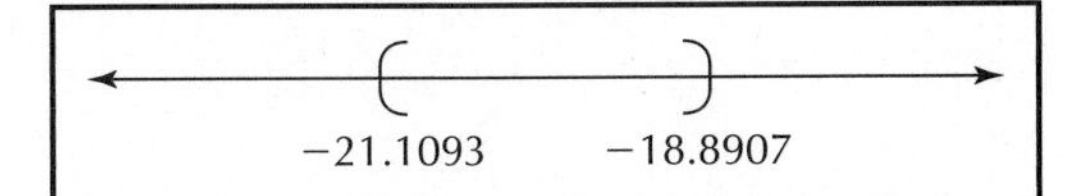

25. Since the distribution of the population is unknown, Case 1 does not apply. Since the sample size of $n = 25$ is small ($n < 30$), Case 2 does not apply. Thus, we cannot construct the indicated confidence interval.

27. Since the distribution of the population is normal, Case 1 applies, so we can use the t interval. df $= n - 1 = 225 - 1 = 224$. The value of 224 for the df does not appear in the df column. The conservative approach would to use the next row with df smaller than 224. That would be df $= 100$. Thus, the "conservative" $t_{\alpha/2}$ is 1.984. $\bar{x} \pm t_{\alpha/2} \cdot (s/\sqrt{n}) = 10 \pm 1.984(5/\sqrt{225}) = 10 \pm 0.6613 = (9.3387, 10.6613)$

29. Since the distribution of the population is unknown, Case 1 does not apply. Since the sample size of $n = 16$ is small ($n < 30$), Case 2 does not apply. Thus, we cannot construct the indicated confidence interval.

31. Since the sample size of $n = 36$ is large ($n \geq 30$), Case 2 applies, so we can use the t interval.

$$df = n - 1 = 36 - 1 = 35,\ t_{\alpha/2} = 2.030$$

$$\bar{x} \pm t_{\alpha/2} \cdot (s/\sqrt{n}) = 50 \pm 2.030(6/\sqrt{36}) = 50 \pm 2.030 = (47.97, 52.03)$$

33. **(a)** $df = n - 1 = 75 - 1 = 74$. The value of 74 for the df does not appear in the df column. The conservative approach would be to use the next row with df smaller than 74. That would be df = 70. Thus, the "conservative" $t_{\alpha/2}$ is 1.994. **(b)** The margin of error $E = t_{\alpha/2} \cdot (s/\sqrt{n}) = 1.994(30/\sqrt{75}) \approx \$6.91$. We can estimate μ, the true mean revenue collected from all parking meters, within \$6.91 with 95% confidence. **(c)** $\bar{x} \pm t_{\alpha/2} \cdot (s/\sqrt{n}) = 120 \pm 1.994(30/\sqrt{75}) = 120 \pm 6.91 = (113.09, 126.91)$. We are 95% confident that the true mean revenue collected from all parking meters μ lies between \$113.09 and \$126.91.

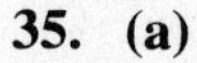
35. **(a)**

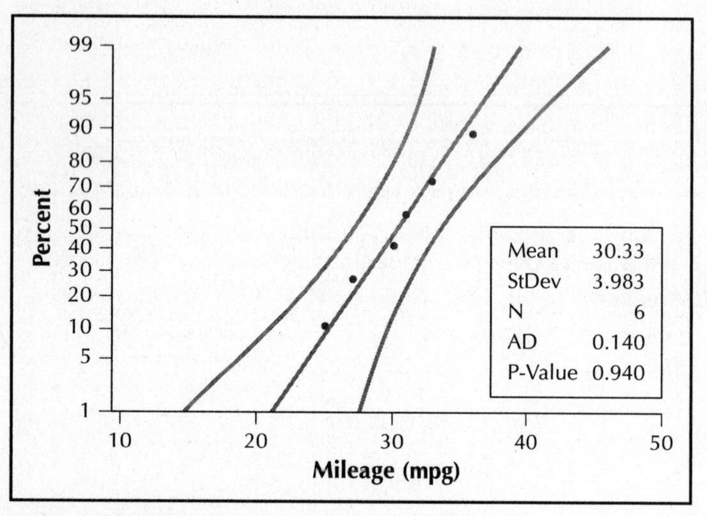

All of the data points lie between the curved lines. In fact, all of the points lie close to the center line. Thus, the distribution appears to be normal. **(b)** $df = n - 1 = 6 - 1 = 5,\ t_{\alpha/2} = 2.015$ **(c)** The margin of error $E = t_{\alpha/2} \cdot (s/\sqrt{n}) = 2.015(3.983/\sqrt{6}) \approx 3.276$ miles per gallon. We can estimate μ, the true mean city gas mileage for hybrid cars, within 3.276 miles per gallon with 90% confidence. **(d)** $\bar{x} \pm t_{\alpha/2} \cdot (s/\sqrt{n}) = 30.33 \pm 2.015(3.983/\sqrt{6}) = 30.33 \pm 3.276 = (27.054, 33.606)$. We are 90% confident that the true mean city gas mileage for hybrid cars μ lies between 27.054 miles per gallon and 33.606 miles per gallon.

37. **(a)**

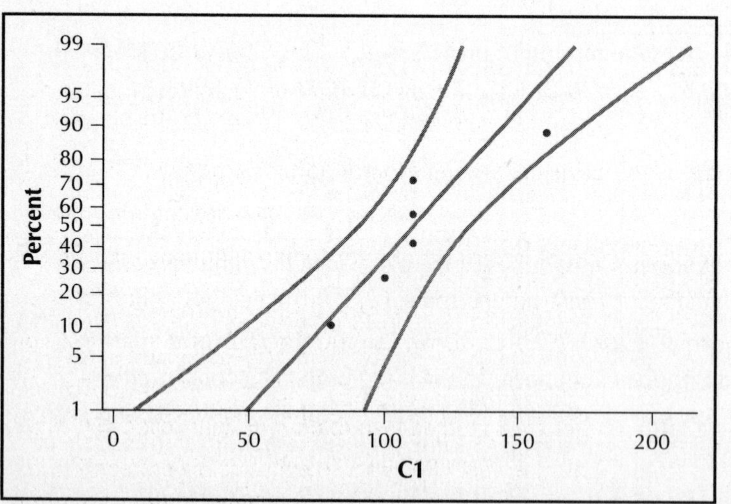

(b) There is no evidence that the distribution is not normal. **(c)** Since the data appear to be normal, Case 1 applies. Thus, a t interval can be used.

39. (a) df $= n - 1 = 30 - 1 = 29$, $t_{\alpha/2} = 2.045$

(b)

Descriptive Statistics: Distance

Variable	N	N*	Mean	SE Mean	StDev	Minimum	Q1	Median	Q3
Distance	30	0	9.933	0.538	2.947	4.000	7.000	10.000	12.000

Variable	Maximum
Distance	15.000

The margin of error is $E = t_{\alpha/2} \cdot (s/\sqrt{n}) = 2.045\,(2.947/\sqrt{30}) \approx 1.1003$ miles. We can estimate μ, the true mean commuting distance, within 1.1003 miles with 95% confidence. **(c)** $\bar{x} \pm t_{\alpha/2} \cdot (s/\sqrt{n}\,) = 9.3333 \pm 2.045\,(2.947/\sqrt{30}) = 9.3333 \pm 1.1003 = (8.2330, 10.4336)$. We are 95% confident that the true mean commuting distance lies between 8.2330 miles and 10.4336 miles.

41. (a) Z interval: $\bar{x} \pm Z_{\alpha/2}(\sigma/\sqrt{n}) = 42 \pm 2.576(12/\sqrt{10}) = 42 \pm 9.7752 = (32.2248, 51.7752)$

t interval: df $= n - 1 = 10 - 1 = 9$

$$\bar{x} \pm t_{\alpha/2} \cdot (s/\sqrt{n}\,) = 42 \pm 3.250(12/\sqrt{10}) = 42 \pm 12.3329 = (29.6671, 54.3329)$$

(b) You can either lower the confidence level or increase the sample size. Lowering the confidence level will give you a shorter confidence interval, but you won't be as confident that your confidence interval contains μ. Increasing the sample size will also give you a shorter confidence interval but it may be difficult to get data for more cities.

43. (a)

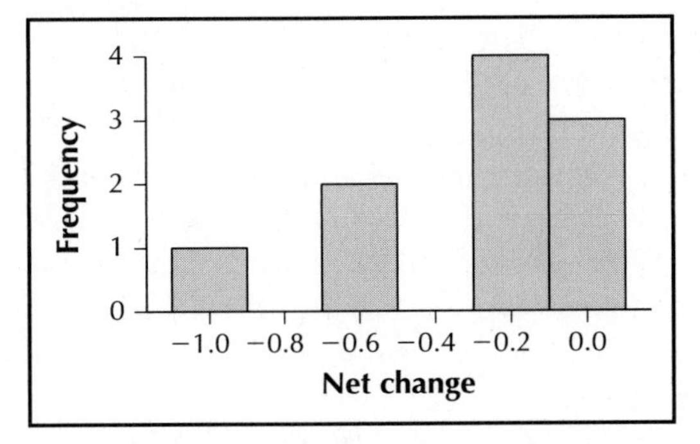

All points lie between the curved lines except one.

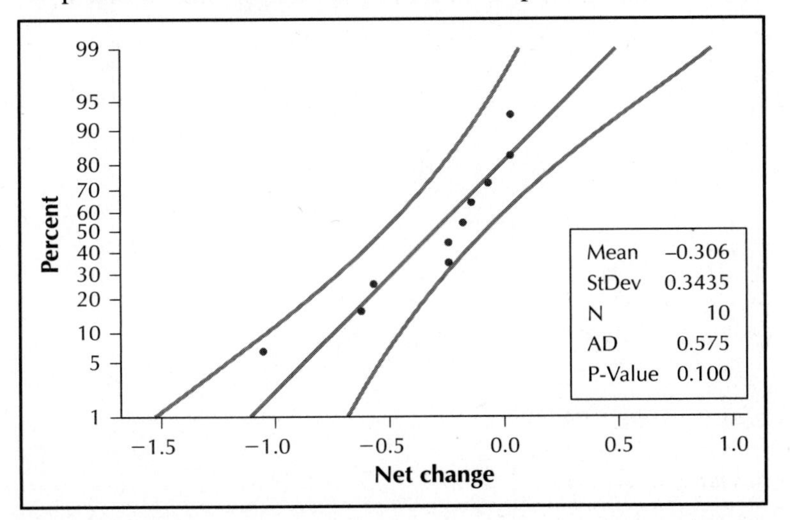

The histogram indicates that the data are left-skewed and therefore not normally distributed. The normal probability plot indicates that the data points all lie between the curved bounds but not along a straight line, so the normality assumption may not be valid. **(b)** $n = 10$, $\bar{x} = -0.306$, $s = 0.3435$ **(c)** df $= n - 1 = 10 - 1 = 9$, $t_{\alpha/2} = 2.262$ **(d)** $\bar{x} \pm t_{\alpha/2} \cdot (s/\sqrt{n}) = -0.306 \pm 2.262(0.3435/\sqrt{10}) = -0.306 \pm 0.2457 = (-0.5517, -0.0603)$ **(e)** The margin of error is $E = t_{\alpha/2} \cdot (s/\sqrt{n}) = 2.262(0.3435/\sqrt{10}) = 0.2457$. We can estimate μ, the true mean net change of all NYSE stocks, within 0.2457 with 95% confidence.

45. **(a)** Except for one outlier on the left, the data appear to be normally distributed.

(b)

Descriptive Statistics: Refusal rate

Variable	N	N*	Mean	SE Mean	StDev	Minimum	Q1	Median	Q3
Refusal rate	10	0	33.35	1.93	6.10	23.20	29.13	33.75	38.67

Variable	Maximum
Refusal rate	42.70

$$\text{df} = n - 1 = 10 - 1 = 9$$
$$\bar{x} \pm t_{\alpha/2} \cdot (s/\sqrt{n}) = 33.35 \pm 2.262(6.10/\sqrt{10})$$
$$= 33.35 \pm 4.3634 = (28.9866, 37.7134).$$

We are 95% confident that the true mean rate of minority mortgage lending refusals for all banks lies between 28.9866% and 37.7134%.

47. **(a)** Since the margin of error is $E = t_{\alpha/2}(s/\sqrt{n})$, a decrease in the confidence level will result in a decrease in $t_{\alpha/2}$, which will result in a decrease in the margin of error. **(b)** Since the width of the confidence interval is 2 E, a decrease in the confidence level will result in a decrease in E, which will result in a decrease in the width of the confidence interval.

49. **(a)** An increase in the sample size will result in an increase in df $= n - 1$, which will result in a decrease in $t_{\alpha/2}$. **(b)** Since the margin of error is $E = t_{\alpha/2}(s/\sqrt{n})$ and the sample size n occurs in the denominator, an increase in the sample size will result in a decrease in $t_{\alpha/2}$ and a decrease in the margin of error. **(c)** Since the width of the confidence interval is 2 E, an increase in the sample size will result in a decrease in E, which will result in a decrease in the width of the confidence interval.

Section 8.3

1. Since the purpose of a confidence interval for p is to estimate p, it would not make sense to construct a confidence interval for p if p is known unless there is some reason to suspect that the value of p has changed.

3. **(a)** We have $n \cdot \hat{p} = 25 \cdot 0.5 = 12.5 \geq 5$ and $n \cdot (1 - \hat{p}) = 25 \cdot (1 - 0.5) = 12.5 \geq 5$.

(b) $Z_{\alpha/2} = 1.96$

(c) $E = Z_{\alpha/2}\sqrt{\dfrac{\hat{p}(1-\hat{p})}{n}} = 1.96\sqrt{\dfrac{0.5(1-0.5)}{25}} = 0.196$

5. **(a)** We have $n \cdot \hat{p} = 100 \cdot 0.95 = 95 \geq 5$ and $n \cdot (1 - \hat{p}) = 100 \cdot (1 - 0.95) = 5 \geq 5$.

(b) $Z_{\alpha/2} = 2.576$

(c) $E = Z_{\alpha/2}\sqrt{\dfrac{\hat{p}(1-\hat{p})}{n}} = 2.576\sqrt{\dfrac{0.95(1-0.95)}{100}} \approx 0.0561$

7. **(a)** We have $n \cdot \hat{p} = 64 \cdot 0.4 = 25.6 \geq 5$ and $n \cdot (1 - \hat{p}) = 64 \cdot (1 - 0.4) = 38.4 \geq 5$.

(b) $Z_{\alpha/2} = 1.96$

(c) $E = Z_{\alpha/2}\sqrt{\dfrac{\hat{p}(1-\hat{p})}{n}} = 1.96\sqrt{\dfrac{0.4(1-0.4)}{64}} \approx 0.1200$

9. **(a)** We have $n \cdot \hat{p} = 49 \cdot 0.3 = 14.7 \geq 5$ and $n \cdot (1 - \hat{p}) = 49 \cdot (1 - 0.3) = 34.3 \geq 5$.

(b) $Z_{\alpha/2} = 1.645$

(c) $E = Z_{\alpha/2}\sqrt{\dfrac{\hat{p}(1-\hat{p})}{n}} = 1.645\sqrt{\dfrac{0.3(1-0.3)}{49}} \approx 0.1077$

11. **(a)** $\hat{p} \pm Z_{\alpha/2}\sqrt{\frac{\hat{p}(1-\hat{p})}{n}} = 0.5 \pm 1.96\sqrt{\frac{0.5(1-0.5)}{25}} = 0.5 \pm 0.196 = (0.304, 0.696)$

(b)

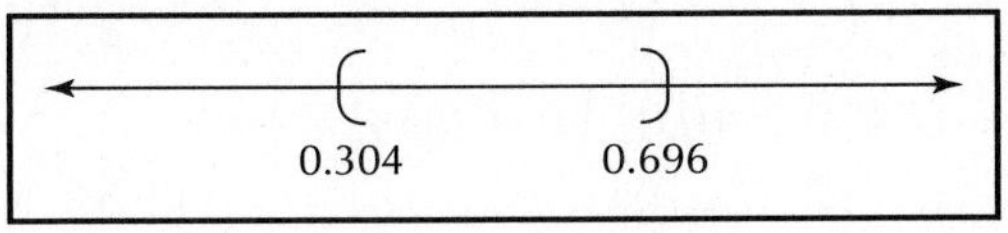

0.304 0.696

13. **(a)** $\hat{p} \pm Z_{\alpha/2}\sqrt{\frac{\hat{p}(1-\hat{p})}{n}} = 0.95 \pm 2.576\sqrt{\frac{0.95(1-0.95)}{100}} \approx 0.95 \pm 0.0561 = (0.8939, 1.0061)$

(b)

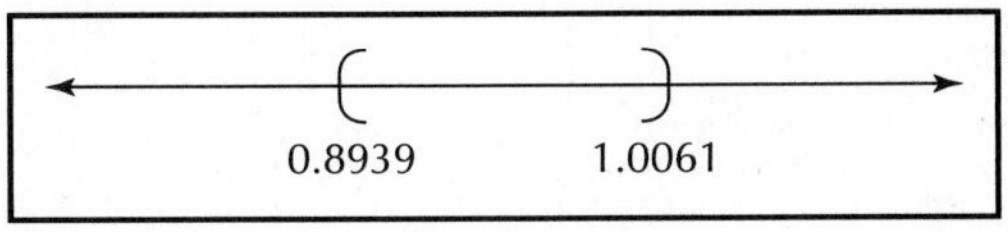

0.8939 1.0061

15. **(a)** $\hat{p} \pm Z_{\alpha/2}\sqrt{\frac{\hat{p}(1-\hat{p})}{n}} = 0.4 \pm 1.96\sqrt{\frac{0.4(1-0.4)}{64}} = 0.4 \pm 0.1200 = (0.280, 0.520)$

(b)

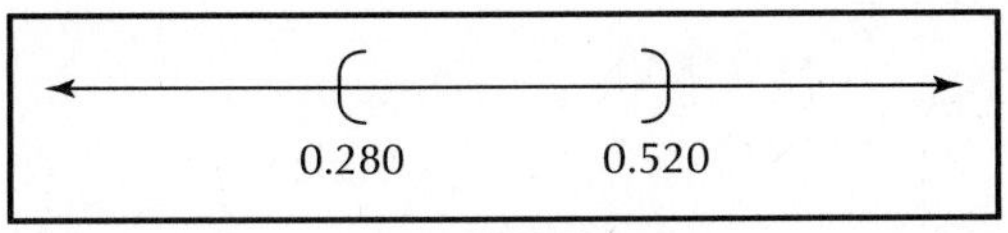

0.280 0.520

17. **(a)** $\hat{p} \pm Z_{\alpha/2}\sqrt{\frac{\hat{p}(1-\hat{p})}{n}} = 0.3 \pm 1.645\sqrt{\frac{0.3(1-0.3)}{49}} = 0.3 \pm 0.1077 = (0.1923, 0.4077)$

(b)

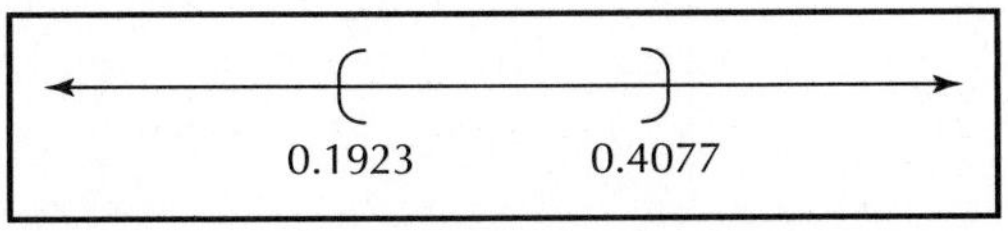

0.1923 0.4077

19. **(a)** $Z_{\alpha/2} = 1.96$ **(b)** We have $n\hat{p} = 121(0.1) = 12.1 \geq 5$ and $n(1-\hat{p}) = 121(1-0.1) = 108.9 \geq 5$, so we may use the Z interval for p.

(c) $E = Z_{\alpha/2}\sqrt{\frac{\hat{p}(1-\hat{p})}{n}} = 1.96\sqrt{\frac{0.1(1-0.1)}{121}} \approx 0.0535$

(d) $\hat{p} \pm Z_{\alpha/2}\sqrt{\frac{\hat{p}(1-\hat{p})}{n}} = 0.1 \pm 1.96\sqrt{\frac{0.1(1-0.1)}{121}} \approx 0.1 \pm 0.0535 = (0.0465, 0.1535)$

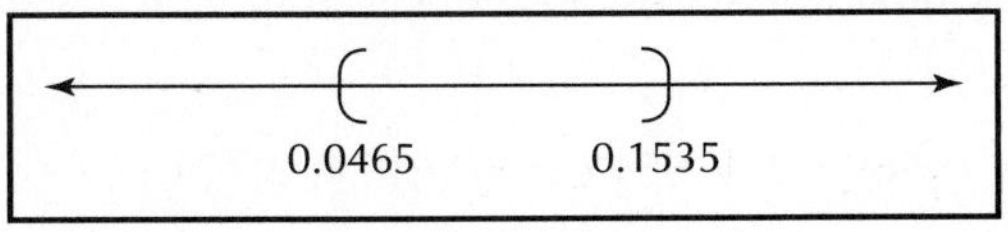

0.0465 0.1535

21. **(a)** $Z_{\alpha/2} = 1.645$ **(b)** We have $n\hat{p} = 16(0.5) = 8 \geq 5$ and $n(1-\hat{p}) = 16(1-0.5) = 8 \geq 5$, so we may use the Z interval for p.

(c) $E = Z_{\alpha/2}\sqrt{\frac{\hat{p}(1-\hat{p})}{n}} = 1.645\sqrt{\frac{0.5(1-0.5)}{16}} \approx 0.2056$

(d) $\hat{p} \pm Z_{\alpha/2}\sqrt{\frac{\hat{p}(1-\hat{p})}{n}} = 0.5 \pm 1.645\sqrt{\frac{0.5(1-0.5)}{16}} \approx 0.5 \pm 0.2056 = (0.2944, 0.7056)$

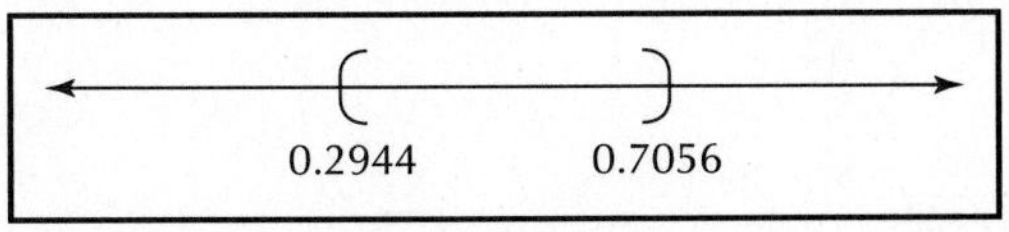

0.2944 0.7056

23. **(a)** $Z_{\alpha/2} = 1.96$ **(b)** We have $n\hat{p} = 25(0.1) = 2.5 < 5$, so we may not use the Z interval for p. **(c)** We have $n\hat{p} = 25(0.1) = 2.5 < 5$, so we may not calculate the margin of error. **(d)** We have $n\hat{p} = 25(0.1) = 2.5 < 5$, so we may not use the Z interval for p.

25. **(a)** $Z_{\alpha/2} = 1.645$ **(b)** We have $n\hat{p} = 36(0.15) = 5.4 \geq 5$ and $n(1 - \hat{p}) = 36(1 - 0.15) = 30.6 \geq 5$, so we may use the Z interval for p.

(c) $E = Z_{\alpha/2}\sqrt{\dfrac{\hat{p}(1-\hat{p})}{n}} = 1.645\sqrt{\dfrac{0.15(1-0.15)}{36}} \approx 0.0979$

(d) $\hat{p} \pm Z_{\alpha/2}\sqrt{\dfrac{\hat{p}(1-\hat{p})}{n}} = 0.15 \pm 1.645\sqrt{\dfrac{0.15(1-0.15)}{36}} \approx 0.15 \pm 0.0979 = (0.0521, 0.2479)$

0.0521 0.2479

27. $Z_{\alpha/2} = 1.96$

(a) $\hat{p} = \dfrac{x}{n} = \dfrac{5}{10} = 0.5$, so $E = Z_{\alpha/2}\sqrt{\dfrac{\hat{p}(1-\hat{p})}{n}} = 1.96\sqrt{\dfrac{0.5(1-0.5)}{10}} \approx 0.3099$

(b) $\hat{p} = \dfrac{x}{n} = \dfrac{50}{100} = 0.5$, so $E = Z_{\alpha/2}\sqrt{\dfrac{\hat{p}(1-\hat{p})}{n}} = 1.96\sqrt{\dfrac{0.5(1-0.5)}{100}} = 0.098$

(c) $\hat{p} = \dfrac{x}{n} = \dfrac{500}{1000} = 0.5$, so $E = Z_{\alpha/2}\sqrt{\dfrac{\hat{p}(1-\hat{p})}{n}} = 1.96\sqrt{\dfrac{0.5(1-0.5)}{1000}} \approx 0.0310$

(d) $\hat{p} = \dfrac{x}{n} = \dfrac{5{,}000}{10{,}000} = 0.5$, so $E = Z_{\alpha/2}\sqrt{\dfrac{\hat{p}(1-\hat{p})}{n}} = 1.96\sqrt{\dfrac{0.5(1-0.5)}{10{,}000}} \approx 0.0098$

29. **(a)** Since the margin of error is

$$E = Z_{\alpha/2}\sqrt{\frac{\hat{p}(1-\hat{p})}{n}},$$

an increase in the sample size while $\hat{p}$ remains constant results in a decrease in the margin of error.
(b) Since the width of the confidence interval is 2 E, an increase in the sample size while $\hat{p}$ remains constant results in a decrease in the width of the confidence interval.

31. **(a)** $Z_{\alpha/2} = 2.576$

(b) $\hat{p} = \dfrac{x}{n} = \dfrac{367}{548} \approx 0.6697$

We have $n\hat{p} = 548(0.6697) = 366.9956 \geq 5$ and $n(1 - \hat{p}) = 548(1 - 0.6697) = 181.0044 \geq 5$, so we may use the Z interval for p.

(c) The margin of error

$$E = Z_{\alpha/2}\sqrt{\frac{\hat{p}(1-\hat{p})}{n}} = 2.576\sqrt{\frac{0.6697(1-0.6697)}{548}} \approx 0.0518.$$

We can estimate p, the true proportion of workers who would be more likely to accept a global assignment, to within 0.0518 with 99% confidence.

(d) $\hat{p} \pm Z_{\alpha/2}\sqrt{\dfrac{\hat{p}(1-\hat{p})}{n}} \approx 0.6697 \pm 2.576\sqrt{\dfrac{0.6697(1-0.6697)}{548}} \approx 0.6697 \pm 0.0518 = (0.6179, 0.7215).$

We are 99% confident that the true proportion of workers who would be more likely to accept a global assignment lies between 0.6179 and 0.7215.

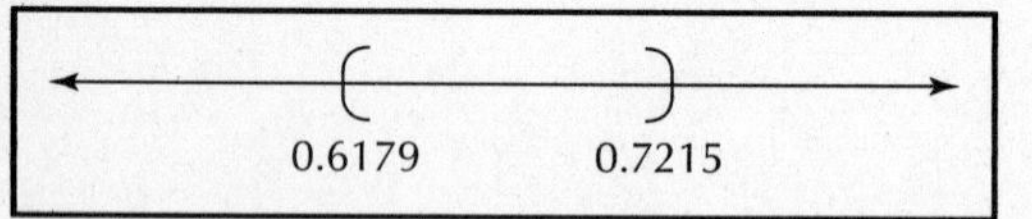

33. **(a)** $Z_{\alpha/2} = 1.96$ **(b)** We have $n\hat{p} = 225(0.64) = 144 \geq 5$ and $n(1 - \hat{p}) = 225(1 - 0.64) = 81 \geq 5$, so we may use the Z interval for p.

(c) The margin of error is

$$E = Z_{\alpha/2}\sqrt{\frac{\hat{p}(1-\hat{p})}{n}} = 1.96\sqrt{\frac{0.64(1-0.64)}{225}} \approx 0.0627.$$

We can estimate p, the true proportion of minorities who oppose race-conscious college admissions, within 0.0627 with 95% confidence.

(d) $\hat{p} \pm Z_{\alpha/2}\sqrt{\frac{\hat{p}(1-\hat{p})}{n}} \approx 0.64 \pm 1.96\sqrt{\frac{0.64(1-0.64)}{225}} \approx 0.64 \pm 0.0627 = (0.5773, 0.7027).$

We are 95% confident that the true proportion of minorities who oppose race-conscious college admissions lies between 0.5773 and 0.7027.

0.5773 0.7027

35. (a) $Z_{\alpha/2} = 1.96$ **(b)** We have $n\hat{p} = 1000(0.40) = 400 \geq 5$ and $n(1-\hat{p}) = 1000(1-0.40) = 600 \geq 5$, so we may use the Z interval for p.

(c) The margin of error is

$$E = Z_{\alpha/2}\sqrt{\frac{\hat{p}(1-\hat{p})}{n}} = 1.96\sqrt{\frac{0.4(1-0.4)}{1{,}000}} \approx 0.0304.$$

We can estimate p, the true proportion of NASCAR racing attendees who own a pickup truck, within 0.0304 with 95% confidence.

(d) $\hat{p} \pm Z_{\alpha/2}\sqrt{\frac{\hat{p}(1-\hat{p})}{n}} \approx 0.4 \pm 1.96\sqrt{\frac{0.4(1-0.4)}{1{,}000}} \approx 0.4 \pm 0.0304 = (0.3696, 0.4304).$

We are 95% confident that the true proportion of NASCAR racing attendees who own a pickup truck lies between 0.3696 and 0.4304.

0.3696 0.4304

37. (a) We have

$$\hat{p} = \frac{x}{n} = \frac{20}{60} \approx 0.3333,$$

so

$$n\hat{p} = 60(0.3333) = 19.998 \geq 5 \text{ and } n(1-\hat{p}) = 60(1-0.3333) = 40.002 \geq 5.$$

Thus, we may use the Z interval for p.

(b) The margin of error is

$$E = Z_{\alpha/2}\sqrt{\frac{\hat{p}(1-\hat{p})}{n}} = 1.645\sqrt{\frac{0.3333(1-0.3333)}{60}} \approx 0.1001.$$

We can estimate p, the true proportion of all anti-terrorist incidents that occurred in 2003 that took place in the Middle East, within 0.1001 with 90% confidence.

(c) $\hat{p} \pm Z_{\alpha/2}\sqrt{\frac{\hat{p}(1-\hat{p})}{n}} \approx 0.3333 \pm 1.645\sqrt{\frac{0.3333(1-0.3333)}{60}} \approx 0.3333 \pm 0.1001 = (0.2332, 0.4334).$

Thus, we are 90% confident that the true proportion of all anti-American terrorist incidents occurring in 2003 that took place in the Middle East lies between 0.2332 and 0.4334.

39. (a) We have $n\hat{p} = 200(0.039) = 7.8 \geq 5$ and $n(1-\hat{p}) = 200(1-0.039) = 192.2 \geq 5$. Thus, we may use the Z interval for p.

(b) The margin of error is

$$E = Z_{\alpha/2}\sqrt{\frac{\hat{p}(1-\hat{p})}{n}} = 2.576\sqrt{\frac{0.039(1-0.039)}{200}} \approx 0.0353.$$

We can estimate p, the true proportion of all white 12th graders who are using ecstasy, within 0.0353 with 99% confidence.

(c) $\hat{p} \pm Z_{\alpha/2}\sqrt{\frac{\hat{p}(1-\hat{p})}{n}} = 0.039 \pm 2.576\sqrt{\frac{0.039(1-0.039)}{200}} \approx 0.039 \pm 0.0353 = (0.0037, 0.0743).$

We are 99% confident that the true proportion of all white 12th graders who are using ecstasy lies between 0.0037 and 0.0743.

41. From Exercise 40 (c), $\hat{p} \approx 0.6735$. We have $n\hat{p} = 49(0.6735) = 33.0015 \geq 5$ and $n(1-\hat{p}) = 49(1-0.6735) = 15.9985 \geq 5$. Thus, we may use the Z interval for p. **(a)** The margin of error is

$$E = Z_{\alpha/2}\sqrt{\frac{\hat{p}(1-\hat{p})}{n}} = 1.96\sqrt{\frac{0.6735(1-0.6735)}{49}} \approx 0.1313.$$

We can estimate p, the true proportion of all high-volume stocks to post an increase in price, within 0.1313 with 95% confidence.

(b) $\hat{p} \pm Z_{\alpha/2}\sqrt{\frac{\hat{p}(1-\hat{p})}{n}} = 0.6735 \pm 1.96\sqrt{\frac{0.6735(1-0.6735)}{49}} \approx 0.6735 \pm 0.1313 = (0.5422, 0.8048)$.

We are 95% confident that the true proportion of all high-volume stocks to post an increase in price lies between 0.5422 and 0.8048.

43. The margin of error is

$$E = Z_{\alpha/2}\sqrt{\frac{\hat{p}(1-\hat{p})}{n}} = 1.96\sqrt{\frac{0.40(1-0.40)}{1006}} \approx 0.0303.$$

45. **(a)** The margin of error is

$$E = Z_{\alpha/2}\sqrt{\frac{\hat{p}(1-\hat{p})}{n}}.$$

Since n is in the denominator of this formula, an increase in the sample size will result in a decrease in the margin of error. **(b)** If we are assuming that the only change is an increase in the sample size, then an increase in sample size will leave the reported sample proportion unchanged. **(c)** Since the confidence level does not depend on the sample size, an increase in sample size will leave the confidence level unchanged.

47. We have

$$\hat{p} = \frac{x}{n} = \frac{482}{1005} \approx 0.4796,$$

so

$$n\hat{p} = 1005(0.4796) = 481.998 \geq 5 \text{ and } n(1-\hat{p}) = 1005(1-0.4796) = 523.002 \geq 5.$$

Thus, we may use the Z interval for p.

(a) $\hat{p} \pm Z_{\alpha/2}\sqrt{\frac{\hat{p}(1-\hat{p})}{n}} \approx 0.4796 \pm 1.645\sqrt{\frac{0.4796(1-0.4796)}{1005}} \approx 0.4796 \pm 0.0259 = (0.4537, 0.5055)$

(b) $\hat{p} \pm Z_{\alpha/2}\sqrt{\frac{\hat{p}(1-\hat{p})}{n}} \approx 0.4796 \pm 2.576\sqrt{\frac{0.4796(1-0.4796)}{1005}} \approx 0.4796 \pm 0.0406 = (0.4390, 0.5202)$

49. **(a)** Since the margin of error is

$$E = Z_{\alpha/2}\sqrt{\frac{\hat{p}(1-\hat{p})}{n}},$$

a decrease in the confidence level will result in a decrease in $Z_{\alpha/2}$, which will result in a decrease in the margin of error. **(b)** A decrease in the confidence level will result in a decrease in $Z_{\alpha/2}$. **(c)** Since the width of the confidence interval is $2E$, a decrease in the confidence level will result in a decrease in E, which will result in a decrease in the width of the confidence interval.

51. We have

$$\hat{p} = \frac{x}{n} = \frac{39}{40} = 0.975,$$

so

$$n\hat{p} = 40(0.975) = 39 \geq 5, \text{ but } n(1-\hat{p}) = 40(1-0.975) = 1 < 5.$$

Thus, we may not use the Z interval for p.

53. **(a)** We have

$$\hat{p} = \frac{x}{n} = \frac{89}{112} \approx 0.7946,$$

so

$$n\hat{p} = 112(0.7946) = 88.9952 \geq 5 \text{ and } n(1-\hat{p}) = 112(1-0.7946) = 23.0048 \geq 5.$$

Thus, we may use the Z interval for p.

$$\hat{p} \pm Z_{\alpha/2}\sqrt{\frac{\hat{p}(1-\hat{p})}{n}} \approx 0.7946 \pm 1.96\sqrt{\frac{0.7946(1-0.7946)}{112}} \approx 0.7946 \pm 0.0748 = (0.7198, 0.8694)$$

(b) **(i)** A decrease in the confidence level will result in a decrease in $Z_{\alpha/2}$ from 1.96 to 1.645.

(ii) Since the margin of error is

$$E = Z_{\alpha/2}\sqrt{\frac{\hat{p}(1-\hat{p})}{n}},$$

a decrease in the confidence level will result in a decrease in $Z_{\alpha/2}$, which will result in a decrease in the margin of error from 0.0748 to 0.0628. **(iii)** Since the width of the confidence interval is $2E$, a decrease in the confidence level will result in a decrease in $Z_{\alpha/2}$, which will result in a decrease in the width of the confidence interval from 0.1496 to 0.1256.

55. **(a)** Since the margin of error is

$$E = Z_{\alpha/2}\sqrt{\frac{\hat{p}(1-\hat{p})}{n}}$$

and n is in the denominator, a decrease in the sample size n will result in an increase in the margin of error. **(b)** If the sample size is decreased and everything else remains constant, a decrease in sample size will leave the reported sample proportion unchanged. **(c)** A decrease in the sample size will leave the confidence level unchanged.

Section 8.4

1. In order to construct a confidence interval for σ^2 or σ, the population must be normal.

3. We can't just use the "point estimate ± margin of error" method used earlier in this chapter because, in order to use this method, the distribution has to be symmetric and the χ^2 curve is not symmetric.

5. True.

7. True.

9. $df = n - 1 = 25 - 1 = 24$, so $\chi^2_{1-\alpha/2} = \chi^2_{0.95} = 13.848$ and $\chi^2_{\alpha/2} = \chi^2_{0.05} = 36.415$.

11. $df = n - 1 = 25 - 1 = 24$, so $\chi^2_{1-\alpha/2} = \chi^2_{0.995} = 9.886$ and $\chi^2_{\alpha/2} = \chi^2_{0.005} = 45.559$.

13. $df = n - 1 = 15 - 1 = 14$, so $\chi^2_{1-\alpha/2} = \chi^2_{0.975} = 5.629$ and $\chi^2_{\alpha/2} = \chi^2_{0.025} = 26.119$.

15. For a given sample size, $\chi^2_{1-\alpha/2}$ decreases and $\chi^2_{\alpha/2}$ increases as the confidence level increases.

17. $df = n - 1 = 25 - 1 = 24$, so $\chi^2_{1-\alpha/2} = \chi^2_{0.95} = 13.848$ and $\chi^2_{\alpha/2} = \chi^2_{0.05} = 36.415$.

$$\text{lower bound} = \frac{(n-1)s^2}{\chi^2_{\alpha/2}} = \frac{(25-1)10}{36.415} \approx 6.59$$

$$\text{upper bound} = \frac{(n-1)s^2}{\chi^2_{1-\alpha/2}} = \frac{(25-1)10}{13.848} \approx 17.33$$

19. $df = n - 1 = 25 - 1 = 24$, so $\chi^2_{1-\alpha/2} = \chi^2_{0.995} = 9.886$ and $\chi^2_{\alpha/2} = \chi^2_{0.005} = 45.559$.

$$\text{lower bound} = \frac{(n-1)s^2}{\chi^2_{\alpha/2}} = \frac{(25-1)10}{45.559} \approx 5.27$$

$$\text{upper bound} = \frac{(n-1)s^2}{\chi^2_{1-\alpha/2}} = \frac{(25-1)10}{9.886} \approx 24.28$$

21. $df = n - 1 = 25 - 1 = 24$, so $\chi^2_{1-\alpha/2} = \chi^2_{0.975} = 12.401$ and $\chi^2_{\alpha/2} = \chi^2_{0.025} = 39.364$.

$$\text{lower bound} = \sqrt{\frac{(n-1)s^2}{\chi^2_{\alpha/2}}} = \sqrt{\frac{(25-1)10}{39.364}} \approx 2.47$$

$$\text{upper bound} = \sqrt{\frac{(n-1)s^2}{\chi^2_{1-\alpha/2}}} = \sqrt{\frac{(25-1)10}{12.401}} \approx 4.40$$

23. As the confidence level increases but the sample size stays the same, the lower bound for the confidence interval for σ^2 decreases and the upper bound for the confidence interval for σ^2 increases

25. $df = n - 1 = 10 - 1 = 9$, so $\chi^2_{1-\alpha/2} = \chi^2_{0.975} = 2.700$ and $\chi^2_{\alpha/2} = \chi^2_{0.025} = 19.023$.

$$\text{lower bound} = \frac{(n-1)s^2}{\chi^2_{\alpha/2}} = \frac{(10-1)10}{19.023} \approx 4.73$$

$$\text{upper bound} = \frac{(n-1)s^2}{\chi^2_{1-\alpha/2}} = \frac{(10-1)10}{2.700} \approx 33.33$$

27. $df = n - 1 = 20 - 1 = 19$, so $\chi^2_{1-\alpha/2} = \chi^2_{0.975} = 8.907$ and $\chi^2_{\alpha/2} = \chi^2_{0.025} = 32.852$.

$$\text{lower bound} = \frac{(n-1)s^2}{\chi^2_{\alpha/2}} = \frac{(20-1)10}{32.852} \approx 5.78$$

$$\text{upper bound} = \frac{(n-1)s^2}{\chi^2_{1-\alpha/2}} = \frac{(20-1)10}{8.907} \approx 21.33$$

29. $df = n - 1 = 15 - 1 = 14$, so $\chi^2_{1-\alpha/2} = \chi^2_{0.975} = 5.629$ and $\chi^2_{\alpha/2} = \chi^2_{0.025} = 26.119$.

$$\text{lower bound} = \sqrt{\frac{(n-1)s^2}{\chi^2_{\alpha/2}}} = \sqrt{\frac{(15-1)10}{26.119}} \approx 2.32$$

$$\text{upper bound} = \sqrt{\frac{(n-1)s^2}{\chi^2_{1-\alpha/2}}} = \sqrt{\frac{(15-1)10}{5.629}} \approx 4.99$$

31. As the sample size increases but the confidence level stays the same, the lower bound of a confidence interval for σ^2 increases and the upper bound of a confidence interval for σ^2 decreases.

33. (a) $df = n - 1 = 5 - 1 = 4$, so $\chi^2_{1-\alpha/2} = \chi^2_{0.975} = 0.484$ and $\chi^2_{\alpha/2} = \chi^2_{0.025} = 11.143$

(b) From Minitab:

```
                  Total
Variable          Count  Variance
Prisoner deaths       5    431.30
```

$$\text{lower bound} = \frac{(n-1)s^2}{\chi^2_{\alpha/2}} = \frac{(5-1)431.30}{11.143} \approx 154.82$$

$$\text{upper bound} = \frac{(n-1)s^2}{\chi^2_{1-\alpha/2}} = \frac{(5-1)431.30}{0.484} \approx 3564.46$$

We are 95% confident that the population variance σ^2 lies between 154.82 prisoners squared and 3564.46 prisoners squared.

(c) $$\text{lower bound} = \sqrt{\frac{(n-1)s^2}{\chi^2_{\alpha/2}}} = \sqrt{\frac{(5-1)431.30}{11.143}} \approx 12.44$$

$$\text{upper bound} = \sqrt{\frac{(n-1)s^2}{\chi^2_{1-\alpha/2}}} = \sqrt{\frac{(5-1)431.30}{0.484}} \approx 59.70$$

We are 95% confident that the population standard deviation σ lies between 12.44 prisoners and 59.70 prisoners.

35. (a) The units used to interpret the confidence interval in 33(b) are prisoners squared. **(b)** Most people would not understand the units of prisoners squared. **(c)** The units used to interpret the confidence interval in 33(c) are prisoners. **(d)** The units of prisoners are more easily understood by most people.

37. (a) $df = n - 1 = 20 - 1 = 19$, so $\chi^2_{1-\alpha/2} = \chi^2_{0.975} = 8.907$ and $\chi^2_{\alpha/2} = \chi^2_{0.025} = 32.852$.

(b) From Minitab:

```
                Total
Variable        Count  Variance
Foot lengths       20     1.638
```

$$\text{lower bound} = \frac{(n-1)s^2}{\chi^2_{\alpha/2}} = \frac{(20-1)1.638}{32.852} \approx 0.95$$

$$\text{upper bound} = \frac{(n-1)s^2}{\chi^2_{1-\alpha/2}} = \frac{(20-1)1.638}{8.907} \approx 3.49$$

We are 95% confident that the population variance σ^2 lies between 0.95 centimeters squared and 3.49 centimeters squared.

(c) $\text{lower bound} = \sqrt{\frac{(n-1)s^2}{\chi^2_{\alpha/2}}} = \sqrt{\frac{(20-1)1.638}{32.852}} \approx 0.97$

$$\text{upper bound} = \sqrt{\frac{(n-1)s^2}{\chi^2_{1-\alpha/2}}} = \sqrt{\frac{(20-1)1.638}{8.907}} \approx 1.87$$

We are 95% confident that the population standard deviation σ lies between 0.97 centimeters and 1.87 centimeters. **(d)** The interpretation in (c) is easier to understand since it is in the same units as our data.

39. **(a)** df $= n - 1 = 10 - 1 = 9$, so $\chi^2_{1-\alpha/2} = \chi^2_{0.995} = 1.735$ and $\chi^2_{\alpha/2} = \chi^2_{0.005} = 23.589$.

(b) From Minitab:

```
                      Total
Variable              Count   Variance
Biomass consumed        10       8.942
```

$$\text{lower bound} = \sqrt{\frac{(n-1)s^2}{\chi^2_{\alpha/2}}} = \sqrt{\frac{(10-1)8.942}{23.589}} \approx 1.85$$

$$\text{upper bound} = \sqrt{\frac{(n-1)s^2}{\chi^2_{1-\alpha/2}}} = \sqrt{\frac{(10-1)8.942}{1.735}} \approx 6.81$$

We are 99% confident that the population standard deviation σ lies between 1.85 trillion BTUs and 6.81 trillion BTUs.

Section 8.5

1. For a fixed sample size, you could not simultaneously maximize your confidence level and minimize the width of your confidence interval because the higher the confidence level, the wider the confidence interval.

3. The formula for the sample size for estimating the population mean μ within a margin of error E with confidence $100(1-\alpha)\%$ is given by

$$n = \left(\frac{(Z_{\alpha/2})\sigma}{E}\right)^2.$$

Since the margin of error E occurs in the denominator, as the margin of error is increased, n decreases, and as the margin of error is decreased, n increases.

5. $n = \left(\frac{(Z_{\alpha/2})\sigma}{E}\right)^2 = \left(\frac{(1.645)10}{32}\right)^2 = 0.2642602539$. Round n up to 1.

7. $n = \left(\frac{(Z_{\alpha/2})\sigma}{E}\right)^2 = \left(\frac{(1.645)10}{8}\right)^2 = 4.228164063$. Round n up to 5.

9. $n = \left(\frac{(Z_{\alpha/2})\sigma}{E}\right)^2 = \left(\frac{(1.645)10}{8}\right)^2 = 4.228164063$. Round n up to 5.

11. $n = \left(\frac{(Z_{\alpha/2})\sigma}{E}\right)^2 = \left(\frac{(2.576)10}{8}\right)^2 = 10.3684$. Round n up to 11.

13. $n = \left(\frac{(Z_{\alpha/2})\sigma}{E}\right)^2 = \left(\frac{(1.645)10}{16}\right)^2 = 1.057041016$. Round n up to 2.

15. $n = \left(\frac{(Z_{\alpha/2})\sigma}{E}\right)^2 = \left(\frac{(1.645)40}{16}\right)^2 = 16.91265625$. Round n up to 17.

17. $n = \left(\frac{0.5 \cdot Z_{\alpha/2}}{E}\right)^2 = \left(\frac{0.5 \cdot 1.96}{0.03}\right)^2 = 1067.111111$. Round n up to 1068.

19. $n = \left(\frac{0.5 \cdot Z_{\alpha/2}}{E}\right)^2 = \left(\frac{0.5 \cdot 1.96}{0.0075}\right)^2 = 17{,}073.77778$. Round n up to 17,074.

21. (a) As the margin of error is halved while the confidence level stays fixed, the sample size required is increased by a factor of approximately 4.

(b)

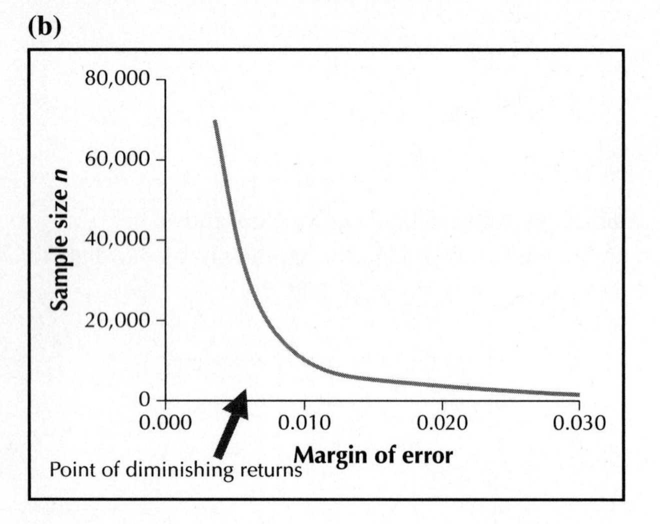

From the graph, we can tell that the point of diminishing returns is somewhere between a margin of error of 0.005 and 0.010.

23. $n = \hat{p}(1-\hat{p})\left(\frac{Z_{\alpha/2}}{E}\right)^2 = (0.1)(1-0.1)\left(\frac{1.96}{0.03}\right)^2 = 384.16$. Round n up to 385.

25. $n = \hat{p}(1-\hat{p})\left(\frac{Z_{\alpha/2}}{E}\right)^2 = (0.001)(1-0.001)\left(\frac{1.96}{0.03}\right)^2 = 4.264176$. Round n up to 5.

27. $n = \left(\frac{0.5 \cdot Z_{\alpha/2}}{E}\right)^2 = \left(\frac{0.5 \cdot 1.645}{0.03}\right)^2 = 751.6736111$. Round n up to 752.

29. $n = \left(\frac{0.5 \cdot Z_{\alpha/2}}{E}\right)^2 = \left(\frac{0.5 \cdot 2.576}{0.03}\right)^2 = 1843.271111$. Round n up to 1844.

31. $n = \hat{p}(1-\hat{p})\left(\frac{Z_{\alpha/2}}{E}\right)^2 = (0.3)(1-0.3)\left(\frac{1.96}{0.03}\right)^2 = 896.3733333$. Round n up to 897.

33. $n = \hat{p}(1-\hat{p})\left(\frac{Z_{\alpha/2}}{E}\right)^2 = (0.7)(1-0.7)\left(\frac{1.96}{0.03}\right)^2 = 896.3733333$. Round n up to 897.

35. From Exercise 55 in Section 8.3, $\hat{p} = 0.5$. Thus,

$$n = \hat{p}(1-\hat{p})\left(\frac{Z_{\alpha/2}}{E}\right)^2 = (0.5)(1-0.5)\left(\frac{2.576}{0.03}\right)^2 = 1843.271111. \text{ Round } n \text{ up to 1844.}$$

37. (a) $n = \left(\frac{(Z_{\alpha/2})\sigma}{E}\right)^2 = \left(\frac{(1.96)20}{25}\right)^2 = 2.458624$. Round n up to 3.

(b) $n = \left(\frac{(Z_{\alpha/2})\sigma}{E}\right)^2 = \left(\frac{(1.96)20}{5}\right)^2 = 61.4656$. Round n up to 62.

39. (a) $n = \left(\frac{0.5 \cdot Z_{\alpha/2}}{E}\right)^2 = \left(\frac{0.5 \cdot 1.645}{0.05}\right)^2 = 270.6025$. Round n up to 271.

(b) $n = \left(\frac{0.5 \cdot Z_{\alpha/2}}{E}\right)^2 = \left(\frac{0.5 \cdot 1.645}{0.01}\right)^2 = 6765.0625$. Round n up to 6766.

41. $n = \hat{p}(1-\hat{p})\left(\frac{Z_{\alpha/2}}{E}\right)^2 = (0.6)(1-0.6)\left(\frac{2.576}{0.03}\right)^2 = 1769.540267$. Round n up to 1770.

43. **(a)** $n = \left(\frac{(Z_{\alpha/2})\sigma}{E}\right)^2 = \left(\frac{(2.576)3}{5}\right)^2 = 2.38887936$. Round n up to 3. **(b)** To lower the required sample size, the psychologist could either lower the confidence level or increase the margin of error.

(c) For a 95% confidence level,

$$n = \left(\frac{(Z_{\alpha/2})\sigma}{E}\right)^2 = \left(\frac{(1.96)3}{5}\right)^2 = 1.382976. \text{ Round } n \text{ up to 2.}$$

For a 90% confidence level,

$$n = \left(\frac{(Z_{\alpha/2})\sigma}{E}\right)^2 = \left(\frac{(1.645)3}{5}\right)^2 = 0.974169. \text{ Round } n \text{ up to 1.}$$

So lowering the confidence level will lower the required sample size.

Suppose the margin of error is increased to 6 episodes.

For a 99% confidence level,

$$n = \left(\frac{(Z_{\alpha/2})\sigma}{E}\right)^2 = \left(\frac{(2.576)3}{6}\right)^2 = 1.658944. \text{ Round } n \text{ up to 2.}$$

Suppose the margin of error is increased to 8 episodes.
For a 99% confidence level,

$$n = \left(\frac{(Z_{\alpha/2})\sigma}{E}\right)^2 = \left(\frac{(2.576)3}{8}\right)^2 = 0.933156. \text{ Round } n \text{ up to 1.}$$

So increasing the margin of error will lower the required sample size.

45. **(a)** $n = \left(\frac{(Z_{\alpha/2})\sigma}{E}\right)^2 = \left(\frac{(1.96)150}{50}\right)^2 = 34.5744$. Round n up to 35.

(b) $n = \left(\frac{(Z_{\alpha/2})\sigma}{E}\right)^2 = \left(\frac{(1.96)150}{10}\right)^2 = 864.36$. Round n up to 865.

Exercises (c) and (d) were worked with the assumptions that the population standard deviation of the amount spent by all American citizens is also $\sigma = \$150$.

(c) $n = \left(\frac{(Z_{\alpha/2})\sigma}{E}\right)^2 = \left(\frac{(1.96)150}{50}\right)^2 = 34.5744$. Round n up to 35.

(d) $n = \left(\frac{(Z_{\alpha/2})\sigma}{E}\right)^2 = \left(\frac{(1.96)150}{10}\right)^2 = 864.36$. Round n up to 865.

(e) The value of the mean does not play any role in calculating the required sample size.

47. Assume a 95% confidence level for (a) and (b).

(a) $n = \left(\frac{(Z_{\alpha/2})\sigma}{E}\right)^2 = \left(\frac{(1.96)500{,}000{,}000}{100{,}000{,}000}\right)^2 = 96.04$. Round n up to 97

(b) $n = \left(\frac{(Z_{\alpha/2})\sigma}{E}\right)^2 = \left(\frac{(1.96)500{,}000{,}000}{10{,}000{,}000}\right)^2 = 9604$; and 9604 days $= \frac{9604}{365} \approx 26.31$ years.

49. $n = \left(\frac{(Z_{\alpha/2})\sigma}{E}\right)^2 = \left(\frac{(1.96)1.6}{0.01}\right)^2 = 98{,}344.96$. Round n up to 98,345.

98,345 days $= \frac{98{,}345}{365} \approx 269.44$ years. Neither the analyst nor the boss will live long enough to collect a sample of this size.

Chapter 8 Review

1. **(a)** Since the population is normal and σ is known, Case 1 applies, so we may use the Z interval.

(b) $\frac{\sigma}{\sqrt{n}} = \frac{10}{\sqrt{25}} = 2$ **(c)** $Z_{\alpha/2} = 1.96$

(d) The margin of error is $E = Z_{\alpha/2}(\sigma/\sqrt{n}) = 1.96(10/\sqrt{25}) = 3.92$. We may estimate μ to within 3.92 with 95% confidence. **(e)** $\bar{x} \pm Z_{\alpha/2}(\sigma/\sqrt{n}) = 50 \pm 1.96(10/\sqrt{25}) = 50 \pm 3.92 = (46.08, 53.92)$. We are 95% confident that the true mean μ lies between 46.08 and 53.92.

3. **(a)** Since the sample size of $n = 100$ is large ($n \geq 30$) and σ is known, Case 2 applies and we may use the Z interval. **(b)** $\sigma/\sqrt{n} = 10/\sqrt{100} = 1$ **(c)** $Z_{\alpha/2} = 1.96$ **(d)** The margin of error is $E = Z_{\alpha/2}(\sigma/\sqrt{n}) = 1.96(10/\sqrt{100}) = 1.96$. We may estimate μ to within 1.96 with 95% confidence. **(e)** $\bar{x} \pm Z_{\alpha/2}(\sigma/\sqrt{n}) = 50 \pm 1.96(10/\sqrt{100}) = 50 \pm 1.96 = (48.04, 51.96)$. We are 95% confident that the true mean μ lies between 48.04 and 51.96.

5. **(a)** $\bar{x} \pm Z_{\alpha/2}(\sigma/\sqrt{n}) = 7 \pm 1.96(2/\sqrt{45}) \approx 70 \pm 0.5844 = (6.4156, 7.5844)$. We are 95% confident that the true mean increase in IQ points for all children after listening to a Mozart piano sonata for about 10 minutes μ lies between 6.4156 points and 7.5844 points. **(b)** $\bar{x} \pm Z_{\alpha/2}(\sigma/\sqrt{n}) = 7 \pm 2.576(2/\sqrt{45}) \approx 70 \pm 0.7680 = (6.2320, 7.7680)$. We are 99% confident that the true mean increase in IQ points for all children after listening to a Mozart piano sonata for about 10 minutes μ lies between 6.2320 points and 7.7680 points. **(c)** The width of the confidence interval increases as the confidence level increases.

7. **(a)** $\bar{x} = \$45{,}000$ **(b)** $\sigma/\sqrt{n} = \$6{,}000/\sqrt{25} = \$1200$ **(c)** $Z_{\alpha/2} = 2.576$ **(d)** Since the distribution is normal and σ is known, Case 1 applies and we may use the Z interval. The margin of error is $E = Z_{\alpha/2}(\sigma/\sqrt{n}) = 2.576(\$6{,}000/\sqrt{25}) = \$3091.20$. We may estimate μ, the mean income of all working women in California, to within \$3091.20 with 99% confidence. **(e)** $\bar{x} \pm Z_{\alpha/2}(\sigma/\sqrt{n}) = \$45{,}000 \pm 2.576(\$6{,}000/\sqrt{25}) = \$45{,}000 \pm \$3091.20 = (41{,}908.80, 48{,}091.20)$. We are 99% confident that the true mean income of all working women in California μ lies between \$41,908.80 and \$48,091.20.

9. Since the mean income of the population of all working women in California is a nonnegative real number, any interval that contains all of the nonnegative real numbers will be a 100% confidence interval.

11. Since the population is normal and σ is unknown, Case 1 applies and we may use the t interval.

$$df = n - 1 = 25 - 1 = 24$$

$$\bar{x} \pm t_{\alpha/2}(s/\sqrt{n}) = 22 \pm 1.711(5/\sqrt{25}) = 22 \pm 1.711 = (20.289, 23.711)$$

13. From Minitab:

Variable	Total Count	Mean	StDev
Number of cigarettes	8	2392.2	274.6

$$df = n - 1 = 8 - 1 = 7$$

(a) $\bar{x} \pm t_{\alpha/2}(s/\sqrt{n}) = 2392.2 \pm 2.365(274.6/\sqrt{8}) \approx 2392.2 \pm 229.61 = (2162.59, 2621.81)$ **(b)** $\bar{x} \pm t_{\alpha/2}(s/\sqrt{n}) = 2392.2 \pm 3.499(274.6/\sqrt{8}) \approx 2392.2 \pm 339.70 = (2052.50, 2731.90)$ **(c)** The interval in (a) is more precise than the interval in (b), but the interval in (b) has higher confidence of containing μ.

15. **(a)** $df = n - 1 = 75 - 1 = 74$. The value of 74 for the df does not appear in the df column. The conservative approach would be to use next row with df smaller than 74. That would be df = 70. Thus, the "conservative" $t_{\alpha/2}$ is 1.667. **(b)** Since the sample size of $n = 75$ is large ($n \geq 30$) and σ is unknown, Case 2 applies and we may use the t interval. The margin of error $E = t_{\alpha/2}(s/\sqrt{n}) = 1.667(\$30/\sqrt{75}) \approx \$5.77$. We can estimate μ, the true mean amount of money collected by all parking meters, within \$5.77 with 90% confidence. **(c)** $\bar{x} \pm t_{\alpha/2}(s/\sqrt{n}) = 120 \pm 1.667(30/\sqrt{75}) \approx 120 \pm 5.77 = (114.23, 125.77)$. We are 90% confident that the true mean amount of money collected by all parking meters lies between \$114.23 and \$125.77.

17. **(a)** A 90% confidence interval will be shorter than a 99% confidence interval, and therefore the estimate will be more precise. But we are more confident that a 99% confidence interval will contain the true value of the mean than we are for a 90% confidence interval. **(b)** $df = n - 1 = 30 - 1 = 29$

$$\text{90\% confidence level: } E = t_{\alpha/2}(s/\sqrt{n}) = 1.699\,(3/\sqrt{30}) \approx 0.9306 \text{ colors}$$

$$\text{99\% confidence level: } E = t_{\alpha/2}(s/\sqrt{n}) = 2.756\,(3/\sqrt{30}) \approx 1.5095 \text{ colors}$$

The margin of error increases as the confidence level increases.

19.

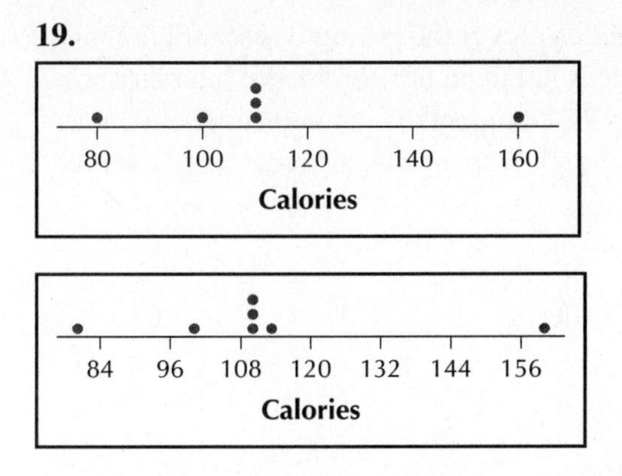

The sample mean of the original six cereals is $\bar{x} = 111.67$ calories. Therefore, the seventh cereal added has to have 111.67 calories per serving. **(a)** If the seventh cereal added has the same mean as the mean of the other six cereals, then the sample mean of the seven cereals would be the same as the sample mean of the original six cereals. **(b)** Since the sample standard deviation is $s = \sqrt{\frac{\Sigma(x-\bar{x})^2}{n-1}}$, adding an x to the data set equal to $\bar{x}$ of the original six cereals means that $x - \bar{x} = 0$ for the added value. Thus, $\Sigma(x-\bar{x})^2$ will remain the same after the seventh value is added, but $n - 1$ will increase by one. Thus, the standard deviation will decrease. **(c)** Since the degrees of freedom is df $= n - 1$, an increase in the sample size n by one will result in an increase in the degrees of freedom from 5 to 6. **(d)** Since $t_{\alpha/2}$ decreases as the degrees of freedom increases, an increase in the sample size n will result in an increase in the degrees of freedom, which will result in a decrease in $t_{\alpha/2}$.

21. **(a)** $Z_{\alpha/2} = 1.96$ **(b)** We have $n\hat{p} = 36(0.1) = 3.6 < 5$, so we may not use the Z interval for p. **(c)** We have $n\hat{p} = 36(0.1) = 3.6 < 5$, so we may not calculate the margin of error. **(d)** We have $n\hat{p} = 36(0.1) = 3.6 < 5$, so we may not use the Z interval for p.

23. **(a)** $Z_{\alpha/2} = 1.96$ **(b)** We have $n\hat{p} = 100(0.99) = 99 \geq 5$ but $n(1-\hat{p}) = 100(1-0.99) = 1 < 5$. Thus, we may not use the Z interval for p. **(c)** We have $n\hat{p} = 100(0.99) = 99 \geq 5$ but $n(1-\hat{p}) = 100(1-0.99) = 1 < 5$. Thus, we may not calculate the margin of error. **(d)** We have $n\hat{p} = 100(0.99) = 99 \geq 5$ but $n(1-\hat{p}) = 100(1-0.99) = 1 < 5$. Thus, we may not use the Z interval for p.

25. **(a)** We have $\hat{p} = \frac{x}{n} = \frac{17}{60} \approx 0.2833$, so $n\hat{p} = 60(0.2833) = 16.998 \geq 5$ and $n(1-\hat{p}) = 60(1-0.2833) = 43.002 \geq 5$. Thus, we may use the Z interval for p.

(b) The margin of error is

$$E = Z_{\alpha/2}\sqrt{\frac{\hat{p}(1-\hat{p})}{n}} = 1.96\sqrt{\frac{0.2833(1-0.2833)}{60}} \approx 0.1140.$$

We can estimate p, the true population proportion, within 0.1140 with 95% confidence.

(c) $\hat{p} \pm Z_{\alpha/2}\sqrt{\frac{\hat{p}(1-\hat{p})}{n}} \approx 0.2833 \pm 1.96\sqrt{\frac{0.2833(1-0.2833)}{60}} \approx 0.2833 \pm 0.1140 = (0.1693, 0.3973)$. We are 95% confident that the true proportion of all anti-American terrorist incidents occurring in 2003 that took place in the Middle East lies between 0.1693 and 0.3973.

27. **(a)** We have $n\hat{p} = 1000(0.89) = 890 \geq 5$ and $n(1-\hat{p}) = 1000(1-0.89) = 110 \geq 5$. Thus, we may use the Z interval for p.

(b) The margin of error is

$$E = Z_{\alpha/2}\sqrt{\frac{\hat{p}(1-\hat{p})}{n}} = 2.576\sqrt{\frac{0.89(1-0.89)}{1000}} \approx 0.0255.$$

We can estimate p, the true proportion of all erasures that resulted in a correct answer at the prestigious school in Fairfield, Connecticut, within 0.0255 with 99% confidence.

(c) $\hat{p} \pm Z_{\alpha/2}\sqrt{\frac{\hat{p}(1-\hat{p})}{n}} \approx 0.89 \pm 2.576\sqrt{\frac{0.89(1-0.89)}{1000}} \approx 0.89 \pm 0.0255 = (0.8645, 0.9155)$. We are 99%

confident that the true proportion of all erasures that result in a correct answer at the prestigious school in Fairfield, Connecticut, lies between 0.8645 and 0.9155. **(d)** Since 0.70 does not lie in our confidence interval and none of the values below 0.70 lie in our confidence interval, we are 99% confident that no number less than or equal to 0.70 is a plausible value for $\hat{p}$. All numbers less than or equal to 0.70 lie below our confidence interval. This is an indication that tampering may have occurred.

For Exercises 29, 31, 33, df $= n - 1 = 36 - 1 = 35$.

29. From the TI-Inspire calculator, $\chi^2_{\alpha/2} = \chi^2_{0.025} = 53.203$ and $\chi^2_{1-\alpha/2} = \chi^2_{0.975} = 20.569$. Thus,

$$\text{lower bound} = \frac{(n-1)s^2}{\chi^2_{\alpha/2}} = \frac{(36-1)100}{53.203} \approx 65.786 \text{ and}$$

$$\text{upper bound} = \frac{(n-1)s^2}{\chi^2_{1-\alpha/2}} = \frac{(36-1)100}{20.569} \approx 170.159.$$

31. From the TI-Inspire calculator, $\chi^2_{\alpha/2} = \chi^2_{0.05} = 49.802$ and $\chi^2_{1-\alpha/2} = \chi^2_{0.95} = 22.465$. Thus,

$$\text{lower bound} = \sqrt{\frac{(n-1)s^2}{\chi^2_{\alpha/2}}} = \sqrt{\frac{(36-1)100}{49.802}} \approx 8.383 \text{ and}$$

$$\text{upper bound} = \sqrt{\frac{(n-1)s^2}{\chi^2_{1-\alpha/2}}} = \sqrt{\frac{(36-1)100}{22.465}} \approx 12.482.$$

33. From the TI-Inspire calculator, $\chi^2_{\alpha/2} = \chi^2_{0.005} = 60.275$ and $\chi^2_{1-\alpha/2} = \chi^2_{0.995} = 17.192$. Thus,

$$\text{lower bound} = \sqrt{\frac{(n-1)s^2}{\chi^2_{\alpha/2}}} = \sqrt{\frac{(36-1)100}{60.275}} \approx 7.620 \text{ and}$$

$$\text{upper bound} = \sqrt{\frac{(n-1)s^2}{\chi^2_{1-\alpha/2}}} = \sqrt{\frac{(36-1)100}{17.192}} \approx 14.268.$$

35. $n = \left(\frac{(Z_{\alpha/2})\sigma}{E}\right)^2 = \left(\frac{(1.96)5}{10}\right)^2 = 0.9604$. Round n up to 1.

37. $n = \left(\frac{(Z_{\alpha/2})\sigma}{E}\right)^2 = \left(\frac{(1.96)1}{10}\right)^2 = 0.038416$. Round n up to 1.

39. $n = \left(\frac{0.5 \cdot Z_{\alpha/2}}{E}\right)^2 = \left(\frac{0.5 \cdot 1.645}{0.03}\right)^2 = 751.6736111$. Round n up to 752.

41. $n = \hat{p}(1-\hat{p})\left(\frac{Z_{\alpha/2}}{E}\right)^2 = (0.9)(1-0.9)\left(\frac{2.576}{0.03}\right)^2 = 663.5776$. Round n up to 664.

43. $n = \hat{p}(1-\hat{p})\left(\frac{Z_{\alpha/2}}{E}\right)^2 = (0.999)(1-0.999)\left(\frac{1.96}{0.03}\right)^2 = 4.264176$. Round n up to 5.

45. **(a)** $n = \hat{p}(1-\hat{p})\left(\frac{Z_{\alpha/2}}{E}\right)^2 = (0.2)(1-0.2)\left(\frac{1.645}{0.03}\right)^2 = 481.0711111$. Round n up to 482.

(b) $n = \hat{p}(1-\hat{p})\left(\frac{Z_{\alpha/2}}{E}\right)^2 = (0.2)(1-0.2)\left(\frac{1.96}{0.03}\right)^2 = 682.9511111$. Round n up to 683.

(c) $n = \hat{p}(1-\hat{p})\left(\frac{Z_{\alpha/2}}{E}\right)^2 = (0.2)(1-0.2)\left(\frac{2.576}{0.03}\right)^2 = 1179.693511$. Round n up to 1180.

Chapter 8 Quiz

1. False. If we take sample after sample for a very long time, then *in the long run*, the proportion of intervals that will contain the parameter μ will equal 95%. The 20 samples in Table 2 are not enough to qualify for the phrase *long run*.

2. True.

3. True.

4. 0

5. 4

6. Less.

7. α is a probability.

8. σ unknown and either the population is normal or the sample size is large ($n \geq 30$).

9. The first method for finding the required sample size if p is unknown is when prior information about p is available. Use the formula $n = \hat{p}(1 - \hat{p})\left(\frac{Z_{\alpha/2}}{E}\right)^2$, where $\hat{p}$ is the sample proportion of successes available from an earlier sample. The second method for finding the required sample size if p is unknown is when no prior information about p is available. Use the formula $n = \left(\frac{0.5 \cdot (Z_{\alpha/2})}{E}\right)^2$. The second method will always deliver the larger sample size.

10. **(a)** $\bar{x} = \$30{,}500$ **(b)** $\sigma/\sqrt{n} = \$3{,}000/\sqrt{49} \approx \$428.57$ **(c)** $Z_{\alpha/2} = 1.645$ **(d)** Since the sample size of $n = 49$ is large ($n \geq 30$) and σ is known, Case 2 applies and we may use the Z interval. The margin of error is $E = Z_{\alpha/2}(\sigma/\sqrt{n}) = 1.645(\$3{,}000/\sqrt{49}) = \$705$. We can estimate μ, the mean cost of a college education, to within \$705 with 90% confidence. **(e)** $\bar{x} \pm Z_{\alpha/2}(\sigma/\sqrt{n}) = \$30{,}500 \pm 1.645(\$3{,}000/\sqrt{49}) = \$30{,}500 \pm \$705 =$ (29,795, 31,205). We are 90% confident that the true mean cost of a college education lies between \$29,795 and \$31,205.

11. **(a)** $\sigma/\sqrt{n} = 210/\sqrt{49} = 30$ pounds **(b)** $Z_{\alpha/2} = 1.645$ **(c)** Since the sample size of $n = 49$ is large ($n \geq 30$) and σ is known, Case 2 applies and we may use the Z interval. The margin of error is $E = Z_{\alpha/2}(\sigma/\sqrt{n}) = 1.645(210/\sqrt{49}) = 49.35$ pounds. We can estimate μ, the mean femur load number in a frontal crash for the passenger in a 2005 Ford Equinox SUV, within 49.35 pounds with 90% confidence. **(d)** $\bar{x} \pm Z_{\alpha/2}(\sigma/\sqrt{n}) = 1003 \pm 1.645(210/\sqrt{49}) = 1003 \pm 49.35 = (953.65, 1052.35)$. We are 90% confident that the true mean femur load number in a frontal crash for the passenger in a 2005 Ford Equinox SUV lies between 953.65 pounds and 1052.35 pounds.

12. **(a)** Since the distribution of the population is normal, Case 1 applies and we may use the t interval.

$$\text{df} = n - 1 = 81 - 1 = 80$$

$\bar{x} \pm t_{\alpha/2}(s/\sqrt{n}) = 57 \pm 1.664\,(10/\sqrt{81}) \approx 57 \pm 1.85 = (55.15, 58.85)$.

We are 90% confident that the true mean day-count of wolves in Isle Royale lies between 55.15 wolves and 58.85 wolves. **(b)** Since we can't decrease the confidence level, the only way we can decrease the margin of error is to increase the sample size.

13. Since the sample size of $n = 64$ is large ($n \geq 30$), Case 2 applies and we may use the t interval. $\text{df} = n - 1 = 64 - 1 = 63$. The value of 63 for the df does not appear in the df column. The conservative approach would be to use the next row with df smaller than 63. That would be df = 60.

(a) Thus, the "conservative" $t_{\alpha/2}$ is 2.000. $\bar{x} \pm t_{\alpha/2}(s/\sqrt{n}) = 9.5 \pm 2.000(0.8/\sqrt{64}) = 9.5 \pm 0.2 = (9.30, 9.70)$. We are 95% confident that the true mean fog index for all of this horror author's books lies between 9.30 and 9.70.
(b) Thus, the "conservative" $t_{\alpha/2}$ is 2.660. $\bar{x} \pm t_{\alpha/2}(s/\sqrt{n}) = 9.5 \pm 2.660(0.8/\sqrt{64}) = 9.5 \pm 0.266 = (9.234, 9.776)$. We are 99% confident that the true mean fog index for all of this horror author's books lies between 9.234 and 9.776. This interval is wider and therefore less precise than the 95% confidence interval in (a).
(c) The 99% confidence interval is wider and less precise than the 95% confidence interval, but we are more confident that the 99% confidence interval contains the true value of the mean. **(d)** Since σ is unknown, it would not be appropriate to use a Z interval.

14. **(a)** Since all of the points lie close to the center line, the normality assumption appears valid. Since σ is unknown, Case 1 applies and we may use the t interval. **(b)** $\text{df} = n - 1 = 15 - 1 = 14$
$\bar{x} \pm t_{\alpha/2}(s/\sqrt{n}) = 75.6 \pm 1.761(6.65/\sqrt{15}) \approx 75.6 \pm 3.0237 = (72.5763, 78.6237)$

We are 90% confident that the true mean heart rate of all women μ lies between 72.5763 beats per minute and 78.6237 beats per minute.

15. We have $\hat{p} = \frac{x}{n} = \frac{991}{3733} \approx 0.2655$. We have $n\hat{p} = 3733(0.2655) = 991.1115 \geq 5$ and $n(1 - \hat{p}) = 3733(1 - 0.2655) = 2741.8885 \geq 5$. Thus, we may use the Z interval for p.

(a) The margin of error is

$$E = Z_{\alpha/2}\sqrt{\frac{\hat{p}(1-\hat{p})}{n}} = 1.96\sqrt{\frac{0.2655(1-0.2655)}{3733}} \approx 0.0142.$$

We can estimate p, the true proportion of all Americans who attended a religious service in response to the attacks on the World Trade Center and the Pentagon, within 0.0142 with 95% confidence.

(b) $\hat{p} \pm Z_{\alpha/2}\sqrt{\frac{\hat{p}(1-\hat{p})}{n}} \approx 0.2655 \pm 1.96\sqrt{\frac{0.2655(1-0.2655)}{3733}} \approx 0.2655 \pm 0.0142 = (0.2513, 0.2797)$.

We are 95% confident that the true proportion of all Americans who attended a religious service in response to the attacks on the World Trade Center and the Pentagon lies between 0.2513 and 0.2797.

16. We have $\hat{p} = \frac{x}{n} = \frac{340}{1000} = 0.34$. We have $n\hat{p} = 1000(0.34) = 340 \geq 5$ and $n(1 - \hat{p}) = 1000(1 - 0.34) = 660 \geq 5$. Thus, we may use the Z interval for p.

(a) The margin of error is

$$E = Z_{\alpha/2}\sqrt{\frac{\hat{p}(1-\hat{p})}{n}} = 2.576\sqrt{\frac{0.34(1-0.34)}{1000}} \approx 0.0386.$$

We can estimate p, the true proportion of all *Quebecois* who favor independence for the Province of Quebec, within 0.0386 with 99% confidence.

(b) $\hat{p} \pm Z_{\alpha/2}\sqrt{\frac{\hat{p}(1-\hat{p})}{n}} = 0.34 \pm 2.576\sqrt{\frac{0.34(1-0.34)}{1000}} \approx 0.34 \pm 0.0386 = (0.3014, 0.3786)$. We are 99% confident that the true proportion of all *Quebecois* who favor independence for the Province of Quebec lies between 0.3014 and 0.3786.

17. (a)

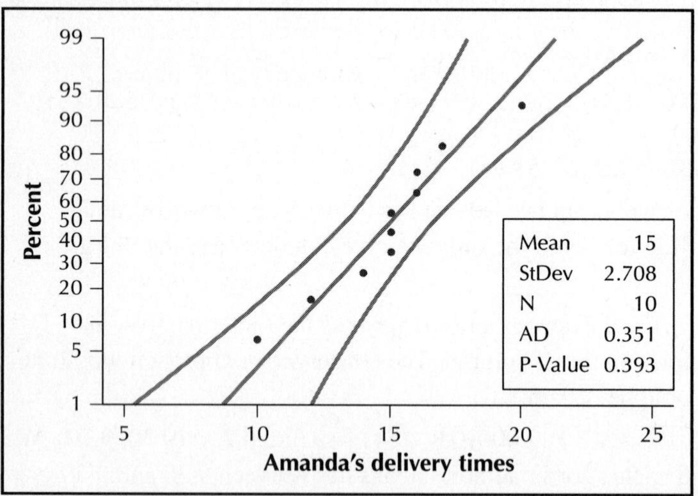

All points lie between the curved lines, so Amanda's delivery times are normally distributed.

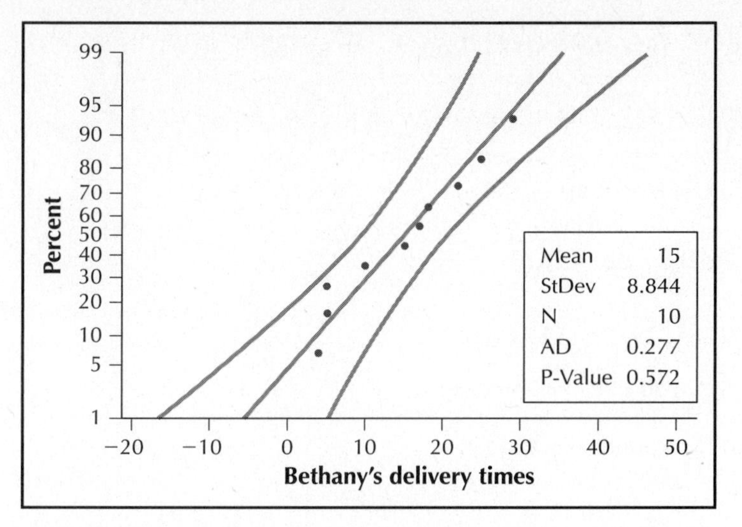

All points lie between the curved lines, so Bethany's delivery times are normally distributed.
df $= n - 1 = 10 - 1 = 9$, so $\chi^2_{\alpha/2} = \chi^2_{0.025} = 19.023$ and $\chi^2_{1-\alpha/2} = \chi^2_{0.975} = 2.700$.

(b) From Minitab:

Descriptive Statistics: Amandas times

Variable	Total Count	Variance
Amandas times	10	7.333

Thus,

$$\text{lower bound} = \sqrt{\frac{(n-1)s^2}{\chi^2_{\alpha/2}}} = \sqrt{\frac{(10-1)7.3333}{19.023}} \approx 1.863 \text{ and}$$

$$\text{upper bound} = \sqrt{\frac{(n-1)s^2}{\chi^2_{1-\alpha/2}}} = \sqrt{\frac{(10-1)7.3333}{2.7000}} \approx 4.944.$$

Therefore, the 95% confidence interval is (1.863, 4.944). We are 95% confident that the population standard deviation of Amanda's delivery times lies between 1.863 minutes and 4.944 minutes.

(c) From Minitab:

Descriptive Statistics: Bethanys times

Variable	Total Count	Variance
Bethanys times	10	78.22

Thus,

$$\text{lower bound} = \sqrt{\frac{(n-1)s^2}{\chi^2_{\alpha/2}}} = \sqrt{\frac{(10-1)78.22}{19.023}} \approx 6.083 \text{ and}$$

$$\text{upper bound} = \sqrt{\frac{(n-1)s^2}{\chi^2_{1-\alpha/2}}} = \sqrt{\frac{(10-1)78.22}{2.700}} \approx 16.147$$

Therefore, the 95% confidence interval is (6.083, 16.147). We are 95% confident that the population standard deviation of Bethany's delivery times lies between 6.083 minutes and 16.147 minutes.

(d) The confidence interval for the population standard deviation of Bethany's delivery times lies above the confidence interval for the population standard deviation of Amanda's delivery times. Thus, the interval estimate for the population standard deviation of Bethany's delivery times is greater than the interval estimate of the population standard deviation of Amanda's delivery times. This shows that there is strong statistical evidence that Bethany's delivery times are more highly variable than Amanda's.

18. **(a)** $n = \hat{p}(1 - \hat{p})\left(\frac{Z_{\alpha/2}}{E}\right)^2 = (0.38)\,(1 - 0.38)\left(\frac{1.96}{0.05}\right)^2 = 362.032384$. Round n up to 363.

(b) $n = \hat{p}(1 - \hat{p})\left(\frac{Z_{\alpha/2}}{E}\right)^2 = (0.38)(1 - 0.38)\left(\frac{2.576}{0.05}\right)^2 = 625.3555302$. Round n up to 626.

19. **(a)** $n = \left(\frac{(Z_{\alpha/2})\sigma}{E}\right)^2 = \left(\frac{(1.96)120}{10}\right)^2 = 553.1904$. Round n up to 554.

(b) $n = \left(\frac{(Z_{\alpha/2})\sigma}{E}\right)^2 = \left(\frac{(1.96)120}{5}\right)^2 = 2212.7616$. Round n up to 2213.

20. **(a)** $n = \left(\frac{0.5 \cdot Z_{\alpha/2}}{E}\right)^2 = \left(\frac{0.5 \cdot 1.645}{0.20}\right)^2 = 16.91265625$. Round n up to 17.

(b) Since the margin of error $E = 0.20$ is so large, it won't be very precise.

(c) $n = \left(\frac{0.5 \cdot Z_{\alpha/2}}{E}\right)^2 = \left(\frac{0.5 \cdot 1.645}{0.03}\right)^2 = 751.6736111$. Round n up to 752.

Chapter 9

Hypothesis Testing

Section 9.1

1. Hypothesis testing is the bedrock of the scientific method. It forms the basis for decision science. Statistical hypothesis testing is a way of formalizing the decision-making process so that a decision can be rendered about the value of a parameter. The researcher draws a conclusion based on the evidence provided by the sample data. Hypothesis testing is the statistical method for resolving conflicting claims about the value of a population parameter.

3. The alternate hypothesis represents an alternative claim about the value of the parameter.

5. There are three forms of the hypotheses in the hypothesis test for the population mean. They are as follows:

Form	Null hypothesis Alternative hypothesis
1	$H_0 : \mu \le \mu_0$ vs. $H_a : \mu > \mu_0$
2	$H_0 : \mu \ge \mu_0$ vs. $H_a : \mu < \mu_0$
3	$H_0 : \mu = \mu_0$ vs. $H_a : \mu \ne \mu_0$

7. The two possible decision errors in a criminal trial are finding the defendant guilty when in reality he did not commit the crime (a Type I error) and finding the defendant not guilty when in reality he did commit the crime (a Type II error).

9. $H_0 : \mu \le 10$ vs. $H_a : \mu > 10$

11. $H_0 : \mu = 0$ vs. $H_a : \mu \ne 0$

13. $H_0 : \mu = 4.0$ vs. $H_a : \mu \ne 4.0$

15. $H_0 : \mu = 36$ vs. $H_a : \mu \ne 36$

17. **(a)** ***Step 1.*** **Find the key words.** The key word "increased" is synonymous with "greater than," >.

Step 2. **Determine the form of the hypotheses.** The symbol > indicates that we should use the hypotheses for the right-tailed test in Table 9.1.

$$H_0 : \mu \le \mu_0 \quad \text{vs.} \quad H_a : \mu > \mu_0$$

Step 3. **Find the value for μ_0 and write your hypotheses.** Asking "greater than what?" (or "increased from what?"), we see that $\mu_0 = 43.9$, so the hypotheses are

$$H_0 : \mu \le 43.9 \quad \text{vs.} \quad H_a : \mu > 43.9$$

(b) Since, in actuality, the population mean referral rate of 45 per 1000 children is greater than the population mean referral rate of 43.9 per 1000 children reported in 2005, the null hypothesis should have been rejected. Since the null hypothesis was not rejected, a Type II error was made.

19. **(a)** ***Step 1.*** **Find the key words.** The key word "increased" is synonymous with "greater than," >.

Step 2. **Determine the form of the hypotheses.** The symbol > indicates that we should use the hypotheses for the right-tailed test in Table 9.1.

$$H_0: \mu \leq \mu_0 \quad \text{vs.} \quad H_a: \mu > \mu_0$$

Step 3. **Find the value for μ_0 and write your hypotheses.** Asking "greater than what?" (or "increased from what?"), we see that no value of μ_0 is given, so the hypotheses are

$$H_0: \mu \leq \mu_0 \quad \text{vs.} \quad H_a: \mu > \mu_0$$

(b) Since the magazine reported that travel costs had increased over the previous two years, they rejected the null hypothesis in favor of the alternative hypothesis. Since the population mean travel costs were actually lower, the null hypothesis is true. Therefore a Type I error was made.

21. (a) ***Step 1.*** **Find the key words.** The key words "in at most" mean "less than or equal to," ≤.
Step 2. **Determine the form of the hypotheses.** The symbol ≤ indicates that we should use the hypotheses for the right-tailed test in Table 9.1.

$$H_0: \mu \leq \mu_0 \quad \text{vs.} \quad H_a: \mu > \mu_0$$

Step 3. **Find the value for μ_0 and write your hypotheses.** Asking "less than or equal to what?" (or "in at most what?"), we see that $\mu_0 = 3$, so the hypotheses are

$$H_0: \mu \leq 3 \quad \text{vs.} \quad H_a: \mu > 3$$

(b) Since the population mean number of years that hybrid vehicles can recoup their initial cost through reduced fuel consumption is 2 years, this indicates that the null hypothesis should not be rejected. Since the study did not reject the null hypothesis, no error was made.

23. (a) ***Step 1.*** **Find the key words.** The key word "changed" is synonymous with "is not equal to," ≠.
Step 2. **Determine the form of the hypotheses.** The symbol ≠ indicates that we should use the hypotheses for the two-tailed test in Table 9.1.

$$H_0: \mu = \mu_0 \quad \text{vs.} \quad H_a: \mu \neq \mu_0$$

Step 3. **Find the value for μ_0 and write your hypotheses.** Asking "is not equal to what?" (or "changed from what?"), we see that $\mu_0 = 339.1$, so the hypotheses are

$$H_0: \mu = 339.1 \quad \text{vs.} \quad H_a: \mu \neq 339.1$$

(b) The two ways to make a correct decision are (1) to conclude that the *mean number* of fatal injury collisions is different from 339.1 per year when the *population mean number* of fatal injury collisions is actually different from 339.1 per year, and (2) to conclude that the *average number* of fatal injury collisions is equal to 339.1 per year when in actuality the *population mean number* of fatal injury collisions is equal to 339.1 per year. **(c)** A Type I error is concluding that the *mean number* of fatal injury collisions is different from 339.1 per year when the *population mean number* of fatal injury collisions is actually equal to 339.1 per year. **(d)** A Type II error is concluding that the *mean number* of fatal injury collisions is equal to 339.1 per year when actually the *population mean number* of fatal injury collisions is not equal to 339.1 per year.

25. (a) ***Step 1.*** **Find the key words.** The key word "changed" is synonymous with "is not equal to," ≠.
Step 2. **Determine the form of the hypotheses.** The symbol ≠ indicates that we should use the hypotheses for the two-tailed test in Table 9.1.

$$H_0: \mu = \mu_0 \quad \text{vs.} \quad H_a: \mu \neq \mu_0$$

Step 3. **Find the value for μ_0 and write your hypotheses.** Asking "is not equal to what?" (or "changed from what?"), we see that $\mu_0 = 175$, so the hypotheses are

$$H_0: \mu = 175 \quad \text{vs.} \quad H_a: \mu \neq 175$$

(b) The two ways that a correct decision could be made are (1) to conclude that the *mean height* of Americans this year is different from 175 centimeters, and (2) to conclude that the *mean height* of Americans is equal to 175 centimeters when the *population mean height* of Americans actually is equal to 175 centimeters.
(c) A Type I error is concluding that the *mean height* of Americans this year has changed from 175 centimeters when it actually is equal to 175 centimeters. **(d)** A Type II error is concluding that the *mean height* of Americans has not changed from 175 centimeters when it actually is not equal to 175 centimeters.

27. **(a)** ***Step 1.* Find the key words.** The key word "increased" is synonymous with "greater than," >.
***Step 2.* Determine the form of the hypotheses.** The symbol > indicates that we should use the hypotheses for the right-tailed test in Table 9.1.

$$H_0: \mu \le \mu_0 \quad \text{vs.} \quad H_a: \mu > \mu_0$$

***Step 3.* Find the value for μ_0 and write your hypotheses.** Asking "greater than what?" (or "increased from what?"), we see that $\mu_0 = 52{,}200$, so the hypotheses are

$$H_0: \mu \le 52{,}200 \quad \text{vs.} \quad H_a: \mu > 52{,}200$$

(b) The two ways that a correct decision could be made are (1) to conclude that the *mean salary* of college graduates is greater than $52,200 when the *population mean salary* of college graduates actually is greater than $52,200, and (2) to conclude that the *mean salary* of college graduates is less than or equal to $52,200 when it actually is less than or equal to $52,200. **(c)** A Type I error is concluding that the *mean salary* of college graduates is greater than $52,200 when it actually is less than or equal to $52,200. **(d)** A Type II error is concluding that the *mean salary* of college graduates is less than or equal to $52,200 when it is actually greater than $52,200.

Section 9.2

1. First we determine our null hypothesis and our alternative hypothesis. Then we calculate the test statistic

$$Z_{data} = \frac{\bar{x} - \mu_0}{\sigma/\sqrt{n}}.$$

Then we calculate the p-value based on the test statistic and the alternative hypothesis. If the p-value $< \alpha$, where α is the level of significance, we conclude that $\bar{x}$ is extreme.

3. The p-value is small when it is less than α.

5. In Example 9.9 the p-value $= 0.0668$. If $\alpha = 0.10$, then p-value $< \alpha$ so we would reject the null hypothesis. We would interpret our conclusion as: "There is sufficient evidence that the population mean user rating for a Dell XPS 410 computer is less than 7.2."

7. $Z_{data} = \dfrac{\bar{x} - \mu_0}{\sigma/\sqrt{n}} = \dfrac{10 - 12}{2/\sqrt{36}} = -6$

9. $Z_{data} = \dfrac{\bar{x} - \mu_0}{\sigma/\sqrt{n}} = \dfrac{14 - 12}{2/\sqrt{36}} = 6$

11. $Z_{data} = \dfrac{\bar{x} - \mu_0}{\sigma/\sqrt{n}} = \dfrac{97.5 - 100}{5/\sqrt{64}} = -4$

13. **(a)** Since p-value < 0.05, we reject the null hypothesis. **(b)** There is evidence that the population mean is less than μ_0.

15. **(a)** Since p-value ≥ 0.05, we do not reject the null hypothesis. **(b)** There is insufficient evidence that the population mean is less than μ_0.

17. **(a)** Since p-value < 0.05, we reject the null hypothesis. **(b)** There is evidence that the population mean is less than μ_0.

19. **(a)**

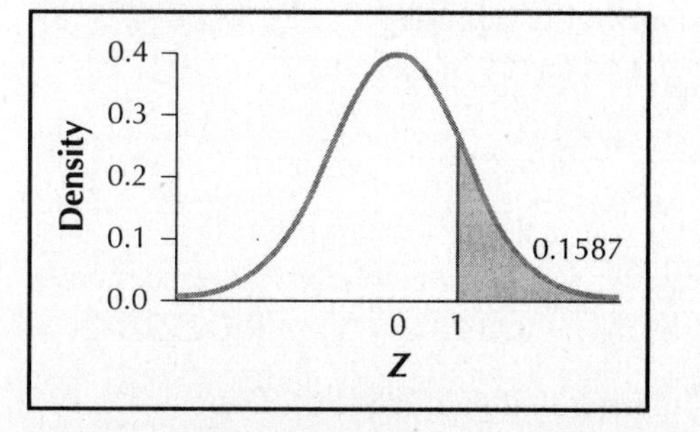

(b) p-value $= P(Z > 1) = 1 - P(Z < 1) = 1 - 0.8413 = 0.1587$

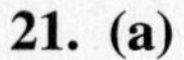

21. (a)

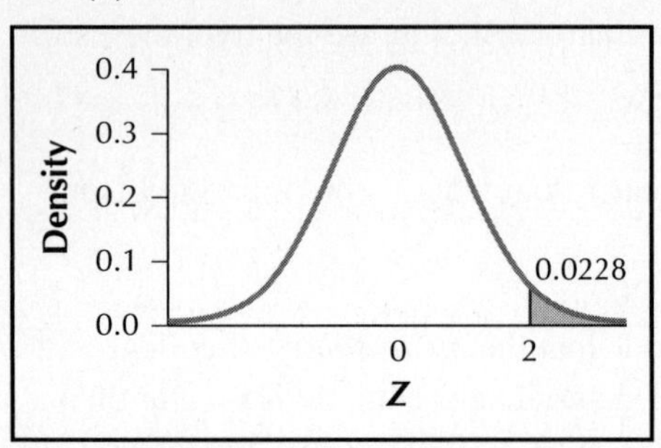

(b) p-value $= P(Z > 2) = 1 - P(Z < 2) = 1 - 0.9772 = 0.0228$

23. (a)

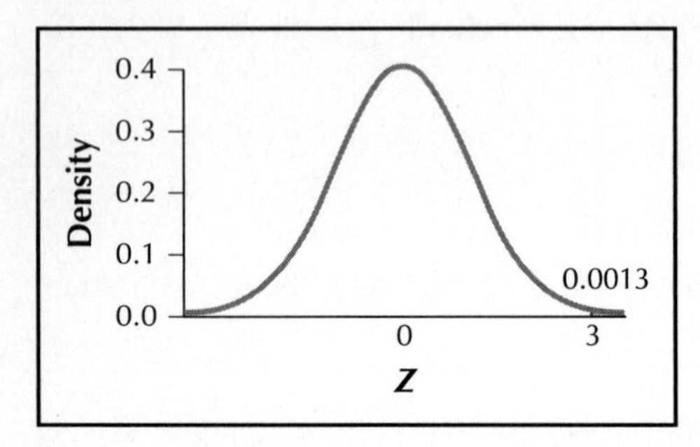

(b) p-value $= P(Z > 3) = 1 - P(Z < 3) = 1 - 0.9987 = 0.0013$

25. (a) $H_0 : \mu \leq 100$ vs. $H_a : \mu > 100$. Reject H_0 if p-value < 0.05.

(b) $Z_{data} = \dfrac{\bar{x} - \mu_0}{\sigma/\sqrt{n}} = \dfrac{102 - 100}{8/\sqrt{64}} = 2$

(c) p-value $= P(Z > 2) = 1 - P(Z < 2) = 1 - 0.9772 = 0.0228$

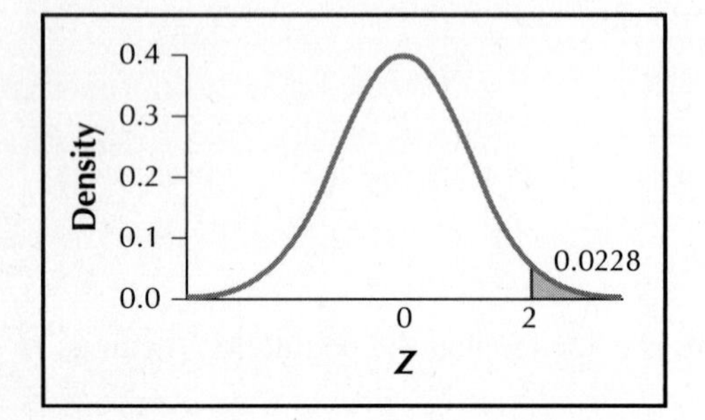

(d) Since p-value < 0.05, reject H_0. There is evidence that the population mean is greater than 100.

27. (a) $H_0 : \mu = 20$ vs. $H_a : \mu \neq 20$. Reject H_0 if p-value < 0.01.

(b) $Z_{data} = \dfrac{\bar{x} - \mu_0}{\sigma/\sqrt{n}} = \dfrac{27 - 20}{5/\sqrt{49}} = 9.8$

(c) p-value $= 2P(Z > 9.8) \approx 0$

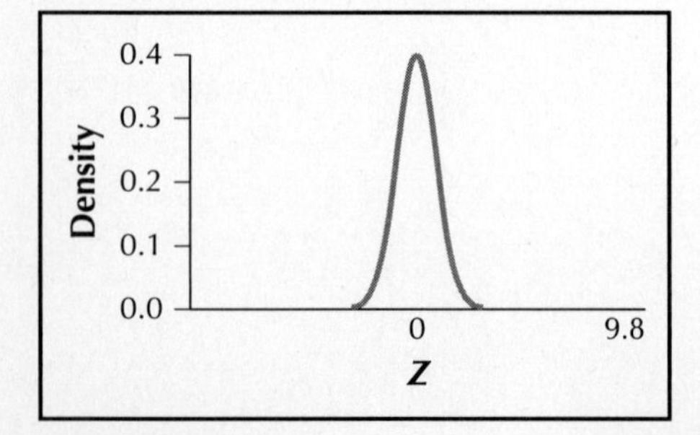

(d) Since p-value < 0.01, reject H_0. There is evidence that the population mean is different from 20.

29. Since $0.001 < p\text{-value} \leq 0.01$, there is very strong evidence against the null hypothesis that the population mean is greater than or equal to μ_0.

31. Since $0.15 < p\text{-value}$, there is no evidence against the null hypothesis that the population mean is greater than or equal to μ_0.

33. Since $p\text{-value} \leq 0.001$, there is extremely strong evidence against the null hypothesis that the population mean is greater than or equal to μ_0.

35. Since $0.15 < p\text{-value}$, there is no evidence against the null hypothesis that the population mean is less than or equal to μ_0.

37. **(a)** $H_0: \mu \leq 1{,}600{,}000{,}000$ vs. $H_a: \mu > 1{,}600{,}000{,}000$. Reject H_0 if $p\text{-value} < 0.05$.

(b) $Z_{data} = \dfrac{\bar{x} - \mu_0}{\sigma/\sqrt{n}} = \dfrac{1{,}500{,}000{,}000 - 1{,}600{,}000{,}000}{5{,}000{,}000{,}000/\sqrt{36}} = -1.2$

(c) $p\text{-value} = P(Z > -1.2) = 1 - P(Z < -1.2) = 1 - 0.1151 = 0.8849$

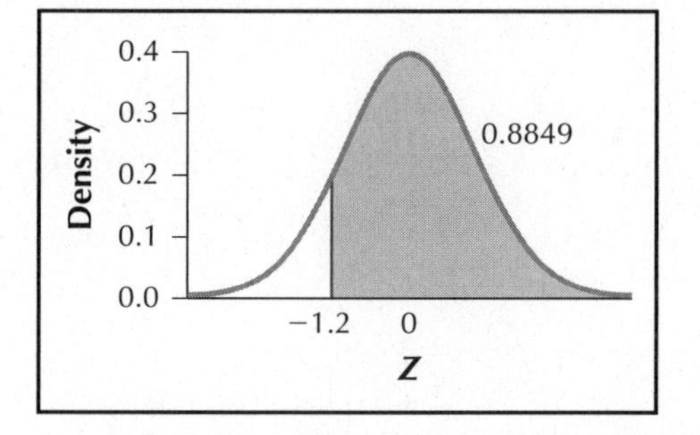

(d) Since $p\text{-value} \geq 0.05$, do not reject H_0. There is insufficient evidence that the population daily mean number of shares traded is greater than 1.6 billion shares.

39. **(a)** $H_0: \mu \leq 2$ vs. $H_a: \mu > 2$. Reject H_0 if $p\text{-value} < 0.05$.

(b) $Z_{data} = \dfrac{\bar{x} - \mu_0}{\sigma/\sqrt{n}} = \dfrac{4 - 2}{0.5/\sqrt{36}} = 24$

(c) $p\text{-value} = P(Z > 24) \approx 0$

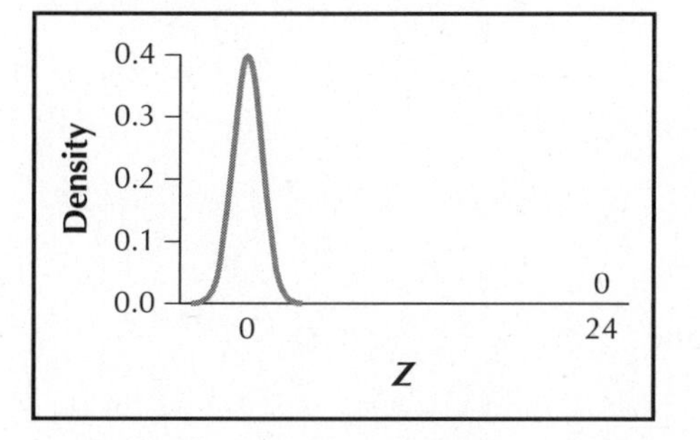

(d) Since $p\text{-value} < 0.05$, reject H_0. There is evidence that the population mean temperature increase in California is greater than 2°F.

41. **(a)** $H_0: \mu \leq 47.2$ vs. $H_a: \mu > 47.2$. Reject H_0 if $p\text{-value} < 0.01$.

(b) $Z_{data} = \dfrac{\bar{x} - \mu_0}{\sigma/\sqrt{n}} = \dfrac{219.7 - 47.2}{12/\sqrt{36}} = 16.60$

(c) p-value $= P(Z > 16.60) \approx 0$

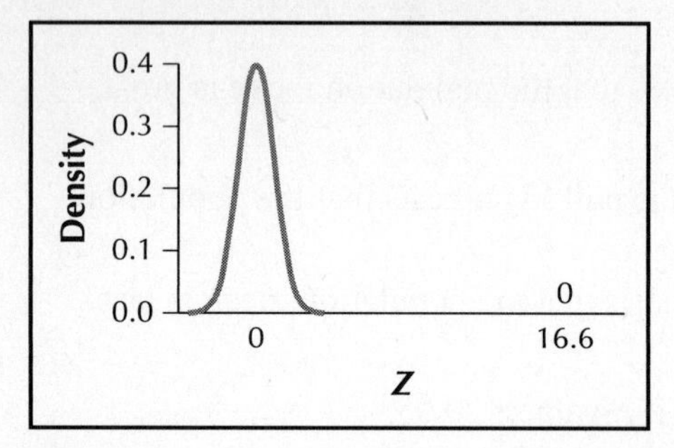

(d) Since p-value < 0.01, reject H_0. There is evidence that the population mean DDT-level in the breast milk of Hispanic women in the Yakima valley is greater than 47.2 parts per billion.

43. (a) $H_0 : \mu \geq 3$ vs. $H_a : \mu < 3$. Reject H_0 if p-value < 0.01.

(b) $Z_{data} = \dfrac{\bar{x} - \mu_0}{\sigma/\sqrt{n}} = \dfrac{2.1 - 3}{0.2/\sqrt{9}} = -13.5$

(c) p-value $= P(Z < -13.5) \approx 0$

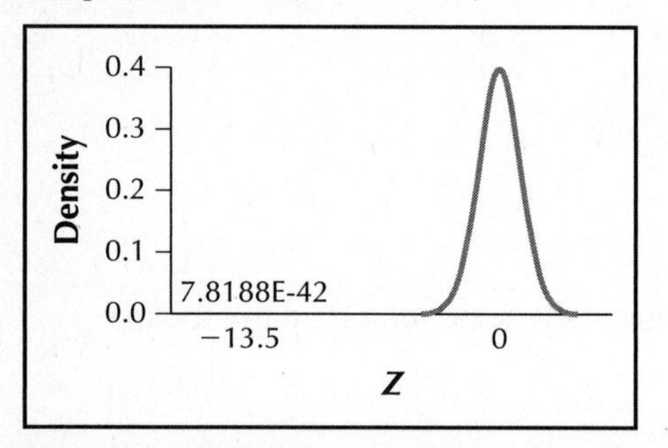

(d) Since p-value < 0.01, reject H_0. There is evidence that the population mean time it takes hybrid cars to recoup their initial cost is less than 3 years.

45. (a) Since all of the points lie within the curved lines, the normal probability plot indicates acceptable normality. Thus Case 1 applies so we may proceed to apply the Z test.

(b) $H_0 : \mu \geq 210$ vs. $H_a : \mu < 210$. Reject H_0 if p-value < 0.01.

$$Z_{data} = \frac{\bar{x} - \mu_0}{\sigma/\sqrt{n}} = \frac{192.39 - 210}{50/\sqrt{23}} = -1.69$$

p-value $= P(Z < -1.69) = 0.0455$

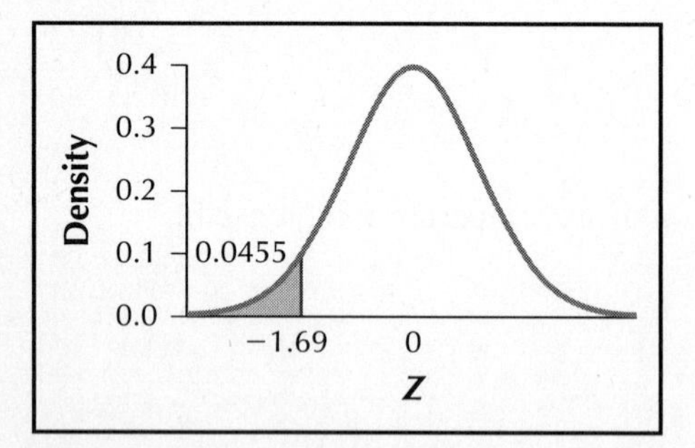

Since p-value ≥ 0.01, do not reject H_0. There is insufficient evidence that the population mean sodium content per serving of breakfast cereal is less than 210 grams.

47. (a) $H_0 : \mu \geq 210$ vs. $H_a : \mu < 210$. Reject H_0 if p-value < 0.05.

$$Z_{data} = \frac{\bar{x} - \mu_0}{\sigma/\sqrt{n}} = \frac{192.39 - 210}{50/\sqrt{23}} = -1.69$$

p-value $= P(Z < -1.69) = 0.0455$

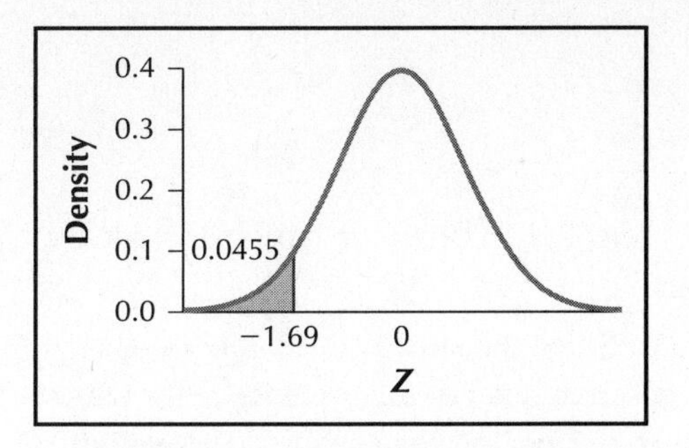

Since p-value < 0.05, reject H_0. There is evidence that the population mean sodium content per serving of breakfast cereal is less than 210 grams. **(b)** The conclusion is different this time because α was increased to a value greater than the p-value. The data have not changed. **(c)** The first alternative would be to report the p-value and assess the strength of the evidence against the null hypothesis. The second alternative is to obtain more data.

49. **(a)** Since the data line up along the center line, the normal probability plot indicates acceptable normality. Therefore, Case 1 applies, so we may apply the Z test. **(b)** $H_0 : \mu \geq 78$ vs. $H_a : \mu < 78$. Reject H_0 if p-value < 0.05.

$$Z_{data} = \frac{\bar{x} - \mu_0}{\sigma/\sqrt{n}} = \frac{75.6 - 78}{9/\sqrt{15}} = -1.03$$

p-value $= P(Z < -1.03) = 0.1515$

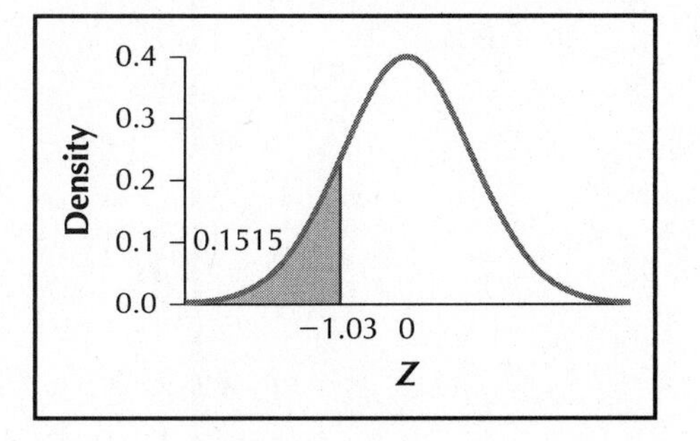

Since p-value ≥ 0.05, do not reject H_0. There is insufficient evidence that the population mean heart rate for all women is less than 78 beats per minute. **(c)** $H_0 : \mu = 78$ vs. $H_a : \mu \neq 78$. Reject H_0 if p-value < 0.05.

$$Z_{data} = \frac{\bar{x} - \mu_0}{\sigma/\sqrt{n}} = \frac{75.6 - 78}{9/\sqrt{15}} = -1.03$$

p-value $= 2P(Z < -1.03) = 2(0.1515) = 0.303$

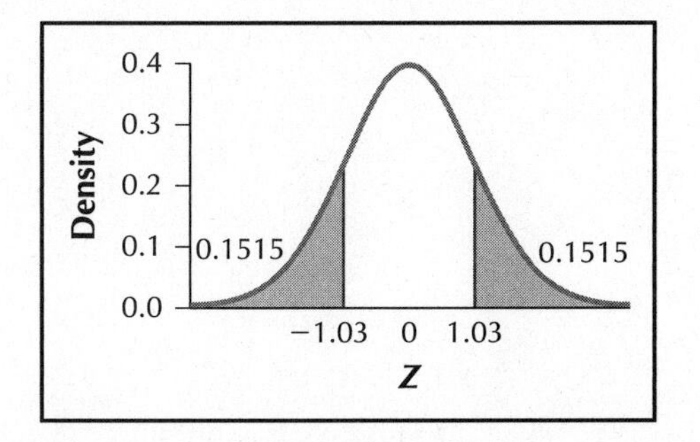

Since p-value ≥ 0.05, do not reject H_0. There is insufficient evidence that the population mean heart rate for all women is different than 78 beats per minute.

51. **(a)** $H_0: \mu \geq 3.15$ vs. $H_a: \mu < 3.15$. Reject H_0 if p-value < 0.05.

(b) $Z_{data} = \dfrac{\bar{x} - \mu_0}{\sigma/\sqrt{n}} = \dfrac{3.05 - 3.15}{1/\sqrt{225}} = -1.5$

(c) p-value $= P(Z < -1.5) = 0.0668$ **(d)** Since p-value ≥ 0.05, do not reject H_0. There is insufficient evidence that the population mean family size in America is less than 3.15 persons.

53. **(a)** We would reject H_0 for any p-value that is less than or equal to α. Since the smallest possible p-value is 0, the smallest p-value for which we would reject H_0 is 0. The smallest possible level of significance α for which we would reject H_0 is the p-value, which is 0.0668. **(b)** The assumption $\sigma = 1$ person seems valid. This value would have been obtained from the sample standard deviation. **(c)** Since our conclusion is do not reject H_0, the only type of error that we might be making is a Type II error. Since we are not rejecting H_0, we know that we are not making a Type I error. **(d)** This headline is not supported by the data and our hypothesis test.

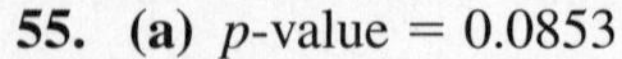

55. **(a)** p-value $= 0.0853$

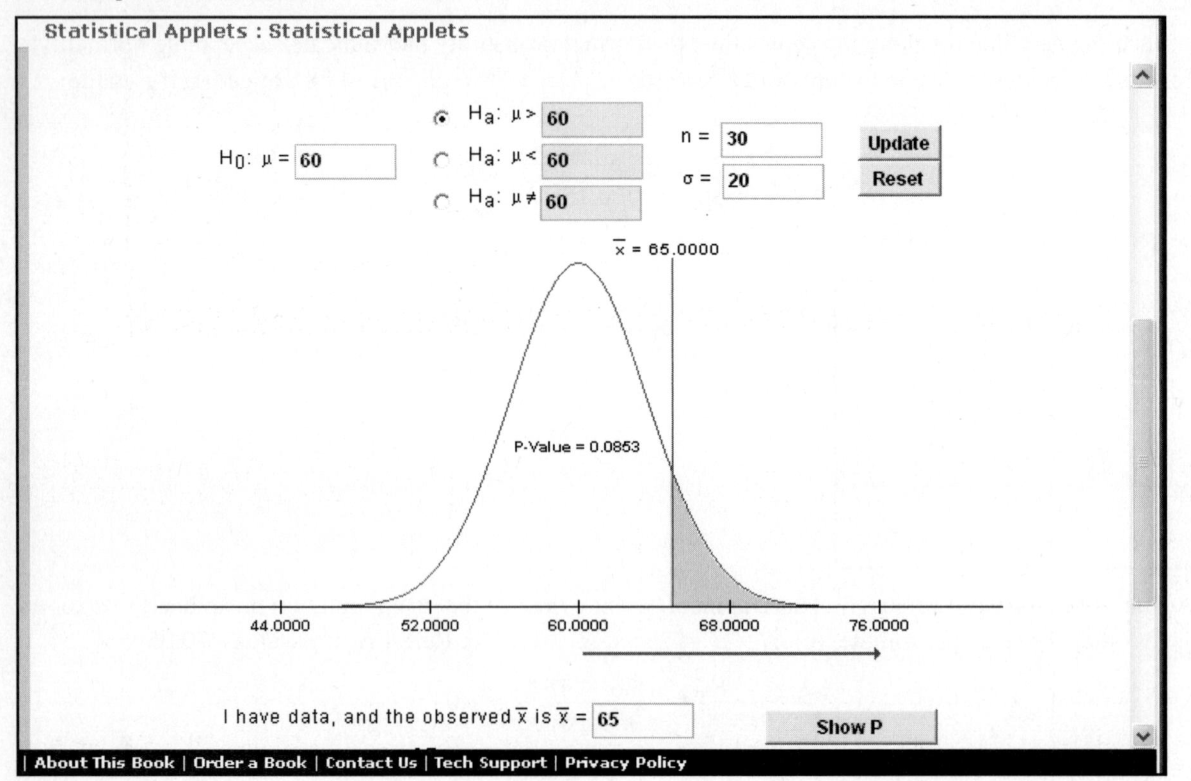

(b) Since p-value ≥ 0.05, we would not reject H_0.

57. **(a)** p-value $= 0.0384$

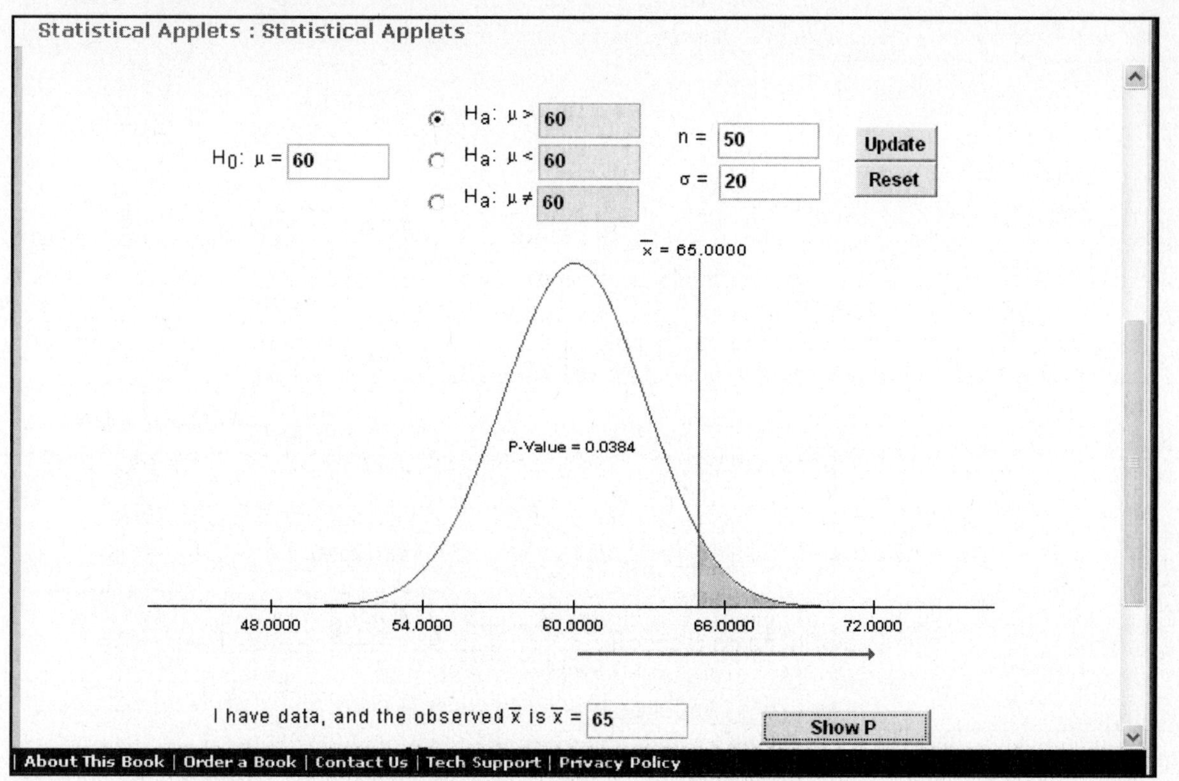

(b) Since p-value < 0.05, we would reject H_0.

Section 9.3

1. The quantity Z_{crit} represents a threshold for significance.

3. The quantity α represents a threshold for significance.

5. The p-value represents a probability.

7. The quantities Z_{data} and Z_{crit} are not directly related.

9. The quantities Z_{data} and the p-value are directly related. For a right-tailed test, the p-value is the area underneath the standard normal curve to the right of Z_{data}, so p-value $= P(Z > Z_1 data)$. For a left-tailed test, the p-value is the area underneath the standard normal curve to the left of Z_{data}, so p-value $= P(Z < Z_1 data)$. For a two-tailed test, the p-value is the area underneath the standard normal curve to the right of $|Z_{data}|$ plus the area underneath the standard normal curve to the left of $-|Z_{data}|$, so p-value $= P(Z > |Z_{data}|) + P(Z < -|Z_{data}|)$.

11. The p-value and Z_{crit} are not directly related.

13. **(a)** For a left-tailed test with $\alpha = 0.10$, $Z_{crit} = -1.28$. **(b)** Reject H_0 if $Z_{data} < -1.28$.

(c)

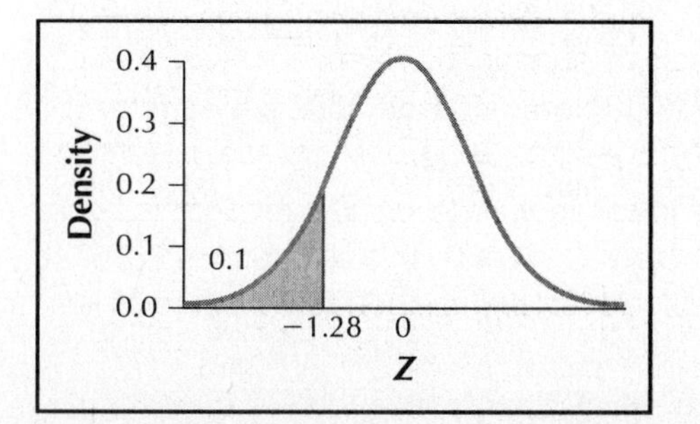

15. **(a)** For a left-tailed test with $\alpha = 0.01$, $Z_{crit} = -2.33$. **(b)** Reject H_0 if $Z_{data} < -2.33$.
(c)

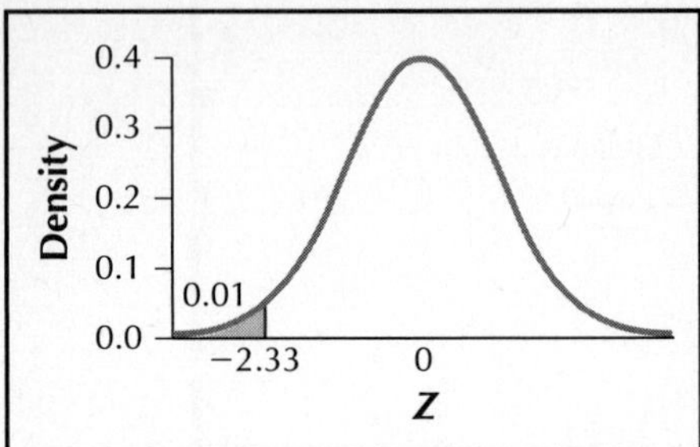

17. **(a)** For a left-tailed test with $\alpha = 0.05$, $Z_{crit} = -1.645$. **(b)** Reject H_0 if $Z_{data} < -1.645$.
(c)

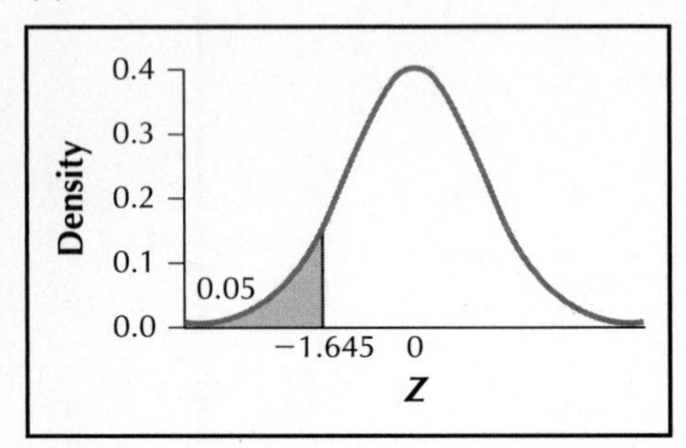

19. **(a)** For a two-tailed test with $\alpha = 0.01$, $Z_{crit} = 2.58$. **(b)** Reject H_0 if $Z_{data} < -2.58$ or $Z_{data} > 2.58$.
(c)

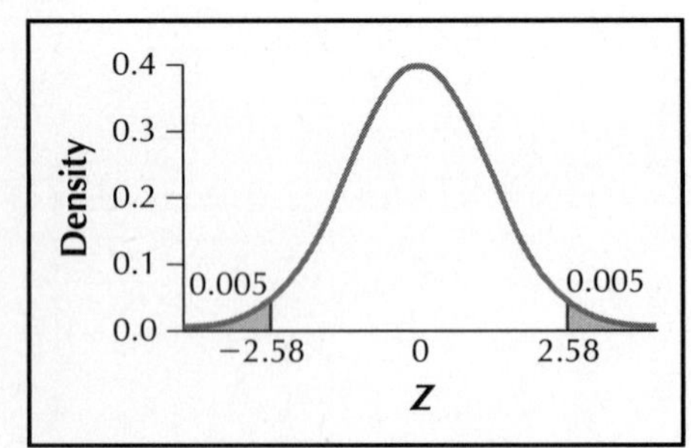

21. **(a)** For a right-tailed test with $\alpha = 0.05$, $Z_{crit} = 1.645$. **(b)** Reject H_0 if $Z_{data} > 1.645$.
(c)

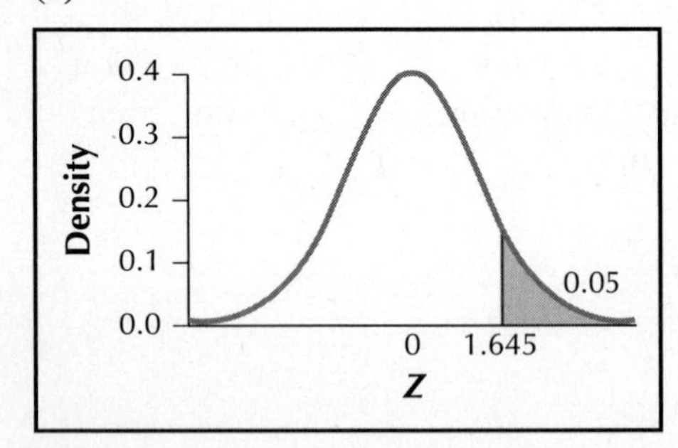

(d) $Z_{data} > 1.645$, so we reject H_0. **(e)** There is evidence that the population mean is greater than μ_0.
23. **(a)** For a right-tailed test with $\alpha = 0.05$, $Z_{crit} = 1.645$. **(b)** Reject H_0 if $Z_{data} > 1.645$.

(c)

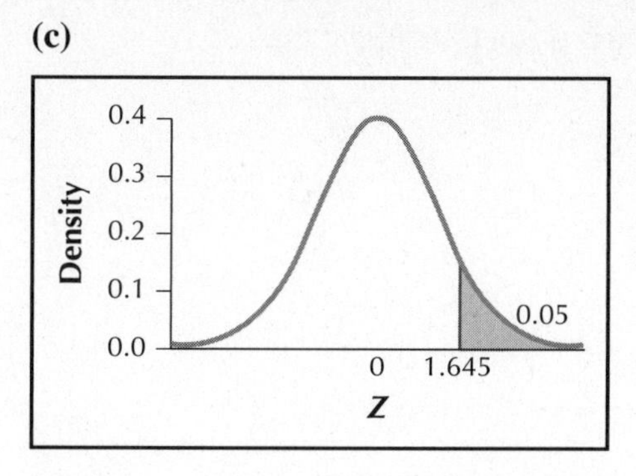

(d) $Z_{data} \leq 1.645$, so we do not reject H_0. **(e)** There is insufficient evidence that the population mean is greater than μ_0.

25. (a) For a left-tailed test with $\alpha = 0.10$, $Z_{crit} = -1.28$. **(b)** Reject H_0 if $Z_{data} < -1.28$.

(c)

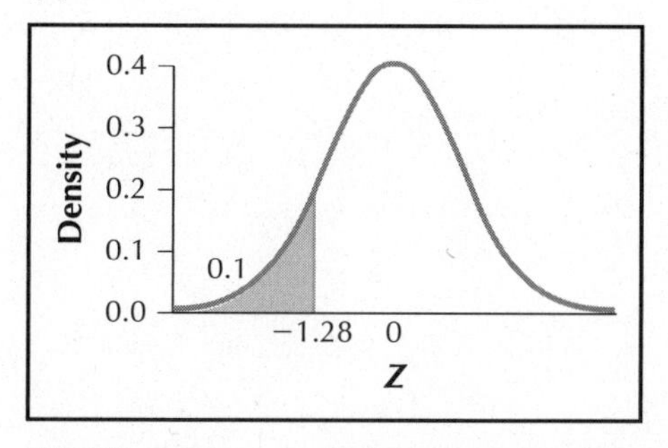

(d) Since $Z_{data} < -1.28$, we reject H_0. **(e)** There is evidence that the population mean is less than μ_0.

27. (a) $H_0 : \mu \leq 100$ vs. $H_a : \mu > 100$ **(b)** For a right-tailed test with $\alpha = 0.05$, $Z_{crit} = 1.645$. Reject H_0 if $Z_{data} > 1.645$.

(c) $Z_{data} = \dfrac{\bar{x} - \mu_0}{\sigma/\sqrt{n}} = \dfrac{102 - 100}{8/\sqrt{64}} = 2$

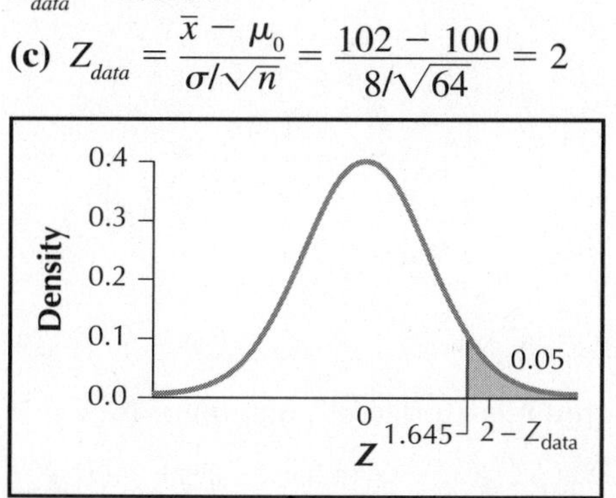

(d) Since $Z_{data} > 1.645$, we reject H_0. There is evidence that the population mean is greater than 100.

29. (a) $H_0 : \mu = 20$ vs. $H_a : \mu \neq 20$ **(b)** For a two-tailed test with $\alpha = 0.01$, $Z_{crit} = 2.58$. Reject H_0 if $Z_{data} < -2.58$ or $Z_{data} > 2.58$.

(c) $Z_{data} = \dfrac{\bar{x} - \mu_0}{\sigma/\sqrt{n}} = \dfrac{27 - 20}{5/\sqrt{49}} = 9.8$

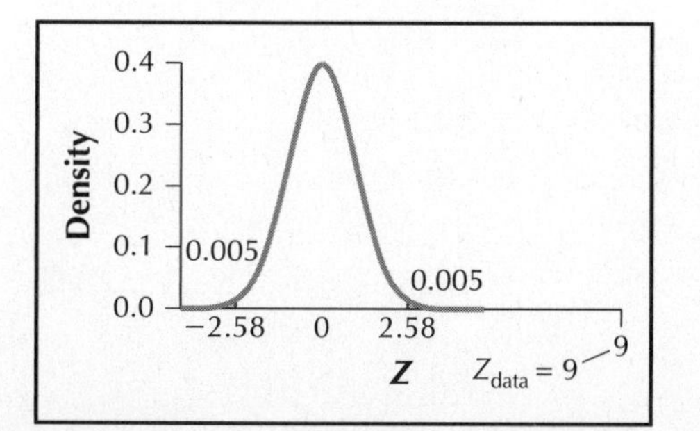

(d) Since $Z_{data} > 2.58$, we reject H_0. There is evidence that the population mean is different than 20.

31. (a) $H_0 : \mu = -3$ vs. $H_a : \mu \neq -3$
Since $\mu_0 = -3$ does not lie in the confidence interval, we reject H_0.
(b) $H_0 : \mu = -2$ vs. $H_a : \mu \neq -2$
Since $\mu_0 = -2$ lies in the confidence interval, we do not reject H_0.
(c) $H_0 : \mu = 0$ vs. $H_a : \mu \neq 0$
Since $\mu_0 = 0$ lies in the confidence interval, we do not reject H_0.
(d) $H_0 : \mu = 5$ vs. $H_a : \mu \neq 5$
Since $\mu_0 = 5$ lies in the confidence interval, we do not reject H_0.
(e) $H_0 : \mu = 7$ vs. $H_a : \mu \neq 7$
Since $\mu_0 = 7$ does not lie in the confidence interval, we reject H_0.

33. (a) $H_0 : \mu \leq 3.24$ vs. $H_a : \mu > 3.24$
(b) For a right-tailed test with $\alpha = 0.05$, $Z_{crit} = 1.645$. Reject H_0 if $Z_{data} > 1.645$.

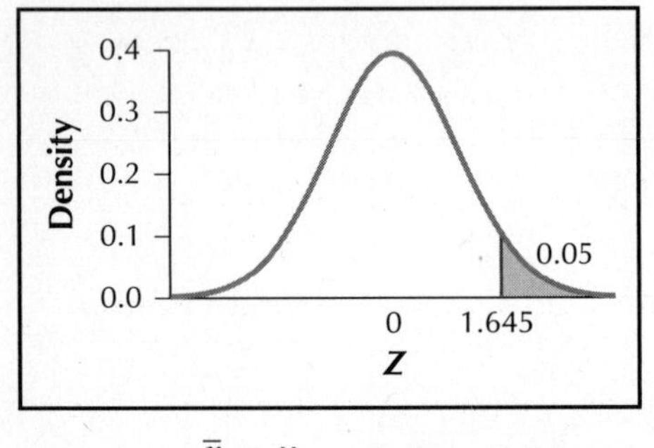

(c) $Z_{data} = \dfrac{\bar{x} - \mu_0}{\sigma/\sqrt{n}} = \dfrac{3.40 - 3.24}{0.25/\sqrt{100}} = 6.4$

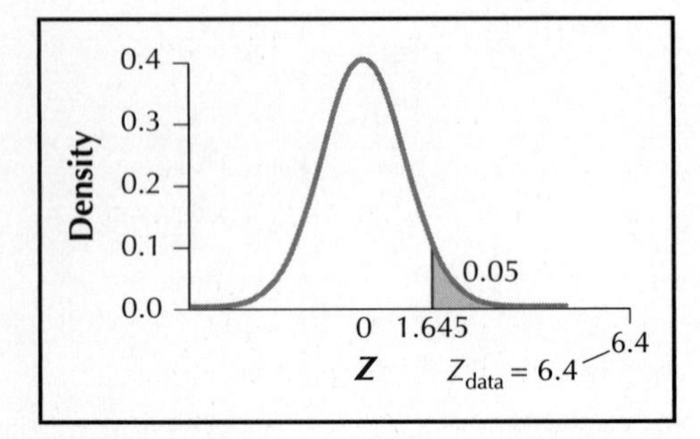

(d) Since $Z_{data} > 1.645$, we reject H_0. There is evidence that the population mean price of a gallon of milk this year is greater than \$3.24 per gallon.

35. (a) $H_0 : \mu \leq 50{,}000$ vs. $H_a : \mu > 50{,}000$ **(b)** For a right-tailed test with $\alpha = 0.01$, $Z_{crit} = 2.33$. Reject H_0 if $Z_{data} > 2.33$.

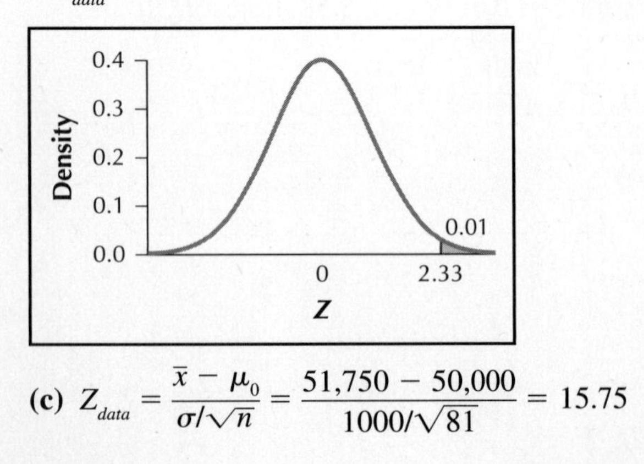

(c) $Z_{data} = \dfrac{\bar{x} - \mu_0}{\sigma/\sqrt{n}} = \dfrac{51{,}750 - 50{,}000}{1000/\sqrt{81}} = 15.75$

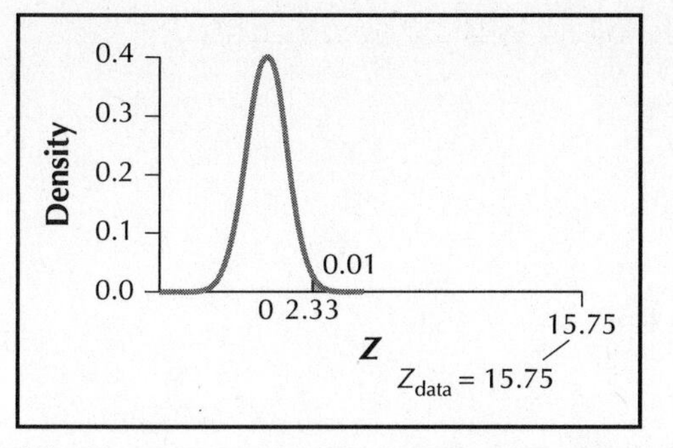

(d) Since $Z_{data} > 2.33$ we reject H_0. There is evidence that the population mean salary of Assistant Professors in Science is greater than \$50,000.

37. **(a)** For a 95% confidence interval, $Z_{\alpha/2} = 1.96$.

$$\bar{x} \pm Z_{\alpha/2}\frac{\sigma}{\sqrt{n}} = 24{,}000 \pm 1.96\left(\frac{3000}{\sqrt{49}}\right) = 24{,}000 \pm 840 = (23{,}160,\ 24{,}840)$$

(b) **(i)** $H_0: \mu = 24{,}000$ vs. $H_a: \mu \neq 24{,}000$. Since $\mu_0 = 24{,}000$ lies in the confidence interval, we do not reject H_0. **(ii)** $H_0: \mu = 23{,}000$ vs. $H_a: \mu \neq 23{,}000$. Since $\mu_0 = 23{,}000$ does not lie in the confidence interval, we reject H_0. **(iii)** $H_0: \mu = 23{,}200$ vs. $H_a: \mu \neq 23{,}200$. Since $\mu_0 = 23{,}200$ lies in the confidence interval, we do not reject H_0. **(iv)** $H_0: \mu = 25{,}000$ vs. $H_a: \mu \neq 25{,}000$. Since $\mu_0 = 25{,}000$ does not lie in the confidence interval, we reject H_0.

39. **(a)** Since the hypotheses are not based on the sample data, they would be unchanged if the sample mean premium was larger than \$11,750. **(b)** Since Z_{crit} is not based on the sample data, it would be unchanged if the sample mean premium was larger than \$11,750. **(c)** Since the critical region is not based on the sample data, it would be unchanged if the sample mean premium was larger than \$11,750.

(d) Since $Z_{data} = \dfrac{\bar{x} - \mu_0}{\sigma/\sqrt{n}}$, a value of $\bar{x}$ larger than \$11,750 would result in a value of Z_{data} greater than 1.67.

(e) From (d), a value of $\bar{x}$ larger than \$11,750 would result in a value of Z_{data} larger than 1.67. Therefore, Z_{data} would still be larger than 1.645, so the conclusion would still be reject H_0.

41. The histogram indicates that the data is extremely right-skewed and therefore not normally distributed. Thus Case 1 does not apply. Since the sample size of $n = 16$ is small ($n < 30$), Case 2 does not apply. Thus it is not appropriate to apply the Z test.

43. **(a)** Since the sample size of $n = 100$ is large ($n \geq 30$), Case 2 applies so it is appropriate to apply the Z test. **(b)** Even though the sample mean $\bar{x} = 6.2$ cents per mile is greater than the hypothesized mean $\mu_0 = 5.9$ cents per mile, this is not enough by itself to reject the null hypothesis. It also depends on the variability of the data and on α. **(c)** $\sigma_{\bar{x}} = \dfrac{\sigma}{\sqrt{n}} = \dfrac{1.5}{\sqrt{100}} = 0.15$ cents per mile

(d) Since

$$\bar{x} = 6.2 \text{ cents per mile is } \frac{\bar{x} - \mu_0}{\sigma_{\bar{x}}} = \frac{6.2 - 5.9}{0.15} = 2$$

standard errors above $\mu_0 = 5.9$ cents per mile, it is mildly extreme.

45. **(a)** $H_0: \mu \leq 5.9$ vs. $H_a: \mu > 5.9$

(b) For a right-tailed test with $\alpha = 0.05$, $Z_{crit} = 1.645$. Reject H_0 if $Z_{data} > 1.645$.

(c) $Z_{data} = \dfrac{\bar{x} - \mu_0}{\sigma/\sqrt{n}} = \dfrac{6.2 - 5.9}{1.5/\sqrt{100}} = 2$

Since $Z_{data} > 1.645$, we reject H_0. There is evidence that the population mean cost of operating an automobile in the United States is greater than 5.9 cents per mile.

Section 9.4

1. The population standard deviation σ is known.
3. Case 1: The population is normal.
   Case 2: The sample size is large ($n \geq 30$).
5. For a left-tailed test with $\alpha = 0.10$ and $df = n - 1 = 12 - 1 = 11$, $t_{crit} = -1.363$.

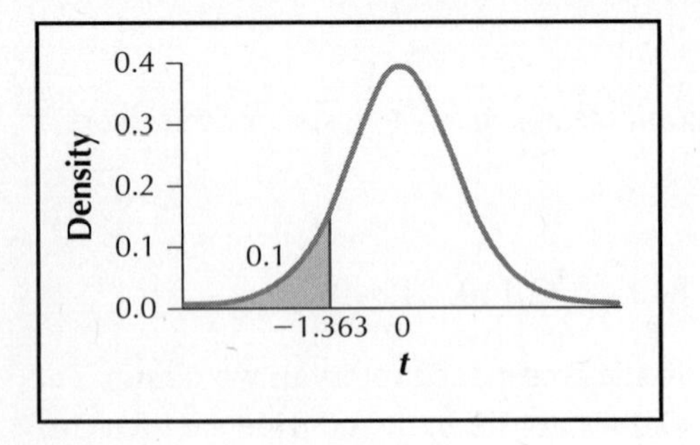

7. For a left-tailed test with $\alpha = 0.01$ and $df = n - 1 = 12 - 1 = 11$, $t_{crit} = -2.718$.

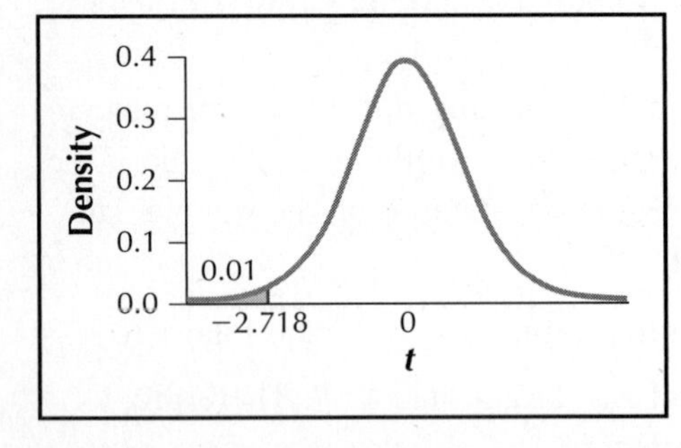

9. For a two-tailed test with $\alpha = 0.10$ and $df = n - 1 = 18 - 1 = 17$, $t_{crit} = 1.740$.

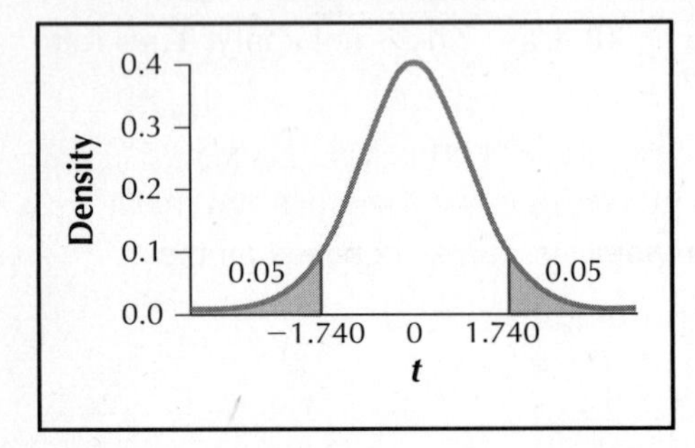

11. For a two-tailed test with $\alpha = 0.01$ and $df = n - 1 = 18 - 1 = 17$, $t_{crit} = 2.898$.

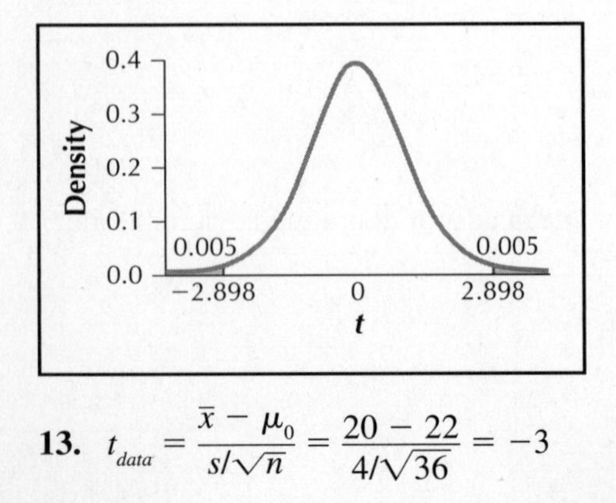

13. $t_{data} = \dfrac{\bar{x} - \mu_0}{s/\sqrt{n}} = \dfrac{20 - 22}{4/\sqrt{36}} = -3$

15. $t_{data} = \dfrac{\bar{x} - \mu_0}{s/\sqrt{n}} = \dfrac{10 - 11}{3/\sqrt{64}} = -2.67$

17. $t_{data} = \dfrac{\bar{x} - \mu_0}{s/\sqrt{n}} = \dfrac{100 - 102}{10/\sqrt{81}} = 3.6$

19. For a one-tailed test with df $= n - 1 = 8 - 1 = 7$ and $t_{data} = 2.5$, $2.365 < t_{data} < 2.998$, so $0.01 < p$-value < 0.025.

21. For a two-tailed test with df $= n - 1 = 18 - 1 = 17$ and $t_{data} = 2.0$, $1.734 < t_{data} < 2.101$, so $0.05 < p$-value < 0.10.

23. For a one-tailed test with df $= n - 1 = 12 - 1 = 11$ and $t_{data} = -2.5$, $-2.718 < t_{data} < -2.201$, so $0.01 < p$-value < 0.025.

25. Solid evidence.

27. Mild evidence.

29. Solid evidence.

31. **(a)** $H_0 : \mu = 0$ vs. $H_a : \mu \neq 0$. Reject H_0 if p-value < 0.05.

(b) $t_{data} = \dfrac{\bar{x} - \mu_0}{s/\sqrt{n}} = \dfrac{1 - 0}{0.5/\sqrt{9}} = 6$

(c) df $= n - 1 = 9 - 1 = 8$, p-value $= 2P(t > 6) = 0.0003233933213$

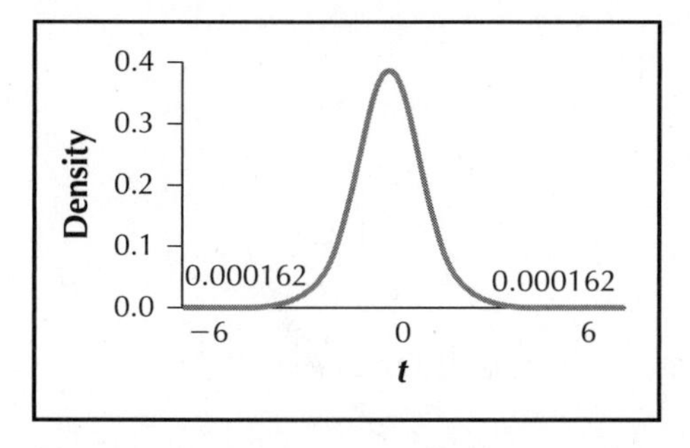

(d) Since p-value < 0.05, we reject H_0. There is evidence that the population mean is different than 0.

33. **(a)** $H_0 : \mu \geq 28$ vs. $H_a : \mu < 28$. Reject H_0 if p-value < 0.05.

(b) $t_{data} = \dfrac{\bar{x} - \mu_0}{s/\sqrt{n}} = \dfrac{27 - 28}{10/\sqrt{100}} = -1$

(c) df $= n - 1 = 100 - 1 = 99$, p-value $= P(t < -1) = 0.1598742373$

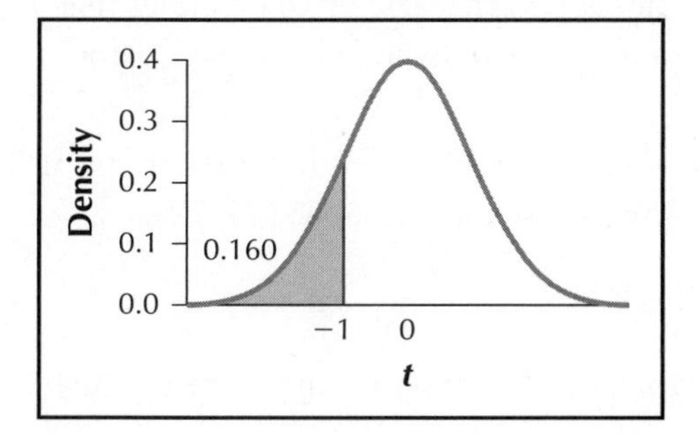

(d) Since p-value ≥ 0.05, we do not reject H_0. There is insufficient evidence that the population mean is less than 28.

35. **(a)** $H_0 : \mu \leq 100$ vs. $H_0 : \mu > 100$ **(b)** For a right-tailed test with $\alpha = 0.01$ and df $= n - 1 = 25 - 1 = 24$, $t_{crit} = 2.492$. Reject H_0 if $t_{data} > 2.492$.

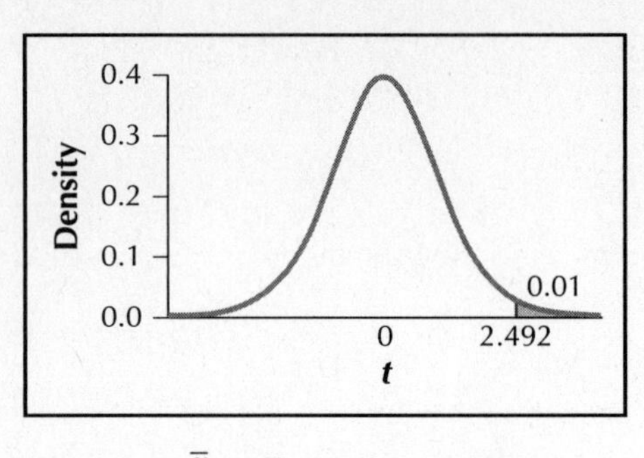

(c) $t_{data} = \dfrac{\bar{x} - \mu_0}{s/\sqrt{n}} = \dfrac{104 - 100}{10/\sqrt{25}} = 2$

(d) Since $t_{data} \le 2.492$, we do not reject H_0. There is insufficient evidence that the population mean is greater than 100.

37. (a) $H_0 : \mu = 9$ vs. $H_0 : \mu \neq 9$ **(b)** For a two-tailed test with $\alpha = 0.10$ and df $= n - 1 = 36 - 1 = 35$, $t_{crit} = 1.690$. Reject H_0 if $t_{data} < -1.690$ or $t_{data} > 1.690$.

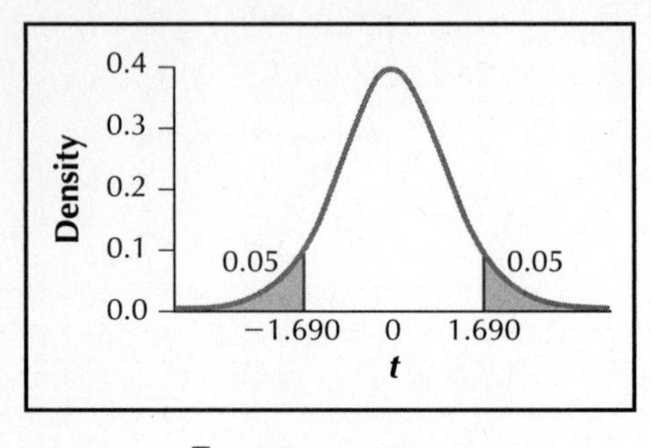

(c) $t_{data} = \dfrac{\bar{x} - \mu_0}{s/\sqrt{n}} = \dfrac{10 - 9}{3/\sqrt{36}} = 2$

(d) Since $t_{data} > 1.690$, we reject H_0. There is evidence that the population mean is different than 9.

39. (a) $s_{\bar{x}} = \dfrac{s}{\sqrt{n}} = \dfrac{8.5021}{\sqrt{14}} = 2.2723$

(b) The sample mean number of home runs $\bar{x} = 13.8571$ lies 0.94 standard errors below the hypothesized mean number of home runs $\mu_0 = 16$.

41. (a) $s_{\bar{x}} = \dfrac{s}{\sqrt{n}} = \dfrac{230}{\sqrt{17}} = 55.7832$ **(b)** The sample mean math journal price $\bar{x} = \$506.82$ lies 1.91 standard errors above the hypothesized mean math journal price $\mu_0 = \$400$.

43. (a) Since the sample size of $n = 13$ is small ($n < 30$), Case 2 does not apply. **(b)** The boxplot shows that the data is right-skewed and therefore not normal. Thus Case 1 does not apply. **(c)** Since neither Case 1 nor Case 2 applies, we may not proceed with the *t* test.

45. (a) $H_0 : \mu \le 60$ vs. $H_a : \mu > 60$ **(b)** For a right-tailed test with $\alpha = 0.05$ and df $= n - 1 = 14 - 1 = 13$, $t_{crit} = 1.771$. Reject H_0 if $t_{data} > 1.771$. **(c)** From the TI-84, $\bar{x} = 70.64285714$ and $s = 15.93065329$. Thus

$$t_{data} = \frac{\bar{x} - \mu_0}{s/\sqrt{n}} = \frac{70.64285714 - 60}{15.93065329/\sqrt{14}} = 2.50.$$

(d) Since $t_{data} > 1.771$, we reject H_0. There is evidence that the population mean response time is greater than 60 milliseconds.

47. (a) Since the prices per bottle for *Obsession* by Calvin Klein are normally distributed, Case 1 applies, so it is appropriate to apply the *t* test. **(b)** Whether or not H_0 is rejected depends not only on whether or not the sample mean is greater than the hypothesized mean but also on the variability of the distribution. Therefore a sample mean that is greater than the hypothesized mean is not enough to reject H_0.

(c) $t_{data} = \dfrac{\bar{x} - \mu_0}{s/\sqrt{n}} = \dfrac{15{,}000 - 12{,}300}{2.50/\sqrt{15}} = 3.87$

(d) $H_0: \mu \le 46.42$ vs. $H_a: \mu > 46.42$. Reject H_0 if p-value < 0.01. From (c), $t_{data} = 3.87$. For a right-tailed test with df $= n - 1 = 15 - 1 = 14$ and $t_{data} = -4.0$, $t_{data} < -2.977$, so p-value < 0.005. Since p-value < 0.01, we reject H_0. There is evidence that the population mean price per bottle of *Obsession* is greater than \$46.42.

49. (a) Since the standard error is $s_{\bar{x}} = \frac{s}{\sqrt{n}}$ and n is in the denominator of the formula, an increase in the sample size will result in a decrease in the standard error. **(b)** Since $t_{data} = \frac{\bar{x} - \mu_0}{s/\sqrt{n}}$, $\bar{x} - \mu_0$ is positive, and $s_{\bar{x}} = \frac{s}{\sqrt{n}}$ is in the denominator, an increase in the sample size n will result in a decrease in $s_{\bar{x}} = \frac{s}{\sqrt{n}}$, which will result in an increase in t_{data}. **(c)** Since t_{crit} depends on the degrees of freedom and the degrees of freedom is df $= n - 1$, an increase in the sample size n will result in an increase in the degrees of freedom, which will result in a decrease in t_{crit}. **(d)** Since p-value $= P(t > t_{data})$, an increase in the sample size n results in an increase in t_{data}, which results in a decrease in the p-value. **(e)** Since an increase in the sample size n results in a decrease in the p-value, and the p-value is already less than $\alpha = 0.01$, the p-value will still be less than $\alpha = 0.01$ after an increase in the sample size n. Thus there is no change in the conclusion. **(f)** Since an increase in the sample size n results in a decrease in the p-value, and the p-value is already less than or equal to 0.001, the p-value will still be less than or equal to 0.001 after an increase in the sample size n. Therefore, the strength of the evidence against the null hypothesis will still be that there is extremely strong evidence against the null hypothesis. **(g)** When the sample size n increases, $\frac{s}{\sqrt{n}}$, t_{crit}, and the p-value decrease, t_{data} increases, and the conclusion and the strength of the evidence against the null hypothesis remain the same.

51. (a) $H_0: \mu \le 30$ vs. $H_a: \mu > 30$. Reject H_0 if p-value < 0.01.

$$t_{data} = \frac{\bar{x} - \mu_0}{s/\sqrt{n}} = \frac{33.7 - 30}{3.301514804/\sqrt{10}} = 3.54$$

For a right-tailed test with df $= n - 1 = 10 - 1 = 9$, p-value $= P(t > 3.54) = 0.0031570524$.

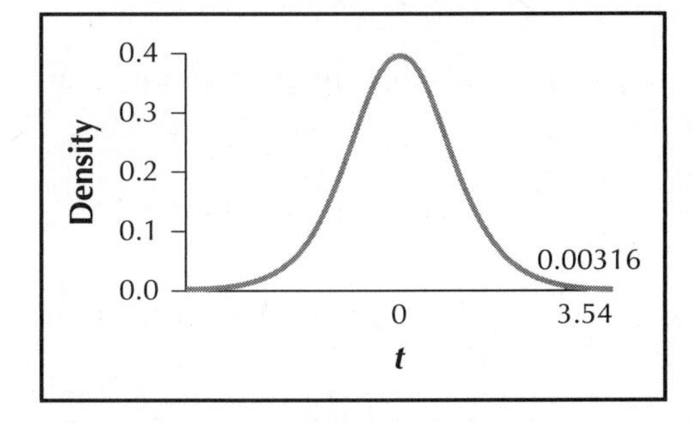

Since p-value < 0.01, we reject H_0. There is evidence that the population mean gas mileage is greater than 30 mpg.

(b) $H_0: \mu \le 30$ vs. $H_a: \mu > 30$. Reject H_0 if p-value < 0.001.

$t_{data} = 3.54$, p-value $= 0.0031570524$

Since p-value ≥ 0.001, we do not reject H_0. There is insufficient evidence that the population mean gas mileage is greater than 30 mpg. **(c)** We could turn to a direct assessment of the strength of the evidence against H_0 or we could obtain more data. **(d)** Since $0.001 < p$-value ≤ 0.01, there is very strong evidence against the null hypothesis that the population mean gas mileage is less than or equal to 30 mpg. This does not change for any value of α we use.

53. (a) Since all of the points lie within the curved lines, the normal probability plot indicates acceptable normality. Thus Case 1 applies, so it is appropriate to apply the t test for the mean.

(b) $n = 10$, $\bar{x} = \$2538.92$, $s = \$404.75$

(c) $s_{\bar{x}} = \frac{s}{\sqrt{n}} = \frac{404.75}{\sqrt{10}} = \$127.99$

(d) $t_{data} = 2.09$. Therefore, the sample mean fees and tuition at community colleges nationwide $\bar{x} = \$2538.92$ is 2.09 standard errors above the hypothesized mean fees and tuition at community colleges nationwide $\mu_0 = \$2272$. Since $2 \le t_{data} < 3$, this is mildly extreme. **(e)** It says "Test of mu = 2272 vs. not = 2272" across the top of the printout. The researchers wanted to test the hypotheses "$H_0 : \mu \le 2272$ vs. $H_a : \mu > 2272$."

55. **(a)** Since the *p*-value for the two-tailed test is twice the *p*-value for the one-tailed test, it is possible to conclude that there is insufficient evidence that the population mean cost has changed, but there is evidence that the population mean cost has increased if α is between the two *p*-values. **(b)** Since $0.01 < p\text{-value} \le 0.05$, there is solid evidence against the null hypothesis that the population mean tuition and fees at community colleges is less than or equal to \$2272. **(c)** Since the test was only done for community colleges and not all colleges, our data and the hypothesis test do not support the headline.

57. $H_0 : \mu \le 26$ vs. $H_a : \mu > 26$. This is a right-tailed test with $\alpha = 0.10$ and $df = n - 1 = 157 - 1 = 156$. Since df = 156 does not appear in our *t*-table, we use the largest degrees of freedom less than 156, which is df = 100. Therefore, we use $t_{crit} = 1.29$. Reject H_0 if $t_{data} > 1.290$. From Exercise 56 (c), $t_{data} = 1.57$. Since $t_{data} > 1.290$, reject H_0. There is evidence that the population mean age at first marriage is greater than 26 years.

59. **(a)** $p\text{-value} = 2P(t > 1.57) = 0.1184412114$

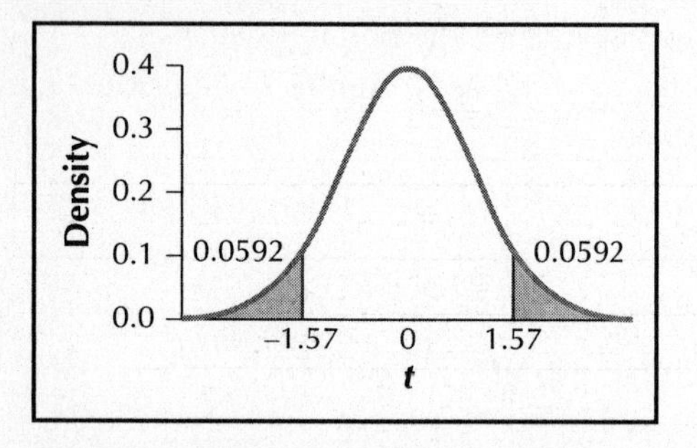

(b) Since $0.10 < p\text{-value} \le 0.15$, there is slight evidence against the null hypothesis that the population mean marriage is 26 years. **(c)** 0.10. Since 0.1184412114 is greater than 0.10, it is closer to 0.10 than it is to 0.05. **(d)** This headline is not supported by the data and the hypothesis test.

61. **(a)** 254 observations, 7 variables

(b)

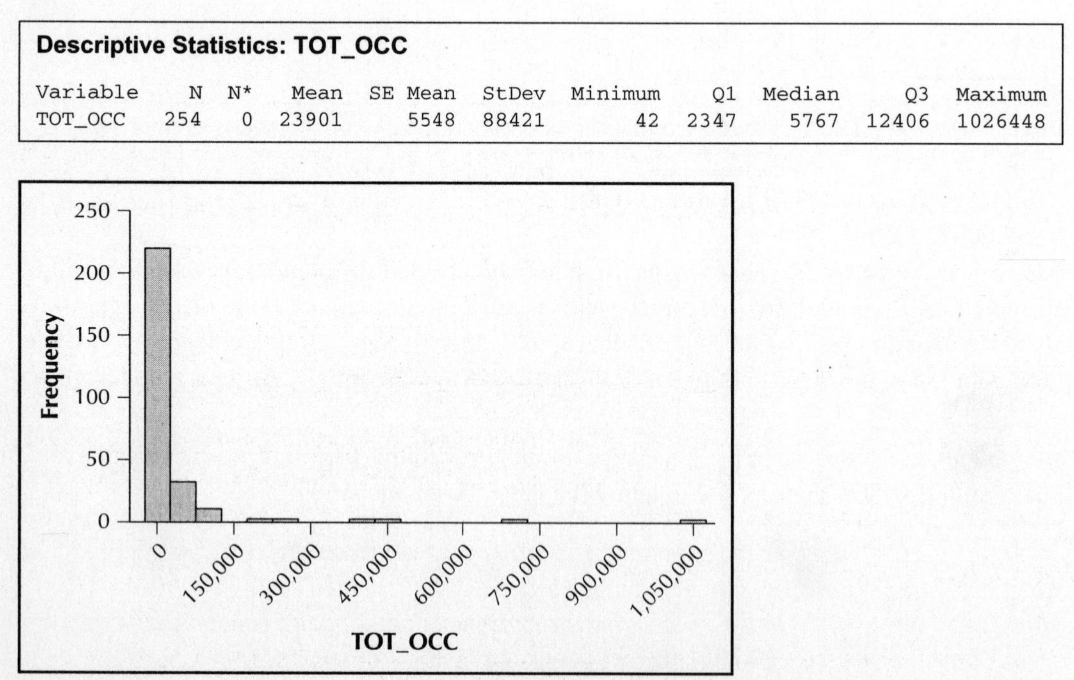

Descriptive Statistics: TOT_OCC

Variable	N	N*	Mean	SE Mean	StDev	Minimum	Q1	Median	Q3	Maximum
TOT_OCC	254	0	23901	5548	88421	42	2347	5767	12406	1026448

$\bar{x} = 23{,}901$, $s = 88{,}421$: The distribution is right-skewed.

(c)

```
One-Sample T: TOT_OCC

Test of mu = 40000 vs not = 40000

Variable    N   Mean  StDev  SE Mean         95% CI           T      P
TOT_OCC   254  23901  88241     5548  (12975, 34828)  -2.83  0.004
```

Since p-value < 0.05, we reject H_0. There is evidence that the population mean total occupied housing units for these counties is different from 40,000.

Section 9.5

1. The difference between $\hat{p}$ and p is that $\hat{p}$ is the sample proportion and p is population proportion.

3. When the sample proportion $\hat{p}$ is unusual or extreme in the sampling distribution of $\hat{p}$ on the assumption that H_0 is correct, then we reject H_0. Otherwise, there is insufficient evidence against H_0 and we should not reject H_0.

5. Between 0 and 1 inclusive: $0 \le p_0 \le 1$.

7. $\sigma_{\hat{p}} = \sqrt{\dfrac{p_0(1-p_0)}{n}} = \sqrt{\dfrac{0.5(1-0.5)}{50}} = 0.0707$

9. $\sigma_{\hat{p}} = \sqrt{\dfrac{p_0(1-p_0)}{n}} = \sqrt{\dfrac{0.5(1-0.5)}{100}} = 0.05$

11. $\sigma_{\hat{p}} = \sqrt{\dfrac{p_0(1-p_0)}{n}} = \sqrt{\dfrac{0.5(1-0.5)}{1000}} = 0.0158$

13. We observe that $\sigma_{\hat{p}}$ for a particular value of p_0 depends only on the sample size and not on the number of successes in the sample. As the sample size increases, $\sigma_{\hat{p}}$ decreases.

15. $\hat{p} = \frac{x}{n} = \frac{25}{50} = 0.5,\ Z_{data} = \dfrac{\hat{p}-p_0}{\sigma_{\hat{p}}} = \dfrac{\hat{p}-p_0}{\sqrt{\dfrac{p_0(1-p_0)}{n}}} = \dfrac{0.5-0.4}{\sqrt{\dfrac{0.4(1-0.4)}{50}}} = 1.44$

17. $\hat{p} = \frac{x}{n} = \frac{35}{50} = 0.7,\ Z_{data} = \dfrac{\hat{p}-p_0}{\sigma_{\hat{p}}} = \dfrac{\hat{p}-p_0}{\sqrt{\dfrac{p_0(1-p_0)}{n}}} = \dfrac{0.7-0.4}{\sqrt{\dfrac{0.4(1-0.4)}{50}}} = 4.33$

19. $\hat{p} = \frac{x}{n} = \frac{20}{80} = 0.25,\ Z_{data} = \dfrac{\hat{p}-p_0}{\sigma_{\hat{p}}} = \dfrac{\hat{p}-p_0}{\sqrt{\dfrac{p_0(1-p_0)}{n}}} = \dfrac{0.25-0.5}{\sqrt{\dfrac{0.5(1-0.5)}{80}}} = -4.47$

21. $\hat{p} = \frac{x}{n} = \frac{40}{80} = 0.5,\ Z_{data} = \dfrac{\hat{p}-p_0}{\sigma_{\hat{p}}} = \dfrac{\hat{p}-p_0}{\sqrt{\dfrac{p_0(1-p_0)}{n}}} = \dfrac{0.5-0.5}{\sqrt{\dfrac{0.5(1-0.5)}{80}}} = 0$

23. $\hat{p} = \frac{x}{n} = \frac{60}{80} = 0.75,\ Z_{data} = \dfrac{\hat{p}-p_0}{\sigma_{\hat{p}}} = \dfrac{\hat{p}-p_0}{\sqrt{\dfrac{p_0(1-p_0)}{n}}} = \dfrac{0.75-0.5}{\sqrt{\dfrac{0.5(1-0.5)}{80}}} = 4.47$

25. **(a)** We have $np_0 = 100(0.4) = 40 \ge 5$ and $n(1-p_0) = 100(1-0.4) = 60 \ge 5$, so we may use the Z test for proportions. **(b)** $H_0: p \le 0.4$ vs. $H_a: p > 0.4$. Reject H_0 if p-value < 0.05.

(c) $\hat{p} = \frac{x}{n} = \frac{44}{100} = 0.44,\ Z_{data} = \dfrac{\hat{p}-p_0}{\sigma_{\hat{p}}} = \dfrac{\hat{p}-p_0}{\sqrt{\dfrac{p_0(1-p_0)}{n}}} = \dfrac{0.44-0.4}{\sqrt{\dfrac{0.4(1-0.4)}{100}}} = 0.82$

(d) p-value $= P(Z > 0.82) = 1 - 0.7939 = 0.2061$ **(e)** Since p-value ≥ 0.05, we do not reject H_0. There is insufficient evidence that the population proportion is greater than 0.4.

27. **(a)** We have $np_0 = 900(0.5) = 450 \geq 5$ and $n(1 - p_0) = 900(1 - 0.5) = 450 \geq 5$, so we may use the Z test for proportions. **(b)** $H_0 : p = 0.5$ vs. $H_a : p \neq 0.5$. Reject H_0 if p-value < 0.05.

(c) $\hat{p} = \frac{x}{n} = \frac{475}{900} = 0.5278, Z_{data} = \frac{\hat{p} - p_0}{\sigma_{\hat{p}}} = \frac{\hat{p} - p_0}{\sqrt{\frac{p_0(1 - p_0)}{n}}} = \frac{0.5278 - 0.5}{\sqrt{\frac{0.5(1 - 0.5)}{900}}} = 1.68$

(d) p-value $= 2P(Z > 1.68) = 2(1 - P(Z < 1.68) = 2(1 - 0.9535) = 0.0930$ **(e)** Since p-value ≥ 0.05, we do not reject H_0. There is insufficient evidence that the population proportion is not equal to 0.5.

29. **(a)** We have $np_0 = 225(0.5) = 112.5 \geq 5$ and $n(1 - p_0) = 225(1 - 0.5) = 112.5 \geq 5$, so we may use the Z test for proportions. **(b)** $H_0 : p \geq 0.5$ vs. $H_a : p < 0.5$ **(c)** For a left-tailed test with $\alpha = 0.05$, $Z_{crit} = -1.645$. Reject H_0 if $Z_{data} < -1.645$.

(d) $\hat{p} = \frac{x}{n} = \frac{100}{225} = 0.4444, Z_{data} = \frac{\hat{p} - p_0}{\sigma_{\hat{p}}} = \frac{\hat{p} - p_0}{\sqrt{\frac{p_0(1 - p_0)}{n}}} = \frac{0.4444 - 0.5}{\sqrt{\frac{0.5(1 - 0.5)}{225}}} = -1.68$

(e) Since $Z_{data} < -1.645$, we reject H_0. There is evidence that the population proportion is less than 0.5.

31. **(a)** We have $np_0 = 400(0.6) = 240 \geq 5$ and $n(1 - p_0) = 400(1 - 0.6) = 160 \geq 5$, so we may use the Z test for proportions. **(b)** $H_0 : p \leq 0.6$ vs. $H_a : p > 0.6$ **(c)** For a right-tailed test with $\alpha = 0.05$, $Z_{crit} = 1.645$. Reject H_0 if $Z_{data} > 1.645$.

(d) $\hat{p} = \frac{x}{n} = \frac{260}{400} = 0.65, Z_{data} = \frac{\hat{p} - p_0}{\sigma_{\hat{p}}} = \frac{\hat{p} - p_0}{\sqrt{\frac{p_0(1 - p_0)}{n}}} = \frac{0.65 - 0.6}{\sqrt{\frac{0.6(1 - 0.6)}{400}}} = 2.04$

(e) Since $Z_{data} > 1.645$, we reject H_0. There is evidence that the population proportion is greater than 0.6.

33. We have $np_0 = 1000(0.368) = 368 \geq 5$ and $n(1 - p_0) = 1000(1 - 0.368) = 632 \geq 5$, so we may use the Z test for proportions.
$H_0 : p \leq 0.368$ vs. $H_a : p > 0.368$. Reject H_0 if p-value < 0.05.

$$\hat{p} = \frac{x}{n} = \frac{380}{1000} = 0.38, Z_{data} = \frac{\hat{p} - p_0}{\sigma_{\hat{p}}} = \frac{\hat{p} - p_0}{\sqrt{\frac{p_0(1 - p_0)}{n}}} = \frac{0.38 - 0.368}{\sqrt{\frac{0.368(1 - 0.368)}{1000}}} = 0.79$$

p-value $= P(Z > 0.79) = 1 - P(Z < 0.79) = 1 - 0.7852 = 0.2148$
Since p-value ≥ 0.05, we do not reject H_0. There is insufficient evidence that the population proportion of births to unmarried women is greater than 0.368.

35. We have $np_0 = 900(0.048) = 43.2 \geq 5$ and $n(1 - p_0) = 900(1 - 0.048) = 856.8 \geq 5$, so we may use the Z test for proportions.
$H_0 : p \leq 0.048$ vs. $H_a : p > 0.048$. Reject H_0 if p-value < 0.01.

$$\hat{p} = \frac{x}{n} = \frac{54}{900} = 0.06, Z_{data} = \frac{\hat{p} - p_0}{\sigma_{\hat{p}}} = \frac{\hat{p} - p_0}{\sqrt{\frac{p_0(1 - p_0)}{n}}} = \frac{0.06 - 0.048}{\sqrt{\frac{0.048(1 - 0.048)}{900}}} = 1.68$$

p-value $= P(Z > 1.68) = 1 - P(Z < 1.68) = 1 - 0.9535 = 0.0465$
Since p-value ≥ 0.01, we do not reject H_0. There is insufficient evidence that the population proportion of persons 12 or older that have used a prescription pain reliever non-medically is greater than 0.048.

37. We have $np_0 = 1000(0.07) = 70 \geq 5$ and $n(1 - p_0) = 1000(1 - 0.07) = 930 \geq 5$, so we may use the Z test for proportions.
$H_0 : p = 0.07$ vs. $H_a : p \neq 0.07$. Reject H_a if p-value < 0.10.

$$\hat{p} = \frac{x}{n} = \frac{80}{1000} = 0.08, Z_{data} = \frac{\hat{p} - p_0}{\sigma_{\hat{p}}} = \frac{\hat{p} - p_0}{\sqrt{\frac{p_0(1 - p_0)}{n}}} = \frac{0.08 - 0.07}{\sqrt{\frac{0.07(1 - 0.07)}{1000}}} = 1.24$$

p-value $= P(Z > 1.24) = 1 - P(Z < 1.24) = 1 - 0.7850 = 0.2150$

Since p-value ≥ 0.10 we do not reject H_0. There is insufficient evidence that the population proportion of hospitalizations of 18- to 44-year-old American women for affective disorders is not equal to 0.07.

39. (a) We have $np_0 = 100(0.153) = 15.3 \geq 5$ and $n(1 - p_0) = 100(1 - 0.153) = 84.7 \geq 5$, so we may use the Z test for proportions. **(b)** If the null hypothesis is correct, the sampling distribution of $\hat{p}$ will be centered at $p_0 = 0.153$. From part (a), the shape of the sampling distribution of $\hat{p}$ will be approximately normal.

(c) $\sigma_{\hat{p}} = \sqrt{\dfrac{p_0(1-p_0)}{n}} = \sqrt{\dfrac{0.153(1-0.153)}{100}} = 0.0360$. This is the standard deviation of the sampling distribution of $\hat{p}$ if H_0 is true. It is a measure of the variability of the sampling distribution of $\hat{p}$ if H_0 is true.

(d) $\hat{p} = \dfrac{x}{n} = \dfrac{23}{100} = 0.23$, $Z_{data} = \dfrac{\hat{p}-p_0}{\sigma_{\hat{p}}} = \dfrac{\hat{p}-p_0}{\sqrt{\dfrac{p_0(1-p_0)}{n}}} = \dfrac{0.23-0.153}{\sqrt{\dfrac{0.153(1-0.153)}{100}}} = 2.14$.

Therefore the sample proportion of Hispanic families that had household income of at least \$75,000 $\hat{p} = 0.23$ lies 2.14 standard deviations above the hypothesized value of the population proportion of Hispanic families that had household income of at least \$75,000 $p_0 = 0.153$. Since $2 \leq Z_{data} < 3$, $\hat{p} = 0.23$ is mildly extreme. **(e)** The value of $\hat{p} = 0.23$ would lie near the right tail of the sampling distribution of $\hat{p}$ because it lies 2.14 standard deviations above $p_0 = 0.153$.

41. (a) $H_0: p = 0.153$ vs. $H_a: p \neq 0.153$ **(b)** Reject H_0 if p-value < 0.01. **(c)** From Exercise 39 (d), $Z_{data} = 2.14$ **(d)** p-value $= 2P(Z > 2.14) = 2(1 - P(Z < 2.14)) = 2(1 - 0.9838) = 0.0324$ **(e)** Since $0.01 < p$-value ≤ 0.05, there is solid evidence against the null hypothesis that the population proportion of Hispanic families that had a household income of at least \$75,000 is equal to 0.153. **(f)** Since p-value ≥ 0.01, we do not reject H_0. There is insufficient evidence that the population proportion of Hispanic families that had a household income of at least \$75,000 is not equal to 0.153.

43. (a) Since $\sigma_{\hat{p}} = \sqrt{\dfrac{p_0(1-p_0)}{n}}$ and n is in the denominator, a decrease in the sample size n will result in an increase in $\sigma_{\hat{p}}$. **(b)** Since $Z_{data} = \dfrac{\hat{p}-p_0}{\sigma_{\hat{p}}}$ and $\hat{p} - p_0$ is negative, a decrease in the sample size n will result in an increase in $\sigma_{\hat{p}}$, which will result in an increase in Z_{data}. **(c)** Since the p-value $= 2P(Z > |Z_{data}|)$ and Z_{data} is negative, a decrease in the sample size results in an increase in Z_{data}, which results in a decrease in $|Z_{data}|$, which results in an increase in the p-value. **(d)** Since α does not depend on the sample size, a decrease in the sample size will leave α unchanged. **(e)** Since the rejection rule is reject H_0 if p-value < 0.05 and the p-value was already larger than 0.05, an increase in the sample size will result in an increase in the p-value, which will result in the same conclusion of do not reject H_0.

45. $H_0: p \geq 0.11$ vs. $H_a: p < 0.11$
For a left-tailed test with $\alpha = 0.05$, $Z_{crit} = -1.645$. Reject H_0 if $Z_{data} < -1.645$. From Exercise 44 (c), $Z_{data} = -1.60$. Since $Z_{data} \geq 1.645$, we do not reject H_0. There is insufficient evidence that the population proportion of children age 6 and under exposed to ETS at home on a regular basis is less than 0.11.

47. (a) $H_0: p \geq 0.11$ vs. $H_a: p < 0.11$. Reject H_0 if p-value < 0.10.
From Exercise 44 (c), $Z_{data} = -1.60$
p-value $= P(Z < -1.60) = 0.0548$
Since p-value < 0.10, we reject H_0. There is evidence that the population proportion of children age 6 and under exposed to environmental tobacco smoke at home on a regular basis is less than 0.11. **(b)** The difference in the conclusions is because we changed the value of α and not because we used different methods for the two different hypothesis tests. Since p-value $= 0.0548$, $0.05 < p$-value < 0.10. Therefore we would reject H_0 for $\alpha = 0.10$ and we would not reject H_0 for $\alpha = 0.05$. **(c)** Since $0.05 < p$-value ≤ 0.10, there is mild evidence against the null hypothesis that the population proportion of children age 6 and under exposed to environmental smoke at home on a regular basis is greater than or equal to 0.11.

49. (a) Since $\sigma_{\hat{p}} = \sqrt{\dfrac{p_0(1-p_0)}{n}}$ and n is in the denominator, doubling the sample size n will result in a decrease in $\sigma_{\hat{p}}$ to $\dfrac{1}{\sqrt{2}}\sigma_{\hat{p}}$. **(b)** Since $Z_{data} = \dfrac{\hat{p}-p_0}{\sigma_{\hat{p}}}$ and $\hat{p} - p_0$ is positive, doubling the sample size n will result in a decrease in $\sigma_{\hat{p}}$, which will result in an increase in Z_{data} by a factor of $\sqrt{2}$ to $\sqrt{2}\,Z_{data}$. **(c)** Since the p-value $= P(Z > Z_{data})$, doubling the sample size results in an increase in Z_{data}, which results in a decrease in the p-value. **(d)** Since α does not depend on the sample size, doubling the sample size will leave α unchanged. **(e)** Doubling the sample size while leaving everything else the same will result in a new p-value of 0.0268. Since p-value < 0.05, the conclusion will now be reject H_0.

Section 9.6

1. One instance where an analyst would be interested in performing a hypothesis test about the population standard deviation is the following: A pharmaceutical company that wishes to ensure the safety of a particular new drug would perform statistical tests to make sure that the drug's effect was consistent and did not vary from patient to patient.

3. Since $\sigma = \sqrt{\sigma^2}$, σ will never be less than 0. Therefore it does not make sense to test whether $\sigma < 0$.

5. The confidence interval for σ can be used to perform a two-tailed test for σ.
If σ_0 lies in the confidence interval we do not reject H_0.
If σ_0 does not lie in the confidence interval we reject H_0.

7. $H_0: \sigma \leq 10$ vs. $H_a: \sigma > 10$

9. $H_0: \sigma = 3$ vs. $H_a: \sigma \neq 3$

11. $\chi^2_{data} = \dfrac{(n-1)s^2}{\sigma_0^2} = \dfrac{(21-1)3}{1^2} = 60$

13. $\chi^2_{data} = \dfrac{(n-1)s^2}{\sigma_0^2} = \dfrac{(16-1)2.5^2}{3^2} = 10.417$

15. $\chi^2_{data} = \dfrac{(n-1)s^2}{\sigma_0^2} = \dfrac{(8-1)350}{20^2} = 6.125$

17. (a)

(b) df $= n - 1 = 21 - 1 = 20$, p-value $= P(\chi^2 > 60) = 7.121750863 \times 10^{-6}$ **(c)** Since p-value < 0.05, we reject H_0. There is evidence that the population standard deviation is greater than 1.

19. (a)

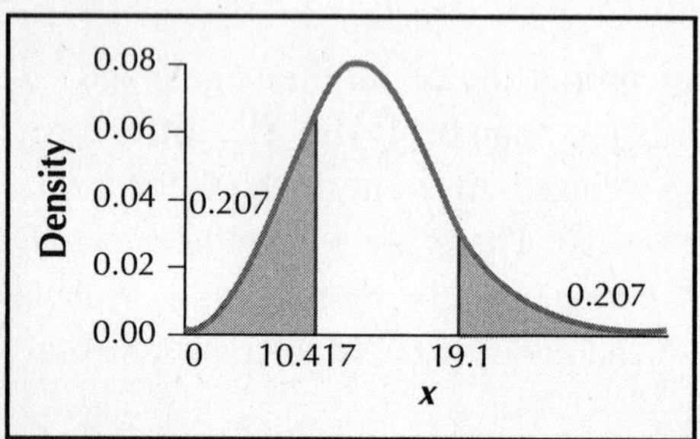

(b) df $= n - 1 = 16 - 1 = 15$, Since $P(\chi^2 > 10.417) = 0.792723378314 > 0.5$, p-value $= 2P(\chi^2 < 10.417) = 0.4145552434$ **(c)** Since p-value ≥ 0.05, we do not reject H_0. There is insufficient evidence that the population standard deviation is different from 3.

21. (a)

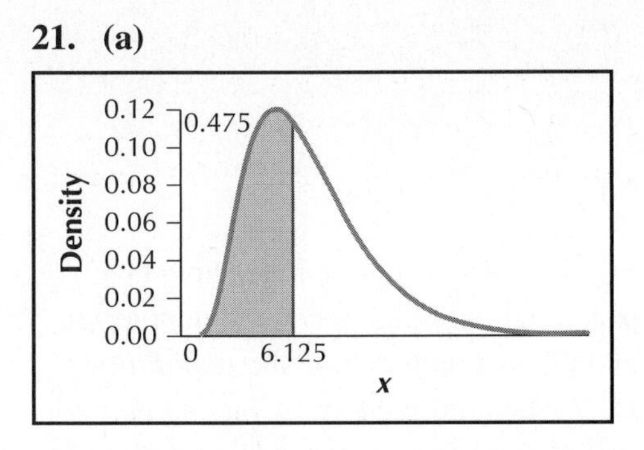

(b) df $= n - 1 = 11 - 1 = 10$, p-value $= P(\chi^2 < 6.125) = 0.4747679539$ **(c)** Since p-value ≥ 0.05, we do not reject H_0. There is insufficient evidence that the population standard deviation is less than 20.

23. df $= n - 1 = 21 - 1 = 20$, $\chi^2_\alpha = \chi^2_{0.05} = 31.410$

25. df $= n - 1 = 16 - 1 = 15$, $\chi^2_{\alpha/2} = \chi^2_{0.025} = 27.488$ and $\chi^2_{1-\alpha/2} = \chi^2_{0.975} = 6.262$

27. df $= n - 1 = 8 - 1 = 7$, $\chi^2_{1-\alpha} = \chi^2_{0.90} = 2.833$

29. (a) Reject H_0 if $\chi^2_{data} > 31.410$. **(b)** Since $\chi^2_{data} > 31.410$ we reject H_0. There is evidence that the population standard deviation is greater than 1.

31. (a) Reject H_0 if $\chi^2_{data} < 6.262$ or $\chi^2_{data} > 27.488$. **(b)** Since χ^2_{data} is not less than 6.262 and χ^2_{data} is not greater than 27.488, we do not reject H_0. There is insufficient evidence that the population standard deviation is different from 3.

33. (a) Reject H_0 if $\chi^2_{data} < 2.833$. **(b)** Since $\chi^2_{data} \geq 2.833$, we do not reject H_0. There is insufficient evidence that the population standard deviation is less than 20.

35. (a) $H_0: \sigma \leq 36.5$ vs. $H_a: \sigma > 36.5$. Reject H_0 if p-value < 0.01.

(b) $\chi^2_{data} = \dfrac{(n-1)s^2}{\sigma_0^2} = \dfrac{(12-1)119.025}{36.5^2} = 982.75$

(c) df $= n - 1 = 12 - 1 = 11$, p-value $= P(\chi^2 > 982.75) \approx 0$ **(d)** Since p-value < 0.01, we reject H_0.

(e) There is evidence that the population standard deviation of DDT-level in the breast milk of Hispanic women in the Yakima valley is greater than 36.5 parts per billion.

37. (a) $H_0: \sigma = 30{,}000$ vs. $H_a: \sigma \neq 30{,}000$. Reject H_0 if p-value < 0.05.

(b) $\chi^2_{data} = \dfrac{(n-1)s^2}{\sigma_0^2} = \dfrac{(7-1)2245.67}{30{,}000^2} = 0.00001497113333$

(c) df $= n - 1 = 7 - 1 = 6$, Since $P(\chi^2 > 0.00001487113333) \approx 1 > 0.5$, p-value $=$ $2P(\chi^2 < 0.00001487113333) \approx 0$ **(d)** Since p-value < 0.05, we reject H_0. **(e)** There is evidence that the population standard deviation of union membership differs from 30,000.

39. Test using the p-value method:

$H_0: \sigma \leq 50$ vs. $H_a: \sigma > 50$. Reject H_0 if p-value < 0.05.

$$\chi^2_{data} = \frac{(n-1)s^2}{\sigma_0^2} = \frac{(101-1)2600}{50^2} = 104$$

$$\text{df} = n - 1 = 101 - 1 = 100,\ p\text{-value} = P(\chi^2 > 104) = 0.3721497012$$

Since p-value ≥ 0.05 we do not reject H_0. There is insufficient evidence that the population standard deviation of test scores for boys is greater than 50 points.

Test using the critical-value method:

$$H_0: \sigma \leq 50 \text{ vs. } H_a: \sigma > 50$$

For a right-tailed test with $\alpha = 0.05$ and df $= n - 1 = 101 - 1 = 100$, $\chi^2_\alpha = \chi^2_{0.05} = 124.342$. Reject H_0 if $\chi^2_{data} > 124.342$.

$$\chi^2_{data} = \frac{(n-1)s^2}{\sigma_0^2} = \frac{(101-1)2600}{50^2} = 104$$

Since $\chi^2_{data} \leq 124.342$, we do not reject H_0. There is insufficient evidence that the population standard deviation of test scores for boys is greater than 50 points.

Chapter 9 Review

1. $H_0 : \mu \geq 12$ vs. $H_a : \mu < 12$

3. $H_0 : \mu \geq 0$ vs. $H_a : \mu < 0$

5. **(a)** $H_0 : \mu \geq 202.7$ vs. $H_a : \mu < 202.7$ **(b)** The two ways that a correct decision could be made are (1) to conclude that the *population mean number* of speeding-related fatalities is less than 202.7 when the *population mean number* of speeding-related fatalities is actually less than 202.7, and (2) to conclude that the *population mean number* of speeding-related fatalities is greater than or equal to 202.7 when the *population mean number* of speeding-related fatalities is actually greater than or equal to 202.7. **(c)** A Type I error would be concluding that the *population mean number* of speeding-related fatalities is less than 202.7 when it actually is greater than or equal to 202.7. **(d)** A Type II error would be concluding that the *population mean number* of speeding-related fatalities is greater than or equal to 202.7 when it actually is less than 202.7.

7. $Z_{data} = \dfrac{\bar{x} - \mu_0}{\sigma/\sqrt{n}} = \dfrac{59 - 60}{10/\sqrt{100}} = -1$

9. $Z_{data} = \dfrac{\bar{x} - \mu_0}{\sigma/\sqrt{n}} = \dfrac{59 - 60}{1/\sqrt{100}} = -10$

11. **(a)** $H_0 : \mu \geq -10$ vs. $H_a : \mu < -10$. Reject H_0 if p-value < 0.01.

(b) $Z_{data} = \dfrac{\bar{x} - \mu_0}{\sigma/\sqrt{n}} = \dfrac{-12 - (-10)}{2/\sqrt{25}} = -5$

(c) p-value $= P(Z < -5) = 2.87105 \times 10^{-7}$

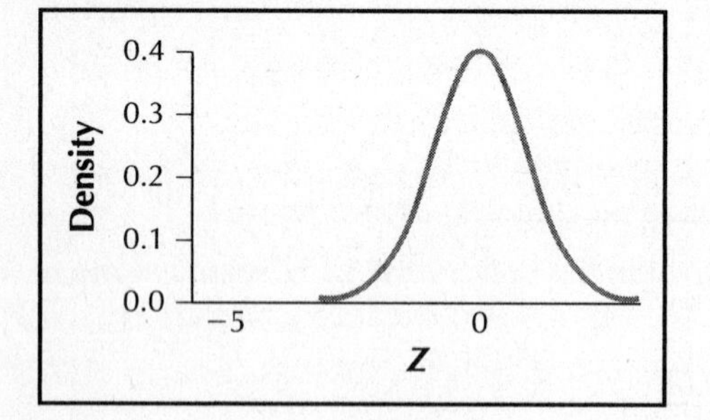

(d) Since p-value < 0.01, reject H_0. There is evidence that the population mean is less than -10.

13. **(a)** For a two-tailed test with $\alpha = 0.01$, $Z_{crit} = 2.58$. **(b)** Reject H_0 if $Z_{data} < -2.58$ or $Z_{data} > 2.58$.
(c)

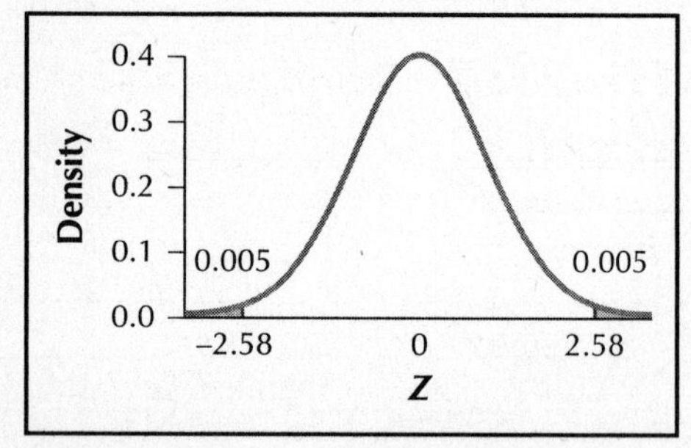

(d) Since Z_{data} is not less than -2.58 and Z_{data} is not greater than 2.58, we do not reject H_0. There is insufficient evidence that the population mean is different from μ_0.

15. **(a)** For a right-tailed test with $\alpha = 0.05$, $Z_{crit} = 1.645$. **(b)** Reject H_0 if $Z_{data} > 1.645$.

(c)

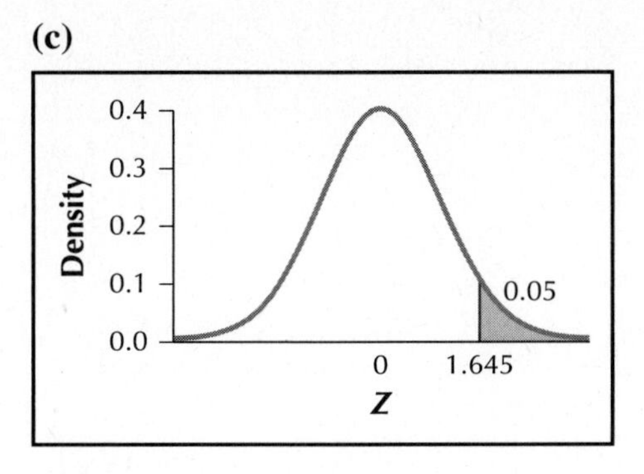

(d) Since $Z_{data} \leq 1.645$, we do not reject H_0. There is insufficient evidence that the population mean is greater than μ_0.

17. **(a)** $H_0 : \mu \leq 52{,}200$ vs. $H_a : \mu > 52{,}200$

(b) For a right-tailed test with $\alpha = 0.10$, $Z_{crit} = 1.28$. Reject H_0 if $Z_{data} > 1.28$.

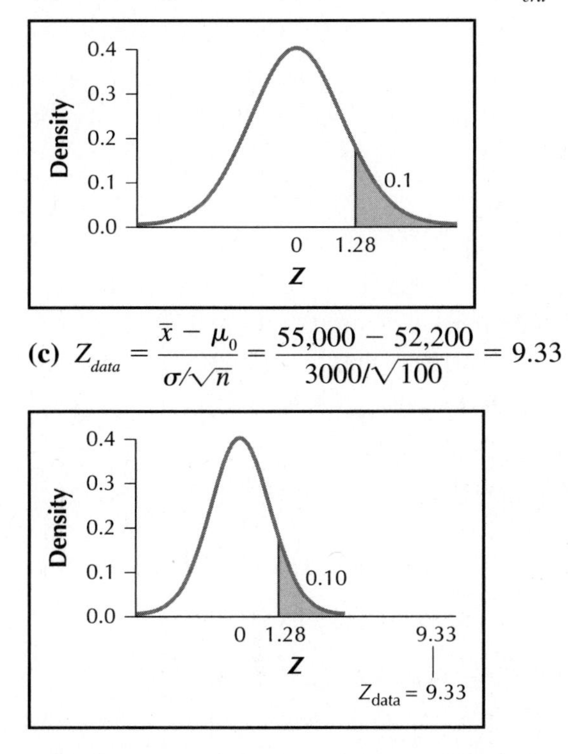

(c) $Z_{data} = \dfrac{\bar{x} - \mu_0}{\sigma/\sqrt{n}} = \dfrac{55{,}000 - 52{,}200}{3000/\sqrt{100}} = 9.33$

(d) Since $Z_{data} > 1.28$, we reject H_0. There is evidence that the population mean salary of college graduates is greater than \$52,200.

19. For a right-tailed test with $\alpha = 0.10$ and $\text{df} = n - 1 = 8 - 1 = 7$, $t_{crit} = 1.415$.

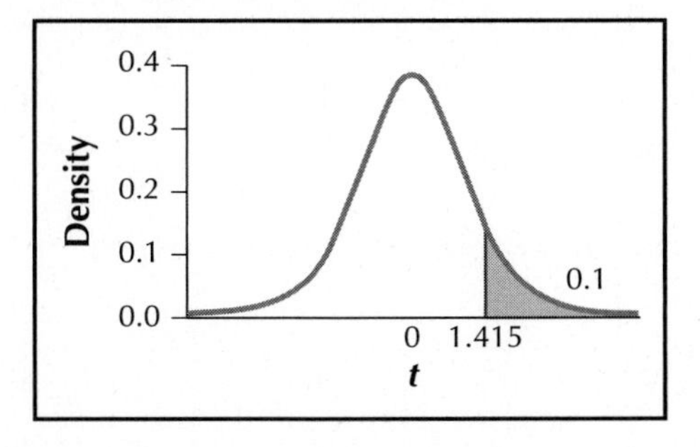

21. For a right-tailed test with $\alpha = 0.01$ and df $= n - 1 = 8 - 1 = 7$, $t_{crit} = 2.998$.

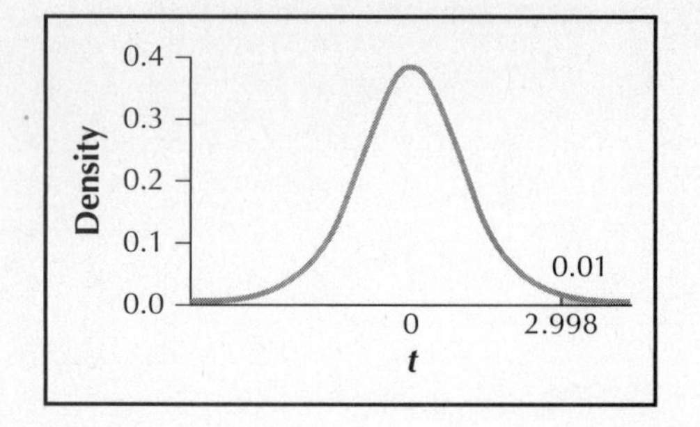

23. $H_0 : \mu = 9$ vs. $H_a : \mu \neq 9$. For a two-tailed test with $\alpha = 0.10$ and df $= n - 1 = 16 - 1 = 15$, $t_{crit} = 1.753$. Reject H_0 if $t_{data} < -1.753$ or $t_{data} > 1.753$.

$$t_{data} = \frac{\bar{x} - \mu_0}{s/\sqrt{n}} = \frac{10 - 9}{3/\sqrt{16}} = 1.33.$$

Since t_{data} is not less than -1.753 and t_{data} is not greater than 1.753, we do not reject H_0. There is insufficient evidence that the population mean is different from 9.

25. **(a)** We have $np_0 = 144(0.7) = 100.8 \geq 5$ and $n(1 - p_0) = 144(1 - 0.7) = 43.2 \geq 5$. **(b)** $H_0 : p = 0.7$ vs. $H_a : p \neq 0.7$. Reject H_0 if p-value < 0.05.

(c) $\hat{p} = \frac{x}{n} = \frac{110}{140} = 0.786$, $Z_{data} = \frac{\hat{p} - p_0}{\sigma_{\hat{p}}} = \frac{\hat{p} - p_0}{\sqrt{\frac{p_0(1 - p_0)}{n}}} = \frac{0.786 - 0.7}{\sqrt{\frac{0.786(1 - 0.786)}{144}}} = 1.67$

(d) p-value $= 2P(Z > 1.67) = 2(1 - P(Z < 1.67)) = 2(1 - 0.9525) = 0.095$ **(e)** Since p-value ≥ 0.05, we do not reject H_0. There is insufficient evidence that the population proportion is different from 0.7.

27. **(a)** We have $np_0 = 1000(0.8) = 800 \geq 5$ and $n(1 - p_0) = 1000(1 - 0.8) = 200 \geq 5$. **(b)** $H_0 : p \leq 0.8$ vs. $H_a : p > 0.8$ **(c)** For a right-tailed test with $\alpha = 0.10$, $Z_{crit} = 1.28$. Reject H_0 if $Z_{data} > 1.28$.

(d) $\hat{p} = \frac{x}{n} = \frac{830}{1000} = 0.83$, $Z_{data} = \frac{\hat{p} - p_0}{\sigma_{\hat{p}}} = \frac{\hat{p} - p_0}{\sqrt{\frac{p_0(1 - p_0)}{n}}} = \frac{0.83 - 0.8}{\sqrt{\frac{0.8(1 - 0.8)}{1000}}} = 2.37$

(e) Since $Z_{data} > 1.28$, we reject H_0. There is evidence that the population proportion is greater than 0.8.

29. **(a)** We have $np_0 = 100(0.4) = 40 \geq 5$ and $n(1 - p_0) = 100(1 - 0.4) = 60 \geq 5$.
(b) $H_0 : p = 0.4$ vs. $H_a : p \neq 0.4$
(c) For a two-tailed test with $\alpha = 0.01$, $Z_{crit} = 2.58$. Reject H_0 if $Z_{data} < -2.58$ or $Z_{data} > 2.58$.

(d) $\hat{p} = \frac{x}{n} = \frac{55}{100} = 0.55$, $Z_{data} = \frac{\hat{p} - p_0}{\sigma_{\hat{p}}} = \frac{\hat{p} - p_0}{\sqrt{\frac{p_0(1 - p_0)}{n}}} = \frac{0.55 - 0.4}{\sqrt{\frac{0.4(1 - 0.4)}{100}}} = 3.06$

(e) Since $Z_{data} > 2.58$, we reject H_0. There is evidence that the population proportion is not equal to 0.4.

31. **(a)** $H_0 : \sigma \geq 35$ vs. $H_a : \sigma < 35$. Reject H_0 if p-value < 0.05.

(b) $\chi^2_{data} = \frac{(n - 1)s^2}{\sigma_0^2} = \frac{(8 - 1)1200}{35^2} = 6.857$

(c) df $= n - 1 = 8 - 1 = 7$, p-value $= P(\chi^2 < 6.857) = 0.5560805474$

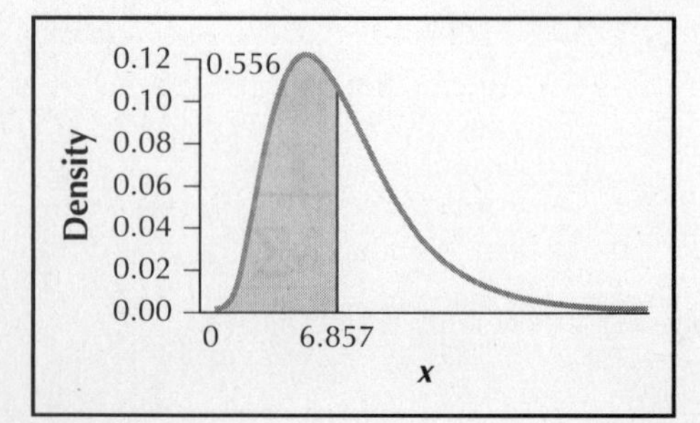

(d) Since p-value ≥ 0.05, we do not reject H_0. There is insufficient evidence that the population standard deviation is less than 35.

33. (a) $H_0: \sigma \leq 6$ vs. $H_a: \sigma > 6$
(b) For a right-tailed test with $\alpha = 0.05$ and df $= n - 1 = 20 - 1 = 19$, $\chi^2_\alpha = \chi^2_{0.05} = 30.144$. Reject H_0 if $\chi^2_{data} > 30.144$.

(c) $\chi^2_{data} = \dfrac{(n-1)s^2}{\sigma_0^2} = \dfrac{(20-1)9^2}{6^2} = 42.75$

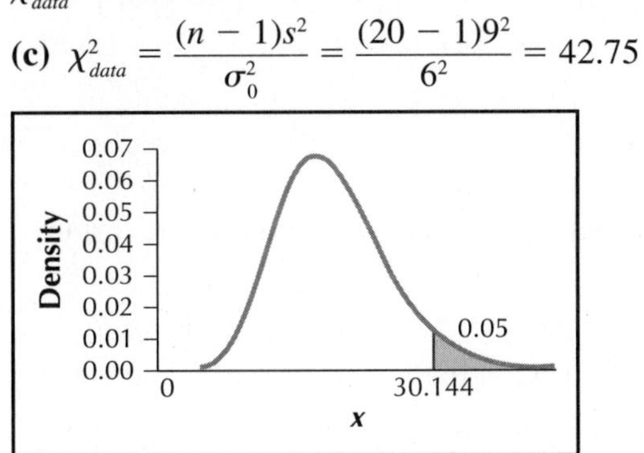

(d) Since $\chi^2_{data} > 30.144$, we reject H_0. There is evidence that the population standard deviation is greater than 6.

35. $H_0: \sigma = 50$ vs. $H_a: \sigma \neq 50$. Reject H_0 if p-value < 0.01.

$$\chi^2_{data} = \frac{(n-1)s^2}{\sigma_0^2} = \frac{(5-1)431.3}{50^2} = 0.690,\ \text{df} = n - 1 = 5 - 1 = 4.$$

Since $P(\chi^2 > 0.690) = 0.9525563754 > 0.5$, p-value $= 2P(\chi^2 < 0.690) = 0.094887$. Since p-value ≥ 0.01, we do not reject H_0. There is insufficient evidence that the population standard deviation differs from 50.

Chapter 9 Quiz

1. False. Only one of them is true and the other one is false.

2. True. The two methods are equivalent.

3. True.

4. I

5. Small.

6. α

7. $np_0 \geq 5$ and $n(1 - p_0) \geq 5$

8. It indicates that there is strong evidence against the null hypothesis. It indicates that there is no evidence against the null hypothesis.

9. No.

10. (a) $H_0: \mu = 47$ vs. $H_a: \mu \neq 47$. Reject H_0 if p-value < 0.10.

(b) $Z_{data} = \dfrac{\bar{x} - \mu_0}{\sigma/\sqrt{n}} = \dfrac{54 - 47}{12/\sqrt{100}} = 5.83$

(c) p-value $= 2P(Z > 5.83) \approx 0$ **(d)** Since p-value < 0.10, we reject H_0. There is evidence that the population mean prison sentence length for felons convicted on drug offenses is different from 47 months. **(e)** Since p-value ≤ 0.001, there is extremely strong evidence against the null hypothesis that the population mean prison sentence length for felons is 47 months.

11. (a) $H_0: \mu \geq 1.14$ vs. $H_a: \mu < 1.14$ **(b)** For a left-tailed test with $\alpha = 0.05$, $Z_{crit} = -1.645$. **(c)** Reject H_0 if $Z_{data} < -1.645$.

(d) $Z_{data} = \dfrac{\bar{x} - \mu_0}{\sigma/\sqrt{n}} = \dfrac{1.07 - 1.14}{0.25/\sqrt{36}} = -1.68$

(e) Since $Z_{data} < -1.645$, we reject H_0. There is evidence that the population mean fee charged by banks when withdrawing funds from an ATM machine not owned by your bank is less than \$1.14.

12. Yes, since the *p*-value method and the critical-value method are equivalent.

13. Type I error, Type II error.

14. **(a)** No, because the population standard deviation is not known. **(b)** $H_0: \mu = 32$ vs. $H_a: \mu \neq 32$ **(c)** For a two-tailed test with $\alpha = 0.10$ and df $= n - 1 = 36 - 1 = 35$, $t_{crit} = 1.690$. **(d)** Reject H_0 if $t_{data} < -1.690$ or $t_{data} > 1.690$.

(e) $t_{data} = \dfrac{\bar{x} - \mu_0}{s/\sqrt{n}} = \dfrac{33.8 - 32}{6/\sqrt{35}} = 1.80$

(f) Since $t_{data} > 1.690$, we reject H_0. There is evidence that the population mean years of potential life lost in alcohol-related fatal automobile accidents is different from 32 years.

15. **(a)** For a two-tailed test, df = 35, $1.690 < t_{data} < 2.030$, so $0.05 < p\text{-value} \leq 0.10$. **(b)** There is mild evidence against the null hypothesis that the population mean years of potential life lost in alcohol-related fatal automobile accidents is 32 years.

16. **(a)** We have $np_0 = 400(0.127) = 50.8 \geq 5$ and $n(1 - p_0) = 400(1 - 0.127) = 349.2 \geq 5$. Therefore it is appropriate to use the *Z* test for the proportion. **(b)** $p_0 = 0.127$, approximately normal

(c) $\sigma_{\hat{p}} = \sqrt{\dfrac{p_0(1 - p_0)}{n}} = \sqrt{\dfrac{0.127(1 - 0.127)}{400}} = 0.0166$. It is a measure of the variability of the distribution of the sample proportions if H_0 is true.

(d) $\hat{p} = \dfrac{x}{n} = \dfrac{57}{400} = 0.1425$, $Z_{data} = \dfrac{\hat{p} - p_0}{\sigma_{\hat{p}}} = \dfrac{\hat{p} - p_0}{\sqrt{\dfrac{p_0(1 - p_0)}{n}}} = \dfrac{0.1425 - 0.127}{\sqrt{\dfrac{0.127(1 - 0.127)}{400}}} = 0.93.$

No, because $-2 < Z_{data} < 2$. **(e)** Near the center.

17. **(a)** $H_0: p \leq 0.127$ vs. $H_a: p > 0.127$ **(b)** Reject H_0 if *p*-value < 0.05. **(c)** From Exercise 16 (d), $Z_{data} = 0.93$. **(d)** $p\text{-value} = P(Z > 0.93) = 1 - P(Z < 0.93) = 1 - 0.0823 = 0.1762$. **(e)** Since $0.15 <$ *p*-value, there is no evidence against the null hypothesis that the population proportion of preterm births is less than or equal to 0.127. **(f)** Since $p\text{-value} \geq 0.05$, we do not reject H_0. There is insufficient evidence that the population proportion of preterm births is greater than 0.127.

18. $H_0: \sigma \geq 0.25$ vs. $H_a: \sigma < 0.25$
For a left-tailed test with $\alpha = 0.10$ and df $= n - 1 = 10 - 1 = 9$, $\chi^2_{1-\alpha} = \chi^2_{0.90} = 4.168$. Reject H_0 if $\chi^2_{data} < 4.168$.

$$\chi^2_{data} = \frac{(n - 1)s^2}{\sigma_0^2} = \frac{(10 - 1)0.1180}{0.25^2} = 16.992$$

Since $\chi^2_{data} \geq 4.168$, we do not reject H_0. There is insufficient evidence that the population standard deviation of net price change is less than 25 cents.

Chapter 10

Two-Sample Inference

Section 10.1

1. Two samples are *independent* when the subjects selected for the first sample do not determine the subjects in the second sample.

3. Matched-pairs or paired samples

5. Since both samples of games were based on the same players, this is an example of dependent sampling.

7. Since the same students are taking both tests, this is an example of dependent sampling.

9.

Subject	1	2	3	4	5
Sample 1	3.0	2.5	3.5	3.0	4.0
Sample 2	2.5	2.5	2.0	2.0	1.5
Differences (Sample 1 minus Sample 2)	0.5	0	1.5	1.0	2.5

(a) $\bar{x}_d = \dfrac{0.5 + 0 + 1.5 + 1.0 + 2.5}{5} = 1.1$

$$s_d = \sqrt{\frac{\sum(x - \bar{x})^2}{n - 1}}$$

$$= \sqrt{\frac{(0.5 - 1.1)^2 + (0 - 1.1)^2 + (1.5 - 1.1)^2 + (1.0 - 1.1)^2 + (2.5 - 1.1)^2}{5 - 1}} \approx 0.9618$$

(b) $\text{df} = n - 1 = 5 - 1 = 4, \bar{x}_d \pm t_{\alpha/2}(s_d/\sqrt{n}) = 1.1 \pm (2.776)(0.9618/\sqrt{5}) =$ $1.1 \pm 1.1940 = (-0.0940, 2.2940)$

11.

Subject	1	2	3	4	5	6	7
Sample 1	20	25	15	10	20	30	15
Sample 2	30	30	20	20	25	35	25
Differences (Sample 1 minus Sample 2)	−10	−5	−5	−10	−5	−5	−10

(a) $\bar{x}_d = \dfrac{(-10) + (-5) + (-5) + (-10) + (-5) + (-5) + (-10)}{7} \approx -7.1429$

To find the standard deviation, we first need to find $\sum(x - \bar{x}_d)^2$.

x	$(x - \bar{x}_d)$	$(x - \bar{x}_d)^2$
−10	$(-10) - (-7.1429) = -2.8571$	8.16302041
−5	$(-5) - (-7.1429) = 2.1429$	4.59202041
−5	$(-5) - (-7.1429) = 2.1429$	4.59202041
−10	$(-10) - (-7.1429) = -2.8571$	8.16302041
−5	$(-5) - (-7.1429) = 2.1429$	4.59202041
−5	$(-5) - (-7.1429) = 2.1429$	4.59202041
−10	$(-10) - (-7.1429) = -2.8571$	8.16302041

$$\sum(x - \bar{x}_d)^2 = 42.85714287$$

Thus,

$$s_d = \sqrt{\frac{\sum(x - \bar{x})^2}{n - 1}} = \sqrt{\frac{42.85714287}{7 - 1}} \approx 2.6726.$$

(b) $\text{df} = n - 1 = 7 - 1 = 6, \bar{x}_d \pm t_{\alpha/2}(s_d/\sqrt{n}) \approx -7.1429 \pm (2.447)(2.6726/\sqrt{7}) = -7.1429 \pm 2.4718 = (-9.6147, -4.6711)$

13. $H_0 : \mu_d \leq 0$ vs. $H_a : \mu_d > 0$

Reject H_0 if p-value < 0.05.

$\text{df} = n - 1 = 5 - 1 = 4, t_{data} = \dfrac{\bar{x}_d}{s_d/\sqrt{n}} = \dfrac{1.1}{0.9618/\sqrt{5}} \approx 2.557$

p-value $= P(t > t_{data}) = P(t > 2.557) = 0.0314187146$

Since p-value < 0.05, we reject H_0. There is evidence that the population mean of the differences is greater than 0.

15. $H_0 : \mu_d \leq 0$ vs. $H_a : \mu_d > 0$

Reject H_0 if p-value < 0.05.

$$t_{data} = \frac{\bar{x}_d}{s_d/\sqrt{n}} = \frac{0.6667}{1.3663/\sqrt{6}} \approx 1.195, \text{ df} = n - 1 = 6 - 1 = 5$$

p-value $= P(t > t_{data}) = P(t > 2.557) = 0.1428363072$

Since p-value ≥ 0.05, we do not reject H_0. There is insufficient evidence that the population mean of the differences is greater than 0.

17. $H_0 : \mu_d = 0$ vs. $H_a : \mu_d \neq 0$

$\text{df} = n - 1 = 7 - 1 = 6, \alpha = 0.10, t_{crit} = 1.943$

Reject H_0 if $t_{data} < -1.943$ or if $t_{data} > 1.943$

$$t_{data} = \frac{\bar{x}_d}{s_d/\sqrt{n}} = \frac{-7.1429}{2.6726/\sqrt{7}} \approx -7.071.$$

Since $t_{data} \leq -1.943$, we reject H_0. There is evidence that the population mean of the differences is different than 0.

19.

	Subaru Forester	**Honda CR-V**	**Toyota RAV-4**	**Nissan Sentra**
2006	\$22,420	\$20,990	\$22,980	\$13,815
2007	\$21,820	\$22,395	\$23,630	\$15,375
Differences (2007 price minus 2006 price)	−\$600	\$1405	\$650	\$1560

(a) From the TI-84 calculator, $\bar{x}_d = \$753.75$ and $s_d \approx \$986.1658$.
(b) df $= n - 1 = 4 - 1 = 3$, $\bar{x}_d \pm t_{\alpha/2}(s_d/\sqrt{n}) = 753.75 \pm (3.182)(986.1658/\sqrt{4}) \approx$
$753.75 \pm 1568.9898 = (-815.2398, 2{,}322.7398)$

21. Critical value method:
$H_0 : \mu_d \leq 0$ vs. $H_a : \mu_d > 0$
df $= n - 1 = 4 - 1 = 3$, $\alpha = 0.05$, $t_{crit} = 2.353$
Reject H_0 if $t_{data} > 2.353$.

$$t_{data} = \frac{\bar{x}_d}{s_d/\sqrt{n}} = \frac{753.75}{986.1658/\sqrt{4}} \approx 1.529.$$

Since $t_{data} \geq -2.353$, we do not reject H_0. There is insufficient evidence that 2007 models are on average more expensive.

***p*-value method:**
$H_0 : \mu_d \leq 0$ vs. $H_a : \mu_d > 0$

Reject H_0 if p-value < 0.05.

$$t_{data} = \frac{\bar{x}_d}{s_d/\sqrt{n}} = \frac{753.75}{986.1658/\sqrt{4}} \approx 1.529, \text{ df} = n - 1 = 4 - 1 = 3$$

p-value $= P(t > t_{data}) = P(t > 1.529) = 0.1119094424$
Since p-value ≥ 0.05, we do not reject H_0. There is insufficient evidence that 2007 models are on average more expensive.

23.

	Northeast	Midwest	South	West
Jan.–Mar. 2007	\$370,300	\$212,800	\$222,900	\$341,500
Jan.–Mar. 2008	\$326,600	\$201,900	\$204,800	\$298,900
Differences (2008 price minus 2007 price)	−\$43,700	−\$10,900	−\$18,100	−\$42,600

(a) From the TI-84, $\bar{x}_d = -\$28{,}825$ and $s_d \approx \$16{,}806.2241$. **(b)** df $= n - 1 = 4 - 1 = 3$, $\bar{x}_d \pm t_{\alpha/2}(s_d/\sqrt{n})$ $= -28{,}825 \pm (2.353)(16{,}806.2241/\sqrt{4}) \approx -28{,}825 \pm 19{,}772.5227 = (-48{,}597.5227, -9052.4773)$.

25. $H_0 : \mu_d = 0$ vs. $H_a : \mu_d \neq 0$
Since $\mu_d = 0$ does not lie in the confidence interval, we reject H_0. There is evidence that the population mean difference between the 1st quarter 2007 median price and the 1st quarter 2008 median price differs from 0.

27.

Country	1995	2003	Differences (2003 minus 1995)
Singapore	523	565	42
Japan	553	543	−10
Hong Kong	508	542	34
England	528	540	12
United States	542	536	−6
Hungary	508	530	22
Latvia	486	530	44
Netherlands	530	525	−5
New Zealand	505	523	18

(*Continued*)

Country	1995	2003	Differences (2003 minus 1995)
Australia	521	521	0
Scotland	514	502	−12
Slovenia	464	490	26
Cyprus	450	480	30
Norway	504	466	−38
Iran	380	414	34

From the TI-84, $\bar{x}_d \approx 12.7333$ and $s_d \approx 23.6899$.

(a) $\text{df} = n - 1 = 15 - 1 = 14$, $\bar{x}_d \pm t_{\alpha/2}(s_d/\sqrt{n}) \approx 12.7333 \pm (1.761)(23.6899/\sqrt{15}) \approx 12.7333 \pm 10.7715 = (1.9618, 23.5048)$

(b) Critical value method:

$H_0: \mu_d \leq 0$ vs. $H_a: \mu_d > 0$

$\text{df} = n - 1 = 15 - 1 = 14$, $\alpha = 0.10$, $t_{crit} = 1.345$

Reject H_0 if $t_{data} > 1.345$.

$$t_{data} = \frac{\bar{x}_d}{s_d/\sqrt{n}} = \frac{12.7333}{23.6899/\sqrt{15}} \approx 2.082.$$

Since $t_{data} > 1.345$, we reject H_0. There is evidence that the 2003 science test scores for fourth graders are higher than the 1995 science test scores for fourth graders.

***p*-value method:**

$H_0: \mu_d \leq 0$ vs. $H_a: \mu_d > 0$

Reject H_0 if p-value < 0.10.

$$t_{data} = \frac{\bar{x}_d}{s_d/\sqrt{n}} = \frac{12.7333}{23.6899/\sqrt{15}} \approx 2.082, \text{ df} = n - 1 = 15 - 1 = 14,$$

$p\text{-value} = P(t > t_{data}) = P(t > 2.082) = 0.0280990904$

Since p-value < 0.10, we reject H_0. There is evidence that the 2003 science test scores for fourth graders are higher than the 1995 science test scores for fourth graders.

29.

Location	Before	After	Differences (After minus Before)
Startup	0.7	0.3	−0.4
Seattle	77.5	43.8	−33.7
Shoreline	63.3	33.6	−29.7
Cashmere	2.1	0.0	−2.1
Othello	7.4	2.0	−5.4
Royal City	1.6	0.7	−0.9
Ephrata	5.7	2.9	−2.8
Oronda	0.5	0.0	−0.5
Chelan	5.1	0.9	−4.2
Manson	1.1	0.6	−0.5
Westport	3.3	2.3	−1.0

(*Continued*)

Location	Before	After	Differences (Before minus After)
Alderton	49.9	40.3	−9.6
Chehalis 1	2.9	0.9	−2.0
Chehalis 2	3.8	1.8	−2.0
Raymond	2.4	3.1	0.7
Yakima	5.5	2.3	−3.2
Ellensburg	14.9	10.7	−4.2
Snoqualmie	19.4	10.4	−9.0
Sunnyside	12.0	11.7	−0.3
Ritzville	39.0	23.7	−15.3
Milton	14.5	11.2	−3.3
Skagit	8.8	17.5	8.7
Spokane	114.7	77.3	−37.4
Kent	25.3	13.8	−11.5
Vancouver	22.4	4.3	−18.1
Raymond	0.7	0.3	−0.4
Poulsbo	5.0	12.8	7.8
Grant	1.5	0.0	−1.5
Covington	2.1	2.9	0.8
Douglas	0.0	0.0	0.0

From the TI-84, $\bar{x}_d \approx -6.0333$ and $s_d \approx 10.8914$.
(a) $\bar{x}_d \approx -6.0333$ collisions per year **(b)** $\text{df} = n - 1 = 30 - 1 = 29$,
$\bar{x}_d \pm t_{\alpha/2}(s_d/\sqrt{n}) \approx -6.0333 \pm (2.045)(10.8914/\sqrt{30}) \approx -6.033 \pm 4.0665 = (1.9665, 10.0995)$

(c) Critical value method:
$H_0 : \mu_d \geq 0$ vs. $H_a : \mu_d < 0$
$\text{df} = n - 1 = 30 - 1 = 29$, $\alpha = 0.01$, $t_{crit} = 2.462$
Reject H_0 if $t_{data} < -2.462$

$$t_{data} = \frac{\bar{x}_d}{s_d/\sqrt{n}} = \frac{-6.0333}{10.8914/\sqrt{30}} \approx -3.034$$

Since $t_{data} < -2.462$, we reject H_0. There is evidence that the improvements lowered the population mean number of collisions per year.

***p*-value method:**
$H_0 : \mu_d \geq 0$ vs. $H_a : \mu_d < 0$
Reject H_0 if p-value < 0.01.

$$t_{data} = \frac{\bar{x}_d}{s_d/\sqrt{n}} = \frac{-6.0333}{10.8914/\sqrt{30}} \approx -3.034, \text{ df} = n - 1 = 30 - 1 = 29,$$

p-value $= P(t < t_{data}) = P(t < -3.034) = 0.002524694$
Since p-value < 0.01, we reject H_0. There is evidence that the improvements lowered the population mean number of collisions per year.

31.

	Detroit Red Wings	Tampa Bay Lightning	Phoenix Coyotes	Atlanta Thrashers	Colorado Avalanche	Dallas Stars
2003–2004 season	3.11	2.99	2.29	2.61	2.88	2.37
2005–2006 season	3.67	3.00	2.95	3.37	3.42	3.08
Differences (2005–2006 season minus 2003–2004 season)	0.56	0.01	0.66	0.76	0.54	0.71

From the TI-84, $\bar{x}_d = 0.54$ and $s_d \approx 0.2731$

Critical value method:

$H_0 : \mu_d \leq 0$ vs. $H_a : \mu_d > 0$

$\text{df} = n - 1 = 6 - 1 = 5$, $\alpha = 0.01$, $t_{crit} = 3.365$

Reject H_0 if $t_{data} > 3.365$

$$t_{data} = \frac{\bar{x}_d}{s_d/\sqrt{n}} = \frac{0.54}{0.2731/\sqrt{6}} \approx 4.843.$$

Since $t_{data} > 3.365$, we reject H_0. There is evidence that the population mean number of goals scored per game has increased.

***p*-value method:**

$H_0 : \mu_d \leq 0$ vs. $H_a : \mu_d > 0$

Reject H_0 if p-value < 0.01.

$$t_{data} = \frac{\bar{x}_d}{s_d/\sqrt{n}} = \frac{0.54}{0.2731/\sqrt{6}} \approx 4.843, \text{ df} = n - 1 = 6 - 1 = 5,$$

p-value $= P(t > t_{data}) = P(t > 4.843) = 0.0023512861$

Since p-value < 0.01, we reject H_0. There is evidence that the population mean number of goals scored per game has increased.

33. If we added a certain amount d to every entry in the 2005–2006 season results, each of the differences would increase by d. Thus $\bar{x}_d$ would increase by d. Since each of the differences is increased by d and $\bar{x}_d$ is increased by d, each of the quantities (difference $- \bar{x}_d$) will remain the same. Since $s_d = \sqrt{\dfrac{\sum(x - \bar{x})^2}{n - 1}}$ and n remains the same, s_d remains the same.

Since $t_{data} = \dfrac{\bar{x}_d}{s_d/\sqrt{n}}$, t_{data} would increase. Since in Exercise 31, $t_{data} > 3.365$, a larger value of t_{data} will also be greater than 3.365. Thus the conclusion would still be reject H_a.

Section 10.2

1. Case 1: The two populations are normally distributed. Case 2: The sample sizes are large (at least 30).

3. It measures the size of the typical error in using $\bar{x}_1 - \bar{x}_2$ to estimate $\mu_1 - \mu_2$.

5. **(a)** Since both sample sizes are large ($n_1 \geq 30$ and $n_2 \geq 30$), it is appropriate to construct a 95% confidence interval for $\mu_1 - \mu_2$. **(b)** $\bar{x}_1 - \bar{x}_2 = 10 - 8 = 2$

(c) $s_{\bar{x}_1 - \bar{x}_2} = \sqrt{\dfrac{s_1^2}{n_1} + \dfrac{s_2^2}{n_2}} = \sqrt{\dfrac{2^2}{36} + \dfrac{2^2}{36}} \approx 0.4714$

(d) $n_1 - 1 = 36 - 1 = 35$ and $n_2 = 36 - 1 = 35$, so df $= 35$. Thus,

$$E = t_{\alpha/2}\sqrt{\frac{s_1^2}{n_1} + \frac{s_2^2}{n_2}} = 2.030\sqrt{\frac{2^2}{36} + \frac{2^2}{36}} \approx 0.9570.$$

(e) $(\bar{x}_1 - \bar{x}_2) \pm E \approx 2 \pm 0.9570 = (1.0430, 2.9570)$. We are 95% confident that the interval (1.0430, 2.9570) captures the difference in population means.

7. (a) Since both sample sizes are large ($n_1 \geq 30$ and $n_2 \geq 30$), it is appropriate to construct a 95% confidence interval for $\mu_1 - \mu_2$. **(b)** $\bar{x}_1 - \bar{x}_2 = -10 - (-5) = -5$

(c) $s_{\bar{x}_1-\bar{x}_2} = \sqrt{\frac{s_1^2}{n_1} + \frac{s_2^2}{n_2}} = \sqrt{\frac{5^2}{30} + \frac{2^2}{30}} \approx 0.9832$

(d) $n_1 - 1 = 30 - 1 = 29$ and $n_2 = 30 - 1 = 29$, so df = 29. Thus,

$$E = t_{\alpha/2}\sqrt{\frac{s_1^2}{n_1} + \frac{s_2^2}{n_2}} = 2.045\sqrt{\frac{5^2}{30} + \frac{2^2}{30}} \approx 2.011$$

(e) $(\bar{x}_1 - \bar{x}_2) \pm E \approx -5 \pm 2.011 = (-7.011, -2.989)$. We are 95% confident that the interval $(-7.011, -2.989)$ captures the difference in population means.

9. (a) Since both sample sizes are large ($n_1 \geq 30$ and $n_2 \geq 30$), it is appropriate to construct a 95% confidence interval for $\mu_1 - \mu_2$. **(b)** $\bar{x}_1 - \bar{x}_2 = 0 - 1 = -1$

(c) $s_{\bar{x}_1-\bar{x}_2} = \sqrt{\frac{s_1^2}{n_1} + \frac{s_2^2}{n_2}} = \sqrt{\frac{3^2}{64} + \frac{1^2}{49}} \approx 0.4013$

(d) $n_1 - 1 = 64 - 1 = 63$ and $n_2 = 49 - 1 = 48$, so df = 48. The value of 48 for the df does not appear in the df column. The conservative approach would be to use next row with df smaller than 48. That would be df = 40. Thus, the "conservative" $t_{\alpha/2}$ is 2.021. Thus,

$$E = t_{\alpha/2}\sqrt{\frac{s_1^2}{n_1} + \frac{s_2^2}{n_2}} = 2.021\sqrt{\frac{3^2}{64} + \frac{1^2}{49}} \approx 0.811.$$

(e) $(\bar{x}_1 - \bar{x}_2) \pm E \approx -1 \pm 0.811 = (-1.811, -0.189)$. We are 95% confident that the interval $(-1.811, -0.189)$ captures the difference in population means.

11. (a) $H_0: \mu_1 = \mu_2$ vs. $H_a: \mu_1 \neq \mu_2$ **(b)** $n_1 - 1 = 36 - 1 = 35$ and $n_2 = 36 - 1 = 35$, so df = 35. $\alpha = 0.10$. $t_{crit} = 1.690$. Reject H_0 if $t_{data} < -1.690$ or $t_{data} > 1.690$.

(c) $t_{data} = \frac{(\bar{x}_1 - \bar{x}_2)}{\sqrt{\frac{s_1^2}{n_1} + \frac{s_2^2}{n_2}}} = \frac{(10 - 8)}{\sqrt{\frac{2^2}{36} + \frac{2^2}{36}}} \approx 4.243$. The difference of the sample means $\bar{x}_1 - \bar{x}_2 = 2$ lies 4.243 standard errors above the difference of the hypothesized means $\mu_1 - \mu_2 = 0$. **(d)** Since $t_{data} > 1.690$, we reject H_0. There is evidence that the population mean for Population 1 is different than the population mean for Population 2.

13. (a) $H_0: \mu_1 = \mu_2$ vs. $H_a: \mu_1 \neq \mu_2$ **(b)** $n_1 - 1 = 30 - 1 = 29$ and $n_2 = 30 - 1 = 29$, so df = 29. Thus, t_{crit} for $\alpha = 0.01$ is 2.756. Reject H_0 if $t_{data} < -2.756$ or $t_{data} > 2.756$.

(c) $t_{data} = \frac{(\bar{x}_1 - \bar{x}_2)}{\sqrt{\frac{s_1^2}{n_1} + \frac{s_2^2}{n_2}}} = \frac{(-10) - (-5)}{\sqrt{\frac{5^2}{30} + \frac{2^2}{30}}} \approx -5.085$. The difference of the sample means $\bar{x}_1 - \bar{x}_2 = -5$ lies 5.085 standard errors below the difference of the hypothesized means $\mu_1 - \mu_2 = 0$. **(d)** Since $t_{data} < -2.756$, we reject H_0. There is evidence that the population mean for Population 1 is different than the population mean for Population 2.

15. (a) $H_0: \mu_1 \leq \mu_2$ vs. $H_a: \mu_1 > \mu_2$ **(b)** $n_1 - 1 = 64 - 1 = 63$ and $n_2 = 49 - 1 = 48$, so df = 48. The value of 48 for the df does not appear in the df column. The conservative approach would be to use next row with df smaller than 48. That would be df = 40. Thus, the "conservative" t_{crit} for $\alpha = 0.01$ is 2.423. Reject H_0 if $t_{data} >$ 2.423.

(c) $t_{data} = \frac{(\bar{x}_1 - \bar{x}_2)}{\sqrt{\frac{s_1^2}{n_1} + \frac{s_2^2}{n_2}}} = \frac{(0 - 1)}{\sqrt{\frac{3^2}{64} + \frac{1^2}{49}}} \approx -2.492$. The difference of the sample means $\bar{x}_1 - \bar{x}_2 = -1$ lies 2.492 standard errors below the difference of the hypothesized means $\mu_1 - \mu_2 = 0$. **(d)** Since $t_{data} \leq 2.423$, we do not reject H_0. There is insufficient evidence that the population mean for Population 1 is greater than the population mean for Population 2.

17. Since neither population is normal, Case 1 does not apply. Since both sample sizes are small ($n_1 < 30$ and $n_2 < 30$), Case 2 does not apply. Therefore, it is not appropriate to perform two-sample t inference.

19. **(a)** Since both sample sizes are large ($n_1 \geq 30$ and $n_2 \geq 30$), Case 2 applies. Thus, it is appropriate to apply two-sample t inference. **(b)** $\bar{x}_1 - \bar{x}_2 = \$31{,}987 - \$33{,}179 = -\$1{,}192$

(c) $s_{\bar{x}_1 - \bar{x}_2} = \sqrt{\frac{s_1^2}{n_1} + \frac{s_2^2}{n_2}} = \sqrt{\frac{5000^2}{36} + \frac{6000^2}{49}} \approx \$1195.4657$ **(d)** $n_1 - 1 = 36 - 1 = 35$ and $n_2 = 49 - 1 = 48$, so df $= 35$. Thus,

$$E = t_{\alpha/2}\sqrt{\frac{s_1^2}{n_1} + \frac{s_2^2}{n_2}} = 2.030\sqrt{\frac{5000^2}{36} + \frac{6000^2}{49}} \approx \$2{,}426.795.$$

(e) $(\bar{x}_1 - \bar{x}_2) \pm E \approx -1{,}192 \pm 2{,}426.795 = (-3{,}618.795, 1{,}234.795)$. We are 95% confident that the interval $(-3{,}618.795, 1{,}234.795)$ captures the difference of the population mean incomes for Sacramento County and Los Angeles County, California.

21. **(a)** $n_1 - 1 = 36 - 1 = 35$ and $n_2 = 49 - 1 = 48$, so df $= 35$. $\alpha = 0.05$. $t_{crit} = 2.030$.

(b) $t_{data} = \dfrac{(\bar{x}_1 - \bar{x}_2)}{\sqrt{\frac{s_1^2}{n_1} + \frac{s_2^2}{n_2}}} = \dfrac{(31{,}987 - 33{,}179)}{\sqrt{\frac{5000^2}{36} + \frac{6000^2}{49}}} \approx -0.997$. The difference in the sample mean incomes for Sacramento County and Los Angeles County, California $\bar{x}_1 - \bar{x}_2 = -1192$ lies 0.997 standard errors below the difference in the hypothesized mean incomes for Sacramento County and Los Angeles County, California, $\mu_1 - \mu_2 = 0$. **(c)** $H_0 : \mu_1 = \mu_2$ vs. $H_a : \mu_1 \neq \mu_2$

Reject H_0 if $t_{data} < -2.030$ or $t_{data} > 2.030$.

Since $t_{data} \geq -2.030$ and $t_{data} \leq 2.030$ we do not reject H_0. There is insufficient evidence that the population mean income for Sacramento County, California, differs from the population mean income for Los Angeles County, California. **(d)** Yes, because it is a two-tail test. Since 0 lies in the interval, we do not reject H_0. There is insufficient evidence that the population mean income for Sacramento County, California, differs from the population mean income for Los Angeles County, California.

23. **(a)** $\bar{x}_1 - \bar{x}_2 = 20.9 - 19.3 = 1.6$

(b) $s_{\bar{x}_1 - \bar{x}_2} = \sqrt{\frac{s_1^2}{n_1} + \frac{s_2^2}{n_2}} = \sqrt{\frac{5^2}{36} + \frac{4^2}{64}} \approx 0.9718$

(c) $n_1 - 1 = 36 - 1 = 35$ and $n_2 = 64 - 1 = 63$, so df $= 35$. Thus,

$$E = t_{\alpha/2}\sqrt{\frac{s_1^2}{n_1} + \frac{s_2^2}{n_2}} = 2.724\sqrt{\frac{5^2}{36} + \frac{4^2}{64}} \approx 2.647.$$

(d) $(\bar{x}_1 - \bar{x}_2) \pm E \approx 1.6 \pm 2.647 = (-1.047, 4.247)$. We are 99% confident that the interval $(-1.047, 4.247)$ captures the difference in the population mean number of children per teacher in the towns of Cupertino, California, and Santa Rosa, California.

25. **(a)** $n_1 - 1 = 36 - 1 = 35$ and $n_2 = 64 - 1 = 63$, so df $= 35$. $\alpha = 0.01$. $t_{crit} = 2.724$.

(b) $t_{data} = \dfrac{(\bar{x}_1 - \bar{x}_2)}{\sqrt{\frac{s_1^2}{n_1} + \frac{s_2^2}{n_2}}} = \dfrac{(20.9 - 19.3)}{\sqrt{\frac{5^2}{36} + \frac{4^2}{64}}} \approx 1.646$. The difference in the sample mean number of children per teacher in the towns of Cupertino, California, and Santa Rosa, California, $\bar{x}_1 - \bar{x}_2 = 1.6$ lies 1.646 standard errors above the difference in the hypothesized mean number of children per teacher in the towns of Cupertino, California, and Santa Rosa, California, $\mu_1 - \mu_2 = 0$.

(c) $H_0 : \mu_1 = \mu_2$ vs. $H_a : \mu_1 \neq \mu_2$

Reject H_0 if $t_{data} < -2.724$ or $t_{data} > 2.724$.

Since $t_{data} \geq -2.724$ and $t_{data} \leq 2.724$ we do not reject H_0. There is insufficient evidence that the population mean number of children per teacher in the town of Cupertino, California, differs from the population mean number of children per teacher in the town of Santa Rosa, California.

27. From the TI-84: North Carolina: $\bar{x}_1 = 159$, $s_1^2 \approx 1015.8182$
Ohio: $\bar{x}_2 \approx 264.6667$, $s_2^2 \approx 3658.6667$

$H_0: \mu_1 = \mu_2$ vs. $H_a: \mu_1 \neq \mu_2$
$n_1 - 1 = 12 - 1 = 11$ and $n_2 = 15 - 1 = 14$, so df = 11. $\alpha = 0.05$. $t_{crit} = 2.201$
Reject H_0 if $t_{data} < -2.201$ or $t_{data} > 2.201$.

$$t_{data} = \frac{(\bar{x}_1 - \bar{x}_2)}{\sqrt{\frac{s_1^2}{n_1} + \frac{s_2^2}{n_2}}} = \frac{(159 - 264.6667)}{\sqrt{\frac{1015.8182}{12} + \frac{3658.6667}{15}}} \approx -5.829$$

Since $t_{data} < -2.201$ we reject H_0. There is evidence that the population mean property tax in Ohio is different than the population mean property tax in North Carolina.

29. **(a)** $n_1 - 1 = 36 - 1 = 35$ and $n_2 = 30 - 1 = 29$, so df = 29.

$$(\bar{x}_1 - \bar{x}_2) \pm t_{\alpha/2}\sqrt{\frac{s_1^2}{n_1} + \frac{s_2^2}{n_2}} = (110 - 150) \pm 2.045\sqrt{\frac{60^2}{36} + \frac{75^2}{30}}$$
$$\approx -40 \pm 34.6747 = (-74.6747, -5.3253).$$

We are 95% confident that our interval $(-74.6747, -5.3253)$ captures the difference in the population mean number of daily visitors to Windvale Park and Cranebrook Park.

(b) $H_0: \mu_1 \geq \mu_2$ vs. $H_a: \mu_1 < \mu_2$
Reject H_0 if p-value < 0.05.

$$t_{data} = \frac{(\bar{x}_1 - \bar{x}_2)}{\sqrt{\frac{s_1^2}{n_1} + \frac{s_2^2}{n_2}}} = \frac{(110 - 150)}{\sqrt{\frac{60^2}{36} + \frac{75^2}{30}}} \approx -2.359$$

p-value $= P(t < t_{data}) = P(t < -2.359) = 0.0126370393$
Since p-value < 0.05 we reject H_0. There is evidence that the population mean number of daily visitors to Windvale Park is less than the population mean number of daily visitors to Cranebrook Park.
(c) No, because confidence intervals can only be used for two-tail tests and this is a one-tail test.

31. **(a)** $\bar{x}_1 - \bar{x}_2 = 29 - 21 = 8$ **(b)** $n_1 - 1 = 100 - 1 = 99$ and $n_2 = 100 - 1 = 99$, so df = 99. The value of 99 for the df does not appear in the df column. The conservative approach would be to use next row with df smaller than 99. That would be df = 90. Thus, the "conservative" $t_{\alpha/2}$ is 2.632.

$$(\bar{x}_1 - \bar{x}_2) \pm t_{\alpha/2}\sqrt{\frac{s_1^2}{n_1} + \frac{s_2^2}{n_2}} = 8 \pm 2.632\sqrt{\frac{59^2}{100} + \frac{52^2}{100}} \approx 8 \pm 20.699 = (-12.699, 28.699)$$

(c) $H_0: \mu_1 = \mu_2$ vs. $H_a: \mu_1 \neq \mu_2$
Reject H_0 if p-value < 0.01.

$$t_{data} = \frac{(\bar{x}_1 - \bar{x}_2)}{\sqrt{\frac{s_1^2}{n_1} + \frac{s_2^2}{n_2}}} = \frac{(29 - 21)}{\sqrt{\frac{59^2}{100} + \frac{52^2}{100}}} \approx 1.017$$

p-value $= 2P(t > t_{data}) = 2P(t > 1.017) = 0.3116324129$
Since p-value ≥ 0.01 we do not reject H_0.
There is insufficient evidence that the mean coached SAT score improvement is different than the mean non-coached SAT score improvement.

33. From the TI-84, $\bar{x}_1 \approx 1324.1429$, $s_1 \approx 345.2996$, $\bar{x}_2 \approx 1822.6667$, and $s_2 \approx 155.6171$.
(a) $\bar{x}_1 - \bar{x}_2 = 1324.1429 - 1822.6667 = -498.5238$
(b) $n_1 - 1 = 7 - 1 = 6$ and $n_2 = 6 - 1 = 5$, so df = 5.

$$(\bar{x}_1 - \bar{x}_2) \pm t_{\alpha/2}\sqrt{\frac{s_1^2}{n_1} + \frac{s_2^2}{n_2}} = -498.5238 \pm 2.105\sqrt{\frac{345.2996^2}{7} + \frac{155.6171^2}{6}}$$
$$\approx -498.5238 \pm 292.4822 = (-791.0060, -206.0416).$$

We are 90% confident that the interval $(-791.0060, -206.0416)$ captures the difference in the population mean daily calorie intake of children in the inner city and the suburbs.
(c) $H_0: \mu_1 \geq \mu_2$ vs. $H_a: \mu_1 < \mu_2$
Reject H_0 if p-value < 0.10.

$$t_{data} = \frac{(\bar{x}_1 - \bar{x}_2)}{\sqrt{\frac{s_1^2}{n_1} + \frac{s_2^2}{n_2}}} = \frac{(1324.1429 - 1822.6667)}{\sqrt{\frac{345.2996^2}{7} + \frac{155.6171^2}{6}}} \approx -3.434$$

p-value $= P(t < t_{data}) = P(t < -3.434) = 0.0092775929$
Since p-value < 0.10 we reject H_0. There is evidence that the population mean daily calorie intake of inner city children is less than that of children from the suburbs. **(d)** Since $0.001 < p$-value ≤ 0.01, there is very strong evidence against the null hypothesis that the population mean daily calorie intake of inner city children is greater than or equal to that of children from the suburbs.

35. **(a)** $n_1 - 1 = 121 - 1 = 120$ and $n_2 = 121 - 1 = 120$, so df $= 120$. The value of 120 for the df does not appear in the df column. The conservative approach would be to use next row with df smaller than 120. That would be df $= 100$. Thus, the "conservative" $t_{\alpha/2}$ is 2.626.

$$(\bar{x}_1 - \bar{x}_2) \pm t_{\alpha/2}\sqrt{\frac{s_1^2}{n_1} + \frac{s_2^2}{n_2}} = (40.4 - 41.7) \pm 2.626\sqrt{\frac{5.5^2}{121} + \frac{5.5^2}{121}} \approx -1.3 \pm 1.857 = (-3.157, 0.557).$$

We are 99% confident that the interval $(-3.157, 0.557)$ captures the difference of the population mean number of hours worked in 2003 and 1994.
(b) $H_0: \mu_1 \geq \mu_2$ vs. $H_a: \mu_1 < \mu_2$
Reject H_0 if p-value < 0.01.

$$t_{data} = \frac{(\bar{x}_1 - \bar{x}_2)}{\sqrt{\frac{s_1^2}{n_1} + \frac{s_2^2}{n_2}}} = \frac{(40.4 - 41.7)}{\sqrt{\frac{5.5^2}{121} + \frac{5.5^2}{121}}} \approx -1.838$$

p-value $= P(t < t_{data}) = P(t < -1.838) = 0.0342678723$
Since p-value ≥ 0.01 we do not reject H_0. There is insufficient evidence that the population mean number of hours worked in 2003 is less than the population mean number of hours worked in 1994.
(c) Since $0.01 < p$-value ≤ 0.05, there is solid evidence against the null hypothesis.

37. **(a)** Since $\bar{x}_1$ is the sample mean number of hours worked in 2003 and $\bar{x}_2$ is the sample mean number of hours worked in 1994, an increase in $\bar{x}_1$ so that it is still smaller than $\bar{x}_2$ will result in an increase in $\bar{x}_1 - \bar{x}_2$, but $\bar{x}_1 - \bar{x}_2$ will still be negative.
(b) Since $t_{data} = \frac{\bar{x}_1 - \bar{x}_2}{\sqrt{\frac{s_1^2}{n_1} + \frac{s_2^2}{n_2}}}$ and an increase in $\bar{x}_1$ so that it is still smaller than $\bar{x}_2$ will result in an increase in $\bar{x}_1 - \bar{x}_2$ but $\bar{x}_1 - \bar{x}_2$ will still be negative, an increase in $\bar{x}_1$ so that it is still smaller than $\bar{x}_2$ will result in an increase in t_{data}, but t_{data} will still be negative. **(c)** Since p-value $= P(t < t_{data})$, an increase in $\bar{x}_1$ so that it is still smaller than $\bar{x}_2$ will result in an increase in t_{data} and therefore an increase in the p-value. **(d)** Since the original p-value ≥ 0.01 and increase in $\bar{x}_1$ so that it is still smaller than $\bar{x}_2$ will result in an increase in the p-value, the p-value will still be greater than or equal to 0.01, so the conclusion will still be do not reject H_0.

39. **(a)** From Minitab:

Descriptive Statistics: CALIF

```
Variable    N  N*   Mean  SE Mean   StDev  Minimum    Q1  Median     Q3
CALIF     858   0  31363     4618  135270     1010  3230    8319  28950

Variable  Maximum
CALIF     3485398
```

$\bar{x}_1 = 31{,}363$, $s_1 = 135{,}270$

(b) From Minitab:

Descriptive Statistics: NEWYORK

Variable	N	N*	Mean	SE Mean	StDev	Minimum	Q1	Median	Q3	Maximum
NEWYORK	790	0	18305	9284	260938	1000	1901	4013	9059	7322564

$\bar{x}_2 = 18{,}305$, $s_2 = 260{,}938$

(c) $H_0: \mu_1 = \mu_2$ vs. $H_a: \mu_1 \neq \mu_2$

Reject H_0 if p-value < 0.05.

$$t_{data} = \frac{(\bar{x}_1 - \bar{x}_2)}{\sqrt{\frac{s_1^2}{n_1} + \frac{s_2^2}{n_2}}} = \frac{(31{,}363 - 18{,}305)}{\sqrt{\frac{135{,}270^2}{858} + \frac{260{,}938^2}{790}}} \approx 1.259$$

$n_1 - 1 = 858 - 1 = 857$ and $n_2 = 790 - 1 = 789$, so df $= 789$. p-value $= 2P(t > t_{data}) = 2P(t > 1.259) = 0.2084026891$. Since p-value ≥ 0.05, we do not reject H_0. There is insufficient evidence that the population mean number of residents in towns and cities in California is different than the population mean number of residents in towns and cities in New York.

Section 10.3

1. $\hat{p}_1$ and $\hat{p}_2$

3. No, because the confidence interval is not based on the assumption that $p_1 = p_2$.

5. Z_{data} measures the distance between sample proportions. Extreme values of Z_{data} indicate evidence against the null hypothesis.

7. (a) We have $x_1 = 80 \geq 5$, $n_1 - x_1 = 20 \geq 5$, $x_2 = 30 \geq 5$, and $n_2 - x_2 = 10 \geq 5$, so it is appropriate to construct a 95% confidence interval for $p_1 - p_2$.

(b) $\hat{p}_1 = \frac{x_1}{n_1} = \frac{80}{100} = 0.80$ and $\hat{p}_2 = \frac{x_2}{n_2} = \frac{30}{40} = 0.75$, so $\hat{p}_1 - \hat{p}_2 = 0.80 - 0.75 = 0.05$

(c) $\hat{q}_1 = 1 - \hat{p}_1 = 1 - 0.80 = 0.20$ and $\hat{q}_2 = 1 - 0.75 = 0.25$.

$$s_{\hat{p}_1 - \hat{p}_2} = \sqrt{\frac{\hat{p}_1 \cdot \hat{q}_1}{n_1} + \frac{\hat{p}_2 \cdot \hat{q}_2}{n_2}} = \sqrt{\frac{(0.80)(0.20)}{100} + \frac{(0.75)(0.25)}{40}} \approx 0.0793.$$

The typical error in estimating the unknown quantity $p_1 - p_2$ is 0.0793.

(d) $E = Z_{\alpha/2} = \sqrt{\frac{\hat{p}_1 \cdot \hat{q}_1}{n_1} + \frac{\hat{p}_2 \cdot \hat{q}_2}{n_2}} = 1.96\sqrt{\frac{(0.80)(0.20)}{100} + \frac{(0.75)(0.25)}{40}} \approx 0.1554.$

The point estimate $\hat{p}_1 - \hat{p}_2 = 0.05$ will lie within 0.1554 of the difference in population proportions $p_1 - p_2$ 95% of the time.

(e) $(\hat{p}_1 - \hat{p}_2) \pm E = 0.05 \pm 0.1554 = (-0.1054, 0.2054)$. We are 95% confident that the difference in population proportions lies between -0.1054 and 0.2054.

9. (a) We have $x_1 = 60 \geq 5$, $n_1 - x_1 = 140 \geq 5$, $x_2 = 40 \geq 5$, and $n_2 - x_2 = 210 \geq 5$, so it is appropriate to construct a 95% confidence interval for $p_1 - p_2$.

(b) $\hat{p}_1 = \frac{x_1}{n_1} = \frac{60}{200} = 0.30$ and $\hat{p}_2 = \frac{x_2}{n_2} = \frac{40}{250} = 0.16$, so $\hat{p}_1 - \hat{p}_2 = 0.30 - 0.16 = 0.14$

(c) $\hat{q}_1 = 1 - \hat{p}_1 = 1 - 0.30 = 0.70$ and $\hat{q}_2 = 1 - 0.16 = 0.84$.

$$s_{\hat{p}_1 - \hat{p}_2} = \sqrt{\frac{\hat{p}_1 \cdot \hat{q}_1}{n_1} + \frac{\hat{p}_2 \cdot \hat{q}_2}{n_2}} = \sqrt{\frac{(0.30)(0.70)}{200} + \frac{(0.16)(0.84)}{250}} \approx 0.0398.$$

The typical error in estimating the unknown quantity $p_1 - p_2$ is 0.0398.

(d) $E = Z_{\alpha/2} = \sqrt{\frac{\hat{p}_1 \cdot \hat{q}_1}{n_1} + \frac{\hat{p}_2 \cdot \hat{q}_2}{n_2}} = 1.96\sqrt{\frac{(0.30)(0.70)}{200} + \frac{(0.16)(0.84)}{250}} \approx 0.078.$

The point estimate $\hat{p}_1 - \hat{p}_2 = 0.14$ will lie within 0.078 of the difference in population proportions $p_1 - p_2$ 95% of the time.

(e) $(\hat{p}_1 - \hat{p}_2) \pm E = 0.14 \pm 0.078 = (0.062, 0.218)$. We are 95% confident that the difference in population proportions lies between 0.062 and 0.218.

11. **(a)** We have $x_1 = 490 \geq 5$, $n_1 - x_1 = 510 \geq 5$, $x_2 = 620 \geq 5$, and $n_2 - x_2 = 380 \geq 5$, so it is appropriate to construct a 95% confidence interval for $p_1 - p_2$.

(b) $\hat{p}_1 = \frac{x_1}{n_1} = \frac{490}{1000} = 0.49$ and $\hat{p}_2 = \frac{x_2}{n_2} = \frac{620}{1000} = 0.62$, so $\hat{p}_1 - \hat{p}_2 = 0.49 - 0.62 = -0.13$

(c) $\hat{q}_1 = 1 - \hat{p}_1 = 1 - 0.49 = 0.51$ and $\hat{q}_2 = 1 - 0.62 = 0.38$.

$$s_{\hat{p}_1 - \hat{p}_2} = \sqrt{\frac{\hat{p}_1 \cdot \hat{q}_1}{n_1} + \frac{\hat{p}_2 \cdot \hat{q}_2}{n_2}} = \sqrt{\frac{(0.49)(0.51)}{1000} + \frac{(0.62)(0.38)}{1000}} \approx 0.0220.$$

The typical error in estimating the unknown quantity $p_1 - p_2$ is 0.0220.

(d) $E = Z_{\alpha/2}\sqrt{\frac{\hat{p}_1 \cdot \hat{q}_1}{n_1} + \frac{\hat{p}_2 \cdot \hat{q}_2}{n_2}} = 1.96\sqrt{\frac{(0.49)(0.51)}{1000} + \frac{(0.62)(0.38)}{1000}} \approx 0.0431.$

The point estimate $\hat{p}_1 - \hat{p}_2 = -0.13$ will lie within 0.0431 of the difference in population proportions $p_1 - p_2$ 95% of the time.

(e) $(\hat{p}_1 - \hat{p}_2) \pm E = -0.13 \pm 0.0431 = (-0.1731, -0.0869)$. We are 95% confident that the difference in population proportions lies between -0.1731 and -0.0869.

13. **(a)** $H_0: p_1 = p_2$ vs. $H_a: p_1 \neq p_2$

$\alpha = 0.10$, $Z_{crit} = 1.645$

Reject H_0 if $Z_{data} < -1.645$ or $Z_{data} > 1.645$.

(b) $\hat{p}_{pooled} = \frac{x_1 + x_2}{n_1 + n_2} = \frac{80 + 30}{100 + 40} = \frac{110}{140} \approx 0.7857$

(c) $Z_{data} = \frac{(\hat{p}_1 - \hat{p}_2)}{\sqrt{\hat{p}_{pooled} \cdot (1 - \hat{p}_{pooled})\left(\frac{1}{n_1} + \frac{1}{n_2}\right)}} \approx \frac{(0.8 - 0.75)}{\sqrt{0.7857 \cdot (1 - 0.7857)\left(\frac{1}{100} + \frac{1}{40}\right)}} \approx 0.65$

(d) Since $Z_{data} \geq -1.645$ and $Z_{data} \leq 1.645$ we do not reject H_0. There is insufficient evidence that the population proportion from Population 1 is different than the population proportion from Population 2.

15. **(a)** $H_0: p_1 \leq p_2$ vs. $H_a: p_1 > p_2$

$\alpha = 0.05$, $Z_{crit} = 1.645$

Reject H_0 if $Z_{data} > 1.645$.

(b) $\hat{p}_{pooled} = \frac{x_1 + x_2}{n_1 + n_2} = \frac{60 + 40}{200 + 250} = \frac{100}{450} \approx 0.2222$

(c) $Z_{data} = \frac{(\hat{p}_1 - \hat{p}_2)}{\sqrt{\hat{p}_{pooled} \cdot (1 - \hat{p}_{pooled})\left(\frac{1}{n_1} + \frac{1}{n_2}\right)}} \approx \frac{(0.30 - 0.16)}{\sqrt{0.2222 \cdot (1 - 0.2222)\left(\frac{1}{200} + \frac{1}{250}\right)}} \approx 3.55$

(d) Since $Z_{data} > 1.645$ we reject H_0. There is evidence that the population proportion from Population 1 is greater than the population proportion from Population 2.

17. **(a)** $H_0: p_1 \leq p_2$ vs. $H_a: p_1 > p_2$

Reject H_0 if p-value < 0.10.

(b) $\hat{p}_{pooled} = \frac{x_1 + x_2}{n_1 + n_2} = \frac{250 + 200}{400 + 400} = \frac{450}{800} = 0.5625$

(c) $Z_{data} = \frac{(\hat{p}_1 - \hat{p}_2)}{\sqrt{\hat{p}_{pooled} \cdot (1 - \hat{p}_{pooled})\left(\frac{1}{n_1} + \frac{1}{n_2}\right)}} \approx \frac{(0.625 - 0.5)}{\sqrt{0.5625 \cdot (1 - 0.5625)\left(\frac{1}{400} + \frac{1}{400}\right)}} \approx 3.56$

(d) p-value $= P(Z > Z_{data}) = P(Z > 3.56) = 0.000185467351$ **(e)** Since p-value < 0.10 we reject H_0. There is evidence that the population proportion from Population 1 is greater than the population proportion from Population 2.

19. **(a)** $H_0: p_1 = p_2$ vs. $H_a: p_1 \neq p_2$
Reject H_0 if p-value < 0.05.
(b) $\hat{p}_{pooled} = \dfrac{x_1 + x_2}{n_1 + n_2} = \dfrac{412 + 498}{527 + 613} = \dfrac{910}{1140} \approx 0.7982$
(c) $Z_{data} = \dfrac{(\hat{p}_1 - \hat{p}_2)}{\sqrt{\hat{p}_{pooled} \cdot (1 - \hat{p}_{pooled})\left(\frac{1}{n_1} + \frac{1}{n_2}\right)}} \approx \dfrac{(0.7818 - 0.8124)}{\sqrt{0.7982 \cdot (1 - 0.7982)\left(\frac{1}{527} + \frac{1}{613}\right)}} \approx -1.28$
(d) p-value $= 2P(Z < Z_{data}) = 2P(Z < -1.28) = 0.2005452669$ **(e)** Since p-value ≥ 0.05 we do not reject H_0. There is insufficient evidence that the population proportion from Population 1 is different than the population proportion from Population 2.

21. **(a)** We have $x_1 = 360.38 \geq 5$, $n_1 - x_1 = 126.62 \geq 5$, $x_2 = 404.21 \geq 5$, and $n_2 - x_2 = 82.79 \geq 5$, so it is appropriate to construct a 95% confidence interval for $p_1 - p_2$. **(b)** $\hat{p}_1 - \hat{p}_2 = 0.74 - 0.83 = -0.09$
(c) $\hat{q}_1 = 1 - \hat{p}_1 = 1 - 0.74 = 0.26$ and $\hat{q}_2 = 1 - \hat{p}_2 = 1 - 0.83 = 0.17$.

$$s_{\hat{p}_1 - \hat{p}_2} = \sqrt{\frac{\hat{p}_1 \cdot \hat{q}_1}{n_1} + \frac{\hat{p}_2 \cdot \hat{q}_2}{n_2}} = \sqrt{\frac{(0.74)(0.26)}{487} + \frac{(0.83)(0.17)}{487}} \approx 0.0262.$$

The typical error in estimating the unknown difference in the population proportion of teenage boys who post their photo on their online profile and the population proportion of teenage girls who do so $p_1 - p_2$ is 0.0262.

(d) $E = Z_{\alpha/2}\sqrt{\dfrac{\hat{p}_1 \cdot \hat{q}_1}{n_1} + \dfrac{\hat{p}_2 \cdot \hat{q}_2}{n_2}} = 1.96\sqrt{\dfrac{(0.74)(0.26)}{487} + \dfrac{(0.83)(0.17)}{487}} \approx 0.0514.$
The point estimate of the difference in the population proportion of teenage boys who post their photo on their online profile and the population proportion of teenage girls who do so $\hat{p}_1 - \hat{p}_2 = -0.09$ will lie within 0.0514 of the difference in population proportions $p_1 - p_2$ 95% of the time.
(e) $(\hat{p}_1 - \hat{p}_2) \pm E \approx -0.09 \pm 0.0514 = (-0.1414, -0.0386)$. We are 95% confident that the difference in the population proportion of teenage boys who post their photo on their online profile and the population proportion of teenage girls who do so lies between -0.1414 and -0.0386.

23. **(a)** $H_0: p_1 = p_2$ vs. $H_a: p_1 \neq p_2$, where $p_1 =$ the population proportion of teenage boys who posted their photo online and $p_2 =$ the population proportion of teenage girls who posted their photo online. **(b)** Zero does not fall within the confidence interval. **(c)** Since 0 does not fall within the confidence interval, we reject H_0. There is evidence that the population proportion of teenage boys who post their photo on their online profile differs from the population proportion of teenage girls who do so.

25. **(a)** $\hat{p}_1 - \hat{p}_2 = 0.42 - 0.30 = 0.12$
(b) $x_1 = n_1 \cdot \hat{p}_1 = (225)(0.42) \approx 95$ and $x_2 = n_2 \cdot \hat{p}_2 = (225)(0.30) \approx 68$

$$\hat{p}_{pooled} = \frac{x_1 + x_2}{n_1 + n_2} = \frac{95 + 68}{225 + 225} = \frac{163}{450} \approx 0.3622$$

(c) $H_0: p_1 = p_2$ vs. $H_a: p_1 \neq p_2$
Reject H_0 if p-value < 0.05.

$$Z_{data} = \frac{(\hat{p}_1 - \hat{p}_2)}{\sqrt{\hat{p}_{pooled} \cdot (1 - \hat{p}_{pooled})\left(\frac{1}{n_1} + \frac{1}{n_2}\right)}} \approx \frac{(0.42 - 0.30)}{\sqrt{0.3622 \cdot (1 - 0.3622)\left(\frac{1}{225} + \frac{1}{225}\right)}} \approx 2.65$$

From the TI-84: p-value $= 2P(Z > Z_{data}) = 2P(Z > 2.65) = 0.0080492611$.
From the table: p-value $= 2P(Z > Z_{data}) = 2P(Z > 2.65) = 2(1 - 0.9960) = 0.008$.
Since p-value < 0.05, we reject H_0. There is evidence of a generation gap in the perception of same-sex marriages.

27. **(a)** $\hat{p}_1 = \dfrac{x_1}{n_1} = \dfrac{14}{54} \approx 0.2593$, $\hat{p}_2 = \dfrac{x_2}{n_2} = \dfrac{25}{45} \approx 0.5556$, $\hat{q}_1 = 1 - \hat{p}_1 \approx 1 - 0.2593 = 0.7407$
and $\hat{q}_2 = 1 - \hat{p}_2 \approx 1 - 0.5556 = 0.4444$.

$$(\hat{p}_1 - \hat{p}_2) \pm Z_{\alpha/2}\sqrt{\frac{\hat{p}_1 \cdot \hat{q}_1}{n_1} + \frac{\hat{p}_2 \cdot \hat{q}_2}{n_2}} \approx (0.2593 - 0.5556) \pm 1.96\sqrt{\frac{(0.2593)(0.7407)}{54} + \frac{(0.5556)(0.4444)}{45}}$$

$$\approx -0.2963 \pm 0.1864 = (-0.4827, -0.1099)$$

(b) $H_0: p_1 = p_2$ vs. $H_a: p_1 \neq p_2$
Since 0 does not fall in the confidence interval, we reject H_0. There is evidence that there is a difference in the population proportions of women who had fetal cells.

29. $H_0: p_1 \le p_2$ vs. $H_a: p_1 > p_2$
Reject H_0 if p-value < 0.05.

$$\hat{p}_{pooled} = \frac{x_1 + x_2}{n_1 + n_2} = \frac{106 + 21}{2000 + 1000} = \frac{127}{3000} \approx 0.0423$$

$$\hat{p}_1 = \frac{x_1}{n_1} = \frac{106}{2000} = 0.053 \text{ and } \hat{p}_2 = \frac{x_2}{n_2} = \frac{21}{1000} = 0.021$$

$$Z_{data} = \frac{(\hat{p}_1 - \hat{p}_2)}{\sqrt{\hat{p}_{pooled} \cdot (1 - \hat{p}_{pooled})\left(\frac{1}{n_1} + \frac{1}{n_2}\right)}} \approx \frac{(0.053 - 0.021)}{\sqrt{0.0423 \cdot (1 - 0.0423)\left(\frac{1}{2000} + \frac{1}{1000}\right)}} \approx 4.11$$

p-value $= P(Z > Z_{data}) = P(Z > 4.11) = 0.00001979384758$
Since p-value < 0.05, we reject H_0. There is evidence that the population proportion of minority-owned businesses in Michigan exceeds that in Minnesota.

31. **(a)** $x_1 = n_1 \cdot \hat{p}_1 = (1000)(0.92) = 920$ and $x_2 = n_2 \cdot \hat{p}_2 = (1000)(0.87) = 870$. We have $x_1 = 920 \ge 5$, $n_1 - x_1 = 80 \ge 5$, $x_2 = 870 \ge 5$, and $n_2 - x_2 = 130 \ge 5$, so it is appropriate to construct a 95% confidence interval for $p_1 - p_2$. **(b)** $\hat{p}_1 - \hat{p}_2 = 0.92 - 0.87 = 0.05$ **(c)** $\hat{q}_1 = 1 - \hat{p}_1 = 1 - 0.92 = 0.08$ and $\hat{q}_2 = 1 - \hat{p}_2 = 1 - 0.87 = 0.13$.

$$s_{\hat{p}_1 - \hat{p}_2} = \sqrt{\frac{\hat{p}_1 \cdot \hat{q}_1}{n_1} + \frac{\hat{p}_2 \cdot \hat{q}_2}{n_2}} = \sqrt{\frac{(0.92)(0.08)}{1000} + \frac{(0.87)(0.13)}{1000}} \approx 0.0137.$$

The typical error in estimating the unknown difference in the population proportion of 18–24-year-old males who listen to the radio each week and the population proportion of males 65 years and older who listen to the radio each week $p_1 - p_2$ is 0.0137.

(d) $E = Z_{\alpha/2}\sqrt{\frac{\hat{p}_1 \cdot \hat{q}_1}{n_1} + \frac{\hat{p}_2 \cdot \hat{q}_2}{n_2}} = 1.96\sqrt{\frac{(0.92)(0.08)}{1000} + \frac{(0.87)(0.13)}{1000}} \approx 0.0268.$

The point estimate of the difference in the population proportion of 18–24-year-old males who listen to the radio each week and the population proportion of males 65 years and older who listen to the radio each week will lie within 0.0268 of the difference in population proportions $p_1 - p_2$ 95% of the time.
(e) $(\hat{p}_1 - \hat{p}_2) \pm E \approx 0.05 \pm 0.0268 = (0.0232, 0.0768)$. We are 95% confident that the difference in the population proportion of 18–24-year-old males who listen to the radio each week and the population proportion of males 65 years and older who listen to the radio each week lies between 0.0232 and 0.0768.

33. **(a)** It will remain 0.05. **(b)** Since $s_{\hat{p}_1 - \hat{p}_2} = \sqrt{\frac{\hat{p}_1 \cdot \hat{q}_1}{n_1} + \frac{\hat{p}_2 \cdot \hat{q}_2}{n_2}}$ and the sample sizes are in the denominators, a decrease in the sample sizes will result in an increase in $s_{\hat{p}_1 - \hat{p}_2}$. This will result in a confidence interval that is less precise. **(c)** Since the width of the 95% confidence interval is $3.92 \cdot s_{\hat{p}_1 - \hat{p}_2}$, a decrease in the sample sizes will result in an increase in $s_{\hat{p}_1 - \hat{p}_2}$, which will result in an increase in the width of the confidence interval. **(d)** Since Z_{crit} does not depend on the sample sizes, a decrease in the sample sizes will leave $Z_1 = 1.96$.
(e) Since p-value $= 2P(Z > Z_{data})$ and

$$Z_{data} = \frac{(\hat{p}_1 - \hat{p}_2)}{\sqrt{\hat{p}_{pooled} \cdot (1 - \hat{p}_{pooled})\left(\frac{1}{n_1} + \frac{1}{n_2}\right)}}$$

and the sample sizes are in the denominator of the denominator, a decrease in the sample sizes will result in $\hat{p}_1$, $\hat{p}_2$, and $\hat{p}_{pooled}$ remaining the same and an increase in the denominator, which will result in a decrease in Z_{data}, which in turn will result in an increase in the p-value to 0.2502. **(f)** Since a decrease in the sample sizes would result in a p-value greater than 0.05, the conclusion would change from reject H_0 to do not reject H_0.

Chapter 10 Review

1.

Subject	1	2	3	4	5	6	7	8
Sample 1	100.7	110.2	105.3	107.1	95.6	109.9	112.3	94.7
Sample 2	104.4	112.5	105.9	111.4	99.8	109.9	115.7	97.7
Differences (Sample 1 minus Sample 2)	−3.7	−2.3	−0.6	−4.3	−4.2	0.0	−3.4	−3.0

(a) From the TI-84, $\bar{x}_d = -2.6875$ and $s_d \approx 1.6146$. **(b)** $df = n - 1 = 8 - 1 = 7, \bar{x}_d \pm t_{\alpha/2}(s_d/\sqrt{n}) = -2.6875 \pm (2.365)(1.6146/\sqrt{8}) \approx -2.6875 \pm 1.3501 = (-4.0376, -1.3374)$

3. $H_0: \mu_d \geq 0$ vs. $H_a: \mu_d < 0$
Reject H_0 if p-value < 0.05.

$$t_{data} = \frac{\bar{x}_d}{s_d/\sqrt{n}} = \frac{-2.6875}{1.6146/\sqrt{8}} \approx -4.708$$

$df = n - 1 = 8 - 1 = 7$, p-value $= P(t < t_{data}) = P(t < -4.708) = 0.0010939869$
Since p-value < 0.05, we reject H_0. There is evidence that the population mean of the differences is less than 0.

5. From the TI-84, $s_d \approx \$49.4733$. $df = n - 1 = 10 - 1 = 9, \bar{x}_d \pm t_{\alpha/2}(s_d/\sqrt{n}) = -39.5 \pm (2.262)(49.4733/\sqrt{10}) \approx -39.5 \pm 35.389 = (-74.889, -4.111)$

7. Since both sample sizes are large ($n_1 \geq 30$ and $n_2 \geq 30$), it is appropriate to construct a 95% confidence interval for $\mu_1 - \mu_2$.

9. $s_{\bar{x}_1 - \bar{x}_2} = \sqrt{\frac{s_1^2}{n_1} + \frac{s_2^2}{n_2}} = \sqrt{\frac{0.01^2}{36} + \frac{0.02^2}{81}} \approx 0.0028$

11. $(\bar{x}_1 - \bar{x}_2) \pm t_{\alpha/2}\sqrt{\frac{s_1^2}{n_1} + \frac{s_2^2}{n_2}} = 0.1 \pm 2.030\sqrt{\frac{0.01^2}{36} + \frac{0.02^2}{81}} \approx 0.1 \pm 0.0056 = (0.0944, 0.1056)$.
We are 95% confident that the interval (0.0944, 0.1056) captures the difference in population means.

13. (a) $5/p_1 = 5/0.16 = 31.25$ and $5/(1 - p_1) = 5/(1 - 0.16) \approx 5.95$, so $n_1 = 32$.
$5/p_2 = 5/0.04 = 125$ and $5/(1 - p_2) = 5/(1 - 0.04) \approx 1.04$, so $n_2 = 125$
(b) $\hat{p}_1 - \hat{p}_2 = 0.16 - 0.04 = 0.12$
(c) $\hat{q}_1 = 1 - \hat{p}_1 = 1 - 0.16 = 0.84$ and $\hat{q}_2 = 1 - \hat{p}_2 = 1 - 0.04 = 0.96$.

$$(\hat{p}_1 - \hat{p}_2) \pm Z_{\alpha/2}\sqrt{\frac{\hat{p}_1 \cdot \hat{q}_1}{n_1} + \frac{\hat{p}_2 \cdot \hat{q}_2}{n_2}} = 0.12 \pm 1.645\sqrt{\frac{(0.16)(0.84)}{32} + \frac{(0.04)(0.96)}{125}}$$

$$\approx 0.12 \pm 0.1104 = (0.0096, 0.2304)$$

(d) $H_0: p_1 = p_2$ vs. $H_a: p_1 \neq p_2$
Reject H_0 if p-value < 0.10.
$x_1 = n_1 \cdot \hat{p}_1 = (32)(0.16) \approx 5$ and $x_2 = n_2 \cdot \hat{p}_2 = (125)(0.04) = 5$

$$\hat{p}_{pooled} = \frac{x_1 + x_2}{n_1 + n_2} = \frac{5 + 5}{32 + 125} = \frac{10}{157} \approx 0.0637$$

$$Z_{data} = \frac{(\hat{p}_1 - \hat{p}_2)}{\sqrt{\hat{p}_{pooled} \cdot (1 - \hat{p}_{pooled})\left(\frac{1}{n_1} + \frac{1}{n_2}\right)}} \approx \frac{(0.16 - 0.04)}{\sqrt{0.0637 \cdot (1 - 0.0637)\left(\frac{1}{32} + \frac{1}{125}\right)}} \approx 2.48$$

From the TI-84, p-value $= 2P(Z > Z_{data}) = 2P(Z > 2.48) = 0.013138259$.
From the table, p-value $= 2P(Z > Z_{data}) = 2P(Z > 2.48) = 2(1 - 0.9934) = 0.0132$.
Since p-value < 0.10, we reject H_0. There is evidence that the population proportion packet loss from Asian websites is different than the population proportion packet loss from North American websites.

15. **(a)** $\hat{p}_1 = \frac{x_1}{n_1} = \frac{641}{823} \approx 0.7789$ and $\hat{p}_2 = \frac{x_2}{n_2} = \frac{658}{824} \approx 0.7985$, so $\hat{p}_1 - \hat{p}_2 = 0.7789 - 0.7985 = -0.0196$.

(b) $\hat{q}_1 = 1 - \hat{p}_1 \approx 1 - 0.7789 = 0.2211$ and $\hat{q}_2 = 1 - \hat{p}_2 \approx 1 - 0.7985 = 0.2015$

$$(\hat{p}_1 - \hat{p}_2) \pm Z_{\alpha/2}\sqrt{\frac{\hat{p}_1 \cdot \hat{q}_1}{n_1} + \frac{\hat{p}_2 \cdot \hat{q}_2}{n_2}} \approx -0.0196 \pm 2.576\sqrt{\frac{(0.7789)(0.2211)}{823} + \frac{(0.7985)(0.2015)}{824}}$$

$$\approx -0.0196 \pm 0.0518 = (-0.0714, 0.0322)$$

(c) $H_0 : p_1 = p_2$ vs. $H_a : p_1 \neq p_2$
Reject H_0 if p-value < 0.01.

$$\hat{p}_{pooled} = \frac{x_1 + x_2}{n_1 + n_2} = \frac{641 + 658}{823 + 824} = \frac{1299}{1647} \approx 0.7887$$

$$Z_{data} = \frac{(\hat{p}_1 - \hat{p}_2)}{\sqrt{\hat{p}_{pooled} \cdot (1 - \hat{p}_{pooled})\left(\frac{1}{n_1} + \frac{1}{n_2}\right)}} \approx \frac{(0.7789 - 0.7985)}{\sqrt{0.7887 \cdot (1 - 0.7887)\left(\frac{1}{823} + \frac{1}{824}\right)}} \approx -0.97$$

From the TI-84, p-value $= 2P(Z < Z_{data}) = 2P(Z < -0.97) = 0.3320464796$.
From the table, p-value $= 2P(Z < Z_{data}) = 2P(Z < -0.97) = 2(0.1660) = 0.3320$.
Since p-value ≥ 0.01, we do not reject H_0. There is insufficient evidence that the population proportion of suspects in Federal courts who were denied bail for violent crimes is different than the population proportion of suspects in Federal courts who were denied bail for property-related crimes.

Chapter 10 Quiz

1. True.
2. True.
3. False. The test statistic Z_{data} measures the distance between the sample proportions.
4. Normal, large (greater than or equal to 30).
5. Margin of error.
6. $\bar{x}_d$
7. $\mu_1 - \mu_2$
8. $s_{\bar{x}_1 - \bar{x}_2}$
9. No difference.
10.

Participant	1	2	3	4	5	6	7	8	9	10
Before	40	20	60	30	50	60	20	40	30	20
After	20	0	40	30	20	60	20	20	0	20
Differences (After minus Before)	−20	−20	−20	0	−30	0	0	−20	−30	0

(a) From the TI-84, $\bar{x}_d = -14$

(b) From the TI-84, $s_d \approx 12.6491$. df $= n - 1 = 10 - 1 = 9$,

$$\bar{x}_d \pm t_{\alpha/2}(s_d/\sqrt{n}) = -14 \pm (1.833)(12.6491/\sqrt{10}) \approx -14 \pm 7.3320 = (-21.3320, -6.6680)$$

(c) $H_0 : \mu_d = 0$ vs. $H_a : \mu_d \neq 0$
Since 0 does not lie in the confidence interval, we reject H_0. There is evidence that the population mean difference in the number of cigarettes smoked before and after attending Butt-Enders is different than 0.

11. (a) $\bar{x}_1 - \bar{x}_2 = \$50{,}000 - \$65{,}000 = -\$15{,}000$
(b) $n_1 - 1 = 40 - 1 = 39$ and $n_2 = 36 - 1 = 35$, so df = 35.

$$(\bar{x}_1 - \bar{x}_2) \pm t_{\alpha/2}\sqrt{\frac{s_1^2}{n_1} + \frac{s_2^2}{n_2}} = -15{,}000 \pm 1.690\sqrt{\frac{15{,}000^2}{40} + \frac{20{,}000^2}{36}}$$
$$\approx -15{,}000 \pm 6913.75 = (-21{,}913.75, -8{,}086.25).$$

We are 90% confident that the interval $(-21{,}913.75, -8{,}086.25)$ captures the difference of the population mean income of Suburb A and the population mean income of Suburb B.

(c) $n_1 - 1 = 40 - 1 = 39$ and $n_2 = 36 - 1 = 35$, so df = 35.

$$(\bar{x}_1 - \bar{x}_2) \pm t_{\alpha/2}\sqrt{\frac{s_1^2}{n_1} + \frac{s_2^2}{n_2}} = -15{,}000 \pm 2.030\sqrt{\frac{15{,}000^2}{40} + \frac{20{,}000^2}{36}}$$
$$\approx -15{,}000 \pm 8{,}304.69 = (-23{,}304.69, -6{,}695.31).$$

We are 95% confident that the interval $(-23{,}304.69, -6{,}695.31)$ captures the difference of the population mean income of Suburb A and the population mean income of Suburb B.

(d) *p*-value method:
$H_0: \mu_1 \geq \mu_2$ vs. $H_a: \mu_1 < \mu_2$
Reject H_0 if *p*-value < 0.10.

$$t_{data} = \frac{(\bar{x}_1 - \bar{x}_2)}{\sqrt{\frac{s_1^2}{n_1} + \frac{s_2^2}{n_2}}} = \frac{(50{,}000 - 65{,}000)}{\sqrt{\frac{15{,}000^2}{40} + \frac{20{,}000^2}{36}}} \approx -3.667$$

p-value $= P(t < t_{data}) = P(t < -3.667) = 0.000404119627$
Since *p*-value < 0.10, we reject H_0. There is evidence that the population mean income of Suburb A is less than the population mean income of Suburb B.

Critical value method:
$H_0: \mu_1 \geq \mu_2$ vs. $H_a: \mu_1 < \mu_2$
$n_1 - 1 = 40 - 1 = 39$ and $n_2 = 36 - 1 = 35$, so df = 35. $\alpha = 0.10$.
$t_{crit} = 1.306$. Reject H_0 if $t_{data} < -1.306$.

$$t_{data} = \frac{(\bar{x}_1 - \bar{x}_2)}{\sqrt{\frac{s_1^2}{n_1} + \frac{s_2^2}{n_2}}} = \frac{(50{,}000 - 65{,}000)}{\sqrt{\frac{15{,}000^2}{40} + \frac{20{,}000^2}{36}}} \approx -3.667$$

Since $t_{data} < -1.306$, we reject H_0. There is evidence that the population mean income of Suburb A is less than the population mean income of Suburb B.

(e) *p*-value method:
$H_0: \mu_1 = \mu_2$ vs. $H_a: \mu_1 \neq \mu_2$
Reject H_0 if *p*-value < 0.05.

$$t_{data} = \frac{(\bar{x}_1 - \bar{x}_2)}{\sqrt{\frac{s_1^2}{n_1} + \frac{s_2^2}{n_2}}} = \frac{(50{,}000 - 65{,}000)}{\sqrt{\frac{15{,}000^2}{40} + \frac{20{,}000^2}{36}}} \approx -3.667$$

p-value $= 2P(t < t_{data}) = 2P(t < -3.667) = 0.0008082392539$
Since *p*-value < 0.05, we reject H_0. There is evidence that the population mean income of Suburb A is different than the population mean income of Suburb B.

Critical value method:
$H_0: \mu_1 = \mu_2$ vs. $H_a: \mu_1 \neq \mu_2$
$n_1 - 1 = 40 - 1 = 39$ and $n_2 = 36 - 1 = 35$, so df = 35. $\alpha = 0.05$.
$t_{crit} = 2.030$. Reject H_0 if $t_{data} < -2.030$ or $t_{data} > 2.030$.

$$t_{data} = \frac{(\bar{x}_1 - \bar{x}_2)}{\sqrt{\frac{s_1^2}{n_1} + \frac{s_2^2}{n_2}}} = \frac{(50{,}000 - 65{,}000)}{\sqrt{\frac{15{,}000^2}{40} + \frac{20{,}000^2}{36}}} \approx -3.667$$

Since $t_{data} < -2.030$ we reject H_0. There is evidence that the population mean income of Suburb A is different than the population mean income of Suburb B.
(f) Since 0 does not fall in the confidence interval in part (c), we would reject H_0. There is evidence that the population mean income of Suburb A is different than the population mean income of Suburb B.
We could have used the confidence interval in (c) to perform the hypothesis test in (e).

12. **(a)** Since both sample sizes are large ($n_1 \geq 30$ and $n_2 \geq 30$), Case 2 applies. Thus, it is appropriate to apply two-sample t inference. **(b)** $\bar{x}_1 - \bar{x}_2 = 200 - 190 = 10$

(c) $s_{\bar{x}_1-\bar{x}_2} = \sqrt{\frac{s_1^2}{n_1} + \frac{s_2^2}{n_2}} = \sqrt{\frac{30^2}{100} + \frac{25^2}{100}} \approx 3.9051$

(d) $n_1 - 1 = 100 - 1 = 99$ and $n_2 = 100 - 1 = 99$, so df $= 99$.
The value of 99 for the df does not appear in the df column. The conservative approach would be to use next row with df smaller than 99. That would be df $= 90$. Thus, the "conservative" $t_{\alpha/2}$ is 1.987.

$$E = t_{\alpha/2}\sqrt{\frac{s_1^2}{n_1} + \frac{s_2^2}{n_2}} = 1.987\sqrt{\frac{30^2}{100} + \frac{25^2}{100}} \approx 7.7595$$

(e) $(\bar{x}_1 - \bar{x}_2) \pm t_{\alpha/2}\sqrt{\frac{s_1^2}{n_1} + \frac{s_2^2}{n_2}} = 10 \pm 1.987\sqrt{\frac{30^2}{100} + \frac{25^2}{100}} \approx 10 \pm 7.7595 = (2.2406, 17.7594)$.

We are 95% confident that the interval (2.2406, 17.7594) captures the difference of the population mean number of bottles processed by the updated machine and the mean number of bottles processed by the non-updated machine.

13. **(a)** $n_1 - 1 = 100 - 1 = 99$ and $n_2 = 100 - 1 = 99$, so df $= 99$. The value of 99 for the df does not appear in the df column. The conservative approach would be to use next row with df smaller than 99. That would be df $= 90$. Thus, the "conservative" t_{crit} is 1.662.

(b) $t_{data} = \dfrac{(\bar{x}_1 - \bar{x}_2)}{\sqrt{\frac{s_1^2}{n_1} + \frac{s_2^2}{n_2}}} = \dfrac{(200 - 190)}{\sqrt{\frac{30^2}{100} + \frac{25^2}{100}}} \approx 2.561$

(c) $H_0: \mu_1 \leq \mu_2$ vs. $H_a: \mu_1 > \mu_2$
$t_{crit} = 1.662$. Reject H_0 if $t_{data} > 1.662$, $t_{data} = 2.561$
Since $t_{data} > 1.662$, we reject H_0. There is evidence that the population mean number of bottles processed by the updated machine is greater than the mean number of bottles processed by the non-updated machine.
(d) Since confidence intervals can only be used to perform two-tail tests and the hypothesis test in (c) is a one-tail test, the confidence interval in Exercise 12 (e) cannot be used to perform the hypothesis test in (c).

14. **(a)** $\bar{x}_1 - \bar{x}_2 = \$13{,}539 - \$19{,}321 = -\$5{,}782$
(b) ***p*-value method:**
$H_0: \mu_1 = \mu_2$ vs. $H_a: \mu_1 \neq \mu_2$
Reject H_0 if p-value < 0.10.

$$t_{data} = \frac{(\bar{x}_1 - \bar{x}_2)}{\sqrt{\frac{s_1^2}{n_1} + \frac{s_2^2}{n_2}}} = \frac{(13{,}539 - 19{,}321)}{\sqrt{\frac{5000^2}{100} + \frac{8000^2}{100}}} \approx -6.129$$

p-value $= 2P(t < t_{data}) = 2P(t < -6.129) \approx 0$
Since p-value < 0.10 we reject H_0. There is evidence that the population mean income of people 18 to 24 years old who never married is different than the population mean income of people 18 to 24 years old who are married.

Critical value method:
$H_0: \mu_1 = \mu_2$ vs. $H_a: \mu_1 \neq \mu_2$
$n_1 - 1 = 100 - 1 = 99$ and $n_2 = 100 - 1 = 99$, so df $= 99$.
The value of 99 for the df does not appear in the df column. The conservative approach would be to use next row with df smaller than 99. That would be df $= 90$. Thus, the "conservative" t_{crit} is 1.662.

Reject H_0 if $t_{data} < -1.662$ or $t_{data} > 1.662$.

$$t_{data} = \frac{(\bar{x}_1 - \bar{x}_2)}{\sqrt{\frac{s_1^2}{n_1} + \frac{s_2^2}{n_2}}} = \frac{(13{,}539 - 19{,}321)}{\sqrt{\frac{5000^2}{100} + \frac{8000^2}{100}}} \approx -6.129$$

Since $t_{data} < -1.662$ we reject H_0. There is evidence that the population mean income of people 18 to 24 years old who never married is different than the population mean income of people 18 to 24 years old who are married. **(c)** Since the conclusion of the two-tail hypothesis test for $\alpha = 0.10$ is reject H_0, the 90% confidence interval would not include 0.

15. (a) $\hat{p}_1 - \hat{p}_2 = 0.385 - 0.379 = 0.006$

(b) $x_1 = n_1 \cdot \hat{p}_1 = (1000)(0.385) = 385$ and $x_2 = n_2 \cdot \hat{p}_2 = (1000)(0.379) = 379$

$$\hat{p}_{pooled} = \frac{x_1 + x_2}{n_1 + n_2} = \frac{385 + 379}{1000 + 1000} = \frac{764}{2000} = 0.382$$

(c) $H_0 : p_1 \leq p_2$ vs. $H_a : p_1 > p_2$
Reject H_0 if p-value < 0.05.

$$Z_{data} = \frac{(\hat{p}_1 - \hat{p}_2)}{\sqrt{\hat{p}_{pooled} \cdot (1 - \hat{p}_{pooled})\left(\frac{1}{n_1} + \frac{1}{n_2}\right)}} = \frac{(0.385 - 0.379)}{\sqrt{0.382 \cdot (1 - 0.382)\left(\frac{1}{1000} + \frac{1}{1000}\right)}} \approx 0.28$$

From the TI-84, p-value $= P(Z > Z_{data}) = P(Z > 0.28) = 0.3897388136$.
From the table, p-value $= P(Z > Z_{data}) = P(Z > 0.28) = 1 - 0.6103 = 0.3897$. Since p-value ≥ 0.05, we do not reject H_0. There is insufficient evidence that the population proportion of 18–20-year-olds who used an illicit drug decreased from 2004 to 2005.

16. *p*-value method:
$H_0 : p_1 \leq p_2$ vs. $H_a : p_1 > p_2$
Reject H_0 if p-value < 0.05.
$\hat{p}_1 - \hat{p}_2 = 0.741 - 0.69 = 0.051$
$x_1 = n_1 \cdot \hat{p}_1 = (1000)(0.741) = 741$ and $x_2 = n_2 \cdot \hat{p}_2 = (1000)(0.69) = 690$

$$\hat{p}_{pooled} = \frac{x_1 + x_2}{n_1 + n_2} = \frac{741 + 690}{1000 + 1000} = \frac{1431}{2000} = 0.7155$$

$$Z_{data} = \frac{(\hat{p}_1 - \hat{p}_2)}{\sqrt{\hat{p}_{pooled} \cdot (1 - \hat{p}_{pooled})\left(\frac{1}{n_1} + \frac{1}{n_2}\right)}} = \frac{(0.741 - 0.69)}{\sqrt{0.7155 \cdot (1 - 0.7155)\left(\frac{1}{1000} + \frac{1}{1000}\right)}} \approx 2.53$$

From the TI-84, p-value $= P(Z > Z_{data}) = P(Z > 2.53) = 0.005703147$. From the table, p-value $= P(Z > Z_{data}) = P(Z > 2.53) = 1 - 0.9943 = 0.0057$. Since p-value < 0.05, we reject H_0. There is evidence that the population proportion of people aged 35 to 44 who were married was lower in 2000 than in 1990.

Critical value method:
$H_0 : p_1 \leq p_2$ vs. $H_a : p_1 > p_2$
$\alpha = 0.05$, $Z_{crit} = 1.645$
Reject H_0 if $Z_{data} > 1.645$.
$\hat{p}_1 - \hat{p}_2 = 0.741 - 0.69 = 0.051$
$x_1 = n_1 \cdot \hat{p}_1 = (1000)(0.741) = 741$ and $x_2 = n_2 \cdot \hat{p}_2 = (1000)(0.69) = 690$

$$\hat{p}_{pooled} = \frac{x_1 + x_2}{n_1 + n_2} = \frac{741 + 690}{1000 + 1000} = \frac{1431}{2000} = 0.7155$$

$$Z_{data} = \frac{(\hat{p}_1 - \hat{p}_2)}{\sqrt{\hat{p}_{pooled} \cdot (1 - \hat{p}_{pooled})\left(\frac{1}{n_1} + \frac{1}{n_2}\right)}} = \frac{(0.741 - 0.69)}{\sqrt{0.7155 \cdot (1 - 0.7155)\left(\frac{1}{1000} + \frac{1}{1000}\right)}} \approx 2.53$$

Since $Z_{data} > 1.645$, we reject H_0. There is evidence that the population proportion of people aged 35 to 44 who were married was lower in 2000 than in 1990.

Chapter 11

Categorical Data Analysis

Section 11.1

1. A random variable is *multinomial* if it satisfies each of the following conditions:

- Each independent trial of the experiment has k possible outcomes, $k = 3, 4, \ldots$
- The ith outcome (category) occurs with probability p_i, where $i = 1, 2, \ldots, k$.
- $\sum_{i=1}^{k} p_i = 1$

3. It is the long run mean of that random variable after an arbitrarily large number of trials.

5. Multinomial.

7. Multinomial.

9. **(a)** Since $\chi^2_{data} = 2 < 4.605$, p-value > 0.10.

(b)

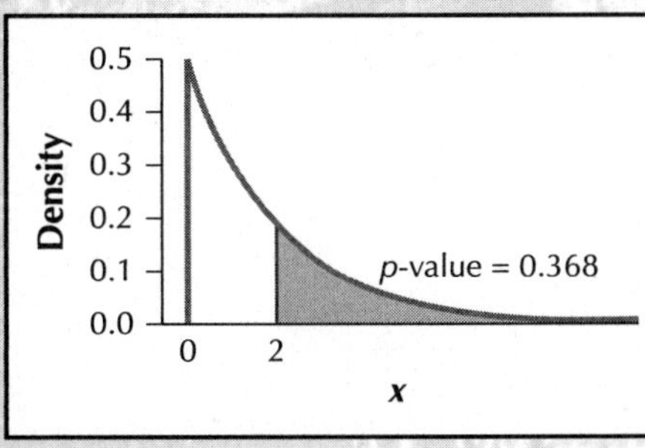

11. **(a)** Since $15.086 < \chi^2_{data} = 16.7 < 16.750$, $0.005 < p$-value < 0.01.

(b)

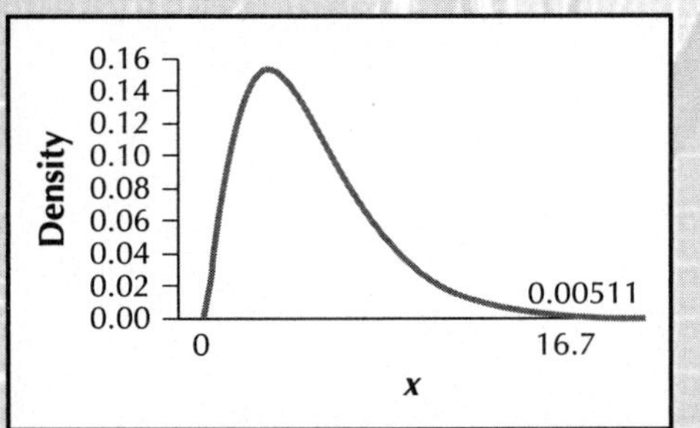

13. **(a)** $E_1 = n \cdot p_1 = (100)(0.50) = 50$, $E_2 = n \cdot p_2 = (100)(0.25) = 25$, $E_3 = n \cdot p_3 = (100)(0.25) = 25$ **(b)** Since none of the expected frequencies is less than 1 and none of the expected frequencies is less than 5, the conditions for performing the χ^2 goodness of fit test are met.

15. **(a)** $E_1 = n \cdot p_1 = (100)(0.9) = 90$, $E_2 = n \cdot p_2 = (100)(0.05) = 5$, $E_3 = n \cdot p_3 = (100)(0.04) = 4$, $E_4 = n \cdot p_4 = (100)(0.01) = 1$ **(b)** Since 50% of the expected frequencies are less than 5, the conditions for performing the χ^2 goodness of fit test are not met.

17.

O_i	E_i	$O_i - E_i$	$(O_i - E_i)^2$	$\frac{(O_i - E_i)^2}{E_i}$
10	12	−2	4	$\frac{4}{12} = 0.3333333333$
12	12	0	0	$\frac{0}{12} = 0$
14	12	2	4	$\frac{4}{12} = 0.3333333333$
				$\chi^2_{data} = 0.6666666666$

Thus, $\chi^2_{data} = \sum \frac{(O_i - E_i)^2}{E_i} = 0.6666666666 \approx 0.667$.

19.

O_i	E_i	$O_i - E_i$	$(O_i - E_i)^2$	$\frac{(O_i - E_i)^2}{E_i}$
20	25	−5	25	$\frac{25}{25} = 1$
30	25	5	25	$\frac{25}{25} = 1$
40	30	10	100	$\frac{100}{30} = 3.333333333$
40	50	−10	100	$\frac{100}{50} = 2$
				$\chi^2_{data} = 7.333333333$

Thus, $\chi^2_{data} = \sum \frac{(O_i - E_i)^2}{E_i} = 7.333333333 \approx 7.333$.

21.

O_i	E_i	$O_i - E_i$	$(O_i - E_i)^2$	$\frac{(O_i - E_i)^2}{E_i}$
1	6	−5	25	$\frac{25}{6} = 4.166666667$
10	6	4	16	$\frac{16}{6} = 2.666666667$
8	6	2	4	$\frac{4}{6} = 0.666666667$
0	6	−6	36	$\frac{36}{6} = 6$
11	6	5	25	$\frac{25}{6} = 4.166666667$
				$\chi^2_{data} = 17.666666667$

Thus, $\chi^2_{data} = \sum \frac{(O_i - E_i)^2}{E_i} = 17.666666667 \approx 17.667$.

23. **(a)** $n = O_1 + O_2 + O_3 = 50 + 25 + 25 = 100$.
$E_1 = n \cdot p_1 = (100)(0.4) = 40$, $E_2 = n \cdot p_2 = (100)(0.3) = 30$,
$E_3 = n \cdot p_3 = (100)(0.3) = 30$.
Since none of the expected frequencies is less than 1 and none of the expected frequencies is less than 5, the conditions for performing the χ^2 goodness of fit test are met.

(b) $\text{df} = k - 1 = 3 - 1 = 2$, $\alpha = 0.05$, $\chi^2_{crit} = \chi^2_{0.05} = 5.991$.
Reject H_0 if $\chi^2_{data} > 5.991$.

(c)

O_i	E_i	$O_i - E_i$	$(O_i - E_i)^2$	$\frac{(O_i - E_i)^2}{E_i}$
50	40	10	100	$\frac{100}{40} = 2.5$
25	30	−5	25	$\frac{25}{30} = 0.8333333333$
25	30	−5	25	$\frac{25}{30} = 0.8333333333$
				$\chi^2_{data} = 4.1666666666$

Thus, $\chi^2_{data} = \sum \frac{(O_i - E_i)^2}{E_i} = 4.1666666666 \approx 4.167.$

(d) Since $\chi^2_{data} \leq 5.991$, we do not reject H_0. There is insufficient evidence that the random variable does not follow the distribution specified in H_0.

25. (a) $n = O_1 + O_2 + O_3 + O_4 + O_5 = 90 + 75 + 15 + 15 + 5 = 200.$
$E_1 = n \cdot p_1 = (200)(0.4) = 80$, $E_2 = n \cdot p_2 = (200)(0.35) = 70$,
$E_3 = n \cdot p_3 = (200)(0.10) = 20$, $E_4 = n \cdot p_4 = (200)(0.10) = 20$,
$E_5 = n \cdot p_5 = (200)(0.05) = 10.$
Since none of the expected frequencies is less than 1 and none of the expected frequencies is less than 5, the conditions for performing the χ^2 goodness of fit test are met.

(b) $\text{df} = k - 1 = 5 - 1 = 4$, $\alpha = 0.10$, $\chi^2_{crit} = \chi^2_{0.10} = 7.779.$
Reject H_0 if $\chi^2_{data} > 7.779$.

(c)

O_i	E_i	$O_i - E_i$	$(O_i - E_i)^2$	$\frac{(O_i - E_i)^2}{E_i}$
90	80	10	100	$\frac{100}{80} = 1.25$
75	70	5	25	$\frac{25}{70} = 0.3571428571$
15	20	−5	25	$\frac{25}{20} = 1.25$
15	20	−5	25	$\frac{25}{20} = 1.25$
5	10	−5	25	$\frac{25}{10} = 2.5$
				$\chi^2_{data} = 6.607142857$

$$\chi^2_{data} = \sum \frac{(O_i - E_i)^2}{E_i} = 6.607142857 \approx 6.607$$

(d) Since $\chi^2_{data} \leq 7.779$, we do not reject H_0. There is insufficient evidence that the random variable does not follow the distribution specified in H_0.

27. (a) Reject H_0 if p-value < 0.05. $n = O_1 + O_2 = 40 + 60 = 100.$
$E_1 = n \cdot p_1 = (100)(0.50) = 50$, $E_2 = n \cdot p_2 = (100)(0.50) = 50.$
Since none of the expected frequencies is less than 1 and none of the expected frequencies is less than 5, the conditions for performing the χ^2 goodness of fit test are met.

(b)

O_i	E_i	$O_i - E_i$	$(O_i - E_i)^2$	$\frac{(O_i - E_i)^2}{E_i}$
40	50	−10	100	$\frac{100}{50} = 2$
60	50	10	100	$\frac{100}{50} = 2$
				$\chi^2_{data} = 4$

$$\chi^2_{data} = \sum \frac{(O_i - E_i)^2}{E_i} = 4$$

(c) Since df $= k - 1 = 2 - 1 = 1$ and $3.841 < \chi^2_{data} = 4 < 5.024$, $0.025 < p\text{-value} < 0.05$.

(d) Since p-value < 0.05, we reject H_0. There is evidence that the random variable does not follow the distribution specified in H_0.

29. **(a)** Reject H_0 if p-value < 0.10.
$n = O_1 + O_2 + O_3 + O_4 = 90 + 55 + 40 + 15 = 200$.
$E_1 = n \cdot p_1 = (200)(0.5) = 100$, $E_2 = n \cdot p_2 = (200)(0.25) = 50$,
$E_3 = n \cdot p_3 = (200)(0.15) = 30$, $E_4 = n \cdot p_4 = (200)(0.1) = 20$.
Since none of the expected frequencies is less than 1 and none of the expected frequencies is less than 5, the conditions for performing the χ^2 goodness of fit test are met.

(b)

O_i	E_i	$O_i - E_i$	$(O_i - E_i)^2$	$\frac{(O_i - E_i)^2}{E_i}$
90	100	−10	100	$\frac{100}{100} = 1$
55	50	5	25	$\frac{25}{50} = 0.5$
40	30	10	100	$\frac{100}{30} = 3.333333333$
15	20	−5	25	$\frac{25}{20} = 0.125$
				$\chi^2_{data} = 6.083333333$

$$\chi^2_{data} = \sum \frac{(O_i - E_i)^2}{E_i} = 6.083333333 \approx 6.083$$

(c) Since df $= k - 1 = 4 - 1 = 3$ and $\chi^2_{data} = 6.083 < 6.251$, p-value > 0.10.

(d) Since p-value ≥ 0.10, we do not reject H_0. There is insufficient evidence that the random variable does not follow the distribution specified in H_0.

31. $H_0 : p_{LHS} = 0.08, p_{HSD} = 0.23, p_{SC} = 0.32, p_{BD} = 0.24, p_{GPG} = 0.13$
H_a : The random variable does not follow the distribution specified in H_0.
$n = O_{LHS} + O_{HSD} + O_{SC} + O_{BD} + O_{GPG} = 12 + 40 + 62 + 54 + 32 = 200$.
$E_{LHS} = n \cdot p_{LHS} = (200)(0.08) = 16$, $E_{HSD} = n \cdot p_{HSD} = (200)(0.23) = 46$,
$E_{SC} = n \cdot p_{SG} = (200)(0.32) = 64$, $E_{BD} = n \cdot p_{BD} = (200)(0.24) = 48$,
$E_{GPG} = n \cdot p_{GPG} = (200)(0.13) = 26$
Since none of the expected frequencies is less than 1 and none of the expected frequencies is less than 5, the conditions for performing the χ^2 goodness of fit test are met.
df $= k - 1 = 5 - 1 = 4$, $\alpha = 0.05$, $\chi^2_{crit} = \chi^2_{0.05} = 9.488$. Reject H_0 if $\chi^2_{data} > 9.488$.

O_i	E_i	$O_i - E_i$	$(O_i - E_i)^2$	$\frac{(O_i - E_i)^2}{E_i}$
12	16	−4	16	$\frac{16}{16} = 1$
40	46	−6	36	$\frac{36}{46} = 0.7826086957$
62	64	−2	4	$\frac{4}{64} = 0.0625$
54	48	6	36	$\frac{36}{48} = 0.75$
32	26	6	36	$\frac{36}{26} = 1.384615385$
				$\chi^2_{data} = 3.979724081$

$$\chi^2_{data} = \sum \frac{(O_i - E_i)^2}{E_i} = 3.979724081 \approx 3.980$$

Since $\chi^2_{data} \leq 9.488$, we do not reject H_0. There is insufficient evidence that the distribution of education levels has changed since 2005.

33. $H_0 : p_{phip} = 0.30, p_{mm} = 0.556, p_{other} = 0.144$
H_a : The random variable does not follow the distribution specified in H_0.
$E_{phip} = n \cdot p_{phip} = (1000)(0.30) = 300,$
$E_{mm} = n \cdot p_{mm} = (1000)(0.556) = 556,$
$E_{other} = n \cdot p_{other} = (1000)(0.144) = 144.$
Since none of the expected frequencies is less than 1 and none of the expected frequencies is less than 5, the conditions for performing the χ^2 goodness of fit test are met.
$\text{df} = k - 1 = 3 - 1 = 2$, $\alpha = 0.05$, $\chi^2_{crit} = \chi^2_{0.05} = 5.991$. Reject H_0 if $\chi^2_{data} > 5.991$.

O_i	E_i	$O_i - E_i$	$(O_i - E_i)^2$	$\frac{(O_i - E_i)^2}{E_i}$
350	300	50	2,500	$\frac{2{,}500}{300} = 8.333333333$
500	556	−56	3,136	$\frac{3{,}136}{556} = 5.64028777$
150	144	6	36	$\frac{36}{144} = 0.25$
				$\chi^2_{data} = 14.2236211$

$$\chi^2_{data} = \sum \frac{(O_i - E_i)^2}{E_i} = 14.2236211 \approx 14.224$$

Since $\chi^2_{data} > 5.991$, we reject H_0. There is evidence that the random variable does not follow the distribution specified in H_0.

35. There were a total of 957,581 + 1,105,547 = 2,063,128 students living in college dormitories in 2000. So in 2000, $p_{males} = \frac{957{,}581}{2{,}063{,}128} = 0.4641403733$ and $p_{females} = \frac{1{,}105{,}547}{2{,}063{,}128} = 0.5358596267$.
$H_0 : p_{males} = 0.4641403733, p_{females} = 0.5358596267$
H_a : The random variable does not follow the distribution specified in H_0.

$n = O_{males} + O_{females} = 1{,}027 + 1{,}406 = 2{,}433.$

$E_{males} = n \cdot p_{males} = (2{,}433)(0.4641403733) \approx 1{,}129,$

$E_{females} = n \cdot p_{males} = (2{,}433)(0.5358596267) \approx 1{,}304.$
Since none of the expected frequencies is less than 1 and none of the expected frequencies is less than 5 the conditions for performing the χ^2 goodness of fit test are met.

$df = k - 1 = 2 - 1 = 1$, $\alpha = 0.05$, $\chi^2_{crit} = \chi^2_{0.05} = 3.841$. Reject H_0 if $\chi^2_{data} > 3.841$.

O_i	E_i	$O_i - E_i$	$(O_i - E_i)^2$	$\frac{(O_i - E_i)^2}{E_i}$
1,027	1,129	−102	10,404	$\frac{10{,}404}{1{,}129} = 9.215234721$
1,406	1,304	102	10,404	$\frac{10{,}404}{1{,}304} = 7.978527607$
				$\chi^2_{data} = 17.19376233$

$$\chi^2_{data} = \sum \frac{(O_i - E_i)^2}{E_i} = 17.19376233 \approx 17.194$$

Since $\chi^2_{data} > 3.841$, we reject H_0. There is evidence that the random variable does not follow the distribution specified in H_0.

37. $H_0: p_{Hertz} = 0.28, p_{Avis} = 0.214, p_{National} = 0.121, p_{Budget} = 0.105, p_{Alamo} = 0.105, p_{Dollar} = 0.085, p_{Others} = 0.09$
H_a: The random variable does not follow the distribution specified in H_0.
$E_{Hertz} = n \cdot p_{Hertz} = (100)(0.28) = 28,$
$E_{Avis} = n \cdot p_{Avis} = (100)(0.214) = 21.4,$
$E_{National} = n \cdot p_{National} = (100)(0.121) = 12.1,$
$E_{Budget} = n \cdot p_{Budget} = (100)(0.105) = 10.5,$
$E_{Alamo} = n \cdot p_{Alamo} = (100)(0.105) = 10.5,$
$E_{Dollar} = n \cdot p_{Dollar} = (100)(0.085) = 8.5,$
$E_{Others} = n \cdot p_{Others} = (100)(0.09) = 9.$
Since none of the expected frequencies is less than 1 and none of the expected frequencies is less than 5, the conditions for performing the χ^2 goodness of fit test are met. Reject H_0 if p-value < 0.05.

O_i	E_i	$O_i - E_i$	$(O_i - E_i)^2$	$\frac{(O_i - E_i)^2}{E_i}$
28	28	0	0	$\frac{0}{28} = 0$
22	21.4	0.6	0.36	$\frac{0.36}{21.4} = 0.0168224299$
12	12.1	−0.1	0.01	$\frac{0.01}{12.1} = 0.000826446281$
10	10.5	−0.5	0.25	$\frac{0.25}{10.5} = 0.0238095238$
10	10.5	−0.5	0.25	$\frac{0.25}{10.5} = 0.0238095238$
8	8.5	−0.5	0.25	$\frac{0.25}{8.5} = 0.0294117647$
10	9	1	1	$\frac{1}{9} = 0.1111111111$
				$\chi^2_{data} = 0.2057907996$

$$\chi^2_{data} = \sum \frac{(O_i - E_i)^2}{E_i} = 0.2057907996 \approx 0.2058$$

Since $df = k - 1 = 7 - 1 = 6$ and $\chi^2_{data} = 0.2058 < 10.645$, p-value > 0.10.
Since p-value > 0.05, we do not reject H_0. There is insufficient evidence that the random variable does not follow the distribution specified in H_0.

39. $H_0: p_{every} = 0.31, p_{almost} = 0.12, p_{onceortwice} = 0.14, p_{few} = 0.24, p_{never} = 0.19$
H_a: The random variable does not follow the distribution specified in H_0.
$E_{every} = n \cdot p_{every} = (100)(0.31) = 31,$
$E_{almost} = n \cdot p_{almost} = (100)(0.12) = 12,$
$E_{onceortwice} = n \cdot p_{onceortwice} = (100)(0.14) = 14,$
$E_{few} = n \cdot p_{few} = (100)(0.24) = 24,$
$E_{never} = n \cdot p_{never} = (100)(0.19) = 19$
Since none of the expected frequencies is less than 1 and none of the expected frequencies is less than 5, the conditions for performing the χ^2 goodness of fit test are met.
df $= k - 1 = 5 - 1 = 4$, $\alpha = 0.10$, $\chi^2_{crit} = \chi^2_{0.10} = 7.779$. Reject H_0 if $\chi^2_{data} > 7.779$.

O_i	E_i	$O_i - E_i$	$(O_i - E_i)^2$	$\frac{(O_i - E_i)^2}{E_i}$
32	31	1	1	$\frac{1}{31} = 0.0322580645$
10	12	−2	4	$\frac{4}{12} = 0.3333333333$
15	14	1	1	$\frac{1}{14} = 0.0714285714$
25	24	1	1	$\frac{1}{24} = 0.0416666667$
18	19	−1	1	$\frac{1}{19} = 0.0526315789$
				$\chi^2_{data} = 0.5313182148$

$$\chi^2_{data} = \sum \frac{(O_i - E_i)^2}{E_i} = 0.5313182148 \approx 0.531$$

Since $\chi^2_{data} \leq 7.779$ we do not reject H_0. There is insufficient evidence that the random variable does not follow the distribution specified in H_0.

41. $H_0: p_{believe} = 0.78, p_{notsure} = 0.12, p_{didnotbelieve} = 0.10$
H_a: The random variable does not follow the distribution specified in H_0.
$E_{believe} = n \cdot p_{believe} = (1000)(0.78) = 780,$
$E_{notsure} = n \cdot p_{notsure} = (1000)(0.12) = 120,$
$E_{didnotbelieve} = n \cdot p_{didnotbelieve} = (1000)(0.10) = 100$
Since none of the expected frequencies is less than 1 and none of the expected frequencies is less than 5, the conditions for performing the χ^2 goodness of fit test are met.
df $= k - 1 = 3 - 1 = 2$, $\alpha = 0.05$, $\chi^2_{crit} = \chi^2_{0.05} = 5.991$. Reject H_0 if $\chi^2_{data} > 5.991$.

O_i	E_i	$O_i - E_i$	$(O_i - E_i)^2$	$\frac{(O_i - E_i)^2}{E_i}$
820	780	40	1,600	$\frac{1,600}{780} = 2.051282051$
110	120	−10	100	$\frac{100}{120} = 0.8333333333$
70	100	−30	900	$\frac{900}{100} = 9$
				$\chi^2_{data} = 11.88461538$

$$\chi^2_{data} = \sum \frac{(O_i - E_i)^2}{E_i} = 11.88461538 \approx 11.885$$

Since $\chi^2_{data} > 5.991$, we reject H_0. There is evidence that the random variable does not follow the distribution specified in H_0.

43. $p_{other} = 1 - p_{personnel} - p_{operations} = 1 - 0.24 - 0.28 = 0.48.$
$H_0 : p_{personnel} = 0.24, p_{operations} = 0.28, p_{other} = 0.48$
H_a : The random variable does not follow the distribution specified in H_0.
$n = O_{personnel} + O_{operations} + O_{other}$ = \$99,000,000,000 + \$117,000,000,000 + \$163,400,000,000 = \$379,400,000,000 = \$379.4 billion.
$E_{personnel} = n \cdot p_{personnel}$ = (\$379.4 billion)(0.24) = \$91.056 billion,
$E_{operations} = n \cdot p_{operations}$ = (\$379.4 billion)(0.28) = \$106.232 billion,
$E_{other} = n \cdot p_{other}$ = (\$379.4 billion)(0.48) = \$182.112 billion.
Since none of the expected frequencies is less than 1 and none of the expected frequencies is less than 5, the conditions for performing the χ^2 goodness of fit test are met. Reject H_0 if p-value < 0.05.

O_i	E_i	$O_i - E_i$	$(O_i - E_i)^2$	$\frac{(O_i - E_i)^2}{E_i}$
99	91.056	7.944	63.107136	$\frac{63.107136}{91.056} = 0.6930585134$
117	106.232	10.768	115.949824	$\frac{115.949824}{106.232} = 1.09147737$
163.4	182.112	−18.712	350.138944	$\frac{350.138944}{182.112} = 1.922657178$
				$\chi^2_{data} = 3.707193061$

$$\chi^2_{data} = \sum \frac{(O_i - E_i)^2}{E_i} = 3.707193061 \approx 3.707$$

Since df $= k - 1 = 3 - 1 = 2$ and $\chi^2_{data} = 3.707 < 4.605$, p-value > 0.10.
Since p-value ≥ 0.05, we do not reject H_0. There is insufficient evidence that the random variable does not follow the distribution specified in H_0.

Section 11.2

1. A contingency table is a tabular summary of the relationship between two categorical variables.

3. The two-sample Z test for the difference in proportions from Chapter 10 is for comparing proportions of two independent populations and the χ^2 test for homogeneity of proportions is for comparing proportions of k independent populations.

5. Table of observed frequencies:

	A1	A2	Total
B1	10	20	30
B2	12	18	30
Total	22	38	60

Table of expected frequencies:

	A1	A2	Total
B1	$\frac{(22)(30)}{60} = 11$	$\frac{(38)(30)}{60} = 19$	30
B2	$\frac{(22)(30)}{60} = 11$	$\frac{(38)(30)}{60} = 19$	30
Total	22	38	60

7. Table of observed frequencies:

	E1	E2	E3	Total
F1	30	20	10	60
F2	35	24	8	67
Total	65	44	18	127

Table of expected frequencies:

	E1	E2	E3	Total
F1	$\frac{(60)(65)}{127} \approx 30.71$	$\frac{(60)(44)}{127} \approx 20.79$	$\frac{(60)(18)}{127} \approx 8.50$	60
F2	$\frac{(67)(65)}{127} \approx 34.29$	$\frac{(67)(44)}{127} \approx 23.21$	$\frac{(67)(18)}{127} \approx 9.50$	67
Total	65	44	18	127

9. Table of observed frequencies:

	I1	I2	I3	Total
J1	100	90	105	295
J2	50	60	55	165
J3	25	15	20	60
Total	175	165	180	520

Table of expected frequencies:

	I1	I2	I3	Total
J1	$\frac{(295)(175)}{520} \approx 99.28$	$\frac{(295)(165)}{520} \approx 93.61$	$\frac{(295)(180)}{520} = 102.12$	295.01
J2	$\frac{(165)(175)}{520} \approx 55.53$	$\frac{(165)(165)}{520} \approx 52.36$	$\frac{(165)(180)}{520} = 57.12$	165.01
J3	$\frac{(60)(175)}{520} \approx 20.19$	$\frac{(60)(165)}{520} \approx 19.04$	$\frac{(60)(180)}{520} = 20.77$	60
Total	175	165.01	180.01	520.02

11. (a) H_0 : Variable A and variable B are independent.

H_a : Variable A and variable B are not independent.

(b)

	A1	A2	Total
B1	$\frac{(22)(30)}{60} = 11$	$\frac{(38)(30)}{60} = 19$	30
B2	$\frac{(22)(30)}{60} = 11$	$\frac{(38)(30)}{60} = 19$	30
Total	22	38	60

Since none of the expected frequencies is less than 1 and none of the expected frequencies is less than 5, the conditions for performing the χ^2 test for independence are met.

(c) df $= (r-1)(c-1) = (2-1)(2-1) = 1$, $\alpha = 0.05$, $\chi^2_{crit} = \chi^2_{0.05} = 3.841$. Reject H_0 if $\chi^2_{data} > 3.841$.

(d) $\chi^2_{data} = \sum \frac{(O_i - E_i)^2}{E_i} = \frac{(10-11)^2}{11} + \frac{(20-19)^2}{19} + \frac{(12-11)^2}{11} + \frac{(18-19)^2}{19} \approx 0.287$

(e) Since $\chi^2_{data} \le 3.841$, we do not reject H_0. There is insufficient evidence that variable A and variable B are not independent.

13. **(a)** H_0: Variable I and variable J are independent.
H_a: Variable I and variable J are not independent.
(b)

	I1	I2	I3	Total
J1	$\frac{(295)(175)}{520} \approx 99.28$	$\frac{(295)(165)}{520} \approx 93.61$	$\frac{(295)(180)}{520} \approx 102.12$	295.01
J2	$\frac{(165)(175)}{520} \approx 55.53$	$\frac{(165)(165)}{520} \approx 52.36$	$\frac{(165)(180)}{520} \approx 57.12$	165.01
J3	$\frac{(60)(175)}{520} \approx 20.19$	$\frac{(60)(165)}{520} \approx 19.04$	$\frac{(60)(180)}{520} \approx 20.77$	60
Total	175	165.01	180.01	520.02

Since none of the expected frequencies is less than 1 and none of the expected frequencies is less than 5, the conditions for performing the χ^2 test for independence are met.
(c) df $= (r-1)(c-1) = (3-1)(3-1) = 4$, $\alpha = 0.01$, $\chi^2_{crit} = \chi^2_{0.01} = 13.277$. Reject H_0 if $\chi^2_{data} > 13.277$.
(d) $\chi^2_{data} = \sum \frac{(O_i - E_i)^2}{E_i} = \frac{(100-99.28)^2}{99.28} + \frac{(90-93.61)^2}{93.61} + \frac{(105-102.12)^2}{102.12} + \frac{(50-55.53)^2}{55.53}$

$+ \frac{(60-52.36)^2}{52.36} + \frac{(55-57.12)^2}{57.12} + \frac{(25-20.19)^2}{20.19} + \frac{(15-19.04)^2}{19.04} + \frac{(20-20.77)^2}{20.77} \approx 4.002$

(e) Since $\chi^2_{data} \le 13.277$, we do not reject H_0. There is insufficient evidence that variable I and variable J are not independent.

15. **(a)** H_0: Variable C and variable D are independent.
H_a: Variable C and variable D are not independent. Reject H_0 if p-value < 0.05.

	C1	C2	Total
D1	$\frac{(110)(150)}{300} = 55$	$\frac{(190)(150)}{300} = 95$	150
D2	$\frac{(110)(150)}{300} = 55$	$\frac{(190)(150)}{300} = 95$	150
Total	110	190	300

Since none of the expected frequencies is less than 1 and none of the expected frequencies is less than 5, the conditions for performing the χ^2 test for independence are met.
(b) $\chi^2_{data} = \sum \frac{(O_i - E_i)^2}{E_i} = \frac{(50-55)^2}{55} + \frac{(100-95)^2}{95} + \frac{(60-55)^2}{55} + \frac{(90-95)^2}{95} \approx 1.435$
(c) Since df $= (r-1)(c-1) = (2-1)(2-1) = 1$ and $\chi^2_{data} = 1.435 < 2.706$, p-value > 0.10.
(d) Since p-value ≥ 0.05, we do not reject H_0. There is insufficient evidence that variable C and variable D are not independent.

17. **(a)** H_0: Variable K and variable L are independent.
H_a: Variable K and variable L are not independent.
Reject H_0 if p-value < 0.01.

	K1	K2	K3	K4	Total
L1	$\frac{(300)(90)}{720} = 37.5$	$\frac{(300)(175)}{720} \approx 72.92$	$\frac{(300)(215)}{720} \approx 89.58$	$\frac{(300)(240)}{720} = 100$	300
L2	$\frac{(190)(90)}{720} = 23.75$	$\frac{(190)(175)}{720} \approx 46.18$	$\frac{(190)(215)}{720} \approx 56.74$	$\frac{(190)(240)}{720} \approx 63.33$	190
L3	$\frac{(230)(90)}{720} = 28.75$	$\frac{(230)(175)}{720} \approx 55.91$	$\frac{(230)(215)}{720} \approx 68.68$	$\frac{(230)(240)}{720} \approx 76.67$	230
Total	90	175	215	240	720

Since none of the expected frequencies is less than 1 and none of the expected frequencies is less than 5, the conditions for performing the χ^2 test for independence are met.

(b) $\chi^2_{data} = \sum \frac{(O_i - E_i)^2}{E_i} = \frac{(40-37.5)^2}{37.5} + \frac{(70-72.92)^2}{72.92} + \frac{(90-89.58)^2}{89.58} + \frac{(100-100)^2}{100}$

$+ \frac{(20-23.75)^2}{23.75} + \frac{(40-46.18)^2}{46.18} + \frac{(60-56.74)^2}{56.74} + \frac{(70-63.33)^2}{63.33} + \frac{(30-28.75)^2}{28.75}$

$+ \frac{(65-55.91)^2}{55.91} + \frac{(65-68.68)^2}{68.68} + \frac{(70-76.67)^2}{76.67} \approx 4.904$

(c) Since df $= (r-1)(c-1) = (3-1)(4-1) = 6$ and $\chi^2_{data} = 4.904 < 10.645$, p-value > 0.10.

(d) Since p-value ≥ 0.01, we do not reject H_0. There is insufficient evidence that variable K and variable L are not independent.

19. (a) $H_0: p_1 = p_2 = p_3$

H_a: Not all the proportions in H_0 are equal.

(b) Observed frequencies:

	Sample 1	Sample 2	Sample 3	Total
Successes	10	20	30	60
Failures	20	45	62	127
Total	30	65	92	187

Expected frequencies:

	Sample 1	Sample 2	Sample 3	Total
Successes	$\frac{(60)(30)}{187} \approx 9.63$	$\frac{(60)(65)}{187} \approx 20.86$	$\frac{(60)(92)}{187} \approx 29.52$	60.01
Failures	$\frac{(127)(30)}{187} \approx 20.37$	$\frac{(127)(65)}{187} \approx 44.14$	$\frac{(127)(92)}{187} \approx 62.48$	126.99
Total	30	65	92	187

Since none of the expected frequencies is less than 1 and none of the expected frequencies is less than 5, the conditions for performing the χ^2 test for homogeneity of proportions are met.

(c) df $= (r-1)(c-1) = (2-1)(3-1) = 2$, $\alpha = 0.05$, $\chi^2_{crit} = \chi^2_{0.05} = 5.991$. Reject H_0 if $\chi^2_{data} > 5.991$.

(d) $\chi^2_{data} = \sum \frac{(O_i - E_i)^2}{E_i} = \frac{(10-9.63)^2}{9.63} + \frac{(20-20.86)^2}{20.86} + \frac{(30-29.52)^2}{29.52}$

$+ \frac{(20-20.37)^2}{20.37} + \frac{(45-44.14)^2}{44.14} + \frac{(62-62.48)^2}{62.48} \approx 0.0846$

(e) Since $\chi^2_{data} \leq 5.991$, we do not reject H_0. There is insufficient evidence that not all of the proportions in H_0 are equal.

21. (a) $H_0: p_1 = p_2 = p_3 = p_4$

H_a: Not all the proportions in H_0 are equal.

(b) Observed frequencies:

	Sample 1	Sample 2	Sample 3	Sample 4	Total
Successes	10	15	20	25	70
Failures	15	24	32	40	111
Total	25	39	52	65	181

Expected frequencies:

	Sample 1	Sample 2	Sample 3	Sample 4	Total
Successes	$\frac{(70)(25)}{181} \approx 9.67$	$\frac{(70)(39)}{181} \approx 15.08$	$\frac{(70)(52)}{181} \approx 20.11$	$\frac{(70)(65)}{181} \approx 25.14$	70
Failures	$\frac{(111)(25)}{181} \approx 15.33$	$\frac{(111)(39)}{181} \approx 23.92$	$\frac{(111)(52)}{181} \approx 31.89$	$\frac{(111)(65)}{181} \approx 39.86$	111
Total	25	39	52	65	181

Since none of the expected frequencies is less than 1 and none of the expected frequencies is less than 5, the conditions for performing the χ^2 test for homogeneity of proportions are met.
(c) df $= (r-1)(c-1) = (2-1)(4-1) = 3$, $\alpha = 0.05$, $\chi^2_{crit} = \chi^2_{0.05} = 7.815$.
Reject H_0 if $\chi^2_{data} > 7.815$.

(d) $$\chi^2_{data} = \sum \frac{(O_i - E_i)^2}{E_i} = \frac{(10-9.67)^2}{9.67} + \frac{(15-15.08)^2}{15.08} + \frac{(20-20.11)^2}{20.11} + \frac{(25-25.14)^2}{25.14}$$
$$+ \frac{(15-15.33)^2}{15.33} + \frac{(24-23.92)^2}{23.92} + \frac{(32-31.89)^2}{31.89} + \frac{(40-39.86)^2}{39.86} \approx 0.0213$$

(e) Since $\chi^2_{data} \le 7.815$, we do not reject H_0. There is insufficient evidence that not all of the proportions in H_0 are equal.

23. (a) $H_0: p_1 = p_2 = p_3$
H_a: Not all the proportions in H_0 are equal.
Reject H_0 if p-value < 0.05.
Observed frequencies:

	Sample 1	Sample 2	Sample 3	Total
Successes	30	60	90	180
Failures	10	25	50	85
Total	40	85	140	265

Expected frequencies:

	Sample 1	Sample 2	Sample 3	Total
Successes	$\frac{(180)(40)}{265} \approx 27.17$	$\frac{(180)(85)}{265} \approx 57.74$	$\frac{(180)(140)}{265} \approx 95.09$	180
Failures	$\frac{(85)(40)}{265} \approx 12.83$	$\frac{(85)(85)}{265} \approx 27.26$	$\frac{(85)(140)}{265} \approx 44.91$	85
Total	40	85	140	265

Since none of the expected frequencies is less than 1 and none of the expected frequencies is less than 5, the conditions for performing the χ^2 test for homogeneity of proportions are met.

(b) $$\chi^2_{data} = \sum \frac{(O_i - E_i)^2}{E_i} = \frac{(30-27.17)^2}{27.17} + \frac{(60-57.74)^2}{57.74} + \frac{(90-95.09)^2}{95.09}$$
$$+ \frac{(10-12.83)^2}{12.83} + \frac{(25-27.26)^2}{27.26} + \frac{(50-44.91)^2}{44.91} \approx 2.044$$

(c) Since df $= (r-1)(c-1) = (2-1)(3-1) = 2$ and $\chi^2_{data} = 2.044 < 4.605$, p-value > 0.10.
(d) Since p-value ≥ 0.05, we do not reject H_0. There is insufficient evidence that not all of the proportions in H_0 are equal.

25. **(a)** $H_0: p_1 = p_2 = p_3 = p_4$
H_a: Not all the proportions in H_0 are equal.
Reject H_0 if p-value < 0.05.
Observed frequencies:

	Sample 1	Sample 2	Sample 3	Sample 4	Total
Successes	10	12	24	32	78
Failures	6	10	15	30	61
Total	16	22	39	62	139

Expected frequencies:

	Sample 1	Sample 2	Sample 3	Sample 4	Total
Successes	$\frac{(78)(16)}{139} \approx 8.98$	$\frac{(78)(22)}{139} \approx 12.35$	$\frac{(78)(39)}{139} \approx 21.88$	$\frac{(78)(62)}{139} \approx 34.79$	78
Failures	$\frac{(61)(16)}{139} \approx 7.02$	$\frac{(61)(22)}{139} \approx 9.65$	$\frac{(61)(39)}{139} \approx 17.12$	$\frac{(61)(62)}{139} \approx 27.21$	61
Total	16	22	39	62	139

Since none of the expected frequencies is less than 1 and none of the expected frequencies is less than 5, the conditions for performing the χ^2 test for homogeneity of proportions are met.

(b) $$\chi^2_{data} = \sum \frac{(O_i - E_i)^2}{E_i} = \frac{(10 - 8.98)^2}{8.98} + \frac{(12 - 12.35)^2}{12.35} + \frac{(24 - 21.88)^2}{21.88}$$
$$+ \frac{(32 - 34.79)^2}{34.79} + \frac{(6 - 7.02)^2}{7.02} + \frac{(10 - 9.65)^2}{9.65} + \frac{(15 - 17.12)^2}{17.12} + \frac{(30 - 27.21)^2}{27.21} \approx 1.264$$

(c) Since df $= (r - 1)(c - 1) = (2 - 1)(4 - 1) = 3$ and $\chi^2_{data} = 1.264 < 6.251$, p-value > 0.10.

(d) Since p-value ≥ 0.05, we do not reject H_0. There is insufficient evidence that not all of the proportions in H_0 are equal.

27. H_0: *Type of stimulus* and *type of mouse* are independent.
H_a: *Type of stimulus* and *type of mouse* are not independent.
Reject H_0 if p-value < 0.05.
Expected frequencies:

	Type of Stimulus		
Type of mouse	**Classical**	**Operant**	**Total**
White	$\frac{(60)(30)}{100} = 18$	$\frac{(60)(70)}{100} = 42$	60
Brown	$\frac{(40)(30)}{100} = 12$	$\frac{(40)(70)}{100} = 28$	40
Total	30	70	100

Since none of the expected frequencies is less than 1 and none of the expected frequencies is less than 5, the conditions for performing the χ^2 test for independence are met.

$$\chi^2_{data} = \sum \frac{(O_i - E_i)^2}{E_i} = \frac{(20 - 18)^2}{18} + \frac{(40 - 42)^2}{42} + \frac{(10 - 12)^2}{12} + \frac{(30 - 28)^2}{28} \approx 0.794$$

Since df $= (r - 1)(c - 1) = (2 - 1)(2 - 1) = 1$ and $\chi^2_{data} = 0.794 < 2.706$, p-value > 0.10. Since p-value ≥ 0.10, we do not reject H_0. There is insufficient evidence that *type of stimulus* and *type of mouse* are not independent.

29. $H_0 : p_{Republicans} = p_{Independents} = p_{Democrats}$
H_a : Not all the proportions in H_0 are equal.
Observed frequencies:

	Restrict	**Not restrict/don't know**	**Total**
Republicans	59	41	100
Independents	52	48	100
Democrats	53	47	100
Total	164	136	300

Expected frequencies:

	Restrict	**Not restrict/don't know**	**Total**
Republicans	$\frac{(100)(164)}{300} \approx 54.67$	$\frac{(100)(136)}{300} \approx 45.33$	100
Independents	$\frac{(100)(164)}{300} \approx 54.67$	$\frac{(100)(136)}{300} \approx 45.33$	100
Democrats	$\frac{(100)(164)}{300} \approx 54.67$	$\frac{(100)(136)}{300} \approx 45.33$	100
Total	164.01	135.99	300

Since none of the expected frequencies is less than 1 and none of the expected frequencies is less than 5, the conditions for performing the χ^2 test for homogeneity of proportions are met.
$\text{df} = (r - 1)(c - 1) = (3 - 1)(2 - 1) = 2$, $\alpha = 0.05$, $\chi^2_{crit} = \chi^2_{0.05} = 5.991$
Reject H_0 if $\chi^2_{data} > 5.991$.

$$\chi^2_{data} = \sum \frac{(O_i - E_i)^2}{E_i} = \frac{(59 - 54.67)^2}{54.67} + \frac{(41 - 45.33)^2}{45.33} + \frac{(52 - 54.67)^2}{54.67}$$

$$+ \frac{(48 - 45.33)^2}{45.33} + \frac{(53 - 54.67)^2}{54.67} + \frac{(47 - 45.33)^2}{45.33} \approx 1.157$$

Since $\chi^2_{data} \leq 5.991$, we do not reject H_0. There is insufficient evidence that not all of the proportions in H_0 are equal.

31. H_0 : *Gender* and *political party preference* are independent.
H_a : *Gender* and *political party preference* are not independent.
Reject H_0 if p-value < 0.01.

	Political Party Preference			
Gender	**Democrat**	**Republican**	**Independent**	**Total**
Female	$\frac{(250)(150)}{500} = 75$	$\frac{(250)(150)}{500} = 75$	$\frac{(250)(200)}{500} = 100$	250
Male	$\frac{(250)(150)}{500} = 75$	$\frac{(250)(150)}{500} = 75$	$\frac{(250)(200)}{500} = 100$	250
Total	150	150	200	500

Since none of the expected frequencies is less than 1 and none of the expected frequencies is less than 5, the conditions for performing the χ^2 test for independence are met.

$$\chi^2_{data} = \sum \frac{(O_i - E_i)^2}{E_i} = \frac{(100 - 75)^2}{75} + \frac{(50 - 75)^2}{75} + \frac{(100 - 100)^2}{100}$$

$$+ \frac{(50 - 75)^2}{75} + \frac{(100 - 75)^2}{75} + \frac{(100 - 100)^2}{100} \approx 33.333$$

df = $(r-1)(c-1) = (2-1)(3-1) = 2$
p-value $= P(\chi^2 > \chi^2_{data}) = P(\chi^2 > 33.333) \approx 0$
Since p-value < 0.01, we reject H_0. There is evidence that *gender* and *political party preference* are not independent.

33. $H_0: p_{Edit} = p_{Arrange}$
H_a: Not all the proportions in H_0 are equal.
Reject H_0 if p-value < 0.05.
Observed frequencies:

	By phone or		
	By e-mail	**In person**	**Total**
Edit or review documents	670	330	1,000
Arrange meetings or appointments	630	370	1,000
Total	1300	700	2,000

Expected frequencies:

	By phone or		
	By e-mail	**In person**	**Total**
Edit or review documents	$\frac{(1,000)(1,300)}{2,000} = 650$	$\frac{(1,000)(700)}{2,000} = 350$	1,000
Arrange meetings or appointments	$\frac{(1,000)(1,300)}{2,000} = 650$	$\frac{(1,000)(700)}{2,000} = 350$	1,000
Total	1,300	700	2,000

Since none of the expected frequencies is less than 1 and none of the expected frequencies is less than 5, the conditions for performing the χ^2 test for homogeneity of proportions are met.

$$\chi^2_{data} = \sum \frac{(O_i - E_i)^2}{E_i} = \frac{(670-650)^2}{650} + \frac{(330-350)^2}{350} + \frac{(630-650)^2}{650} + \frac{(370-350)^2}{350} \approx 3.516$$

Since df $= (r-1)(c-1) = (2-1)(2-1) = 1$ and $2.706 < \chi^2_{data} = 3.516 < 3.841$, $0.05 < p$-value < 0.10. Since p-value ≥ 0.05, we do not reject H_0. There is insufficient evidence that not all of the proportions in H_0 are equal.

35. $H_0: p_{Work} = p_{Personal}$
H_a: Not all the proportions in H_0 are equal.
Reject H_0 if p-value < 0.01.
Observed frequencies:

	None	**Some**	**A lot**	**Total**
Work e-mail	53	36	11	100
Personal e-mail	22	48	30	100
Total	75	84	41	200

Expected frequencies:

	None	**Some**	**A lot**	**Total**
Work e-mail	$\frac{(100)(75)}{200} = 37.5$	$\frac{(100)(84)}{200} = 42$	$\frac{(100)(41)}{200} = 20.5$	100
Personal e-mail	$\frac{(100)(75)}{200} = 37.5$	$\frac{(100)(84)}{200} = 42$	$\frac{(100)(41)}{200} = 20.5$	100
Total	75	84	41	200

Since none of the expected frequencies is less than 1 and none of the expected frequencies is less than 5, the conditions for performing the χ^2 test for homogeneity of proportions are met.

$$\chi^2_{data} = \sum \frac{(O_i - E_i)^2}{E_i} = \frac{(53 - 37.5)^2}{37.5} + \frac{(36 - 42)^2}{42} + \frac{(11 - 20.5)^2}{20.5} + \frac{(22 - 37.5)^2}{37.5} + \frac{(48 - 42)^2}{42} + \frac{(30 - 20.5)^2}{20.5} \approx 23.332$$

Since df $= (r - 1)(c - 1) = (2 - 1)(3 - 1) = 2$ and $\chi^2_{data} = 23.332 > 10.597$, p-value < 0.005. Since p-value < 0.01, we reject H_0. There is evidence that not all of the proportions in H_0 are equal.

37. $H_0 : p_{Urban} = p_{Suburban} = p_{Rural}$
H_a : Not all the proportions in H_0 are equal.
Reject H_0 if p-value < 0.05.
Observed frequencies:

	Use online dating	**Don't use online dating**	**Total**
Urban	(0.13)(1000) = 130	1000 − 130 = 870	1000
Suburban	(0.10)(1000) = 100	1000 − 100 = 900	1000
Rural	(0.09)(1000) = 90	1000 − 90 = 910	1000
Total	320	2680	3000

Expected frequencies:

	Use online dating	**Don't use online dating**	**Total**
Urban	$\frac{(1000)(320)}{3000} \approx 106.67$	$\frac{(1000)(2680)}{3000} \approx 893.33$	1000
Suburban	$\frac{(1000)(320)}{3000} \approx 106.67$	$\frac{(1000)(2680)}{3000} \approx 893.33$	1000
Rural	$\frac{(1000)(320)}{3000} \approx 106.67$	$\frac{(1000)(2680)}{3000} \approx 893.33$	1000
Total	320.01	2679.99	3000

Since none of the expected frequencies is less than 1 and none of the expected frequencies is less than 5, the conditions for performing the χ^2 test for homogeneity of proportions are met.

$$\chi^2_{data} = \sum \frac{(O_i - E_i)^2}{E_i} = \frac{(130 - 106.67)^2}{106.67} + \frac{(870 - 893.33)^2}{893.33} + \frac{(100 - 106.67)^2}{106.67} + \frac{(900 - 893.33)^2}{893.33} + \frac{(90 - 106.67)^2}{106.67} + \frac{(910 - 893.33)^2}{893.33} \approx 9.095$$

Since df $= (r - 1)(c - 1) = (3 - 1)(2 - 1) = 2$ and $7.738 < \chi^2_{data} = 9.095 < 9.210$, $0.01 < p$-value < 0.025. Since p-value < 0.05, we reject H_0. There is evidence that not all of the proportions in H_0 are equal.

39. **(a)** Dependent.
(b) H_0 : *Gender* and *goals* are independent.
H_a : *Gender* and *goals* are not independent.

Reject H_0 if p-value < 0.05.

```
Tabulated statistics: GENDER, GOALS

Rows: GENDER   Columns: GOALS

        Grades  Popular  Sports  All

boy        117       50      60  227
girl       130       91      30  251
All        247      141      90  478

Cell Contents:       Count

Pearson Chi-Square = 21.455, DF = 2, P-Value = 0.000
Likelihood Ratio Chi-Square = 21.769, DF = 2, P-Value = 0.000
```

Since p-value ≈ 0, p-value < 0.05. Thus, we reject H_0. There is evidence that *gender* and *goals* are not independent.

41. **(a)** Dependent.

(b) H_0 : *Goals* and *urb_rur* are independent.

H_a : *Goals* and *urb_rur* are not independent.

Reject H_0 if p-value < 0.10.

```
Tabulated statistics: URB_RUR, GOALS

Rows: URB_RUR   Columns: GOALS

            Grades  Popular  Sports  All

Rural           57       50      42  149
Suburban        87       42      22  151
Urban          103       49      26  178
All            247      141      90  478

Cell Contents:       Count


Pearson Chi-Square = 18.828, DF = 4, P-Value = 0.001
Likelihood Ratio Chi-Square = 18.571, DF = 4, P-Value = 0.001
```

Since p-value 0.001, p-value < 0.10. Thus, we reject H_0. There is evidence that *urb_rural* and *goals* are not independent.

43. $H_0 : p_{Jan} = p_{Feb} = p_{Mar} = p_{Apr} = p_{May} = p_{June} = p_{July} = p_{Aug} = p_{Sept} = p_{Oct} = p_{Nov} = p_{Dec}$

H_a : Not all the proportions in H_0 are equal.

Reject H_0 if p-value < 0.01.

Expected frequencies:

Month	Dates not drafted	Dates drafted	All
Jan.	$\frac{(31)(171)}{366} \approx 14.48$	$\frac{(31)(195)}{366} \approx 16.52$	31
Feb.	$\frac{(29)(171)}{366} \approx 13.55$	$\frac{(29)(195)}{366} \approx 15.45$	29
Mar.	$\frac{(31)(171)}{366} \approx 14.48$	$\frac{(31)(195)}{366} \approx 16.52$	31
Apr.	$\frac{(30)(171)}{366} \approx 14.02$	$\frac{(30)(195)}{366} \approx 15.98$	30
May	$\frac{(31)(171)}{366} \approx 14.48$	$\frac{(31)(195)}{366} \approx 16.52$	31

(Continued)

Month	Dates not drafted	Dates drafted	All
June	$\frac{(30)(171)}{366} \approx 14.02$	$\frac{(30)(195)}{366} \approx 15.98$	30
July	$\frac{(31)(171)}{366} \approx 14.48$	$\frac{(31)(195)}{366} \approx 16.52$	31
Aug.	$\frac{(31)(171)}{366} \approx 14.48$	$\frac{(31)(195)}{366} \approx 16.52$	31
Sept.	$\frac{(30)(171)}{366} \approx 14.02$	$\frac{(30)(195)}{366} \approx 15.98$	30
Oct.	$\frac{(31)(171)}{366} \approx 14.48$	$\frac{(31)(195)}{366} \approx 16.52$	31
Nov.	$\frac{(30)(171)}{366} \approx 14.02$	$\frac{(30)(195)}{366} \approx 15.98$	30
Dec.	$\frac{(31)(171)}{366} \approx 14.48$	$\frac{(31)(195)}{366} \approx 16.52$	31
All	170.99	195.01	366

$$\begin{aligned}\chi^2_{data} = \sum\frac{(O_i - E_i)^2}{E_i} &= \frac{(17-14.48)^2}{14.48} + \frac{(14-16.52)^2}{16.52} + \frac{(16-13.55)^2}{13.55} \\ &+ \frac{(13-15.45)^2}{15.45} + \frac{(21-14.48)^2}{14.48} + \frac{(10-16.52)^2}{16.52} \\ &+ \frac{(18-14.02)^2}{14.02} + \frac{(12-15.98)^2}{15.98} + \frac{(17-14.48)^2}{14.48} \\ &+ \frac{(14-16.52)^2}{16.52} + \frac{(16-14.02)^2}{14.02} + \frac{(14-15.98)^2}{15.98} \\ &+ \frac{(13-14.48)^2}{14.48} + \frac{(18-16.52)^2}{16.52} + \frac{(12-14.48)^2}{14.48} \\ &+ \frac{(19-16.52)^2}{16.52} + \frac{(11-14.02)^2}{14.02} + \frac{(19-15.98)^2}{15.98} \\ &+ \frac{(17-14.48)^2}{14.48} + \frac{(14-16.52)^2}{16.52} + \frac{(8-14.02)^2}{14.02} \\ &+ \frac{(22-15.98)^2}{15.98} + \frac{(5-14.48)^2}{14.48} + \frac{(26-16.52)^2}{16.52} \approx 30.257\end{aligned}$$

Since $\text{df} = (r-1)(c-1) = (12-1)(2-1) = 11$ and $\chi^2_{data} = 30.257 > 26.757$, p-value < 0.005. Since p-value < 0.01, we reject H_0. There is evidence that not all of the proportions in H_0 are equal.

Chapter 11 Review

1. $H_0: p_{USCan} = 0.32, p_{USMex} = 0.22, p_{CanUS} = 0.31, p_{MexUS} = 0.15$
H_a: The random variable does not follow the distribution specified in H_0.
$n = O_{USCan} + O_{USMex} + O_{CanUS} + O_{MexUS} = \$25 \text{ billion} + \$15 \text{ billion} + \$20 \text{ billion} + \$10 \text{ billion} = \$70 \text{ billion}$
$E_{USCan} = n \cdot p_{USCan} = (\$70 \text{ billion})(0.32) = \$22.4 \text{ billion}$
$E_{USMex} = n \cdot p_{USMex} = (\$70 \text{ billion})(0.22) = \$15.4 \text{ billion}$
$E_{CanUS} = n \cdot p_{CanUS} = (\$70 \text{ billion})(0.31) = \$21.7 \text{ billion}$
$E_{MexUS} = n \cdot p_{MexUS} = (\$70 \text{ billion})(0.15) = \$10.5 \text{ billion}$
Since none of the expected frequencies is less than 1 and none of the expected frequencies is less than 5, the conditions for performing the χ^2 goodness of fit test are met. Reject H_0 if p-value < 0.05.

O_i	E_i	$O_i - E_i$	$(O_i - E_i)^2$	$\frac{(O_i - E_i)^2}{E_i}$
25	22.4	2.6	6.76	$\frac{6.76}{22.4} = 0.3017857143$
15	15.4	−0.4	0.16	$\frac{0.16}{15.4} = 0.0103896104$
20	21.7	−1.7	2.89	$\frac{2.89}{21.7} = 0.1331797235$
10	10.5	−0.5	0.25	$\frac{0.25}{10.5} = 0.0238095238$
				$\chi^2_{data} = 0.469164572$

$$\chi^2_{data} = \sum \frac{(O_i - E_i)^2}{E_i} = 0.469164572 \approx 0.469$$

Since df $= k - 1 = 4 - 1 = 3$ and $\chi^2_{data} = 0.469 < 6.251$, p-value > 0.10.

From the TI-84, p-value $= 0.9989565445$. Since p-value ≥ 0.05, we do not reject H_0. There is insufficient evidence that the random variable does not follow the distribution specified in H_0.

3. $H_0 : p_{rock} = 0.40, p_{\frac{rap}{hiphop}} = 0.25, p_{country} = 0.15, p_{blues} = 0.08,$

$p_{jazz} = 0.07, p_{classical} = 0.05$

H_a : The random variable does not follow the distribution specified in H_0.

$E_{rock} = n \cdot p_{rock} = (200)(0.40) = 80$

$E_{\frac{rap}{hiphop}} = n \cdot p_{\frac{rap}{hiphop}} = (200)(0.25) = 50$

$E_{country} = n \cdot p_{country} = (200)(0.15) = 30$

$E_{blues} = n \cdot p_{blues} = (200)(0.08) = 16$

$E_{jazz} = n \cdot p_{jazz} = (200)(0.07) = 14$

$E_{classical} = n \cdot p_{classical} = (200)(0.05) = 10$

Since none of the expected frequencies is less than 1 and none of the expected frequencies is less than 5, the conditions for performing the χ^2 goodness of fit test are met.

df $= k - 1 = 6 - 1 = 5$, $\alpha = 0.05$, $\chi^2_{crit} = \chi^2_{0.05} = 11.071$. Reject H_0 if $\chi^2_{data} > 11.071$.

O_i	E_i	$O_i - E_i$	$(O_i - E_i)^2$	$\frac{(O_i - E_i)^2}{E_i}$
60	80	−20	400	$\frac{400}{80} = 5$
60	50	10	100	$\frac{100}{50} = 2$
35	30	5	25	$\frac{25}{30} = 0.8333333333$
12	16	−4	16	$\frac{16}{16} = 1$
18	14	4	16	$\frac{16}{14} = 1.142857143$
15	10	5	25	$\frac{25}{10} = 2.5$
				$\chi^2_{data} = 12.47619048$

$$\chi^2_{data} = \sum \frac{(O_i - E_i)^2}{E_i} = 12.47619048 \approx 12.476$$

Since $\chi^2_{data} > 11.071$, we reject H_0. There is evidence that the random variable does not follow the distribution specified in H_0.

5. $H_0: p_0 = 0.10, p_1 = 0.10, p_2 = 0.10, p_3 = 0.10, p_4 = 0.10, p_5 = 0.10, p_6 = 0.10, p_7 = 0.10, p_8 = 0.10, p_9 = 0.10$

H_a : Not all the proportions in H_0 are equal.

Reject H_0 if p-value < 0.05.

$E_0 = n \cdot p_0 = (218)(0.10) = 21.8$

$E_1 = n \cdot p_1 = (218)(0.10) = 21.8$

$E_2 = n \cdot p_2 = (218)(0.10) = 21.8$

$E_3 = n \cdot p_3 = (218)(0.10) = 21.8$

$E_4 = n \cdot p_4 = (218)(0.10) = 21.8$

$E_5 = n \cdot p_5 = (218)(0.10) = 21.8$

$E_6 = n \cdot p_6 = (218)(0.10) = 21.8$

$E_7 = n \cdot p_7 = (218)(0.10) = 21.8$

$E_8 = n \cdot p_8 = (218)(0.10) = 21.8$

$E_9 = n \cdot p_9 = (218)(0.10) = 21.8$

Since none of the expected frequencies is less than 1 and none of the expected frequencies is less than 5, the conditions for performing the χ^2 goodness of fit test are met.

O_i	E_i	$O_i - E_i$	$(O_i - E_i)^2$	$\frac{(O_i - E_i)^2}{E_i}$
26	21.8	4.2	17.64	$\frac{17.64}{21.8} = 0.8091743119$
12	21.8	−9.8	96.04	$\frac{96.04}{21.8} = 4.405504587$
26	21.8	4.2	17.64	$\frac{17.64}{21.8} = 0.8091743119$
18	21.8	−3.8	14.44	$\frac{14.44}{21.8} = 0.6623853211$
23	21.8	1.2	1.44	$\frac{1.44}{21.8} = 0.0660550459$
19	21.8	−2.8	7.84	$\frac{7.84}{21.8} = 0.3596330275$
18	21.8	−3.8	14.44	$\frac{14.44}{21.8} = 0.6623853211$
27	21.8	5.2	27.04	$\frac{27.04}{21.8} = 1.240366972$
30	21.8	8.2	67.24	$\frac{67.24}{21.8} = 3.08440367$
19	21.8	−2.8	7.84	$\frac{7.84}{21.8} = 0.3596330275$
				$\chi^2_{data} = 12.4587156$

$$\chi^2_{data} = \sum \frac{(O_i - E_i)^2}{E_i} = 12.4587156 \approx 12.459$$

Since df $= k - 1 = 10 - 1 = 9$ and $\chi^2_{data} = 12.459 < 14.684$, p-value > 0.10.

From the TI-84, p-value $= 0.18866727$. Since p-value ≥ 0.05, we do not reject H_0. There is insufficient evidence that the random variable does not follow the distribution specified in H_0.

7. (a) A higher proportion of the females with high GPAs take the SAT exam than the proportion of the females with lower GPAs.

(b) $H_0: p_{A+} = p_A = p_{A-} = p_B = p_C = p_{D/F}$
H_a: Not all the proportions in H_0 are equal.
Reject H_0 if p-value < 0.05.
Observed frequencies:

	High School Grade Point Average						
Gender	**A+**	**A**	**A−**	**B**	**C**	**D–F**	**Total**
Female	60	62	59	53	43	43	320
Male	40	38	41	47	57	57	280
Total	100	100	100	100	100	100	600

Expected frequencies:

	High School Grade Point Average						
Gender	**A+**	**A**	**A−**	**B**	**C**	**D–F**	**Total**
Female	$\frac{(320)(100)}{600} \approx 53.33$	$\frac{(320)(100)}{600} \approx 53.33$	$\frac{(320)(100)}{600} \approx 53.33$	$\frac{(320)(100)}{600} \approx 53.33$	$\frac{(320)(100)}{600} \approx 53.33$	$\frac{(320)(100)}{600} \approx 53.33$	319.98
Male	$\frac{(280)(100)}{600} \approx 46.67$	$\frac{(280)(100)}{600} \approx 46.67$	$\frac{(280)(100)}{600} \approx 46.67$	$\frac{(280)(100)}{600} \approx 46.67$	$\frac{(280)(100)}{600} \approx 46.67$	$\frac{(280)(100)}{600} \approx 46.67$	280.02
Total	100	100	100	100	100	100	600

Since none of the expected frequencies is less than 1 and none of the expected frequencies is less than 5, the conditions for performing the χ^2 test for homogeneity of proportions are met.

$$\chi^2_{data} = \sum \frac{(O_i - E_i)^2}{E_i} = \frac{(60 - 53.33)^2}{53.33} + \frac{(62 - 53.33)^2}{53.33} + \frac{(59 - 53.33)^2}{53.33}$$
$$+ \frac{(53 - 53.33)^2}{53.33} + \frac{(43 - 53.33)^2}{20.38} + \frac{(43 - 53.33)^2}{53.33}$$
$$+ \frac{(40 - 46.67)^2}{46.67} + \frac{(38 - 46.67)^2}{46.67} + \frac{(41 - 46.67)^2}{46.67}$$
$$+ \frac{(47 - 46.67)^2}{46.67} + \frac{(57 - 46.67)^2}{46.67} + \frac{(57 - 46.67)^2}{46.67} \approx 14.678$$

Since df $= (r - 1)(c - 1) = (2 - 1)(6 - 1) = 5$ and $12.833 < \chi^2_{data} = 14.678 < 15.086$, $0.01 < p$-value < 0.025. Since p-value < 0.05, we reject H_0. There is evidence that not all of the proportions in H_0 are equal.

9. $H_0: p_{Northeast} = p_{Midwest} = p_{South} = p_{West}$
H_a: Not all the proportions in H_0 are equal.
Reject H_0 if p-value < 0.01.
Observed frequencies:

	Northeast	**Midwest**	**South**	**West**	**Total**
Had HIV test	(1000)(0.568) = 568	(1000)(0.493) = 493	(1000)(0.585) = 585	(1000)(0.502) = 502	2148
Did not have HIV test	1000 − 568 = 432	1000 − 493 = 507	1000 − 585 = 415	1000 − 502 = 498	1852
Total	1000	1000	1000	1000	4000

Expected frequencies:

	Northeast	Midwest	South	West	Total
Had HIV test	$\frac{(2148)(1000)}{4000} = 537$	$\frac{(2148)(1000)}{4000} = 537$	$\frac{(2148)(1000)}{4000} = 537$	$\frac{(2148)(1000)}{4000} = 537$	2148
Did not Have HIV test	$\frac{(1852)(1000)}{4000} = 463$	$\frac{(1852)(1000)}{4000} = 463$	$\frac{(1852)(1000)}{4000} = 463$	$\frac{(1852)(1000)}{4000} = 463$	1852
Total	1000	1000	1000	1000	4000

Since none of the expected frequencies is less than 1 and none of the expected frequencies is less than 5, the conditions for performing the χ^2 test for homogeneity of proportions are met.

$$\chi^2_{data} = \sum\frac{(O_i - E_i)^2}{E_i} = \frac{(568-537)^2}{537} + \frac{(493-537)^2}{537} + \frac{(585-537)^2}{537}$$
$$+ \frac{(502-537)^2}{537} + \frac{(432-463)^2}{463} + \frac{(507-463)^2}{463}$$
$$+ \frac{(415-463)^2}{463} + \frac{(498-463)^2}{463} \approx 25.846$$

Since df $= (r-1)(c-1) = (2-1)(4-1) = 3$ and $\chi^2_{data} = 25.846 > 12.838$, p-value < 0.005. Since p-value ≤ 0.01, we reject H_0. There is evidence that not all of the proportions in H_0 are equal.

11. H_0 : Age and radio station type are independent.
H_a : Age and radio station type are not independent.
Reject H_0 if p-value < 0.05.
Observed frequencies:

	Pop contemporary hit radio stations	Alternative radio stations	Total
12–17 years old	240	170	410
18–24 years old	250	260	510
Total	490	430	920

Expected frequencies:

	Pop contemporary hit radio stations	Alternative radio stations	Total
12–17 years old	$\frac{(410)(490)}{920} \approx 218.37$	$\frac{(410)(430)}{930} \approx 191.63$	410
18–24 years old	$\frac{(510)(490)}{920} \approx 271.63$	$\frac{(510)(430)}{920} \approx 238.37$	510
Total	490	430	920

Since none of the expected frequencies is less than 1 and none of the expected frequencies is less than 5, the conditions for performing the χ^2 test for independence are met.

$$\chi^2_{data} = \sum\frac{(O_i - E_i)^2}{E_i} = \frac{(240-218.37)^2}{218.37} + \frac{(170-191.63)^2}{191.63} + \frac{(250-271.63)^2}{271.63}$$
$$+ \frac{(260-238.37)^2}{238.37} \approx 8.269$$

Since df $= (r-1)(c-1) = (2-1)(2-1) = 1$ and $\chi^2_{data} = 8.269 > 7.879$, p-value < 0.005. Since p-value < 0.05, we reject H_0. There is evidence that age and radio station type are not independent.

13. $H_0: p_{Whites} = p_{Blacks} = p_{Hispanics}$
H_a: Not all the proportions in H_0 are equal.
Reject H_0 if p-value < 0.05.
Observed frequencies:

	Whites	Blacks	Hispanics	Total
Uses the Internet	$(400)(0.82) = 328$	$(400)(0.65) = 260$	$(400)(0.82) = 328$	916
Does not use the Internet	$400 - 328 = 72$	$400 - 260 = 140$	$400 - 328 = 72$	284
Total	400	400	400	1200

Expected frequencies:

	Whites	Blacks	Hispanics	Total
Uses the Internet	$\frac{(916)(400)}{1200} \approx 305.33$	$\frac{(916)(400)}{1200} \approx 305.33$	$\frac{(916)(400)}{1200} \approx 305.33$	915.99
Does not use the Internet	$\frac{(284)(400)}{1200} \approx 94.67$	$\frac{(284)(400)}{1200} \approx 94.67$	$\frac{(284)(400)}{1200} \approx 94.67$	284.01
Total	400	400	400	1200

Since none of the expected frequencies is less than 1 and none of the expected frequencies is less than 5, the conditions for performing the χ^2 test for homogeneity of proportions are met.

$$\chi^2_{data} = \sum \frac{(O_i - E_i)^2}{E_i} = \frac{(328 - 305.33)^2}{305.33} + \frac{(260 - 305.33)^2}{305.33} + \frac{(328 - 305.33)^2}{305.33}$$

$$+ \frac{(72 - 94.67)^2}{94.67} + \frac{(140 - 94.67)^2}{94.67} + \frac{(72 - 94.67)^2}{94.67} \approx 42.658$$

Since df $= (r - 1)(c - 1) = (2 - 1)(3 - 1) = 2$ and $\chi^2_{data} = 42.658 > 10.597$, p-value < 0.005. Since p-value < 0.05, we reject H_0. There is evidence that not all of the proportions in H_0 are equal.

Chapter 11 Quiz

1. True.

2. False. In a test for independence, the degrees of freedom equal $(r - 1)(c - 1)$, where r = the number of rows and c = the number of columns.

3. False. In the test for homogeneity of proportions, the alternative hypothesis states that the random variable does not follow the distribution specified in H_0.

4. 1, 5

5. Equal.

6. Expected frequency.

7. **(a)** The critical-value method, **(b)** the exact p-value method, or **(c)** the estimated p-value method.

8. H_a, the alternative hypothesis.

9. Degrees of freedom $= (r - 1)(c - 1)$, where r = the number of categories in the row variable and c = the number of categories in the column variable.

10. **(a)** $n = O_1 + O_2 + O_3 + O_4 + O_5 = 8 + 9 + 10 + 11 + 12 = 50$.
$E_1 = n \cdot p_1 = (50)(0.2) = 10$, $E_2 = n \cdot p_2 = (50)(0.2) = 10$.
$E_3 = n \cdot p_3 = (50)(0.2) = 10$, $E_4 = n \cdot p_4 = (50)(0.2) = 10$.
$E_5 = n \cdot p_5 = (50)(0.2) = 10$.

(b) Since none of the expected frequencies is less than 1 and none of the expected frequencies is less than 5, the conditions for performing the χ^2 goodness of fit test are met.

(c) df $= k - 1 = 5 - 1 = 4$, $\alpha = 0.05$, $\chi^2_{crit} = \chi^2_{0.05} = 9.488$. Reject H_0 if $\chi^2_{data} > 9.488$.

(d)

O_i	E_i	$O_i - E_i$	$(O_i - E_i)^2$	$\frac{(O_i - E_i)^2}{E_i}$
8	10	−2	4	$\frac{4}{10} = 0.4$
9	10	−1	1	$\frac{1}{10} = 0.1$
10	10	0	0	$\frac{0}{10} = 0$
11	10	1	1	$\frac{1}{10} = 0.1$
12	10	2	4	$\frac{4}{10} = 0.4$
				$\chi^2_{data} = 1$

(e) Since $\chi^2_{data} \leq 9.488$, we do not reject H_0. There is insufficient evidence that the random variable does not follow the distribution specified in H_0.

11. (a) $n = O_1 + O_2 + O_3 + O_4 + O_5 + O_6 = 50 + 40 + 30 + 20 + 10 + 10 = 160$.
$E_1 = n \cdot p_1 = (160)(0.3) = 48$, $E_2 = n \cdot p_2 = (160)(0.25) = 40$,
$E_3 = n \cdot p_3 = (160)(0.2) = 32$, $E_4 = n \cdot p_4 = (160)(0.15) = 24$,
$E_5 = n \cdot p_5 = (160)(0.06) = 9.6$, $E_6 = n \cdot p_6 = (160)(0.04) = 6.4$.
(b) Since none of the expected frequencies is less than 1 and none of the expected frequencies is less than 5, the conditions for performing the χ^2 goodness of fit test are met.
(c) df $= k - 1 = 6 - 1 = 5$, $\alpha = 0.05$, $\chi^2_{crit} = \chi^2_{0.05} = 11.071$. Reject H_0 if $\chi^2_{data} > 11.071$.

(d)

O_i	E_i	$O_i - E_i$	$(O_i - E_i)^2$	$\frac{(O_i - E_i)^2}{E_i}$
50	48	2	4	$\frac{4}{48} = 0.0833333333$
40	40	0	0	$\frac{0}{40} = 0$
30	32	−2	4	$\frac{4}{32} = 0.125$
20	24	−4	16	$\frac{16}{24} = 0.6666666667$
10	9.6	0.4	0.16	$\frac{0.16}{9.6} = 0.0166666667$
10	6.4	3.6	12.96	$\frac{12.96}{6.4} = 2.025$
				$\chi^2_{data} = 2.916666667$

$$\chi^2_{data} = \sum \frac{(O_i - E_i)^2}{E_i} = 2.916666667 \approx 2.917$$

(e) Since $\chi^2_{data} \leq 11.071$, we do not reject H_0. There is insufficient evidence that the random variable does not follow the distribution specified in H_0.

12. (a) $n = O_1 + O_2 + O_3 + O_4 + O_5 = 18 + 19 + 21 + 22 + 20 = 100$.
$E_1 = n \cdot p_1 = (100)(0.2) = 20$, $E_2 = n \cdot p_2 = (100)(0.2) = 20$,
$E_3 = n \cdot p_3 = (100)(0.2) = 20$, $E_4 = n \cdot p_4 = (100)(0.2) = 20$,
$E_5 = n \cdot p_5 = (100)(0.2) = 20$.

(b) Since none of the expected frequencies is less than 1 and none of the expected frequencies is less than 5, the conditions for performing the χ^2 goodness of fit test are met.

(c) df $= k - 1 = 5 - 1 = 4$, $\alpha = 0.01$, $\chi^2_{crit} = \chi^2_{0.01} = 13.277$. Reject H_0 if $\chi^2_{data} > 13.277$.

(d)

O_i	E_i	$O_i - E_i$	$(O_i - E_i)^2$	$\frac{(O_i - E_i)^2}{E_i}$
18	20	−2	4	$\frac{4}{20} = 0.2$
19	20	−1	1	$\frac{1}{20} = 0.05$
21	20	1	1	$\frac{1}{20} = 0.05$
22	20	2	4	$\frac{4}{20} = 0.2$
20	20	0	0	$\frac{0}{20} = 0$
				$\chi^2_{data} = 0.5$

(e) Since $\chi^2_{data} \le 13.277$, we do not reject H_0. There is insufficient evidence that the random variable does not follow the distribution specified in H_0.

13. **(a)** $n = O_1 + O_2 + O_3 + O_4 + O_5 + O_6 = 65 + 55 + 30 + 25 + 15 + 10 = 200$.
$E_1 = n \cdot p_1 = (200)(0.3) = 60$, $E_2 = n \cdot p_2 = (200)(0.25) = 50$,
$E_3 = n \cdot p_3 = (200)(0.20) = 40$, $E_4 = n \cdot p_4 = (200)(0.15) = 30$,
$E_5 = n \cdot p_5 = (200)(0.06) = 12$, $E_6 = n \cdot p_6 = (200)(0.04) = 8$.

(b) Since none of the expected frequencies is less than 1 and none of the expected frequencies is less than 5, the conditions for performing the χ^2 goodness of fit test are met.

(c) df $= k - 1 = 6 - 1 = 5$, $\alpha = 0.05$, $\chi^2_{crit} = \chi^2_{0.05} = 11.071$. Reject H_0 if $\chi^2_{data} > 11.071$.

(d)

O_i	E_i	$O_i - E_i$	$(O_i - E_i)^2$	$\frac{(O_i - E_i)^2}{E_i}$
65	60	5	25	$\frac{25}{60} = 0.4166666667$
55	50	5	25	$\frac{25}{50} = 0.5$
30	40	−10	100	$\frac{100}{40} = 2.5$
25	30	−5	25	$\frac{25}{30} = 0.8333333333$
15	12	3	9	$\frac{9}{12} = 0.75$
10	8	2	4	$\frac{4}{8} = 0.5$
				$\chi^2_{data} = 5.5$

(e) Since $\chi^2_{data} \le 11.071$, we do not reject H_0. There is insufficient evidence that the random variable does not follow the distribution specified in H_0.

14. **(a)** The higher the grade level, the higher the proportion of students who have used an illicit drug.

(b) $H_0 : p_{8thgraders} = p_{10thgraders} = p_{12thgraders}$
H_a : Not all the proportions in H_0 are equal.
Reject H_0 if p-value < 0.01.

Observed frequencies:

	8th-graders	10th-graders	12th-graders	Total
Have used an illicit drug	3,655	6,527	7,461	17,643
Have never used an illicit drug	13,345	9,873	7,139	30,357
Total	17,000	16,400	14,600	48,000

Expected frequencies:

	8th-graders	10th-graders	12th-graders	Total
Have used an illicit drug	$\frac{(17{,}643)(17{,}000)}{48{,}000} \approx 6248.56$	$\frac{(17{,}643)(16{,}400)}{48{,}000} \approx 6028.03$	$\frac{(17{,}643)(14{,}600)}{48{,}000} \approx 5366.41$	17,643
Have never used an illicit drug	$\frac{(30{,}357)(17{,}000)}{48{,}000} \approx 10{,}751.44$	$\frac{(30{,}357)(16{,}400)}{48{,}000} \approx 10{,}371.98$	$\frac{(30{,}357)(14{,}600)}{48{,}000} \approx 9233.59$	30,357.01
Total	17,000	16,400.01	14,600	48,000.01

Since none of the expected frequencies is less than 1 and none of the expected frequencies is less than 5, the conditions for performing the χ^2 test for homogeneity of proportions are met.

$$\chi^2_{data} = \sum\frac{(O_i - E_i)^2}{E_i} = \frac{(3{,}655 - 6248.56)^2}{6{,}248.56} + \frac{(6{,}527 - 6028.03)^2}{6028.03} + \frac{(7{,}461 - 5366.41)^2}{5366.41}$$
$$+ \frac{(13{,}345 - 10{,}751.44)^2}{10{,}751.44} + \frac{(9{,}873 - 10{,}371.98)^2}{10{,}371.98} + \frac{(7{,}139 - 9233.59)^2}{9233.59}$$
$$\approx 3{,}060.142$$

Since df $= (r - 1)(c - 1) = (2 - 1)(3 - 1) = 2$ and $\chi^2_{data} = 3{,}060.142 > 10.597$, p-value < 0.005. Since p-value < 0.01, we reject H_0. There is evidence that not all of the proportions in H_0 are equal.

15. $H_0 : p_{Whites} = p_{Blacks} = p_{Hispanics}$
H_a : Not all the proportions in H_0 are equal.
Reject H_0 if p-value < 0.01.
Observed frequencies:

	Whites	Blacks	Hispanics	Total
Uses the Internet	$(400)(0.83) = 332$	$(400)(0.76) = 304$	$(400)(0.87) = 348$	984
Does not use the Internet	$400 - 332 = 68$	$400 - 304 = 96$	$400 - 348 = 52$	216
Total	400	400	400	1200

Expected frequencies:

	Whites	Blacks	Hispanics	Total
Uses the Internet	$\frac{(984)(400)}{1200} = 328$	$\frac{(984)(400)}{1200} = 328$	$\frac{(984)(400)}{1200} = 328$	984
Does not use the Internet	$\frac{(216)(400)}{1200} = 72$	$\frac{(216)(400)}{1200} = 72$	$\frac{(216)(400)}{1200} = 72$	216
Total	400	400	400	1200

Since none of the expected frequencies is less than 1 and none of the expected frequencies is less than 5, the conditions for performing the χ^2 test for homogeneity of proportions are met.

$$\chi^2_{data} = \sum \frac{(O_i - E_i)^2}{E_i} = \frac{(332 - 328)^2}{328} + \frac{(304 - 328)^2}{328} + \frac{(348 - 328)^2}{328}$$

$$+ \frac{(68 - 72)^2}{72} + \frac{(96 - 72)^2}{72} + \frac{(52 - 72)^2}{72} \approx 16.802$$

Since df $= (r - 1)(c - 1) = (2 - 1)(3 - 1) = 2$ and $\chi^2_{data} = 16.802 > 10.597$, p-value < 0.005. Since p-value < 0.01, we reject H_0. There is evidence that not all of the proportions in H_0 are equal.

16. H_0 : Gender and sport preference are independent.
H_a : Gender and sport preference are not independent. Reject H_0 if p-value < 0.05.
Expected frequencies:

	Sport Preference			
Gender	**Basketball**	**Soccer**	**Swimming**	**Total**
Female	$\frac{(100)(80)}{200} = 40$	$\frac{(100)(50)}{200} = 25$	$\frac{(100)(70)}{200} = 35$	100
Male	$\frac{(100)(80)}{200} = 40$	$\frac{(100)(50)}{200} = 25$	$\frac{(100)(70)}{200} = 35$	100
Total	80	50	70	200

Since none of the expected frequencies is less than 1 and none of the expected frequencies is less than 5, the conditions for performing the χ^2 test for independence are met.

$$\chi^2_{data} = \sum \frac{(O_i - E_i)^2}{E_i} = \frac{(30 - 40)^2}{40} + \frac{(20 - 25)^2}{25} + \frac{(50 - 35)^2}{35}$$

$$+ \frac{(50 - 40)^2}{40} + \frac{(30 - 25)^2}{25} + \frac{(20 - 35)^2}{35} \approx 19.857$$

Since df $= (r - 1)(c - 1) = (2 - 1)(3 - 1) = 2$ and $\chi^2_{data} = 19.857 > 10.597$, p-value < 0.005. Since p-value < 0.05, we reject H_0. There is evidence that gender and sport preference are not independent.

17. $H_0 : p_{Texas} = p_{Oklahoma} = p_{Pennsylvania}$
H_a : Not all the proportions in H_0 are equal.
Reject H_0 if p-value < 0.05.
Observed frequencies:

	Texas	**Oklahoma**	**Pennsylvania**	**Total**
Beef cattle on smaller scale operations	103,000	3,600	11,400	118,000
Beef cattle on operations that are not smaller scale	28,000	44,400	600	73,000
Total	131,000	48,000	12,000	191,000

Expected frequencies:

	Texas	Oklahoma	Pennsylvania	Total
Beef cattle on smaller scale operations	$\frac{(118{,}000)(131{,}000)}{191{,}000} \approx 80{,}931.94$	$\frac{(118{,}000)(48{,}000)}{191{,}000} \approx 29{,}654.45$	$\frac{(118{,}000)(12{,}000)}{191{,}000} \approx 7413.61$	118,000
Beef cattle on operations that are not smaller scale	$\frac{(73{,}000)(131{,}000)}{191{,}000} \approx 50{,}068.06$	$\frac{(73{,}000)(48{,}000)}{191{,}000} \approx 18{,}345.55$	$\frac{(73{,}000)(12{,}000)}{191{,}000} \approx 4586.39$	73,000
Total	131,000	48,000	12,000	191,000

Since none of the expected frequencies is less than 1 and none of the expected frequencies is less than 5, the conditions for performing the χ^2 test for homogeneity of proportions are met.

$$\chi^2_{data} = \sum \frac{(O_i - E_i)^2}{E_i} = \frac{(103{,}000 - 80{,}931.94)^2}{80{,}931.94} + \frac{(3{,}600 - 29{,}654.45)^2}{29{,}654.45}$$
$$+ \frac{(11{,}400 - 7413.61)^2}{7413.61} + \frac{(28{,}000 - 50{,}068.06)^2}{50{,}068.06}$$
$$+ \frac{(44{,}400 - 18{,}345.55)^2}{18{,}345.55} + \frac{(600 - 4586.39)^2}{4586.39} \approx 81{,}246.708$$

Since df $= (r - 1)(c - 1) = (2 - 1)(3 - 1) = 2$ and $\chi^2_{data} = 81{,}246.708 > 10.597$, p-value < 0.005. Since p-value < 0.01, we reject H_0. There is evidence that not all of the proportions in H_0 are equal.

Chapter 12

Analysis of Variance

Section 12.1

1. True.

3. True.

5. True.

7. MSTR measures the variability in the sample means. MSE measures the variability within the samples.

9. ANOVA works by comparing (a) the variability in the sample means with (b) the variability within each sample. (a) larger than (b) is evidence that the population means are not all equal and that we should reject the null hypothesis.

11. Each of the *k* populations is normally distributed, the variances of the populations are all equal, and the samples are independently drawn.

13. **(a)** $F_{crit} = 5.59$

(b)

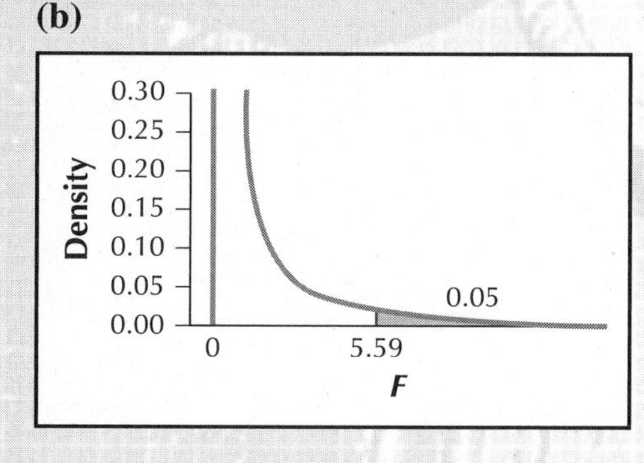

15. **(a)** $F_{crit} = 6.36$

(b)

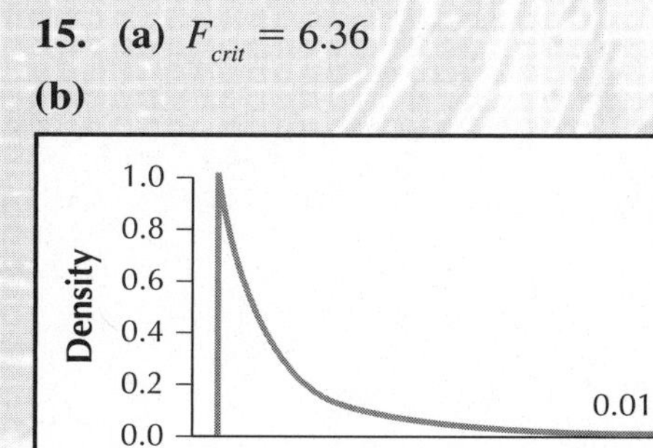

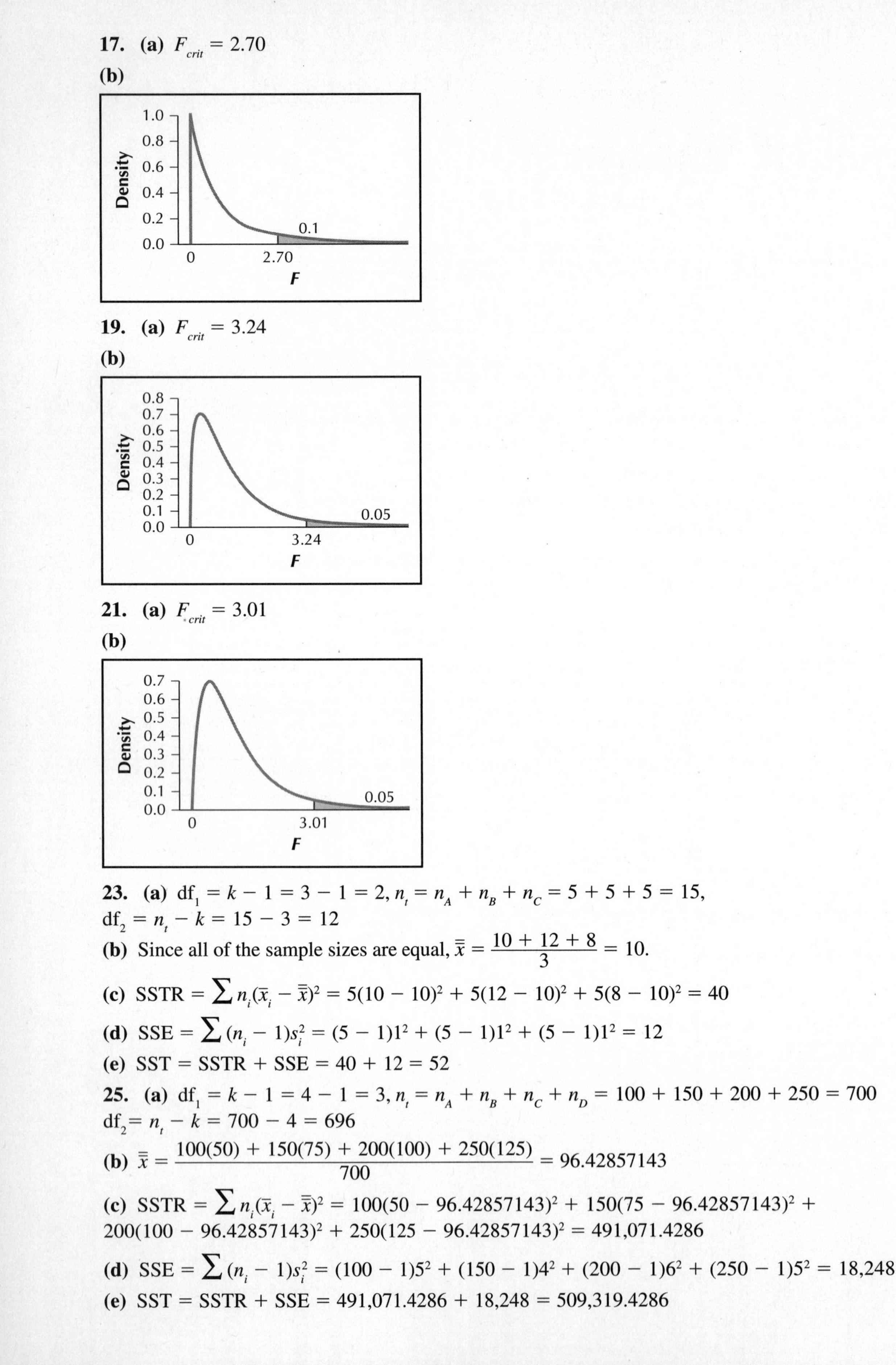

17. **(a)** $F_{crit} = 2.70$

(b)

19. **(a)** $F_{crit} = 3.24$

(b)

21. **(a)** $F_{crit} = 3.01$

(b)

23. **(a)** $df_1 = k - 1 = 3 - 1 = 2$, $n_t = n_A + n_B + n_C = 5 + 5 + 5 = 15$,
$df_2 = n_t - k = 15 - 3 = 12$

(b) Since all of the sample sizes are equal, $\bar{\bar{x}} = \dfrac{10 + 12 + 8}{3} = 10$.

(c) $SSTR = \sum n_i(\bar{x}_i - \bar{\bar{x}})^2 = 5(10 - 10)^2 + 5(12 - 10)^2 + 5(8 - 10)^2 = 40$

(d) $SSE = \sum (n_i - 1)s_i^2 = (5 - 1)1^2 + (5 - 1)1^2 + (5 - 1)1^2 = 12$

(e) $SST = SSTR + SSE = 40 + 12 = 52$

25. **(a)** $df_1 = k - 1 = 4 - 1 = 3$, $n_t = n_A + n_B + n_C + n_D = 100 + 150 + 200 + 250 = 700$
$df_2 = n_t - k = 700 - 4 = 696$

(b) $\bar{\bar{x}} = \dfrac{100(50) + 150(75) + 200(100) + 250(125)}{700} = 96.42857143$

(c) $SSTR = \sum n_i(\bar{x}_i - \bar{\bar{x}})^2 = 100(50 - 96.42857143)^2 + 150(75 - 96.42857143)^2 +$
$200(100 - 96.42857143)^2 + 250(125 - 96.42857143)^2 = 491{,}071.4286$

(d) $SSE = \sum (n_i - 1)s_i^2 = (100 - 1)5^2 + (150 - 1)4^2 + (200 - 1)6^2 + (250 - 1)5^2 = 18{,}248$

(e) $SST = SSTR + SSE = 491{,}071.4286 + 18{,}248 = 509{,}319.4286$

27. **(a)** $\text{MSTR} = \dfrac{\text{SSTR}}{\text{df}_1} = \dfrac{40}{2} = 20$ **(b)** $\text{MSE} = \dfrac{\text{SSE}}{\text{df}_2} = \dfrac{12}{12} = 1$ **(c)** $F_{data} = \dfrac{\text{MSTR}}{\text{MSE}} = \dfrac{20}{1} = 20$

29. **(a)** $\text{MSTR} = \dfrac{\text{SSTR}}{\text{df}_1} = \dfrac{491{,}071.4286}{3} = 163{,}690.4762$

(b) $\text{MSE} = \dfrac{\text{SSE}}{\text{df}_2} = \dfrac{18{,}248}{696} = 26.2183908$

(c) $F_{data} = \dfrac{\text{MSTR}}{\text{MSE}} = \dfrac{163{,}690.4762}{26.2183908} = 6243.345652 \approx 6243.3457$

31. **(a)** $S_{Online} = 15.0555 < 25.2982 = 2S_{Hybrid}$

(b) **(i)** $\text{df}_1 = k - 1 = 3 - 1 = 2, n_t = 6 + 6 + 6 = 18, \text{df}_2 = n_t - k = 18 - 3 = 15$

(ii) Since all of the sample sizes are equal, $\bar{\bar{x}} = \dfrac{71.6667 + 74.1667 + 80}{3} = 75.2778$

(iii) $\text{SSTR} = \sum n_i(\bar{x}_i - \bar{\bar{x}})^2 = 6(71.6667 - 75.2778)^2 + 6(74.1667 - 75.2778)^2 + 6(80 - 75.2778)^2 = 219.4426$

(iv) $\text{SSE} = \sum (n_i - 1)s_i^2 = (6 - 1)15.0555^2 + (6 - 1)13.1972^2 + (6 - 1)12.6491^2 = 2804.1695$

(v) $\text{SST} = \text{SSTR} + \text{SSE} = 219.4426 + 2804.1695 = 3023.6121$

(vi) $\text{MSTR} = \dfrac{\text{SSTR}}{\text{df}_1} = \dfrac{219.4426}{2} = 109.7213$

(vii) $\text{MSE} = \dfrac{\text{SSE}}{\text{df}_2} = \dfrac{2804.1695}{15} = 186.944633$

(viii) $F_{data} = \dfrac{\text{MSTR}}{\text{MSE}} = \dfrac{109.7213}{3186.944633} \approx 0.5869$

(c)

Source of variation	Sum of squares	Degrees of freedom	Mean square	*F*-test statistic
Treatment	SSTR = 219,4426	$\text{df}_1 = 2$	MSTR = 109.7213	$F_{data} = 0.5869$
Error	SSE = 2804.1695	$\text{df}_2 = 15$	MSE = 186.944633	
Total	SST = 3023.6121			

33. **(a)** $S_{Female} = 0.743 < 1.398 = 2S_{Male}$

(b) **(i)** $\text{df}_1 = k - 1 = 2 - 1 = 1, n_t = 65 + 65 = 130,$
$\text{df}_2 = n_t - k = 130 - 2 = 128$

(ii) Since all of the sample sizes are equal, $\bar{\bar{x}} = \dfrac{98.384 + 98.104}{2} = 98.244.$

(iii) $\text{SSTR} = \sum n_i(\bar{x}_i - \bar{\bar{x}})^2 = 65(98.384 - 98.244)^2 + 65(98.104 - 98.244)^2 = 2.548$

(iv) $\text{SSE} = \sum (n_i - 1)s_i^2 = (65 - 1)\, 0.743^2 + (65 - 1)\, 0.699^2 = 66.6016$

(v) $\text{SST} = \text{SSTR} + \text{SSE} = 2.548 + 66.6016 = 69.1496$

(vi) $\text{MSTR} = \dfrac{\text{SSTR}}{\text{df}_1} = \dfrac{2.548}{1} = 2.548$

(vii) $\text{MSE} = \dfrac{\text{SSE}}{\text{df}_2} = \dfrac{66.6016}{128} = 0.520325$

(viii) $F_{data} = \dfrac{\text{MSTR}}{\text{MSE}} = \dfrac{2.548}{0.520325} \approx 4.896939413$

(c)

Source of variation	Sum of squares	Degrees of freedom	Mean square	*F*-test statistic
Treatment	SSTR = 2.548	$\text{df}_1 = 1$	MSTR = 2.548	$F_{data} = 4.896939413$
Error	SSE = 66.6016	$\text{df}_2 = 128$	MSE = 0.520325	
Total	SST = 69.1496			

The method of inference from an earlier chapter that we could also use to solve this problem is inference about two independent means from Chapter 10.

35. **(a)** $S_{None} = 3.22 < 5.48 = 2S_{Catholic}$

(b) **(i)** $\text{df}_1 = k - 1 = 5 - 1 = 4$, $n_t = 1660 + 683 + 68 + 339 + 141 = 2891$,
$\text{df}_2 = n_t - k = 2891 - 5 = 2886$

(ii) $\bar{\bar{x}} = \dfrac{1660(13.10) + 683(13.51) + 68(15.37) + 339(13.52) + 141(14.46)}{2891} = 13.36583535$

(iii) $\text{SSTR} = \sum n_i(\bar{x}_i - \bar{\bar{x}})^2 = 1660(13.10 - 13.36583535)^2 + 683(13.51 - 13.36583535)^2 + 68(15.37 - 13.36583535)^2 + 339(13.52 - 13.36583535)^2 + 141(14.46 - 13.36583535)^2 = 581.5002576$

(iv) $\text{SSE} = \sum (n_i - 1)s_i^2 = (1660 - 1)2.87^2 + (683 - 1)2.74^2 + (68 - 1)2.80^2 + (339 - 1)3.22^2 + (141 - 1)3.18^2 = 24{,}230.7355$

(v) $\text{SST} = \text{SSTR} + \text{SSE} = 581.5002576 + 24{,}230.7355 = 24{,}812.23576$

(vi) $\text{MSTR} = \dfrac{\text{SSTR}}{\text{df}_1} = \dfrac{581.5002576}{4} = 145.3750644$

(vii) $\text{MSE} - \dfrac{\text{SSE}}{\text{df}_2} - \dfrac{24{,}230.7355}{2886} = 8.395958247$

(viii) $F_{data} = \dfrac{\text{MSTR}}{\text{MSE}} = \dfrac{145.3750644}{8.395958247} \approx 17.31488654$

(c)

Source of variation	Sum of squares	Degrees of freedom	Mean square	*F*-test statistic
Treatment	SSTR = 581.5002576	$\text{df}_1 = 4$	MSTR = 145.3750644	$F_{data} = 17.31488654$
Error	SSE = 24,230.7355	$\text{df}_2 = 2886$	MSE = 8.395958247	
Total	SST = 24,812.23576			

37. From the TI-84:

Before: $n = 12, \bar{x} = 10.9167, s = 3.6199$

During: $n = 12, \bar{x} = 13.4167, s = 5.5014$

After: $n = 12, \bar{x} = 11.4583, s = 3.1094$

(a) $S_{During} = 5.5014 < 6.2188 = 2S_{After}$

(b) **(i)** $\text{df}_1 = k - 1 = 3 - 1 = 2$, $n_t = 12 + 12 + 12 = 36$,
$\text{df}_2 = n_t - k = 36 - 3 = 33$

(ii) Since all of the sample sizes are equal, $\bar{\bar{x}} = \dfrac{10.9167 + 13.4167 + 11.4583}{3} = 11.93056667.$

(iii) $\text{SSTR} = \sum n_i(\bar{x}_i - \bar{\bar{x}})^2 = 12(10.9167 - 11.93056667)^2 + 12(13.4167 - 11.93056667)^2 + 12(11.4583 - 11.93056667)^2 = 41.51464448$

(iv) $\text{SSE} = \sum (n_i - 1)s_i^2 = (12 - 1)3.6199^2 + (12 - 1)5.5014^2 + (12 - 1)3.1094^2 = 583.4119096$

(v) $\text{SST} = \text{SSTR} + \text{SSE} = 41.51464448 + 583.4119096 = 624.9265541$

(vi) $\text{MSTR} = \dfrac{\text{SSTR}}{\text{df}_1} = \dfrac{41.5146448}{2} = 20.7573224$

(vii) $\text{MSE} = \dfrac{\text{SSE}}{\text{df}_2} = \dfrac{583.4119096}{33} = 17.67914878$

(viii) $F_{data} = \dfrac{\text{MSTR}}{\text{MSE}} = \dfrac{20.7573224}{17.67914878} \approx 1.1741$

(c)

Source of variation	Sum of squares	Degrees of freedom	Mean square	F-test statistic
Treatment	SSTR = 41.5146	$\text{df}_1 = 2$	MSTR = 20.7573	$F_{data} = 1.1741$
Error	SSE = 583.4119	$\text{df}_2 = 33$	MSE = 17.6791	
Total	SST = 624.9265			

Section 12.2

1. Use the closest value.

3. Against.

5. **(a)** $H_0: \mu_A = \mu_B = \mu_C$
H_a: Not all of the population means are equal.
Reject H_0 if p-value < 0.05.
(b) From Exercise 27, Section 12.1, $F_{data} = 20$.
(c) Since $\text{df}_1 = 2$, $\text{df}_2 = 12$, and $F_{data} = 20 > 12.97$, p-value < 0.001.
(d) Since p-value < 0.05, we reject H_0. There is evidence that not all of the population means are equal.

7. **(a)** $H_0: \mu_A = \mu_B = \mu_C = \mu_D$
H_a: Not all of the population means are equal.
(b) $\text{df}_1 = 3$, $\text{df}_2 = 696$, $\alpha = 0.05$, $F_{crit} = 2.61$. Reject H_0 if $F_{data} > 2.61$. **(c)** From Exercise 29, Section 12.1, $F_{data} = 6243.3457$. **(d)** Since $F_{data} > 2.61$, we reject H_0. There is evidence that not all of the population means are equal.

9. **(a)** $\text{SST} = \text{SSTR} + \text{SSE} = 120 + 315 = 435$, $\text{df}_1 = k - 1 = 7 - 1 = 6$,

$n_t = 7(10) = 70$, $\text{df}_2 = n_t - k = 70 - 7 = 63$, $\text{MSTR} = \dfrac{\text{SSTR}}{\text{df}_1} = \dfrac{120}{6} = 20$,

$\text{MSE} = \dfrac{\text{SSE}}{\text{df}_2} = \dfrac{315}{63} = 5$, $F_{data} = \dfrac{\text{MSTR}}{\text{MSE}} = \dfrac{20}{5} = 4$.
Missing values are bold.

Source of variation	Sum of squares	Degrees of freedom	Mean square	F-test statistic
Treatment	SSTR = 120	$\text{df}_1 = 6$	MSTR = 20	$F_{data} = 4$
Error	SSE = 315	$\text{df}_2 = 63$	MSE = 5	
Total	SST = 435			

(b) $H_0: \mu_1 = \mu_2 = \mu_3 = \mu_4 = \mu_5 = \mu_6 = \mu_7$
H_a: Not all of the population means are equal.
Reject H_0 if p-value < 0.05. $F_{data} = 4$. Since $\text{df}_1 = 6$, $\text{df}_2 = 63$, and $3.19 < F_{data} = 4 < 4.51$, $0.001 < p$-value < 0.01. Since p-value < 0.05, we reject H_0. There is evidence that not all of the population means are equal.

11. **(a)** $F_{data} = \dfrac{\text{MSTR}}{\text{MSE}} = \dfrac{10}{\text{MSE}} = 1.0$, so MSE = 10. $\text{MSTR} = \dfrac{\text{SSTR}}{\text{df}_1} = \dfrac{\text{SSTR}}{4} = 10$, so SSTR = 4(10) = 40.
SSE = SST − SSTR = 440 − 40 = 400

$\text{MSE} = \dfrac{\text{SSE}}{\text{df}_2} = \dfrac{400}{\text{df}_2} = 10$, so $\text{df}_2 = \dfrac{400}{10} = 40$.

Missing values are bold.

Source of variation	Sum of squares	Degrees of freedom	Mean square	F-test statistic
Treatment	SSTR = 40	$df_1 = 4$	MSTR = 10	$F_{data} = 1.0$
Error	SSE = 400	$df_2 = 40$	MSE = 10	
Total	SST = 440			

(b) $H_0: \mu_1 = \mu_2 = \mu_3 = \mu_4 = \mu_5$
H_a: Not all of the population means are equal.
$df_1 = 4$, $df_2 = 40$, $\alpha = 0.10$, $F_{crit} = 2.06$. Reject H_0 if $F_{data} > 2.06$.

$F_{data} = 1.0$. Since $F_{data} \le 2.06$, we do not reject H_0. There is insufficient evidence that not all of the population means are equal.

13. (a) $H_0: \mu_{Protestant} = \mu_{Catholic} = \mu_{Jewish} = \mu_{None} = \mu_{Other}$
H_a: Not all of the population means are equal.
$\mu_{Protestant}$ = the population mean number of years of education of people whose religious preference is Protestant.
$\mu_{Catholic}$ = the population mean number of years of education of people whose religious preference is Catholic.
μ_{Jewish} = the population mean number of years of education of people whose religious preference is Jewish.
μ_{None} = the population mean number of years of education of people whose religious preference is None.
μ_{Other} = the population mean number of years of education of people whose religious preference is Other.
Reject H_0 if p-value < 0.05.

(b) From Exercise 35, Section 12.1, $F_{data} = 17.3149$. **(c)** Since $df_1 = 4$, $df_2 = 2886$, and $F_{data} = 17.3149 > 4.65$, p-value < 0.001. **(d)** Since p-value < 0.05, we reject H_0. There is evidence that not all of the population mean numbers of years of education are equal.

15. (a) $H_0: \mu_{Before} = \mu_{During} = \mu_{After}$
H_a: Not all of the population means are equal.
μ_{Before} = the population mean number of emergency room visits before a full moon.
μ_{During} = the population mean number of emergency room visits during a full moon.
μ_{After} = the population mean number of emergency room visits after a full moon.
(b) $df_1 = 2$, $df_2 = 33$, $\alpha = 0.05$, $F_{crit} = 3.39$. Reject H_0 if $F_{data} > 3.39$. **(c)** From Exercise 37, Section 12.1, $F_{data} = 1.1741$. **(d)** Since $F_{data} \le 3.39$, we do not reject H_0. There is insufficient evidence that not all of the population mean numbers of emergency room visits are equal.

17. (a) Since one of the boxplots only overlaps 2 of the 5 other boxplots, the comparison boxplot of the nutritional ratings may be considered as evidence against the null hypothesis that all population mean nutritional ratings were equal. **(b)** We would expect the null hypothesis to be rejected.

19. The independence assumption is validated since the cereals were selected randomly from the manufacturers so that the selection of a cereal from one manufacturer did not affect the selection of cereals from other manufacturers.

21. $H_0: \mu_{Online} = \mu_{Traditional} = \mu_{Hybrid}$
H_a: Not all of the population means are equal
$df_1 = 2$, $df_2 = 15$, $\alpha = 0.01$, $F_{crit} = 6.36$. Reject H_0 if $F_{data} > 6.36$.
From Exercise 31, Section 12.1, $F_{data} = 0.5869$. Since $F_{data} \le 6.36$, we do not reject H_0. There is insufficient evidence that not all of the population mean numbers of emergency room visits are equal.

23. (a) (i) n_t would stay the same. We are neither adding nor deleting data values. Instead, we are adjusting existing data values. So the total sample size remains the same. **(ii)** k would stay the same. The number of treatments is unaffected by the shift in the data values. **(iii)** Recall that SSTR measures the variability among the sample means. Since the sample mean for the online scores is already the smallest, decreasing each of the online scores by 10 points will decrease the sample mean of the online scores by 10 points to $\bar{x} = 61.6667$. This would increase the variability in sample means. Thus, SSTR would increase. **(iv)** Recall that SSE measures the within-sample variability. Since each online score is decreased by 10, the decrease would have no effect on the spread of the sample. Therefore, SSE stays the same. **(v)** Use parts (iii) and (iv), as well as the identity SST = SSTR + SSE. If SSTR is increased and SSE stays the same, then SST will increase. **(vi)** Use (ii) and (iii), as well as the identity $= \frac{\text{SSTR}}{\text{df}_1}$. If SSTR increases while $\text{df}_1 = k - 1$ stays the same, then MSTR would also increase. **(vii)** Use parts (i), (ii), and (iv), as well as the identity $\text{MSE} = \frac{\text{SSE}}{\text{df}_2}$. If SSE and $\text{df}_2 = n_t - k$ stay the same, then MSE will stay the same. **(viii)** Use parts (vi) and (vii), as well as the identity $F_{data} = \frac{\text{MSTR}}{\text{MSE}}$. If MSTR is increases while MSE stays the same, then F_{data} would also increase.

(b) Since the original F_{data} was less than 6.36, our conclusion was to not reject H_0. From part (a) (viii), we know that F_{data} is increased. The new value for F_{data} is 2.346449094. This is still less than 6.36, so our conclusion would still be to not reject H_0.

25. (a) Since all of the sample means are the same, there is no variability among the sample means. Thus, SSTR = 0, MSTR = 0, and $F_{data} = 0$. **(b)** p-value $= P(F > F_{data}) = P(F > 0) = 1$ **(c)** Since p-value ≥ 0.05, we would not reject H_0.

27. (a) (i) $\text{df} = n_1 - 1 = 68 - 1 = 67, \bar{x}_1 \pm t_{\alpha/2}\left(\frac{s_1}{\sqrt{n_1}}\right) = 27.603 \pm 2.660\left(\frac{6.58}{\sqrt{68}}\right) \approx 27.603 \pm 2.1225 =$ (25.4805, 29.7255)

(ii) $\text{df} = n_2 - 1 = 79 - 1 = 78, \bar{x}_2 \pm t_{\alpha/2}\left(\frac{s_2}{\sqrt{n_2}}\right) = 30.451 \pm 2.648\left(\frac{6.09}{\sqrt{79}}\right) \approx 30.451 \pm 1.8144 =$ (28.6366, 32.2654)

(iii) $\text{df} = n_3 - 1 = 245 - 1 = 244, \bar{x}_3 \pm t_{\alpha/2}\left(\frac{s_3}{\sqrt{n_3}}\right) = 20.033 \pm 2.626\left(\frac{6.440}{\sqrt{245}}\right) \approx 20.033 \pm 1.0804 =$ (18.9526, 21.1134)

(b) The confidence interval for the population mean gas mileage of American cars does not overlap the other two confidence intervals. This is evidence against the null hypothesis that all of the population means are equal.

29. Since some of the boxplots do not overlap each other, there is evidence for differences in the population means.

31. From Minitab:

One-way ANOVA: CHEST_IN versus SIZE2

```
Source   DF       SS     MS     F      P
SIZE2     7   5001.2  714.5  9.07  0.000
Error   333  26236.5   78.8
Total   340  31237.7

S = 8.876   R-Sq = 16.01%   R-Sq(adj) = 14.24%


                                Individual 95% CIs For Mean Based on
                                Pooled StDev
Level   N    Mean   StDev   -------+---------+---------+---------+--
1      84  46.429   8.697               (--*---)
2      70  46.229   7.355               (--*---)
3      60  46.583   8.674               (---*--)
4      15  45.733   7.382         (------*-------)
5      13  42.462   6.306   (-------*-------)
6      30  54.233  12.094                             (----*-----)
7      36  50.583   8.936                    (----*----)
8      33  56.909  10.474                                  (----*----)
                            -------+---------+---------+---------+--
                                42.0      48.0      54.0      60.0

Pooled StDev = 8.876
```

(a) $H_0: \mu_1 = \mu_2 = \mu_3 = \mu_4 = \mu_5 = \mu_6 = \mu_7 = \mu_8$
H_a: Not all of the population means are equal.

(b) Reject H_0 if p-value < 0.05. **(c)** $F_{data} = 9.07$ **(d)** p-value ≈ 0 **(e)** Since p-value < 0.05, we reject H_0. **(f)** There is evidence that not all of the population means are equal.

(g)

Source	DF	SS	MS	F	P
SIZE2	7	5001.2	714.5	9.07	0.000
Error	333	26236.5	78.8		
Total	340	31237.7			

33. From Minitab:

One-way ANOVA: CHEST_IN versus PROTECT2

```
Source     DF       SS     MS     F      P
PROTECT2    3    844.4  281.5  3.12  0.026
Error     334  30168.7   90.3
Total     337  31013.1

S = 9.504   R-Sq = 2.72%   R-Sq(adj) = 1.85%


                                 Individual 99% CIs For Mean Based on Pooled StDev
Level    N    Mean   StDev   -+---------+---------+---------+--------
1      193  49.611  10.301                          (------*-------)
2       41  47.707   8.044               (---------------*--------------)
3       46  48.022   9.614                  (-------------*--------------)
4       58  45.362   7.314    (-----------*------------)
                             -+---------+---------+---------+--------
                            42.5      45.0      47.5      50.0

Pooled StDev = 9.504
```

(a) $H_0: \mu_1 = \mu_2 = \mu_3 = \mu_4$
H_a : Not all of the population means are equal.
(b) Reject H_0 if p-value < 0.01. **(c)** $F_{data} = 3.12$ **(d)** p-value $= 0.026$ **(e)** Since p-value ≥ 0.01, we do not reject H_0. **(f)** There is insufficient evidence that not all of the population means are equal.

(g)

Source	DF	SS	MS	F	P
PROTECT2	3	844.4	281.5	3.12	0.026
Error	334	30168.7	90.3		
Total	337	31013.1			

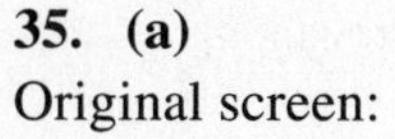

35. (a)

Original screen:

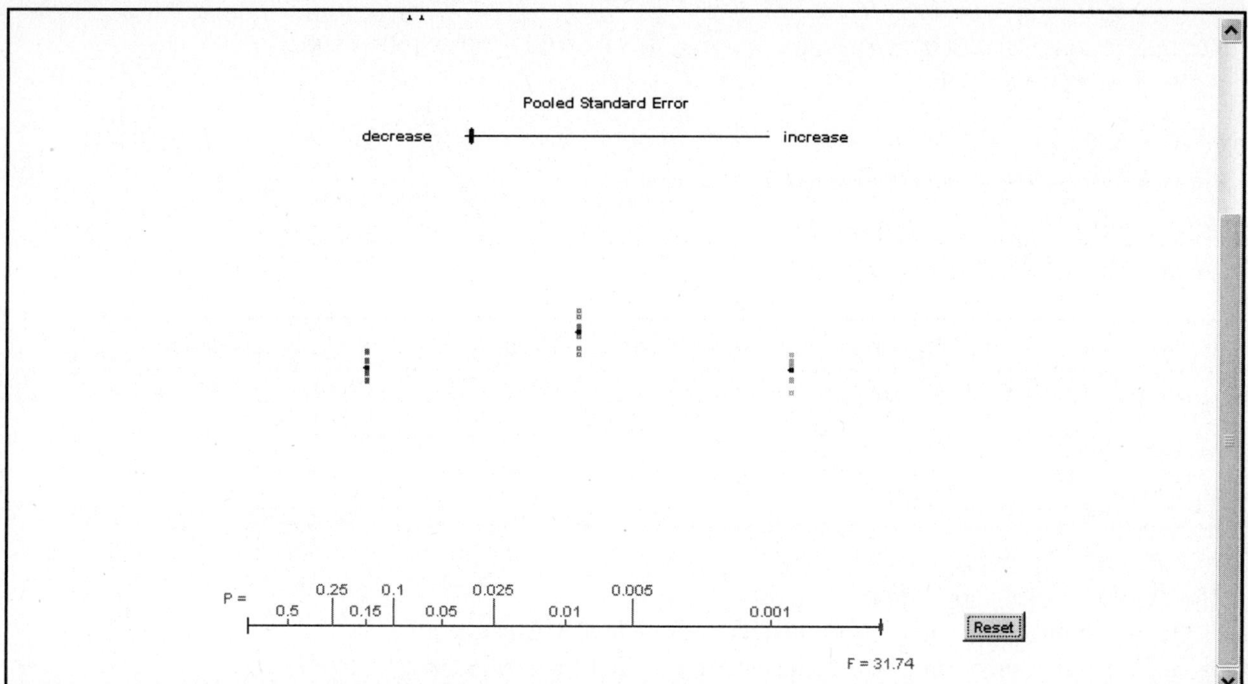

Screen after black dots are made almost level horizontally:

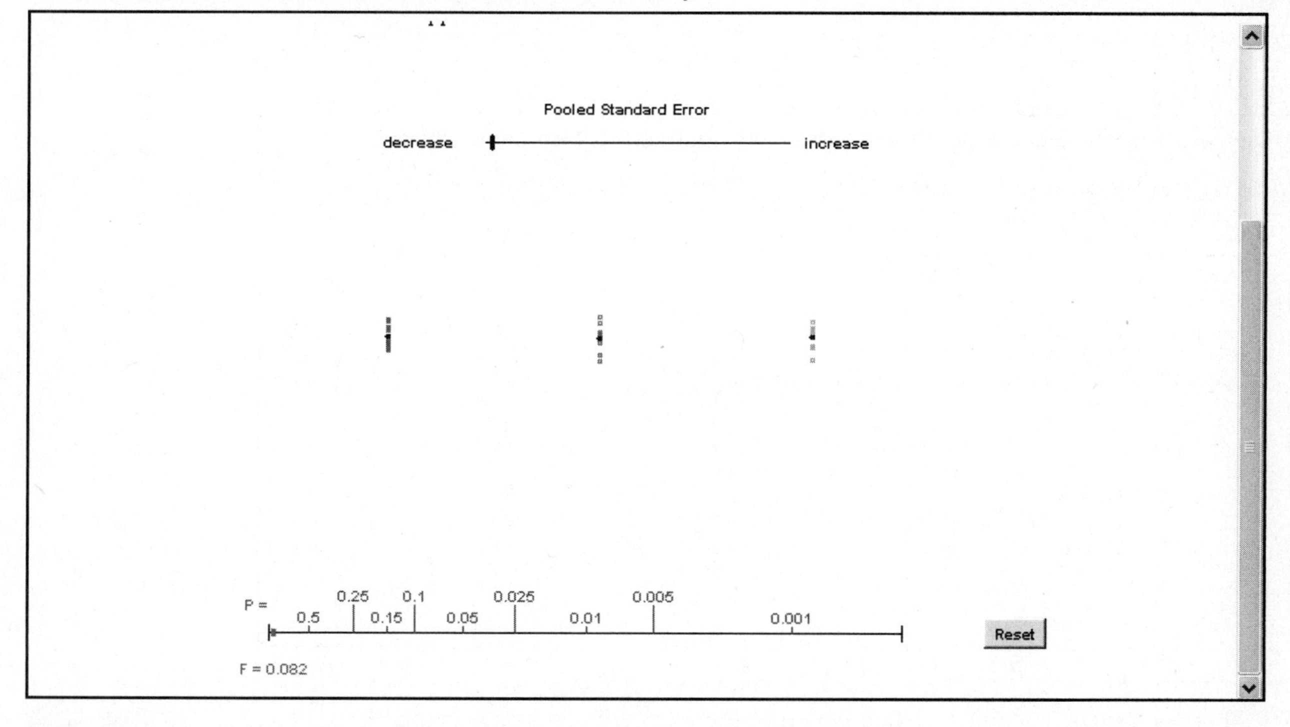

F decreases.

(b) If all of the sample means are about the same, then the between sample variability is small and all of the sample means are close to $\bar{\bar{x}}$. Therefore, SSTR $= \sum n_i(\bar{x}_i - \bar{\bar{x}})^2$ decreases. Since the number of observations in each sample remains the same, the total number of observations remains the same. The number of treatments also remains the same. Therefore, MSTR $= \dfrac{\text{SSTR}}{n_t - k}$ decreases. Since the within sample variability does not change, SSE does not change. Therefore, MSE $= \dfrac{\text{SSE}}{k - 1}$ does not change. Thus, $F_{data} = \dfrac{\text{MSTR}}{\text{MSE}}$ decreases.

Chapter 12 Review

1. $df_1 = k - 1 = 4 - 1 = 3$, $n_t = n_A + n_B + n_C + n_D = 50 + 100 + 50 + 100 = 300$,
$df_2 = n_t - k = 300 - 4 = 296$

3. $SSTR = \sum n_i(\bar{x}_i - \bar{\bar{x}})^2 = 50(0 - 10)^2 + 100(10 - 10)^2 + 50(20 - 10)^2 + 100(10 - 10)^2 = 10{,}000$

5. $SST = SSTR + SSE = 10{,}000 + 1157.5 = 11{,}157.5.$

7. $MSE = \frac{SSE}{df_2} = \frac{1157.5}{296} = 3.910472973$

9.

Source of variation	Sum of squares	Degrees of freedom	Mean square	*F*-test statistic
Treatment	SSTR = 10,000	$df_1 = 3$	MSTR = 3333.3333	$F_{data} = 852.4117985$
Error	SSE = 1157.5	$df_2 = 296$	MSE = 3.910472973	
Total	SST = 11,157.5			

11. **(a)** $H_0: \mu_1 = \mu_2 = \mu_3$
H_a: Not all of the population means are equal.
where μ_1 = population mean level of satisfaction for Medical Treatment 1
μ_2 = population mean level of satisfaction for Medical Treatment 2
μ_3 = population mean level of satisfaction for Medical Treatment 3

(b) $df_1 = 2$, $df_2 = 18$, $\alpha = 0.05$, $F_{crit} = 3.49$

(c) Reject H_0 if $F_{data} > 3.49$.

(d) From Minitab:

One-way ANOVA: Medical Treatment 1, Medical Treatment 2, Medical Treatment 3

```
Source  DF     SS    MS     F      P
Factor   2   4114  2057  3.19  0.065
Error   18  11600   644
Total   20  15714

S = 25.39   R-Sq = 26.18%   R-Sq(adj) = 17.98%

                                               Individual 95% CIs For Mean Based on
                                               Pooled StDev
Level                  N   Mean  StDev   ------+---------+---------+---------+---
Medical Treatment 1    7  48.57  32.50   (---------*---------)
Medical Treatment 2    7  65.71  23.70            (---------*---------)
Medical Treatment 3    7  82.86  17.76                     (----------*----------)
                                         ------+---------+---------+---------+---
                                              40        60        80       100

Pooled StDev = 25.39
```

(e) From (d), $F_{data} = 3.19$. Since $F_{data} \le 3.49$, we do not reject H_0.

(f) There is insufficient evidence that not all of the population means are equal.

(g)

```
Source  DF     SS    MS     F      P
Factor   2   4114  2057  3.19  0.065
Error   18  11600   644
Total   20  15714
```

Or using a calculator:
$\bar{x}_1 = 48.5714, s_1 = 32.4954, n_1 = 7$
$\bar{x}_2 = 65.7143, s_2 = 23.7045, n_2 = 7$
$\bar{x}_3 = 82.8571, s_3 = 17.7616, n_3 = 7$
(i) $df_1 = k - 1 = 3 - 1 = 2, n_t = 7 + 7 + 7 = 21$
$df_2 = n_t - k = 21 - 3 = 18$
(ii) $\bar{\bar{x}} = \dfrac{48.5714 + 65.7143 + 82.8571}{3} = 65.7142667$
(iii) $\text{SSTR} = \sum n_i(\bar{x}_i - \bar{\bar{x}})^2 = 7(48.5714 - 65.7142667)^2 + 7(65.7143 - 65.7142667)^2 + 7(82.8571 - 65.7142667)^2 = 4114.282286$
(iv) $\text{SSE} = \sum(n_i - 1)s_i^2 = (7 - 1)32.4954^2 + (7 - 1)23.7045^2 + (7 - 1)17.7616^2 = 11{,}599.97266$
(v) $\text{SST} = \text{SSTR} + \text{SSE} = 4114.282286 + 11{,}599.97266 = 15{,}714.25494$
(vi) $\text{MSTR} = \dfrac{\text{SSTR}}{df_1} = \dfrac{4114.282286}{2} = 2057.141143$
(vii) $\text{MSE} = \dfrac{\text{SSE}}{df_2} = \dfrac{11{,}599.97266}{18} = 644.4429256$
(viii) $F_{data} = \dfrac{\text{MSTR}}{\text{MSE}} = \dfrac{2057.141143}{644.4429256} \approx 3.1921$

Source	df	SS	MS	*F*	*P*
Factor	2	4114	2057	3.19	0.065
Error	18	11600	644		
Total	20	15714			

13. (a) $H_0: \mu_A = \mu_B = \mu_C = \mu_D = \mu_E$
H_a: Not all of the population means are equal.
μ_A = the population mean price per pound (in dollars) of apples from Supermarket A
μ_B = the population mean price per pound (in dollars) of apples from Supermarket B
μ_C = the population mean price per pound (in dollars) of apples from Supermarket C
μ_D = the population mean price per pound (in dollars) of apples from Supermarket D
μ_E = the population mean price per pound (in dollars) of apples from Supermarket E
Reject H_0 if p-value < 0.05.
(b) From Minitab:

One-way ANOVA: A, B, C, D, E

```
Source  DF      SS     MS      F      P
Factor   4  10.395  2.599  17.53  0.000
Error   22   3.262  0.148
Total   26  13.656

S = 0.3850   R-Sq = 76.12%   R-Sq(adj) = 71.77%
```

$F_{data} = 17.53$
(c) p-value ≈ 0
(d) Since p-value < 0.05, we reject H_0. There is evidence that not all of the population means are equal.
(e)

```
Source  DF      SS     MS      F      P
Factor   4  10.395  2.599  17.53  0.000
Error   22   3.262  0.148
Total   26  13.656
```

Or using a calculator:
$\bar{x}_A = 1.6333$, $s_A = 0.4719$, $n_A = 6$
$\bar{x}_B = 2.28$, $s_B = 0.4207$, $n_B = 5$
$\bar{x}_C = 2.4833$, $s_C = 0.3764$, $n_C = 6$
$\bar{x}_D = 3$, $s_D = 0.3240$, $n_D = 5$
$\bar{x}_E = 3.44$, $s_E = 0.2793$, $n_E = 5$
(i) $\text{df}_1 = k - 1 = 5 - 1 = 4$, $n_t = 6 + 5 + 6 + 5 + 5 = 27$
$\text{df}_2 = n_t - k = 27 - 5 = 22.$

(ii) $\bar{\bar{x}} = \dfrac{6(1.6333) + 5(2.28) + 6(2.4833) + 5(3) + 5(3.44)}{27} = 2.529614815$

(iii) $\text{SSTR} = \sum n_i(\bar{x}_i - \bar{\bar{x}})^2 = 6(1.6333 - 2.529614815)^2 + 5(2.28 - 2.529614815)^2 + 6(2.4833 - 2.529614815)^2 + 5(3 - 2.529614815)^2 + 5(3.44 - 2.529614815)^2 = 10.39500667$

(iv) $\text{SSE} = \sum(n_i - 1)s_i^2 = (6 - 1)\,0.4719^2 + (5 - 1)\,0.4207^2 + (6 - 1)\,0.3764^2 + (5 - 1)\,0.3240^2 + (5 - 1)\,0.2793^2 = 3.26172477$

(v) $\text{SST} = \text{SSTR} + \text{SSE} = 10.39500667 + 3.26172477 = 13.65673144$

(vi) $\text{MSTR} = \dfrac{\text{SSTR}}{\text{df}_1} = \dfrac{10.39500667}{4} = 2.598766675$

(vii) $\text{MSE} = \dfrac{\text{SSE}}{\text{df}_2} = \dfrac{3.26172477}{22} = 0.1482602168$

(viii) $F_{data} = \dfrac{\text{MSTR}}{\text{MSE}} = \dfrac{2.598766675}{0.1482602168} \approx 17.5284$

Source of variation	Sum of squares	Degrees of freedom	Mean square	F-test statistic
Treatment	SSTR = 10.39500667	$\text{df}_1 = 4$	MSTR = 2.598766675	$F_{data} = 17.5284$
Error	SSE = 3.26172477	$\text{df}_2 = 22$	MSE = 0.1482602168	
Total	SST = 13.65673144			

Chapter 12 Quiz

1. False.
2. True.
3. False. If we reject the null hypothesis in an ANOVA, we conclude that not all of the population means are different. This means that at least one of the population means is different from the rest.
4. Mean square.
5. Mean square treatment.
6. Mean square error.
7. $\bar{\bar{x}}$
8. Mean square equals the sum of squares divided by the degrees of freedom.
9. F_{data}
10. **(a)** **(i)** $\text{df}_1 = k - 1 = 3 - 1 = 2$, $n_t = 199 + 83 + 103 = 385$
$\text{df}_2 = n_t - k = 385 - 3 = 382$

(ii) $\bar{\bar{x}} = \dfrac{199(29.3) + 83(20.0) + 103(15.0)}{385} = 23.46935065$

(iii) $\text{SSTR} = \sum n_i(\bar{x}_i - \bar{\bar{x}})^2 = 199(29.3 - 23.46935065)^2 + 83(20.0 - 23.46935065)^2 + 103(15.0 - 23.46935065)^2 = 15{,}152.49833$

(iv) $\text{SSE} = \sum(n_i - 1)s_i^2 = (199 - 1)5.7^2 + (83 - 1)3.8^2 + (103 - 1)2.9^2 = 8474.92$

(v) $\text{SST} = \text{SSTR} + \text{SSE} = 15{,}152.49833 + 8474.92 = 23{,}627.41833$

(vi) $\text{MSTR} = \dfrac{\text{SSTR}}{\text{df}_1} = \dfrac{15{,}152.49833}{2} = 7576.249165.$

(vii) $\text{MSE} = \dfrac{\text{SSE}}{\text{df}_2} = \dfrac{8474.92}{382} = 22.1856445$

(viii) $F_{data} = \dfrac{\text{MSTR}}{\text{MSE}} = \dfrac{7576.249165}{22.1856445} \approx 341.4931564$

(b)

Source of variation	Sum of squares	Degrees of freedom	Mean square	*F*-test statistic
Treatment	SSTR = 15,152.49833	$\text{df}_1 = 2$	MSTR = 7576.249165	$F_{data} = 341.4931564$
Error	SSE = 8474.92	$\text{df}_2 = 382$	MSE = 22.1856445	
Total	SST = 23,627.41833			

11. **(a)** **(i)** $\text{df}_1 = k - 1 = 5 - 1 = 4$, $n_t = 964 + 72 + 342 + 79 + 478 = 1935$
$\text{df}_2 = n_t - k = 1935 - 5 = 1930$

(ii) $\bar{\bar{x}} = \dfrac{964(42.76) + 72(40.13) + 342(43.69) + 79(41.66) + 478(41.03)}{1935} = 42.35424289$

(iii) $\text{SSTR} = \sum n_i(\bar{x}_i - \bar{\bar{x}})^2 = 964(42.76 - 42.35424289)^2 + 72(40.13 - 42.35424289)^2 +$
$342(43.69 - 42.35424289)^2 + 79(41.66 - 42.35424289)^2 + 478(41.03 - 42.35424289)^2 = 2001.432666$

(iv) $\text{SSE} = \sum(n_i - 1)s_i^2 = (964 - 1)\,14.08^2 + (72 - 1)\,14.28^2 + (342 - 1)\,13.93^2 + (79 - 1)\,15.71^2 +$
$(478 - 1)\,14.03^2 = 384{,}702.6296.$

(v) $\text{SST} = \text{SSTR} + \text{SSE} = 2001.432666 + 384{,}702.6296 = 386{,}704.0623$

(vi) $\text{MSTR} = \dfrac{\text{SSTR}}{\text{df}_1} = \dfrac{2001.432666}{4} = 500.3581665$

(vii) $\text{MSE} = \dfrac{\text{SSE}}{\text{df}_2} = \dfrac{386{,}704.0623}{1930} = 199.3277874$

(viii) $F_{data} = \dfrac{\text{MSTR}}{\text{MSE}} = \dfrac{7576.249165}{22.1856445} \approx 2.510227871$

(b)

Source of variation	Sum of squares	Degrees of freedom	Mean square	*F*-test statistic
Treatment	SSTR = 2001.432669	$\text{df}_1 = 4$	MSTR = 500.3581674	$F_{data} = 2.510227871$
Error	SSE = 384,702.6296	$\text{df}_2 = 1930$	MSE = 199.3277874	
Total	SST = 386,704.0623			

12. **(a)** **(i)** $\text{df}_1 = k - 1 = 3 - 1 = 2$, $n_t = 23 + 8 + 8 = 39$
$\text{df}_2 = n_t - k = 39 - 3 = 36$

(ii) $\bar{\bar{x}} = \dfrac{23(109) + 8(95) + 8(115)}{39} = 107.3589744$

(iii) $\text{SSTR} = \sum n_i(\bar{x}_i - \bar{\bar{x}})^2 = 23(109 - 107.3589744)^2 + 8(95 - 107.3589744)^2 +$
$8(115 - 107.3589744)^2 = 1750.974357$

(iv) $\text{SSE} = \sum(n_i - 1)s_i^2 = (23 - 1)\,22^2 + (8 - 1)\,29^2 + (8 - 1)\,23^2 = 20{,}238$

(v) $\text{SST} = \text{SSTR} + \text{SSE} = 1750.974357 + 20{,}238 = 21{,}988.97436$

(vi) $\text{MSTR} = \dfrac{\text{SSTR}}{\text{df}_1} = \dfrac{1750.974357}{2} = 875.4871785$

(vii) $\text{MSE} = \dfrac{\text{SSE}}{\text{df}_2} = \dfrac{20{,}238}{36} = 562.1666667$

(viii) $F_{data} = \dfrac{\text{MSTR}}{\text{MSE}} = \dfrac{875.4871785}{562.1666667} \approx 1.557344523$

(b)

Source of variation	Sum of squares	Degrees of freedom	Mean square	*F*-test statistic
Treatment	SSTR = 1750.974359	$\text{df}_1 = 2$	MSTR = 875.4871795	$F_{data} = 1.557344523$
Error	SSE = 20,238	$\text{df}_2 = 36$	MSE = 562.1666667	
Total	SST = 21,988.97436			

13. (a) (i) $\text{df}_1 = k - 1 = 5 - 1 = 4$, $n_t = 6 + 5 + 6 + 5 + 5 = 27$
$\text{df}_2 = n_t - k = 27 - 5 = 22$

(ii) $\bar{\bar{x}} = \dfrac{6(1.5) + 5(2.0) + 6(2.5) + 5(3.0) + 5(3.5)}{27} = 2.462962963$

(iii) $\text{SSTR} = \sum n_i(\bar{x}_i - \bar{\bar{x}})^2 = 6(1.5 - 2.462962963)^2 + 5(2.0 - 2.462962963)^2 + 6(2.5 - 2.462962963)^2 + 5(3.0 - 2.462962963)^2 + 5(3.5 - 2.462962963)^2 = 13.46296296$

(iv) $\text{SSE} = \sum(n_i - 1)s_i^2 = (6 - 1)\,0.5477^2 + (5 - 1)\,0.3082^2 + (6 - 1)\,0.3950^2 + (5 - 1)\,0.3240^2 + (5 - 1)\,0.3082^2 = 3.45980337$

(v) $\text{SST} = \text{SSTR} + \text{SSE} = 13.46296296 + 3.45980337 = 16.92276633$

(vi) $\text{MSTR} = \dfrac{\text{SSTR}}{\text{df}_1} = \dfrac{13.46296296}{4} = 3.36574074$

(vii) $\text{MSE} = \dfrac{\text{SSE}}{\text{df}_2} = \dfrac{3.45980337}{22} = 0.1572637895$

(viii) $F_{data} = \dfrac{\text{MSTR}}{\text{MSE}} = \dfrac{3.36574074}{0.1572637895} \approx 21.40187993$

(b)

Source of variation	Sum of squares	Degrees of freedom	Mean square	*F*-test statistic
Treatment	SSTR = 13.46296296	$\text{df}_1 = 4$	MSTR = 3.365740741	$F_{data} = 21.40187993$
Error	SSE = 3.45980337	$\text{df}_2 = 22$	MSE = 0.1572637895	
Total	SST = 16.92276633			

14. (a) $H_0: \mu_4 = \mu_6 = \mu_8$
H_a : Not all of the population means are equal.
μ_4 = the population mean miles per gallon for 4-cylinder cars
μ_6 = the population mean miles per gallon for 6-cylinder cars
μ_8 = the population mean miles per gallon for 8-cylinder cars
Reject H_0 if p-value < 0.05.
(b) From Exercise 10, $F_{data} = 341.4932$.
(c) Since $\text{df}_1 = 3$, $\text{df}_2 = 126$, and $F_{data} = 341.4932 > 5.86$, p-value < 0.001.
(d) Since p-value < 0.05, we reject H_0. There is evidence that not all of the population mean gas mileages are equal.

15. (a) $H_0: \mu_{Married} = \mu_{Widowed} = \mu_{Divorced} = \mu_{Separated} = \mu_{Nevermarried}$
H_a : Not all of the population means are equal.

$\mu_{Married}$ = the population mean number of hours worked by people who are married
$\mu_{Widowed}$ = the population mean number of hours worked by people who are widowed
$\mu_{Divorced}$ = the population mean number of hours worked by people who are divorced
$\mu_{Separated}$ = the population mean number of hours worked by people who are separated
$\mu_{Nevermarried}$ = the population mean number of hours worked by people who have never been married
Reject H_0 if p-value < 0.05.
(b) From Exercise 11, $F_{data} = 2.5102$.
(c) Since $df_1 = 4$, $df_2 = 1930$, and $2.38 < F_{data} = 2.5102 < 2.80$, $0.025 < p\text{-value} < 0.05$.
(d) Since p-value < 0.05, we reject H_0. There is evidence that not all of the population mean numbers of hours worked are equal.

16. (a) $H_0: \mu_{Kelloggs} = \mu_{Quaker} = \mu_{RalstonPurina}$
H_a : Not all of the population means are equal.
$\mu_{Kelloggs}$ = the population mean number of calories per serving in breakfast cereals made by Kellogg's
μ_{Quaker} = the population mean number of calories per serving in breakfast cereals made by Quaker
$\mu_{RalstonPurina}$ = the population mean number of calories per serving in breakfast cereals made by Ralston Purina
(b) $df_1 = 2$, $df_2 = 36$, $\alpha = 0.05$, $F_{crit} = 3.39$. Reject H_0 if $F_{data} > 3.39$.
(c) From Exercise 12, $F_{data} = 1.5573$.
(d) Since $F_{data} \le 3.39$, we do not reject H_0. There is insufficient evidence that not all of the population mean numbers of calories per serving are equal.

17. (a) $H_0: \mu_{Freshmen} = \mu_{Sophomores} = \mu_{Juniors} = \mu_{Seniors} = \mu_{GraduateStudents}$
H_a : Not all of the population means are equal.
$\mu_{Freshmen}$ = the population mean GPA of freshmen at this university
$\mu_{Sophomores}$ = the population mean GPA of sophomores at this university
$\mu_{Juniors}$ = the population mean GPA of juniors at this university
$\mu_{Seniors}$ = the population mean GPA of seniors at this university
$\mu_{GraduateStudents}$ = the population mean GPA of graduate students at this university
(b) $df_1 = 4$, $df_2 = 22$, $\alpha = 0.05$, $F_{crit} = 2.87$.
Reject H_0 if $F_{data} > 2.87$.
(c) From Exercise 13, $F_{data} = 21.4019$.
(d) Since $F_{data} > 2.87$, we reject H_0. There is evidence that not all of the population mean GPAs are equal.

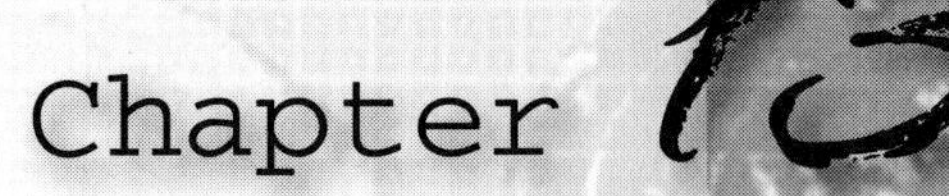

Regression Analysis

Section 13.1

1. The y-intercept b_0 is interpreted as "the estimated y when x equals zero." The slope b_1 is interpreted as "the estimated change in y for a unit increase in x."

3. Out of all possible straight lines, the least squares criterion chooses the line with the smallest SSE (Sum of Squared Error).

5. The standard error of the estimate s is a measure of the size of the typical difference between the predicted value of y and the observed value of y.

7. (a)

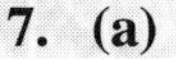

(b)

x	y	xy	x^2	y^2
1	15	15	1	225
2	20	40	4	400
3	20	60	9	400
4	25	100	16	625
5	25	125	25	625
$\sum x = 15$	$\sum y = 105$	$\sum xy = 340$	$\sum x^2 = 55$	$\sum y^2 = 2{,}275$

(c) $b_1 = \dfrac{\sum xy - \left(\sum x\right)\left(\sum y\right)/n}{\sum x^2 - \left(\sum x\right)^2/n} = \dfrac{340 - (15)(105)/5}{55 - (15)^2/5} = \dfrac{25}{10} = 2.5$

Now, $\bar{x} = \dfrac{\sum x}{n} = \dfrac{15}{5} = 3$ and $\bar{y} = \dfrac{\sum y}{n} = \dfrac{105}{5} = 21$ giving,

$b_0 = \bar{y} - (b_1 \cdot \bar{x}) = 21 - (2.5)(3) = 13.5$

So $b_0 = 13.5$, $b_1 = 2.5$, and $\hat{y} = 13.5 + 2.5\,x$.

9. (a)

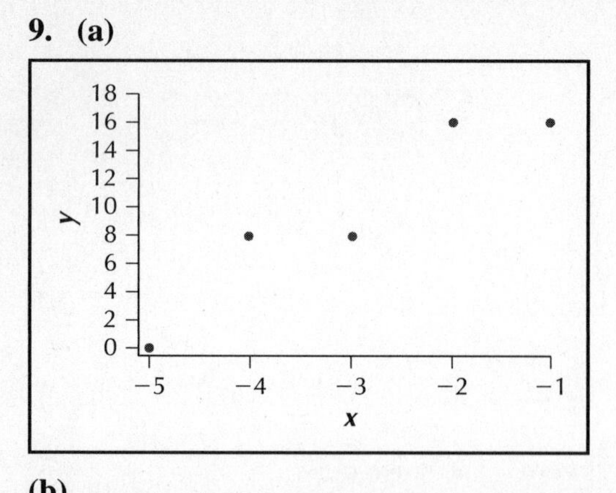

(b)

x	y	xy	x^2	y^2
−5	0	0	25	0
−4	8	−32	16	64
−3	8	−24	9	64
−2	16	−32	4	256
−1	16	−16	1	256
$\sum x = -15$	$\sum y = 48$	$\sum xy = -104$	$\sum x^2 = 55$	$\sum y^2 = 640$

(c) $b_1 = \dfrac{\sum xy - \left(\sum x\right)\left(\sum y\right)/n}{\sum x^2 - \left(\sum x\right)^2/n} = \dfrac{-104 - (-15)(48)/5}{55 - (-15)^2/5} = \dfrac{40}{10} = 4$

Now, $\bar{x} = \dfrac{\sum x}{n} = \dfrac{-15}{5} = -3$ and $\bar{y} = \dfrac{\sum y}{n} = \dfrac{48}{5} = 9.6$ giving,
$b_0 = \bar{y} - (b_1 \cdot \bar{x}) = 9.6 - (4)(-3) = 21.6$
So $b_0 = 21.6$, $b_1 = 4$, and $\hat{y} = 21.6 + 4x$.

11. (a)

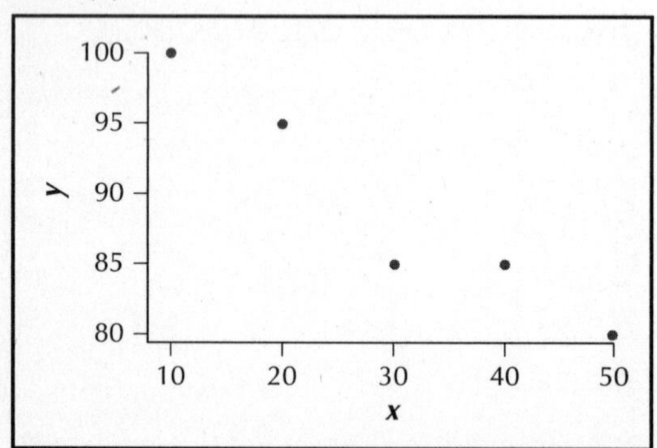

(b)

x	y	xy	x^2	y^2
10	100	1,000	100	10,000
20	95	1,900	400	9,025
30	85	2,550	900	7,225
40	85	3,400	1,600	7,225
50	80	4,000	2,500	6,400
$\sum x = 150$	$\sum y = 445$	$\sum xy = 12{,}850$	$\sum x^2 = 5{,}500$	$\sum y^2 = 39{,}875$

(c) $b_1 = \dfrac{\sum xy - \left(\sum x\right)\left(\sum y\right)/n}{\sum x^2 - \left(\sum x\right)^2/n} = \dfrac{12{,}850 - (150)(445)/5}{5{,}500 - (150)^2/5} = \dfrac{-500}{1000} = -0.5$

Now, $\bar{x} = \dfrac{\sum x}{n} = \dfrac{150}{5} = 30$ and $\bar{y} = \dfrac{\sum y}{n} = \dfrac{445}{5} = 89$ giving, $b_0 = \bar{y} - (b_1 \cdot \bar{x}) = 89 - (-0.5)(30) = 104$

So $b_0 = 104$, $b_1 = -0.5$, and $\hat{y} = 104 - 0.5\,x$.

13. (a)

x	y	**Predicted value** $\hat{y} = 13.5 + 2.5\,x$	**Residual** $(y - \hat{y})$	**(Residual)²** $(y - \hat{y})^2$
1	15	16	−1	1
2	20	18.5	1.5	2.25
3	20	21	−1	1
4	25	23.5	1.5	2.25
5	25	26	−1	1

(b) SSE $= \sum(y - \hat{y})^2 = 7.5$. The quantity SSE measures the sum of the squared distances between the predicted value $\hat{y}$ and the actual value y.

(c) $n = 5$, $s = \sqrt{\dfrac{\text{SSE}}{n-2}} = \sqrt{\dfrac{7.5}{5-2}} \approx 1.5811$. The typical error in prediction is 1.5811.

15. (a)

x	y	**Predicted value** $\hat{y} = 21.6 + 4x$	**Residual** $(y - \hat{y})$	**(Residual)²** $(y - \hat{y})^2$
−5	0	1.6	−1.6	2.56
−4	8	5.6	2.4	5.76
−3	8	9.6	−1.6	2.56
−2	16	13.6	2.4	5.76
−1	16	17.6	−1.6	2.56

(b) SSE $= \sum(y - \hat{y})^2 = 19.2$. The quantity SSE measures the sum of the squared distances between the predicted value $\hat{y}$ and the actual value y.

(c) $n = 5$, $s = \sqrt{\dfrac{\text{SSE}}{n-2}} = \sqrt{\dfrac{19.2}{5-2}} \approx 2.5298$. The typical error in prediction is 2.5298.

17. (a)

x	y	**Predicted value** $\hat{y} = 104 - 0.5x$	**Residual** $(y - \hat{y})$	**(Residual)²** $(y - \hat{y})^2$
10	100	99	1	1
20	95	94	1	1
30	85	89	−4	16
40	85	84	1	1
50	80	79	1	1

(b) SSE $= \sum(y - \hat{y})^2 = 20$. The quantity SSE measures the sum of the squared distances between the predicted value $\hat{y}$ and the actual value y.

(c) $s = \sqrt{\dfrac{\text{SSE}}{n-2}} = \sqrt{\dfrac{20}{5-2}} \approx 2.5820$. The typical error in prediction is 2.5820.

19. (a)

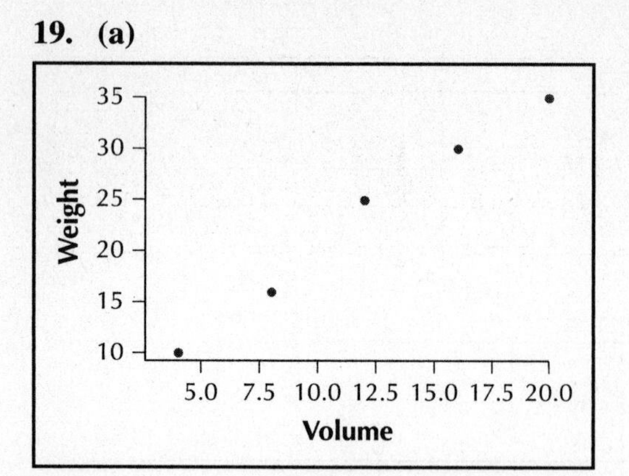

(b)

x	y	xy	x^2	y^2
4	10	40	16	100
8	16	128	64	256
12	25	300	144	625
16	30	480	256	900
20	35	700	400	1,225
$\sum x = 60$	$\sum y = 116$	$\sum xy = 1{,}648$	$\sum x^2 = 880$	$\sum y^2 = 3{,}106$

$$b_1 = \frac{\sum xy - \left(\sum x\right)\left(\sum y\right)/n}{\sum x^2 - \left(\sum x\right)^2/n} = \frac{1{,}648 - (60)(116)/5}{880 - (60)^2/5} = \frac{256}{160} = 1.6$$

Now, $\bar{x} = \frac{\sum x}{n} = \frac{60}{5} = 12$ and $\bar{y} = \frac{\sum y}{n} = \frac{116}{5} = 23.2$ giving, $b_0 = \bar{y} - (b_1 \cdot \bar{x}) = 23.2 - (1.6)(12) = 4$.
So $b_0 = 4$, $b_1 = 1.6$, and $\hat{y} = 4 + 1.6x$.

(c) $b_0 = 4$ means that the estimated weight of a package that has a volume of 0 cubic meters is 4 kg.
$b_1 = 1.6$ means that for every additional cubic meter of volume the estimated weight of the package is increased by 1.6 kg.

(d)

x	y	**Predicted value** $\hat{y} = 4 + 1.6x$	**Residual** $(y - \hat{y})$	**(Residual)^2** $(y - \hat{y})^2$
4	10	10.4	−0.4	0.16
8	16	16.8	−0.8	0.64
12	25	23.2	1.8	3.24
16	30	29.6	0.4	0.16
20	35	36	−1	1

$$\text{SSE} = \sum(y - \hat{y})^2 = 5.2,\ s = \sqrt{\frac{\text{SSE}}{n-2}} = \sqrt{\frac{5.2}{5-2}} \approx 1.3166 \text{ kg.}$$

If we know the volume (x) of the package, then our estimate of the weight of the package will typically differ from the actual weight of the package by 1.3166 kg.

21. (a)

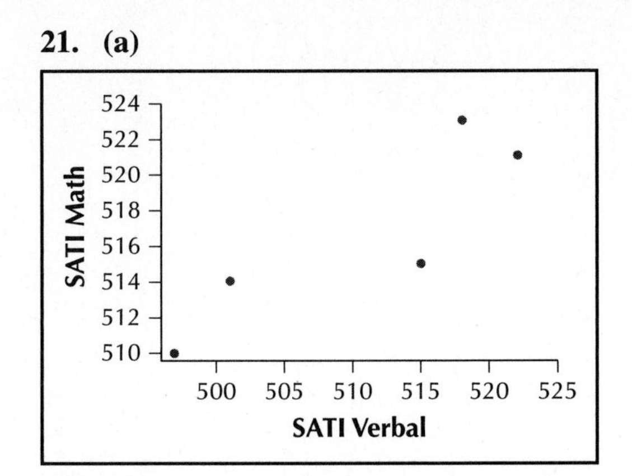

(b)

x	y	xy	x^2	y^2
497	510	253,470	247,009	260,100
515	515	265,225	265,225	265,225
518	523	270,914	268,324	273,529
501	514	257,514	251,001	264,196
522	521	271,962	272,484	271,441
$\Sigma x = 2{,}553$	$\Sigma y = 2{,}583$	$\Sigma xy = 1{,}319{,}085$	$\Sigma x^2 = 1{,}304{,}043$	$\Sigma y^2 = 1{,}334{,}491$

$$b_1 = \frac{\sum xy - \left(\sum x\right)\left(\sum y\right)/n}{\sum x^2 - \left(\sum x\right)^2/n} = \frac{1{,}319{,}085 - (2553)(2583)/5}{1{,}304{,}043 - (2553)^2/5} = \frac{205.2}{481.2} = 0.4264339152 \approx 0.4264$$

Now, $\bar{x} = \frac{\sum x}{n} = \frac{2553}{5} = 510.6$ and $\bar{y} = \frac{\sum y}{n} = \frac{2583}{5} = 516.6$ giving,

$b_0 = \bar{y} - (b_1 \cdot \bar{x}) = 516.6 - (0.4264339152)\,(510.6) = 298.8628429 \approx 298.8628.$

So $b_0 = 298.8628$, $b_1 = 0.4264$, and $\hat{y} = 298.8628 + 0.4264x$.

(c) $b_0 = 298.8628$ means that the estimated SAT I Math score for a student that has a SAT I Verbal score of 0 is 298.8628. $b_1 = 0.4264$ means that for every additional point in the SAT I Verbal score the estimated SAT I Math score is increased by 0.4264 points.

(d)

x	y	**Predicted Value** $\hat{y} = 298.8628 + 0.4264x$	**Residual** $(y - \hat{y})$	**(Residual)^2** $(y - \hat{y})^2$
497	510	510.7836	−0.7836	0.61402896
515	515	518.4588	−3.4588	11.96329744
518	523	519.738	3.262	10.640644
501	514	512.4892	1.5108	2.28251664
522	521	521.4436	−0.4436	0.19678096

$$\text{SSE} = \sum (y - \hat{y})^2 = 25.697268,\ s = \sqrt{\frac{\text{SSE}}{n-2}} = \sqrt{\frac{25.697268}{5-2}} \approx 2.9267 \text{ points.}$$

If we know the SAT I Verbal score (x), then our estimate of the SAT I Math score will typically differ from the actual SAT I Math score by 2.9267 points.

23. (a)

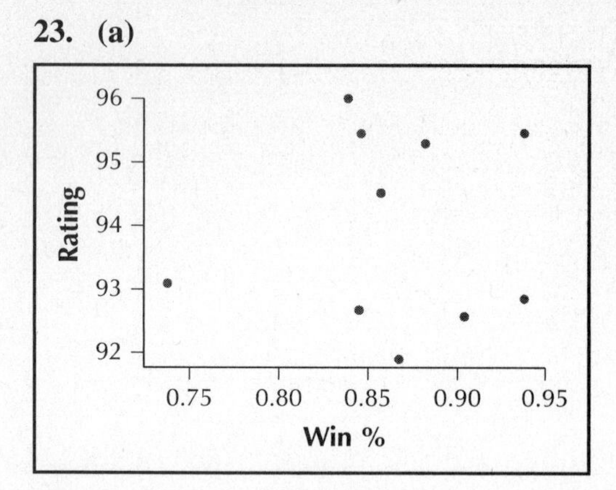

(b)

x	y	xy	x^2	y^2
0.838	96.020	80.46476	0.702244	9219.8404
0.938	95.493	89.572434	0.879844	9118.913049
0.846	95.478	80.774388	0.715716	9116.048484
0.882	95.320	84.07224	0.777924	9085.9024
0.857	94.541	81.021637	0.734449	8938.000681
0.737	93.091	68.608067	0.543169	8665.934281
0.938	92.862	87.104556	0.879844	8623.351044
0.844	92.692	78.232048	0.712336	8591.806864
0.903	92.609	83.625927	0.815409	8576.426881
0.867	91.912	79.687704	0.751689	8447.815744
$\sum x = 8.65$	$\sum y = 940.018$	$\sum xy = 813.163761$	$\sum x^2 = 7.512624$	$\sum y^2 = 88{,}384.03983$

$$b_1 = \frac{\sum xy - \left(\sum x\right)\left(\sum y\right)/n}{\sum x^2 - \left(\sum x\right)^2/n} = \frac{813.163761 - (8.65)(940.018)/10}{7.512624 - (8.65)^2/10} = \frac{0.048191}{0.030374} = 1.586587213 \approx 1.5866$$

Now, $\bar{x} = \frac{\sum x}{n} = \frac{8.65}{10} = 0.865$ and $\bar{y} = \frac{\sum y}{n} = \frac{940.018}{10} = 94.0018$ giving,

$b_0 = \bar{y} - (b_1 \cdot \bar{x}) = 94.0018 - (1.586587213)(0.865) = 92.62940206 \approx 92.6294.$

So $b_0 = 92.6294$, $b_1 = 1.5866$, and $\hat{y} = 92.6294 + 1.5866\,x$.

(c) $b_0 = 92.6294$ means that the estimated rating for a team that has an 0% win percentage is 92.6294. $b_1 = 1.5866$ means that for every additional percentage point of wins the estimated rating increases by 1.5866 percentage points.

(d)

x	y	**Predicted Value** $\hat{y} = 92.6294 + 1.5866\,x$	**Residual** $(y - \hat{y})$	**(Residual)²** $(y - \hat{y})^2$
0.838	96.020	93.9589708	2.0610292	4.247841363
0.938	95.493	94.1176308	1.3753692	1.891640436
0.846	95.478	93.9716636	1.5063364	2.26904935
0.882	95.320	94.0287812	1.2912188	1.667245989
0.857	94.541	93.9891162	0.5518838	0.3045757287
0.737	93.091	93.7987242	−0.7077242	0.5008735433
0.938	92.862	94.1176308	−1.2556308	1.576608706

(Continued)

x	y	Predicted Value $\hat{y} = 92.6294 + 1.5866\,x$	Residual $(y - \hat{y})$	$(\text{Residual})^2$ $(y - \hat{y})^2$
0.844	92.692	93.9684904	−1.2764904	1.629427741
0.903	92.609	94.0620998	−1.4530998	2.111499029
0.867	91.912	94.0049822	−2.0929822	4.38057449

$$\text{SSE} = \sum(y - \hat{y})^2 = 20.57933638,\ s = \sqrt{\frac{\text{SSE}}{n-2}} = \sqrt{\frac{20.57933638}{10-2}} \approx 1.6039.$$

If we know the percentage of wins, then our estimate of the rating will typically differ from the actual rating by 1.6039.

25. (a)

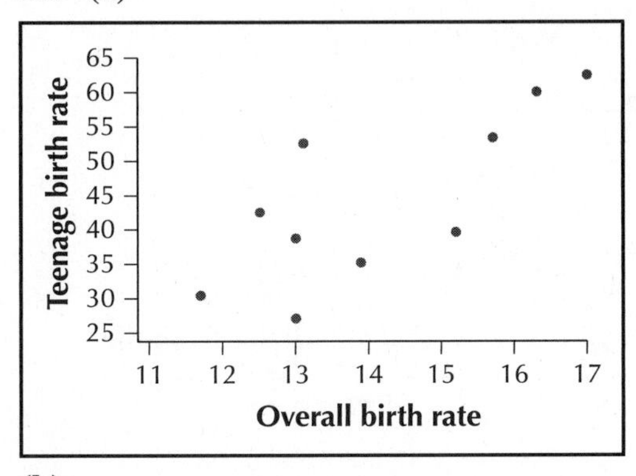

(b)

x	y	xy	x^2	y^2
13.1	52.4	686.44	171.61	2745.76
16.3	60.1	979.63	265.69	3612.01
15.2	39.5	600.4	231.04	1560.25
12.5	42.4	530	156.25	1797.76
15.7	53.4	838.38	246.49	2851.56
13.0	26.9	349.7	169	723.61
13.0	38.5	500.5	169	1482.25
11.7	30.5	356.85	136.89	930.25
17.0	62.6	1064.2	289	3918.76
13.9	35.2	489.28	193.21	1329.04
$\sum x = 141.4$	$\sum y = 441.5$	$\sum xy = 6395.38$	$\sum x^2 = 2028.18$	$\sum y^2 = 20{,}861.25$

$$b_1 = \frac{\sum xy - \left(\sum x\right)\left(\sum y\right)/n}{\sum x^2 - \left(\sum x\right)^2/n} = \frac{6395.38 - (141.4)(441.5)/10}{2028.18 - (141.4)^2/10} = \frac{152.57}{28.784} = 5.300514175 \approx 5.3005$$

Now, $\bar{x} = \dfrac{\sum x}{n} = \dfrac{141.4}{10} = 14.14$ and $\bar{y} = \dfrac{\sum y}{n} = \dfrac{441.5}{10} = 44.15$ giving,

$b_0 = \bar{y} - (b_1 \cdot \bar{x}) = 44.15 - (5.300514175)\,(14.14) = -30.79927043 \approx -30.7993.$

So $b_0 = -30.7993$, $b_1 = 5.3005$, and $\hat{y} = -30.7993 + 5.3005\,x$.

(c) $b_0 = -30.7993$ means that the estimated teenage birth rate for a state with an overall birth rate (x) of 0 is -30.7993.

$b_1 = 5.3005$ means that for every additional birth per 1000 women in the overall birth rate the teenage birth rate increases by 5.3005 births per 1000 women.

(d)

x	y	Predicted Value $\hat{y} = -30.7993 + 5.3005\,x$	Residual $(y - \hat{y})$	(Residual)^2 $(y - \hat{y})^2$
13.1	52.4	38.63725	13.76275	189.4132876
16.3	60.1	55.59885	4.50115	20.26035132
15.2	39.5	49.7683	−10.2683	105.4379849
12.5	42.4	35.45695	6.94305	48.2059433
15.7	53.4	52.41855	0.98145	1.018900606
13.0	26.9	38.1072	−11.2072	125.6013318
13.0	38.5	38.1072	0.3928	0.15429184
11.7	30.5	31.21655	−0.71655	0.5134439025
17.0	62.6	59.3092	3.2908	10.82936464
13.9	35.2	42.87765	−7.67765	58.94630952

$$\text{SSE} = \sum (y - \hat{y})^2 = 560.3812094,\ s = \sqrt{\frac{\text{SSE}}{n-2}} = \sqrt{\frac{560.3812094}{10-2}} \approx 8.3694.$$

If we know the overall birth rate, then our estimate of the teenage birth rate will typically differ from the actual teenage birth rate by 8.3694.

27. (a)

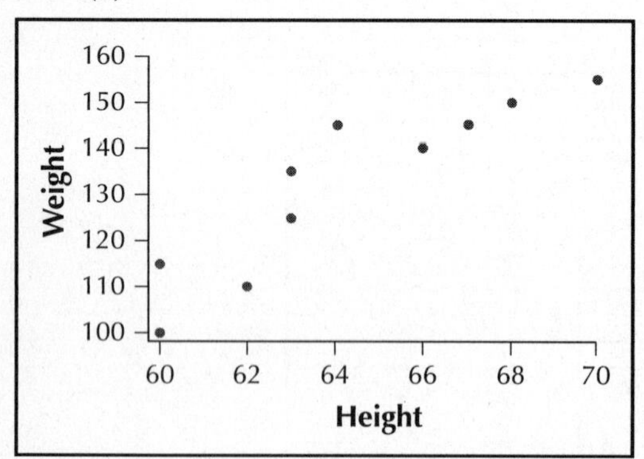

(b)

x	y	xy	x^2	y^2
60	100	6,000	3,600	10,000
60	115	6,900	3,600	13,225
62	110	6,820	3,844	12,100
63	125	7,875	3,969	15,625
63	135	8,505	3,969	18,225
64	145	8,550	4,096	21,025
66	140	9,280	4,356	19,600
67	145	9,240	4,489	21,025
68	150	9,715	4,624	22,500
70	155	10,850	4,900	24,025
$\sum x = 643$	$\sum y = 1{,}320$	$\sum xy = 85{,}385$	$\sum x^2 = 41{,}447$	$\sum y^2 = 177{,}350$

$$b_1 = \frac{\sum xy - \left(\sum x\right)\left(\sum y\right)/n}{\sum x^2 - \left(\sum x\right)^2/n} = \frac{85{,}385 - (643)(1320)/10}{41{,}447 - (643)^2/10} = \frac{509}{102.1} = 4.985308521 \approx 4.9853$$

Now, $\bar{x} = \frac{\sum x}{n} = \frac{643}{10} = 64.3$ and $\bar{y} = \frac{\sum y}{n} = \frac{1320}{10} = 132$ giving,

$b_0 = \bar{y} - (b_1 \cdot \bar{x}) = 132 - (4.985308521)(64.3) = -188.5553379 \approx -188.5553.$

So $b_0 = -188.5553$, $b_1 = 4.9853$, and $\hat{y} = -188.5553 + 4.9853\,x$.

(c) $b_0 = -188.5553$ means that the estimated weight (x) for a female student whose height is 0 inches is -188.5553 pounds. $b_1 = 4.9853$ means that for every additional inch in height of a female student her weight increases by 4.9853 pounds.

(d)

x	y	**Predicted Value** $\hat{y} = -188.5553 + 4.9853\,x$	**Residual** $(y - \hat{y})$	**(Residual)^2** $(y - \hat{y})^2$
60	100	110.5627	−10.5627	111.57063129
60	115	110.5627	4.4373	19.68963129
62	110	120.5333	−10.5333	110.9504089
63	125	125.5186	−0.5186	0.26894596
63	135	125.5186	9.4814	89.89694596
64	145	130.5039	14.4961	210.1369152
66	140	140.4745	−0.4745	0.22515025
67	145	145.4598	−0.4598	0.21141604
68	150	150.4451	−0.4451	0.19811401
70	155	160.4157	−5.4157	29.32980649

$$\text{SSE} = \sum (y - \hat{y})^2 = 572.4779654,\ s = \sqrt{\frac{\text{SSE}}{n-2}} = \sqrt{\frac{572.4779654}{10-2}} \approx 8.4593 \text{ pounds.}$$

If we know the female student's height, then the predicted weight will differ from the actual weight by 8.4593 pounds.

29. **(a)** If $x = 5$, then $\hat{y} = 11.9470 + 1.0466\,(5) = 17.18°$F. Since $x = 5$ does not lie between 7 and 70, this estimate represents extrapolation. It may not be appropriate to use extrapolation because the relationship between the variables may no longer be linear outside of the range of x. **(b)** If $x = 50$, then $\hat{y} = 11.9470 + 1.0466\,(50) = 64.277°$F.

31. **(a)** If $x = 100$, then $\hat{y} = 15.9859 + 0.7711\,(100) = 93.0959$. Since $x = 100$ does not lie between 50 and 90, this estimate represents extrapolation. It may not be appropriate to use extrapolation because the relationship between the variables may no longer be linear outside of the range of x. **(b)** If $x = 75$, then $\hat{y} = 15.9859 + 0.7711\,(75) = 73.8184$.

33. **(a)** If $x = 0.50$, then $\hat{y} = 0.3406 + 0.3416\,(0.50) = 0.5114$ dollars per pound. **(b)** If $x = 0.35$, then $\hat{y} = 0.3406 + 0.3416\,(0.35) = 0.46016$ dollars per pound. Since $x = 0.35$ does not lie between 0.374 and 0.516, this estimate represents extrapolation. It may not be appropriate to use extrapolation because the relationship between the variables may no longer be linear outside of the range of x.

35. **(a)**

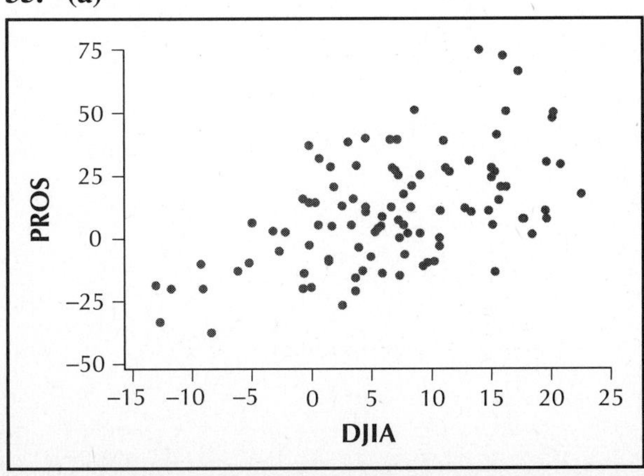

(b) $\hat{y} = 0.832 + 1.4890\,x$ **(c)** $s = 18.8545$. If we know the Dow Jones Industrial Average, then our estimate of the percent increase or decrease in the stock portfolio chosen by the pros will differ from the actual percent increase or decrease by 18.8545.

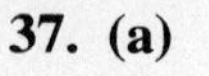

37. (a)

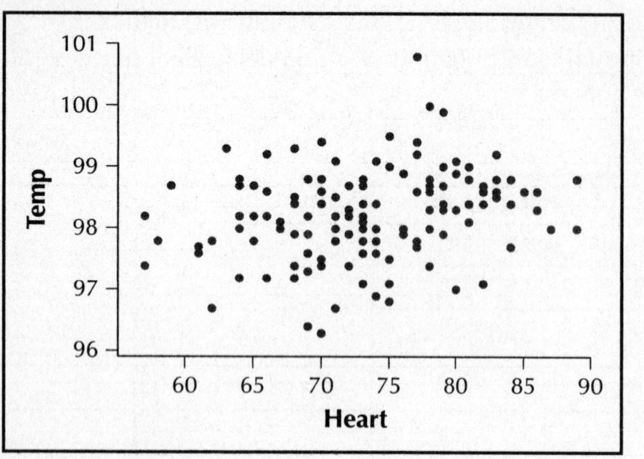

(b) $\hat{y} = 0.963071 + 0.026328\,x$ **(c)** $s = 0.711986$. If we know a person's heart rate, then our estimate of the person's temperature will differ from the person's actual temperature by 0.711986°F.

39. Answers will vary.

41. Answers will vary.

43. (a) Answers will vary. One possible answer.

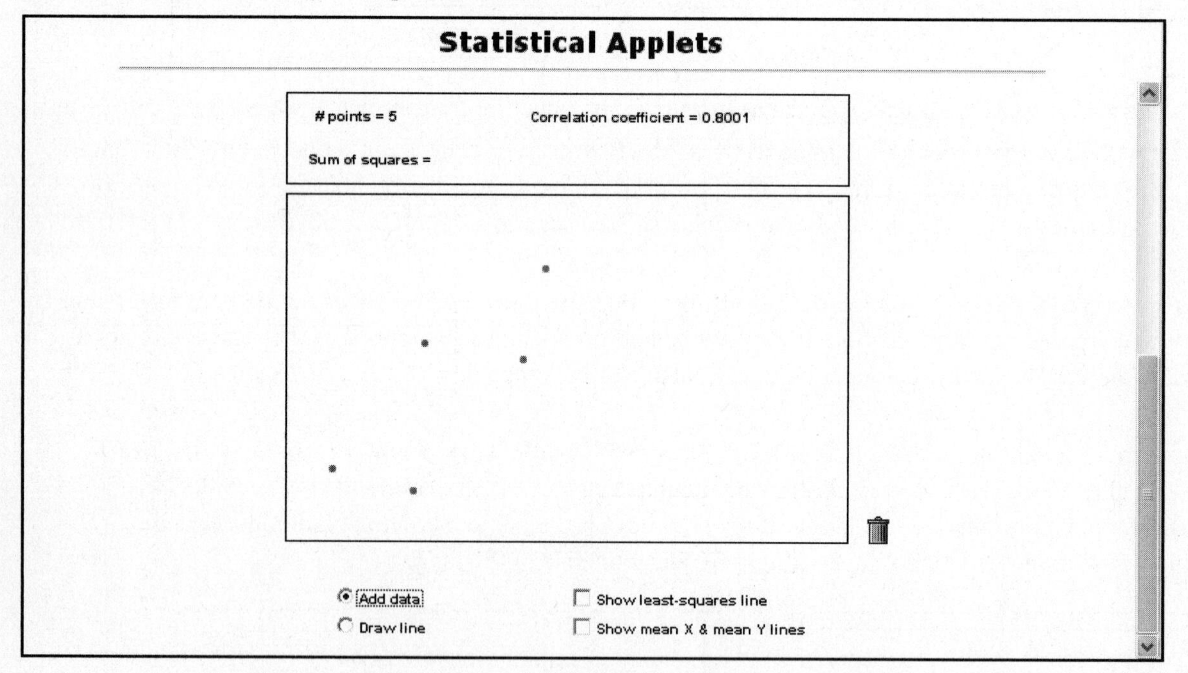

(b) Answers will vary. One possible answer.

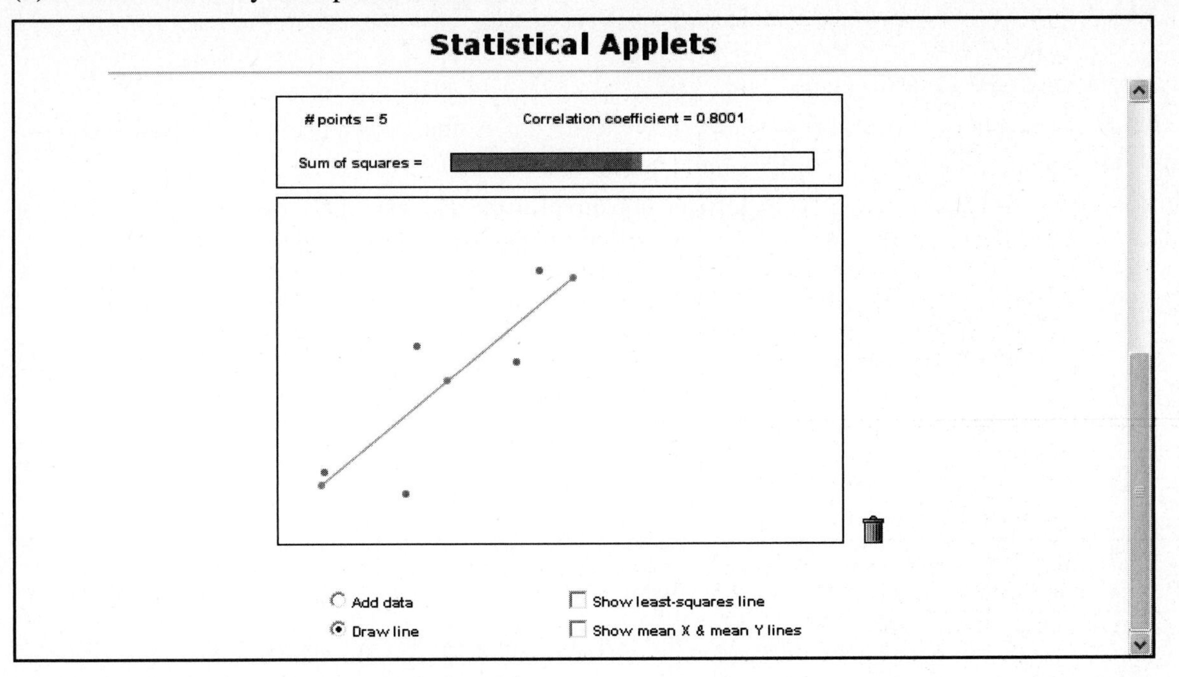

45. Answers will vary. One possible answer.

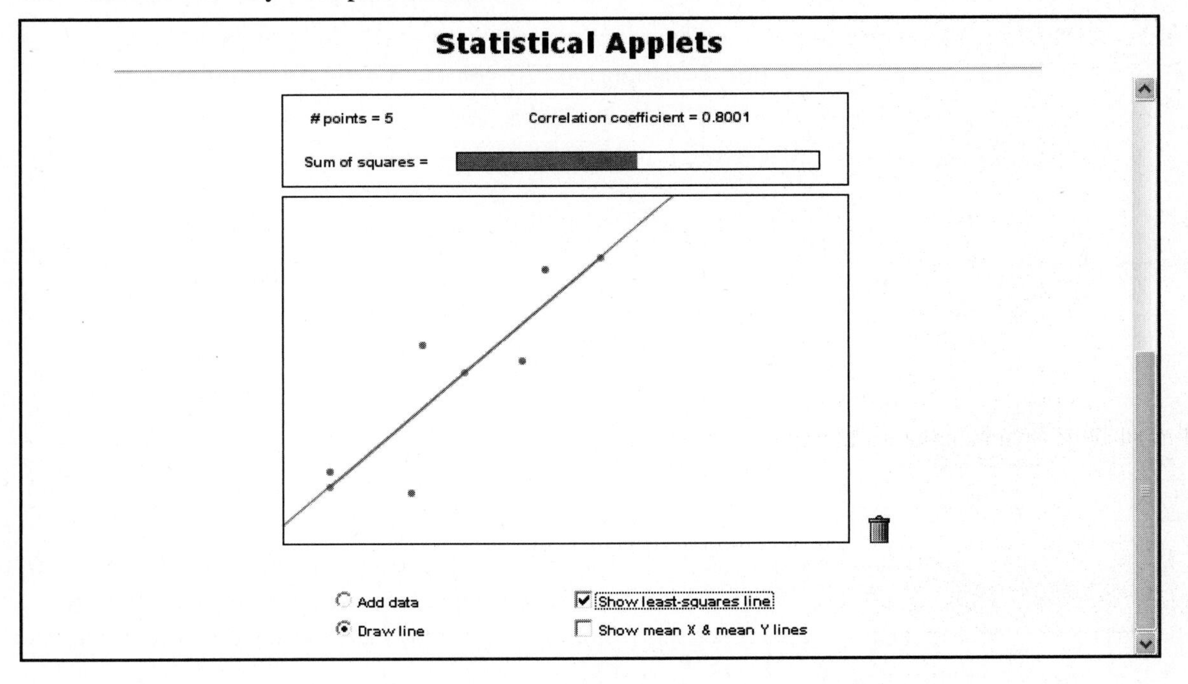

Section 13.2

1. SST is a measure of the variability in y.

3. SSR measures the amount of improvement in the accuracy of our estimate when using the regression equation compared with relying only on the y-values and ignoring the x information.

5. A value of r^2 close to 1 indicates that the regression equation fits the data extremely well. A value of r^2 close to 0 indicates that the regression equation fits the data extremely poorly.

7. (a) $\bar{y} = \frac{\sum y}{n} = \frac{105}{5} = 21$

y	$y - \bar{y}$	$(y - \bar{y})^2$
15	−6	36
20	−1	1
20	−1	1
25	4	16
25	4	16

(b) $\text{SST} = \sum(y - \bar{y})^2 = 70$. This quantity measures the variability in y. **(c)** From Exercise 13, Section 13.1, $\text{SSE} = 7.5$. Thus, $\text{SSR} = \text{SST} - \text{SSE} = 70 - 7.5 = 62.5$. This quantity measures the amount of improvement in the accuracy of our estimates when using the regression equation compared with relying only on the y-values and ignoring the x information. **(d)** $r^2 = \frac{\text{SSR}}{\text{SST}} = \frac{62.5}{70} = 0.8928571429$. Thus, 0.8928571429 of the variability in y is accounted for by the linear relationship between y and x. **(e)** From Exercise 7, Section 13.1, the slope $b_1 = 2.5$, which is positive and tells us that the sign of r is positive. Hence, $r = \sqrt{r^2} = \sqrt{0.8928571429} = 0.9449111825$. Thus, y and x are strongly positively correlated.

9. (a) $\bar{y} = \frac{\sum y}{n} = \frac{48}{5} = 9.6$

y	$y - \bar{y}$	$(y - \bar{y})^2$
0	−9.6	92.16
8	−1.6	2.56
8	−1.6	2.56
16	6.4	40.96
16	6.4	40.96

(b) $\text{SST} = \sum(y - \bar{y})^2 = 179.2$. This quantity measures the variability in y. **(c)** From Exercise 15, Section 13.1, $\text{SSE} = 19.2$. Thus $\text{SSR} = \text{SST} - \text{SSE} = 179.2 - 19.2 = 160$. This quantity measures the amount of improvement in the accuracy of our estimates when using the regression equation compared with relying only on the y-values and ignoring the x information. **(d)** $r^2 = \frac{\text{SSR}}{\text{SST}} = \frac{160}{179.2} = 0.8928571429$. Thus 0.8928571429 of the variability in y is accounted for by the linear relationship between y and x. **(e)** From Exercise 9, Section 13.1, the slope $b_1 = 4$, which is positive and tells us that the sign of r is positive. Hence, $r = \sqrt{r^2} = \sqrt{0.8928571429} = 0.9449111825$. Thus, y and x are strongly positively correlated.

11. (a) $\bar{y} = \frac{\sum y}{n} = \frac{445}{5} = 89$

y	$y - \bar{y}$	$(y - \bar{y})^2$
100	11	121
95	6	36
85	−4	16
85	−4	16
80	9	81

(b) $SST = \sum(y - \bar{y})^2 = 270$. This quantity measures the variability in y. **(c)** From Exercise 17, Section 13.1, $SSE = 20$. Thus, $SSR = SST - SSE = 270 - 20 = 250$. This quantity measures the amount of improvement in the accuracy of our estimates when using the regression equation compared with relying only on the y-values and ignoring the x information.
(d) $r^2 = \frac{SSR}{SST} = \frac{250}{270} = 0.9259259259$. Thus, 0.9259259259 of the variability in y is accounted for by the linear relationship between y and x. **(e)** From Exercise 11, Section 13.1, the slope $b_1 = -0.5$, which is negative and tells us that the sign of r is negative. Hence, $r = -\sqrt{r^2} = -\sqrt{0.9259259259} = -0.9622504486$. Thus, y and x are strongly negatively correlated.

13. **(a)** $\bar{y} = \frac{\sum y}{n} = \frac{116}{5} = 23.2$

y	$y - \bar{y}$	$(y - \bar{y})^2$
10	−13.2	174.24
16	−7.2	51.84
25	1.8	3.24
30	6.8	46.24
35	11.8	139.24

$SST = \sum(y - \bar{y})^2 = 414.8$
From Exercise 19, Section 13.1, $SSE = 5.2$. Thus, $SSR = SST - SSE = 414.8 - 5.2 = 409.6$.
$r^2 = \frac{SSR}{SST} = \frac{409.6}{414.8} = 0.987463838$. Thus, 0.987463838 of the variability in *weight* (y) is accounted for by the linear relationship between *weight* (y) and *volume* (x).
(b) From Exercise 19, Section 13.1, the slope $b_1 = 1.6$, which is positive and tells us that the sign of r is positive. Hence, $r = \sqrt{r^2} = \sqrt{0.987463838} = 0.9937121505$. Thus, *weight* ($y$) and *volume* ($x$) are strongly positively correlated.

15. **(a)** $\bar{y} = \frac{\sum y}{n} = \frac{2583}{5} = 516.6$

y	$y - \bar{y}$	$(y - \bar{y})^2$
510	−6.6	43.56
515	−1.6	2.56
523	6.4	40.96
514	−2.6	6.76
521	4.4	19.36

$SST = \sum(y - \bar{y})^2 = 113.2$

From Exercise 21, Section 13.1, $SSE = 25.697268$. Thus, $SSR = SST - SSE = 113.2 - 25.697268 = 87.502732$.
$r^2 = \frac{SSR}{SST} = \frac{87.502732}{113.2} = 0.7729923322$. Thus, 0.7729923322 of the variability in *SAT I Math scores* (y) is accounted for by the linear relationship between *SAT I Math scores* (y) and *SAT I Verbal scores* (x).

(b) From Exercise 21, Section 13.1, the slope $b_1 = 0.4264$, which is positive and tells us that the sign of r is positive. Hence, $r = \sqrt{r^2} = \sqrt{0.7729923322} = 0.8791998249$. Thus, *SAT I Math scores* (y) and *SAT I Verbal scores* (x) are strongly positively correlated.

17. (a) $\bar{y} = \frac{\sum y}{n} = \frac{940.018}{10} = 94.0018$

y	$y - \bar{y}$	$(y - \bar{y})^2$
96.020	2.0182	4.07313124
95.493	1.4912	2.22367744
95.478	1.4762	2.17916644
95.320	1.3182	1.73765124
94.541	0.5392	0.29073664
93.091	−0.9108	0.82955664
92.862	−1.1398	1.29914404
92.692	−1.3098	1.71557604
92.609	−1.3928	1.93989184
91.912	−2.0898	4.36726404

$SST = \sum(y - \bar{y})^2 = 20.6557956$

From Exercise 23, Section 13.1, SSE = 20.57933638. Thus, SSR = SST − SSE = 20.6557956 − 20.57933638 = 0.07645922.

$r^2 = \frac{SSR}{SST} = \frac{0.07645922}{20.6557956} = 0.0037015868$. Thus, 0.0037015868 of the variability in the rating (y) is accounted for by the linear relationship between *rating* (y) and *win percentage* (x).

(b) From Exercise 23, Section 13.1, the slope $b_1 = 1.5866$, which is positive and tells us that the sign of r is positive. Hence, $r = \sqrt{r^2} = \sqrt{0.0037015868} = 0.0608406672$. Thus, *rating* ($y$) and *win percentage* ($x$) are not correlated.

19. (a)

x	y	xy	x^2	y^2
7,246	3,744	27,129,024	52,504,516	14,017,536
4,750	3,509	16,667,750	22,562,500	12,313,081
15,223	9,629	146,582,267	231,739,729	92,717,641
1,915	1,292	2,474,180	3,667,225	1,669,264
3,463	2,561	8,868,743	11,992,369	6,558,721
8,134	4,901	39,864,734	66,161,956	24,019,801
2,614	2,277	5,952,078	6,832,996	5,184,729
2,327	2,122	4,937,894	5,414,929	4,502,884
3,701	2,697	9,981,597	13,697,401	7,273,809
8,881	5,642	50,106,602	78,872,161	31,832,164
1,387	889	1,233,043	1,923,769	790,321
3,222	1,554	5,006,988	10,381,284	2,414,916
3,254	2,343	7,624,122	10,588,516	5,489,649
2,626	1,407	3,694,782	6,895,876	1,979,649
2,528	1,745	4,411,360	6,390,784	3,045,025
13,631	7,848	106,976,088	185,804,161	61,591,104
$\sum x = 84{,}902$	$\sum y = 54{,}160$	$\sum xy = 441{,}511{,}252$	$\sum x^2 = 715{,}430{,}172$	$\sum y^2 = 275{,}400{,}294$

$$SST = \sum y^2 - \frac{\left(\sum y\right)^2}{n} = 275{,}400{,}294 - \frac{54{,}160^2}{16} = 92{,}068{,}694$$

$$SSR = \frac{\left(\sum xy - \frac{\left(\sum x\right)\left(\sum y\right)}{n}\right)^2}{\sum x^2 - \frac{\left(\sum x\right)^2}{n}} = \frac{\left(441{,}511{,}252 - \frac{(84{,}902)(54{,}160)}{16}\right)^2}{715{,}430{,}172 - \frac{84{,}902^2}{16}} = \frac{(154{,}117{,}982)^2}{264{,}908{,}321.8} = 89{,}662{,}537.64$$

$r^2 = \frac{SSR}{SST} = \frac{89{,}662{,}537.64}{92{,}068{,}094} = 0.9738656404$. Thus, 0.9738656404 of the variability in the *number of veterans 65 and over* (y) is accounted for by the linear relationship between the *number of veterans 65 and over* (y) and the *number of veterans under 65* (x).

(b) $b_1 = \frac{\sum xy - \left(\sum x\right)\left(\sum y\right)/n}{\sum x^2 - \left(\sum x\right)^2/n} = \frac{441{,}511{,}252 - (84{,}902)(54{,}160)/16}{715{,}430{,}172 - (84.902)^2/16} = \frac{154{,}117{,}982}{264{,}908{,}321.8} = 0.58177856$

Since the slope is positive, this tells us that the sign of r is positive.

Hence, $r = \sqrt{r^2} = \sqrt{0.9738656404} = 0.9868463104$. Thus, the *number of veterans 65 and over* (y) and the *number of veterans under 65* (x) are strongly positively correlated.

21. (a)

x	y	xy	x^2	y^2
900	1.5	1,350	810,000	2.25
925	2.6	2,405	855,625	6.76
950	1.9	1,805	902,500	3.61
975	2.7	2,632.5	950,625	7.29
1,000	2.0	2,000	1,000,000	4
1,000	2.5	2,500	1,000,000	6.25
1,025	3.0	3,075	1,050,625	9
1,025	2.4	2,460	1,050,625	5.76
1,050	2.1	2,205	1,102,500	4.41
1,050	2.9	3,045	1,102,500	8.41
1,075	2.7	2,902.5	1,155,625	7.29
1,075	3.1	3,332.5	1,155,625	9.61
1,100	3.0	3,300	1,210,000	9
1,125	3.3	3,712.5	1,265,625	10.89
1,150	3.2	3,680	1,322,500	10.24
$\Sigma x = 15{,}425$	$\Sigma y = 38.9$	$\Sigma xy = 40{,}405$	$\Sigma x^2 = 15{,}934{,}375$	$\Sigma y^2 = 104.77$

$$SST = \sum y^2 - \frac{\left(\sum y\right)^2}{n} = 104.77 - \frac{38.9^2}{15} = 3.889333333$$

$$SSR = \frac{\left(\sum xy - \frac{\left(\sum x\right)\left(\sum y\right)}{n}\right)^2}{\sum x^2 - \frac{\left(\sum x\right)^2}{n}} = \frac{\left(40{,}405 - \frac{(15{,}425)(38.9)}{15}\right)^2}{15{,}934{,}375 - \frac{15{,}425^2}{15}} = \frac{(402.8333333)^2}{72{,}333.33333} = 2.243428955$$

$r^2 = \frac{SSR}{SST} = \frac{2.243428955}{3.889333333} = 0.5768158096$. Thus, 0.5768158096 of the variability in *GPAs* (y) is accounted for by the linear relationship between *GPAs* (y) and the *combined SAT scores* (x).

(b) $b_1 = \frac{\sum xy - \left(\sum x\right)\left(\sum y\right)/n}{\sum x^2 - \left(\sum x\right)^2/n} = \frac{40{,}405 - (15{,}425)(38.9)/15}{15{,}934{,}375 - (15{,}425)^2/15} = \frac{402.8333333}{72{,}333.33333} = 0.0055691244$

Since the slope is positive, this tells us that the sign of r is positive.

Hence, $r = \sqrt{r^2} = \sqrt{0.576818096} = 0.75948391$.

Thus, *GPAs* (y) and the combined *SAT scores* (x) are strongly positively correlated.

23. (a)

x	y	xy	x^2	y^2
3.86	0.04	0.1544	14.8996	0.0016
3.41	−0.56	−1.9096	11.6281	0.3136
32.02	0.78	24.9756	1025.2804	0.6084
3.04	−0.03	−0.0912	9.2416	0.0009
13.96	−1.96	−27.3616	194.8816	3.8416
10.36	−0.82	−8.4952	107.3296	0.6724
14.74	1.62	23.8788	217.2676	2.6244
33.19	0.03	0.9957	1101.5761	0.0009
39.15	0.46	18.009	1532.7225	0.2116
33.05	−0.21	−6.9405	1092.3025	0.0441
$\sum x = 186.78$	$\sum y = -0.65$	$\sum xy = 23.2154$	$\sum x^2 = 5307.1296$	$\sum y^2 = 8.3195$

$$\text{SST} = \sum y^2 - \frac{\left(\sum y\right)^2}{n} = 8.3195 - \frac{(-0.65)^2}{10} = 8.27725$$

$$\text{SSR} = \frac{\left(\sum xy - \frac{\left(\sum x\right)\left(\sum y\right)}{n}\right)^2}{\sum x^2 - \frac{\left(\sum x\right)^2}{n}} = \frac{\left(23.2154 - \frac{(186.78)(-0.65)}{10}\right)^2}{5307.1296 - \frac{186.78^2}{10}} = \frac{(35.3561)^2}{1818.45276} = 0.6874271549$$

$r^2 = \frac{\text{SSR}}{\text{SST}} = \frac{0.6874271549}{8.27725} = 0.0830501863$. Thus, 0.0830501863 of the variability in the *change in stock price* (y) is accounted for by the linear relationship between the *change in stock price* (y) and *the stock price* (x).

(b) $b_1 = \frac{\sum xy - \left(\sum x\right)\left(\sum y\right)/n}{\sum x^2 - \left(\sum x\right)^2/n} = \frac{23.2154 - (186.78)(-0.65)/10}{5307.1296 - (186.78)^2/10} = \frac{35.3561}{1818.45276} = 0.0194429576$

Since the slope is positive, this tells us that the sign of r is positive.

Hence, $r = \sqrt{r^2} = \sqrt{0.0830501863} = 0.2881842923$.

Thus, *change in stock price* (y) and the *stock price* (x) are not correlated.

25. (a)

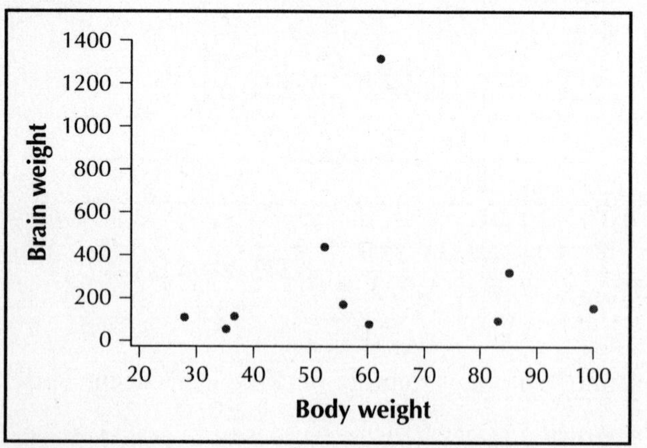

Yes, humans.

(b)

Regression Analysis: Brain weight versus Body weight

```
The regression equation is
Brain weight = 194 + 1.59 Body weight


Predictor      Coef  SE Coef     T      P
Constant      193.5    360.7  0.54  0.606
Body weight   1.595    5.656  0.28  0.785


S = 402.732   R-Sq = 1.0%   R-Sq(adj) = 0.0%


Analysis of Variance

Source           DF       SS      MS     F      P
Regression        1    12896   12896  0.08  0.785
Residual Error    8  1297544  162193
Total             9  1310441


Unusual Observations

       Body   Brain
Obs  weight  weight  Fit  SE Fit  Residual  St Resid
  8      62    1320  292     128      1028     2.69R

R denotes an observation with a large standardized
residual.
```

(c)

Regression Analysis: Brain weight versus Body weight

```
The regression equation is
Brain weight = 111 + 1.07 Body weight


Predictor      Coef  SE Coef     T      P
Constant      110.6    119.1  0.93  0.384
Body weight   1.068    1.862  0.57  0.584


S = 132.469   R-Sq = 4.5%   R-Sq(adj) = 0.0%


Analysis of Variance

Source           DF      SS     MS     F      P
Regression        1    5780   5780  0.33  0.584
Residual Error    7  122837  17548
Total             8  128617


Unusual Observations

       Body   Brain
Obs  weight  weight    Fit  SE Fit  Residual  St Resid
  1      52   440.0  166.3    46.2     273.7     2.20R

R denotes an observation with a large standardized residual.
```

(i) b_0 decreased from 193.5 to 110.6.
(ii) b_1 decreased from 1.595 to 1.068.
(iii) r^2 increased from 0.010 to 0.045.
(iv) s decreased from 402.732 to 132.469.

27. (a)

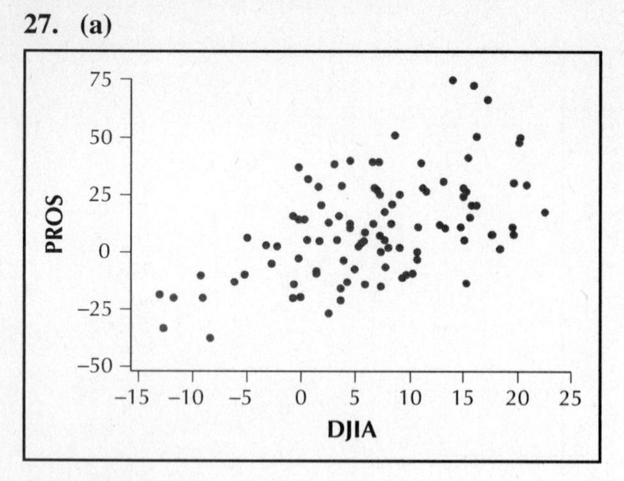

From Minitab:

Regression Analysis: PROS versus DJIA

```
The regression equation is
PROS = 0.83 + 1.49 DJIA


Predictor      Coef   SE Coef      T      P
Constant      0.832     2.475   0.34  0.737
DJIA         1.4890    0.2359   6.31  0.000


S = 18.8545   R-Sq = 28.9%   R-Sq(adj) = 28.2%


Analysis of Variance

Source          DF     SS     MS      F      P
Regression       1  14158  14158  39.83  0.000
Residual Error  98  34838    355
Total           99  48996
```

(b) $\hat{y} = 0.832 + 1.4890\,x$ **(c)** $r^2 = 0.289$. Thus, 0.289 of the variability in the performance of the stock portfolios chosen by the pros (y) is accounted for by the linear relationship between the performance of the stock portfolios chosen by the pros (y) and the Dow Jones Industrial Average (x). **(d)** $s = 18.8545$. If we know the Dow Jones Industrial Average, then our estimate of the performance of the stock portfolio chosen by the pros will differ from the actual performance by 18.8545. **(e)** Since $b_1 = 1.4890$ is positive, r is positive. Thus, $r = \sqrt{r^2} = \sqrt{0.289} = 0.5376$. Thus, the performance of the stock portfolios chosen by the pros (y) and the Dow Jones Industrial Average (x) are mildly positively correlated.

29. (a)

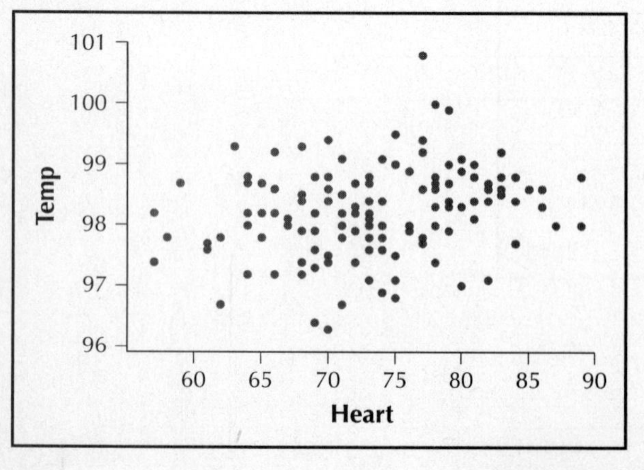

From Minitab:

```
Regression Analysis: TEMP versus HEART

The regression equation is
TEMP = 96.3 + 0.0263 HEART


Predictor       Coef    SE Coef       T      P
Constant     96.3068     0.6577  146.43  0.000
HEART       0.026335   0.008876    2.97  0.004


S = 0.711969   R-Sq = 6.4%   R-Sq(adj) = 5.7%


Analysis of Variance

Source          DF       SS      MS     F      P
Regression       1   4.4618  4.4618  8.80  0.004
Residual Error 128  64.8832  0.5069
Total          129  69.3449
```

(b) $\hat{y} = 96.3068 + 0.026335\,x$

(c) $r^2 = 0.064$. Thus, 0.064 of the variability in people's body temperatures (y) is accounted for by the linear relationship between people's body temperatures (x) and people's heart rates (x)

(d) $s = 0.711969$. If we know a person's heart rate, then our estimate of the person's body temperature will differ from the person's actual body temperature by 0.711969 °F.

(e) Since $b_1 = 0.026335$ is positive, r is positive. Thus, $r = \sqrt{r^2} = \sqrt{0.064} = 0.253$. Thus, a person's body temperature (y) and a person's heart rate (x) are not correlated.

Section 13.3

1. The regression equation is calculated from a sample and is only valid for values of x in the range of the sample data. The population regression equation may be used to approximate the relationship between the predictor variable x and the response variable y for the entire population of (x, y) pairs.

3. We construct a scatterplot of the residuals against the fitted values and a normal probability plot of the residuals. We must make sure that the scatter plot contains no strong evidence of any unhealthy patterns and that the normal probability plot indicates no evidence of departures from normality in residuals.

5. It means that there is no relationship between x and y.

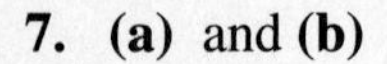

7. **(a)** and **(b)**

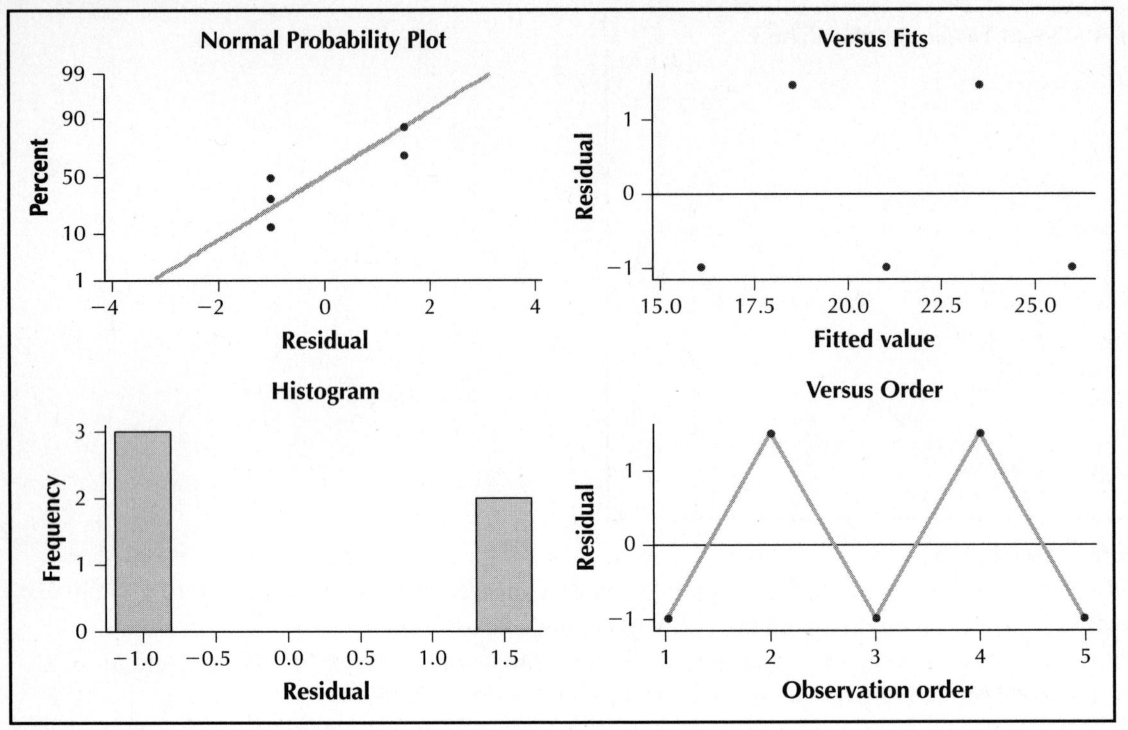

(c) The scatterplot of the residuals contains an unhealthy pattern, so the regression assumptions are not verified.

9. **(a)** and **(b)**

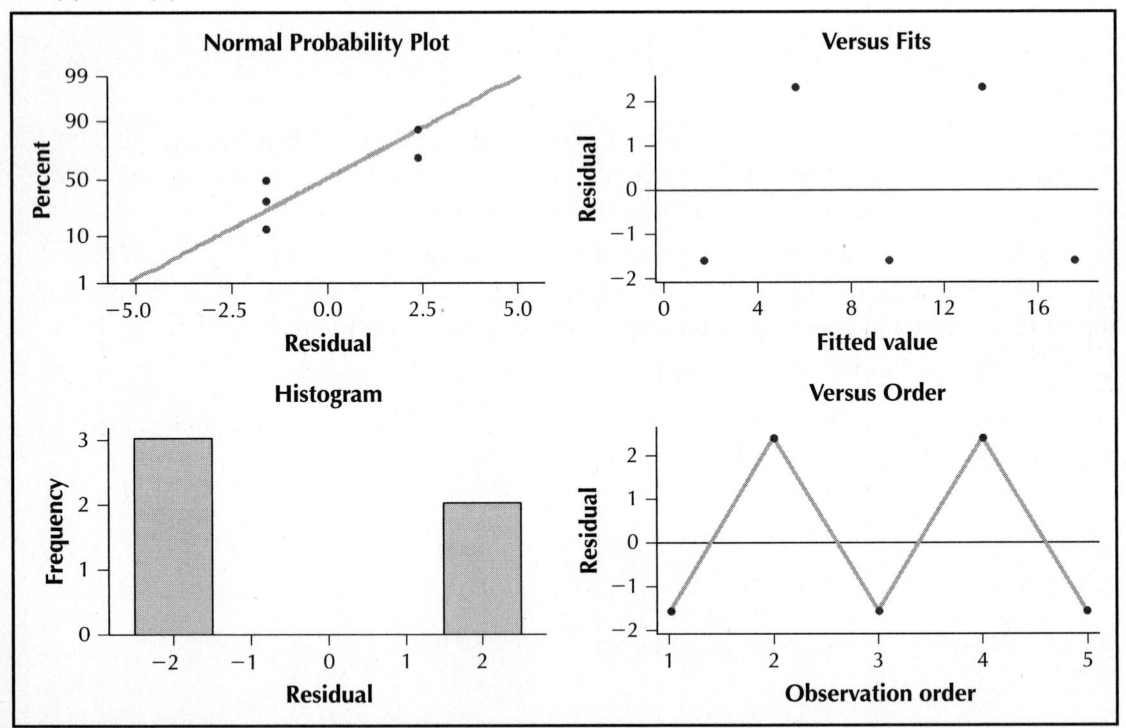

(c) The scatterplot of the residuals contains an unhealthy pattern, so the regression assumptions are not verified.

11. **(a)** and **(b)**

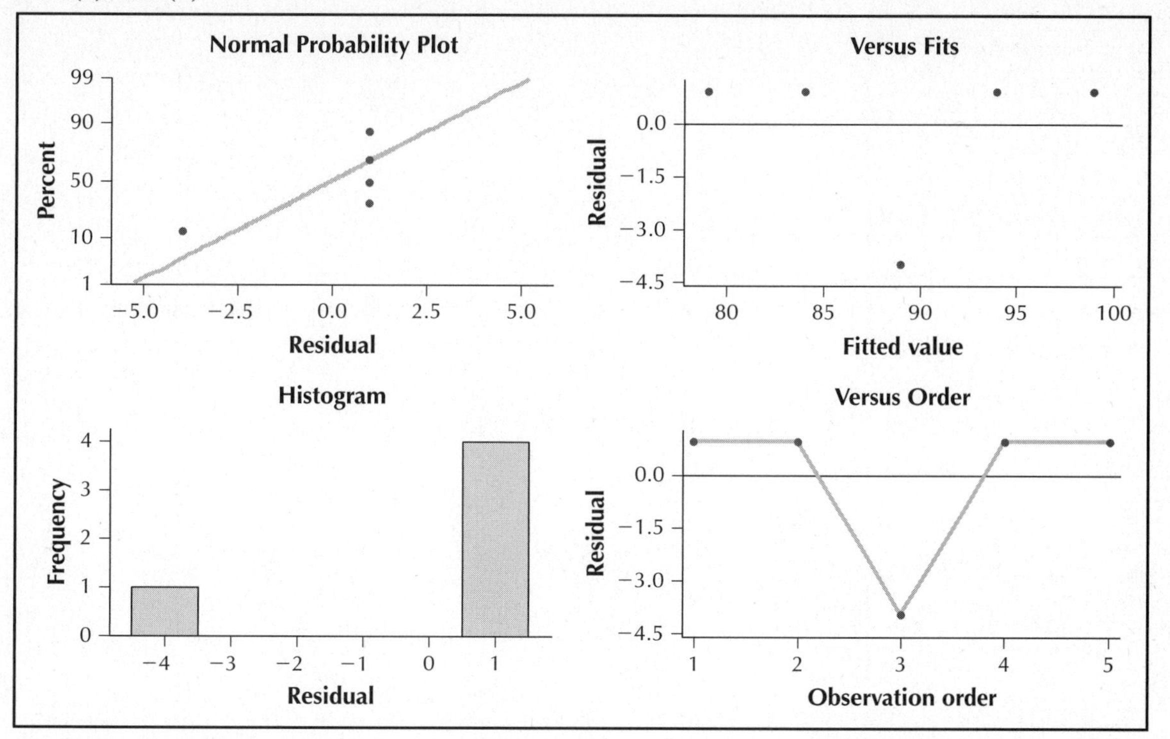

(c) The scatterplot of the residuals contains an unhealthy pattern, so the regression assumptions are not verified.

13. **(a)** Calculating $\sum(x - \bar{x})^2$

$$\bar{x} = \frac{\sum x}{n} = \frac{1 + 2 + 3 + 4 + 5}{5} = \frac{15}{5} = 3$$

x	$(x - \bar{x})^2$
1	$(1 - 3)^2 = (-2)^2 = 4$
2	$(2 - 3)^2 = (-1)^2 = 1$
3	$(3 - 3)^2 = (0)^2 = 0$
4	$(4 - 3)^2 = (1)^2 = 1$
5	$(5 - 3)^2 = (2)^2 = 4$
	$\sum(x - \bar{x})^2 = 10$

From Exercise 13, Section 13.1, $s \approx 1.5811$. Thus, $s_{b_1} = \dfrac{s}{\sqrt{\sum(x - \bar{x})^2}} = \dfrac{1.5811}{\sqrt{10}} = 0.5$. The typical error in using b_1 to estimate β_1 is $s_{b_1} = 0.5$.

(b) $\text{df} = n - 2 = 5 - 2 = 3$, $t_{\alpha/2} = 3.182$

(c) $b_1 \pm t_{\alpha/2}\, s_{b_1} = 2.5 \pm (3.182)(0.5) = 2.5 \pm 1.591 = (0.909, 4.091)$

15. **(a)** Calculating $\sum(x - \bar{x})^2$

$$\bar{x} = \frac{\sum x}{n} = \frac{(-1) + (-2) + (-3) + (-4) + (-5)}{5} = \frac{-15}{5} = -3$$

x	$(x-\bar{x})^2$
−1	$((-1)-(-3))^2 = (2)^2 = 4$
−2	$((-2)-(-3))^2 = (1)^2 = 1$
−3	$((-3)-(-3))^2 = (0)^2 = 0$
−4	$((-4)-(-3))^2 = (-1)^2 = 1$
−5	$((-5)-(-3))^2 = (-2)^2 = 4$
	$\sum(x-\bar{x})^2 = 10$

From Exercise 15, Section 13.1, $s \approx 2.5298$. Thus, $s_{b_1} = \dfrac{s}{\sqrt{\sum(x-\bar{x})^2}} = \dfrac{2.5298}{\sqrt{10}} = 0.8$. The typical error in using b_1 to estimate β_1 is $s_{b_1} = 0.8$.

(b) $\text{df} = n - 2 = 5 - 2 = 3$, $t_{\alpha/2} = 3.182$

(c) $b_1 \pm t_{\alpha/2}\, s_{b_1} = 4.0 \pm (3.182)(0.8) \approx 4.0 \pm 2.5456 = (1.4544, 6.5456)$

17. (a) Calculating $\sum(x-\bar{x})^2$

$$\bar{x} = \frac{\sum x}{n} = \frac{10+20+30+40+50}{5} = \frac{150}{5} = 30$$

x	$(x-\bar{x})^2$
10	$(10-30)^2 = (-20)^2 = 400$
20	$(20-30)^2 = (-10)^2 = 100$
30	$(30-30)^2 = (0)^2 = 0$
40	$(40-30)^2 = (10)^2 = 100$
50	$(50-30)^2 = (20)^2 = 400$
	$\sum(x-\bar{x})^2 = 1{,}000$

From Exercise 17, Section 13.1, $s \approx 2.5820$. Thus, $s_{b_1} = \dfrac{s}{\sqrt{\sum(x-\bar{x})^2}} = \dfrac{2.5820}{\sqrt{1000}} \approx 0.08165$. The typical error in using b_1 to estimate β_1 is $s_{b_1} = 0.08165$.

(b) $\text{df} = n - 2 = 5 - 2 = 3$, $t_{\alpha/2} = 3.182$

(c) $b_1 \pm t_{\alpha/2}\, s_{b_1} = -0.5 \pm (3.182)(0.08165) \approx -0.5 \pm 0.2598 = (-0.7598, -0.2402)$

19. (a) From Exercise 13, $s_{b_1} = 0.5$. Thus, $t_{data} = \dfrac{b_1}{s_{b_1}} = \dfrac{2.5}{0.5} = 5$.

(b) $\text{df} = n - 2 = 5 - 2 = 3$, p-value $= 2 \cdot P(t > |t_{data}|) = 2 \cdot P(t > |5|) = 0.0153924381$

(c) $H_0: \beta_1 = 0$. There is no relationship between x and y.
$H_a: \beta_1 \neq 0$. There is a linear relationship between x and y.
Reject H_0 if p-value < 0.05. Since p-value < 0.05, we reject H_0. There is evidence for a linear relationship between x and y.

21. (a) From Exercise 15, $s_{b_1} = 0.8$. Thus, $t_{data} = \dfrac{b_1}{s_{b_1}} = \dfrac{4.0}{0.8} = 5$

(b) $\text{df} = n - 2 = 5 - 2 = 3$, p-value $= 2 \cdot P(t > |t_{data}|) = 2 \cdot P(t > |5|) = 0.0153924381$

(c) $H_0: \beta_1 = 0$. There is no relationship between x and y.
$H_a: \beta_1 \neq 0$. There is a linear relationship between x and y.
Reject H_0 if p-value < 0.05. Since p-value < 0.05, we reject H_0. There is evidence for a linear relationship between x and y.

23. (a) $\text{df} = n - 2 = 5 - 2 = 3$, $\alpha = 0.05$, $t_{crit} = 3.182$

(b) From Exercise 17, $s_{b_1} = 0.08165$. Thus, $t_{data} = \dfrac{b_1}{s_{b_1}} = \dfrac{-0.5}{0.08165} = -6.12$.

(c) $H_0: \beta_1 = 0$. There is no relationship between x and y.
$H_a: \beta_1 \neq 0$. There is a linear relationship between x and y.

Reject H_0 if $t_{data} < -3.182$ or $t_{data} > 3.182$. Since $t_{data} < -3.182$, we reject H_0. There is evidence for a linear relationship between x and y.

25. **(a)** Calculating $\sum(x - \bar{x})^2$

$$\bar{x} = \frac{\sum x}{n} = \frac{4 + 8 + 12 + 16 + 20}{5} = \frac{60}{5} = 12$$

x	$(x - \bar{x})^2$
4	$(4 - 12)^2 = (-8)^2 = 64$
8	$(8 - 12)^2 = (-4)^2 = 16$
12	$(12 - 12)^2 = (0)^2 = 0$
16	$(16 - 12)^2 = (4)^2 = 16$
20	$(20 - 12)^2 = (8)^2 = 64$
	$\sum(x - \bar{x})^2 = 160$

From Exercise 19, Section 13.1, $s \approx 1.3166$. Thus, $s_{b_1} = \dfrac{s}{\sqrt{\sum(x - \bar{x})^2}} = \dfrac{1.3166}{\sqrt{160}} \approx 0.1041$. The typical error in using b_1 to estimate β_1 is $s_{b_1} = 0.1041$.

(b) df $= n - 2 = 5 - 2 = 3$, $t_{\alpha/2} = 3.182$

(c) From Exercise 19, Section 13.1, $b_1 = 1.6$. Thus, $b_1 \pm t_{\alpha/2}\, s_{b_1} = 1.6 \pm (3.182)(0.1041) \approx 1.6 \pm 0.3312 = (1.2688, 1.9312)$.

(d) We are 95% confident that the interval (1.2688, 1.9312) captures the population slope β_1 of the relationship between *volume* and *weight*.

27. **(a)** Calculating $\sum(x - \bar{x})^2$

$$\bar{x} = \frac{\sum x}{n} = \frac{497 + 515 + 518 + 501 + 522}{5} = \frac{2553}{5} = 510.6$$

x	$(x - \bar{x})^2$
497	$(497 - 510.6)^2 = (-13.6)^2 = 184.96$
515	$(515 - 510.6)^2 = (4.4)^2 = 19.36$
518	$(518 - 510.6)^2 = (7.4)^2 = 54.76$
501	$(501 - 510.6)^2 = (-9.6)^2 = 92.16$
522	$(522 - 510.6)^2 = (11.4)^2 = 129.96$
	$\sum(x - \bar{x})^2 = 481.2$

From Exercise 21, Section 13.1, $s \approx 2.9267$. Thus, $s_{b_1} = \dfrac{s}{\sqrt{\sum(x - \bar{x})^2}} = \dfrac{2.9267}{\sqrt{481.2}} \approx 0.1334$. The typical error in using b_1 to estimate β_1 is $s_{b_1} = 0.1334$.

(b) df $= n - 2 = 5 - 2 = 3$, $t_{\alpha/2} = 3.182$

(c) From Exercise 21, Section 13.1, $b_1 = 0.4264$. Thus, $b_1 \pm t_{\alpha/2}\, s_{b_1} = 0.4264 \pm (3.182)(0.1334) \approx 0.4264 \pm 0.4245 = (0.0019, 0.8509)$

(d) We are 95% confident that the interval (0.0019, 0.8509) captures the population slope β_1 of the relationship between *SAT I Verbal* and *SAT I Math*.

29. **(a)** Calculating $\sum(x - \bar{x})^2$

$$\bar{x} = \frac{\sum x}{n} = \frac{0.838 + 0.938 + 0.846 + 0.882 + 0.857 + 0.737 + 0.938 + 0.844 + 0.903 + 0.867}{10}$$

$$= \frac{8.65}{10} = 0.865$$

x	$(x-\bar{x})^2$
0.838	$(0.838-0.865)^2=(-0.027)^2=0.000729$
0.938	$(0.938-0.865)^2=(0.073)^2=0.005329$
0.846	$(0.846-0.865)^2=(-0.019)^2=0.000361$
0.882	$(0.882-0.865)^2=(0.017)^2=0.000289$
0.857	$(0.857-0.865)^2=(-0.008)^2=0.000064$
0.737	$(0.737-0.865)^2=(-0.128)^2=0.016384$
0.938	$(0.938-0.865)^2=(0.073)^2=0.005329$
0.844	$(0.844-0.865)^2=(-0.21)^2=0.000441$
0.903	$(0.903-0.865)^2=(0.038)^2=0.001444$
0.867	$(0.867-0.865)^2=(0.002)^2=0.000004$
	$\sum(x-\bar{x})^2=0.030374$

From Exercise 23, Section 13.1, $s \approx 1.6039$. Thus, $s_{b_1} = \dfrac{s}{\sqrt{\sum(x-\bar{x})^2}} = \dfrac{1.6039}{\sqrt{0.030374}} \approx 9.2029$. The typical error in using b_1 to estimate β_1 is $s_{b_1} = 9.2029$.

(b) $\text{df} = n - 2 = 10 - 2 = 8$, $t_{\alpha/2} = 2.306$

(c) From Exercise 23, Section 13.1, $b_1 = 1.5866$. Thus, $b_1 \pm t_{\alpha/2}\, s_{b_1} = 1.5866 \pm (2.306)(9.2029) \approx 1.5866 \pm 21.2219 = (-19.6353, 22.8085)$

(d) We are 95% confident that the interval $(-19.6353, 22.8085)$ captures the population slope β_1 of the relationship between *win%* and *rating*.

31. (a) From Exercise 21, Section 13.2, $\sum x = 15{,}425$, SST $= 3.889333333$, and SSR $= 2.243428955$.

Thus, $\bar{x} = \dfrac{\sum x}{n} = \dfrac{15{,}425}{15} = 1028.333333$ and SSE = SST − SSR = 3.889333333 − 2.243428955 = 1.645904378. Therefore, $s = \sqrt{\dfrac{\text{SSE}}{n-2}} = \sqrt{\dfrac{1.645904378}{15-2}} \approx 0.3558$. Calculating $\sum(x-\bar{x})^2$

x	$(x-\bar{x})^2$
900	$(900-1028.333333)^2=(-128.333333)^2=16{,}469.44436$
925	$(925-1028.333333)^2=(-103.333333)^2=10{,}677.77771$
950	$(950-1028.333333)^2=(-78.333333)^2=6{,}136.111059$
975	$(975-1028.333333)^2=(-53.333333)^2=2{,}844.444409$
1000	$(1000-1028.333333)^2=(-28.333333)^2=802.7777589$
1000	$(1000-1028.333333)^2=(-28.333333)^2=802.7777589$
1025	$(1025-1028.333333)^2=(-3.333333)^2=11.11110889$
1025	$(1025-1028.333333)^2=(-3.333333)^2=11.11110889$
1050	$(1050-1028.333333)^2=(21.666667)^2=469.4444589$
1050	$(1050-1028.333333)^2=(21.666667)^2=469.4444589$
1075	$(1075-1028.333333)^2=(46.666667)^2=2{,}177.777809$
1075	$(1075-1028.333333)^2=(46.666667)^2=2{,}177.777809$
1100	$(1100-1028.333333)^2=(71.666667)^2=5{,}136.111159$
1125	$(1125-1028.333333)^2=(96.666667)^2=9{,}344.444509$
1150	$(1150-1028.333333)^2=(121.666667)^2=14{,}802.77786$
	$\sum(x-\bar{x})^2=72{,}333.33334$

Thus, $s_{b_1} = \dfrac{s}{\sqrt{\sum(x-\bar{x})^2}} = \dfrac{0.3558}{\sqrt{72{,}333.33334}} \approx 0.001323$. The typical error in using b_1 to estimate β_1 is $s_{b_1} = 0.001323$.

(b) df $= n - 2 = 15 - 2 = 13$, $t_{\alpha/2} = 2.160$

(c) From Exercise 21, Section 13.2, $b_1 = 0.0056$. Thus, $b_1 \pm t_{\alpha/2}\, s_{b_1} = 0.0056 \pm (2.160)\,(0.001323) \approx 0.0056 \pm 0.0029 = (0.0027, 0.0085)$.

(d) We are 95% confident that the interval (0.0027, 0.0085) captures the population slope β_1 of the relationship between *combined SAT score* and *grade point average.*

33. There is evidence against the null hypothesis that no linear relationship exists. The points appear to lie near a line with a positive slope.

35. $H_0 : \beta_1 = 0$. There is no relationship between *SAT I Verbal* (x) and *SAT I Math* (y).
$H_a : \beta_1 \neq 0$. There is a linear relationship between *SAT I Verbal* (x) and *SAT I Math* (y).

37. $s_{b_1} = \dfrac{s}{\sqrt{\sum(x-\bar{x})^2}} = \dfrac{2.92665}{\sqrt{481.2}} \approx 0.1334$

$$t_{data} = \frac{b_1}{s_{b_1}} = \frac{0.4264}{0.1334} \approx 3.196$$

39. (a)

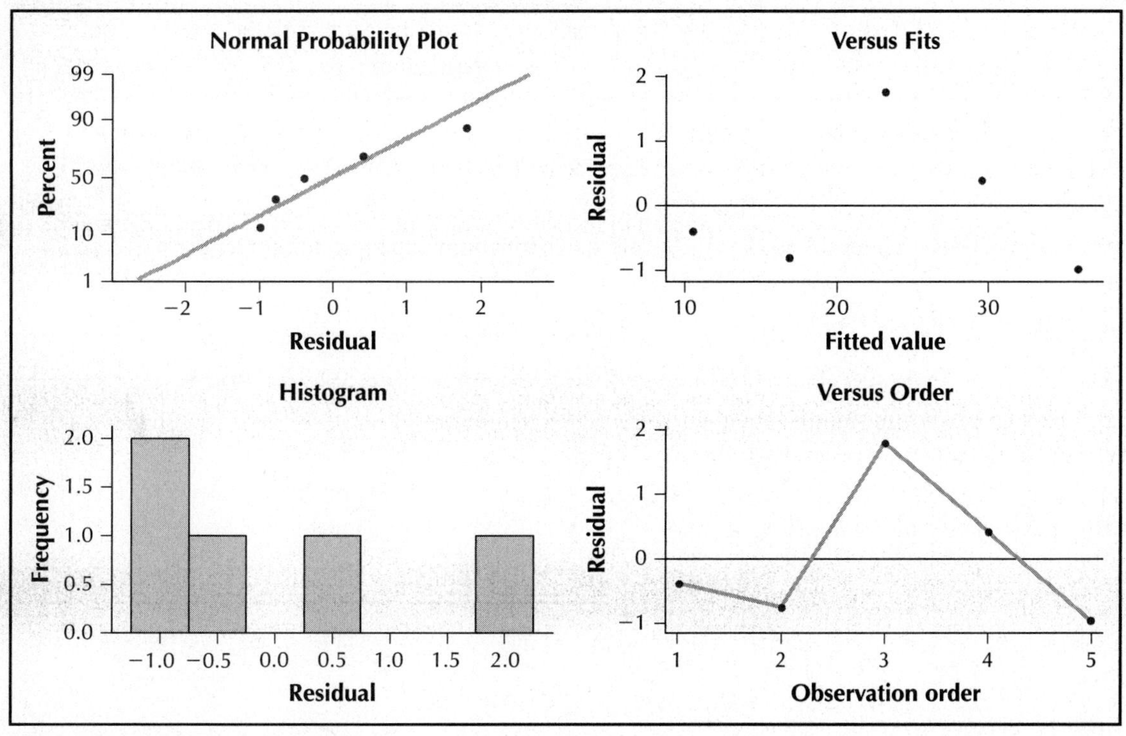

The scatterplot of the residuals contains an unhealthy pattern. Therefore, we conclude that the regression assumptions are not verified.

(b) Since 0 is not in the confidence interval (1.2688, 1.9312), we would expect to reject the null hypothesis that $\beta_1 = 0$.

(c) From Exercise 25, $s_{b_1} = 0.1041$. Thus, $t_{data} = \dfrac{b_1}{s_{b_1}} = \dfrac{1.6}{0.1041} \approx 15.370$

(d) df $= n - 2 = 5 - 2 = 3$, p-value $= 2 \cdot P(t > |t_{data}|) = 2 \cdot P(t > |15.370|) = 0.0005982328458$.

(e) $H_0 : \beta_1 = 0$. There is no relationship between *volume* (x) and *weight* (y).
$H_a : \beta_1 \neq 0$. There is a linear relationship between *volume* (x) and *weight* (y).
Reject H_0 if p-value < 0.05. Since p-value < 0.05, we reject H_0. There is evidence for a linear relationship between *volume* (x) and *weight* (y).

41. (a)

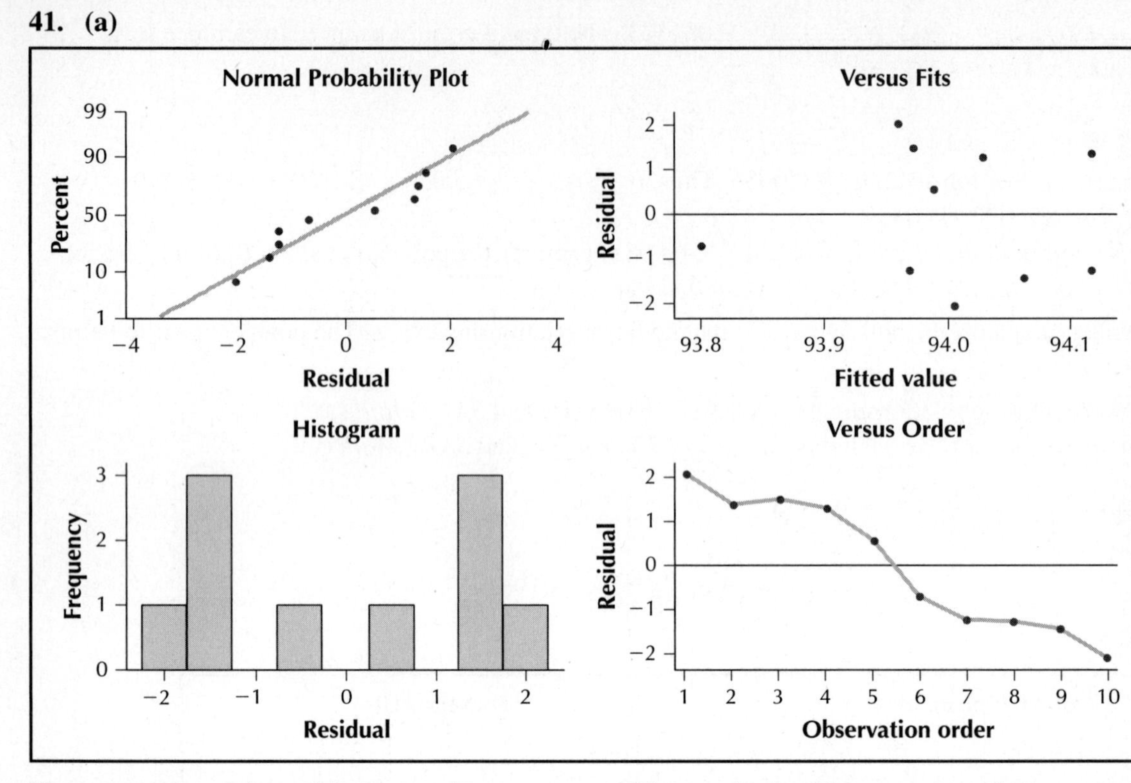

The scatterplot of the residuals contains no strong evidence of unhealthy patterns and the normal probability plot indicates no evidence of departures from normality in the residuals. Therefore, we conclude that the regression assumptions are verified.

(b) Since 0 lies in the confidence interval $(-19.6353, 22.8085)$, we would expect to not reject the null hypothesis that $\beta_1 = 0$.

(c) From Exercise 29, $s_{b_1} = 9.2029$. Thus, $t_{data} = \frac{b_1}{s_{b_1}} = \frac{1.587}{9.2029} \approx 0.1724$

(d) $\text{df} = n - 2 = 10 - 2 = 8$, p-value $= 2 \cdot P(t > |t_{data}|) = 2 \cdot P(t > |0.1724|) = 0.8674046954$

(e) $H_0 : \beta_1 = 0$. There is no relationship between *win%* (x) and *rating* (y).

$H_a : \beta_1 \neq 0$. There is a linear relationship between *win%* (x) and *rating* (y).

Reject H_0 if p-value < 0.05. Since p-value ≥ 0.05, we do not reject H_0. There is insufficient evidence for a linear relationship between *win%* (x) and *rating* (y).

43. (a)

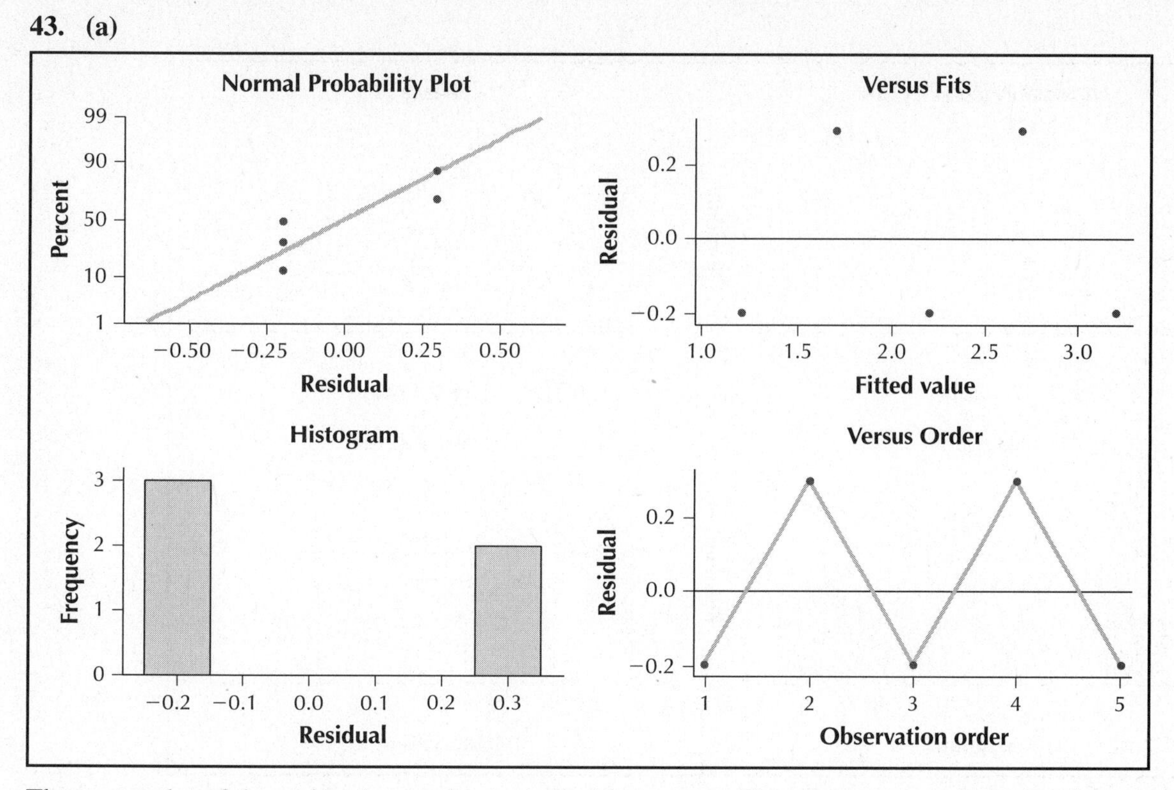

The scatterplot of the residuals contains an unhealthy pattern. Therefore, we conclude that the regression assumptions are not verified.

(b) Since 0 does not lie in the confidence interval (0.1818, 0.8182), we would expect to reject the null hypothesis that $\beta_1 = 0$.

(c) $\text{df} = n - 2 = 5 - 2 = 3$, $\alpha = 0.05$, $t_{crit} = 3.182$

(d) From Exercise 26, $s_{b_1} = 0.1000$. Thus, $t_{data} = \dfrac{b_1}{s_{b_1}} = \dfrac{0.5}{0.1000} \approx 5$.

(e) $H_0 : \beta_1 = 0$. There is no relationship between *family size* (x) and *pets* (y).
$H_a : \beta_1 \neq 0$. There is a linear relationship between *family size* (x) and *pets* (y).
Reject H_0 if $t_{data} < -3.182$ or $t_{data} > 3.182$. Since $t_{data} > 3.182$, we reject H_0. There is evidence for a linear relationship between *family size* (x) and *pets* (y).

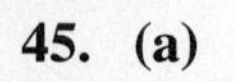

45. (a)

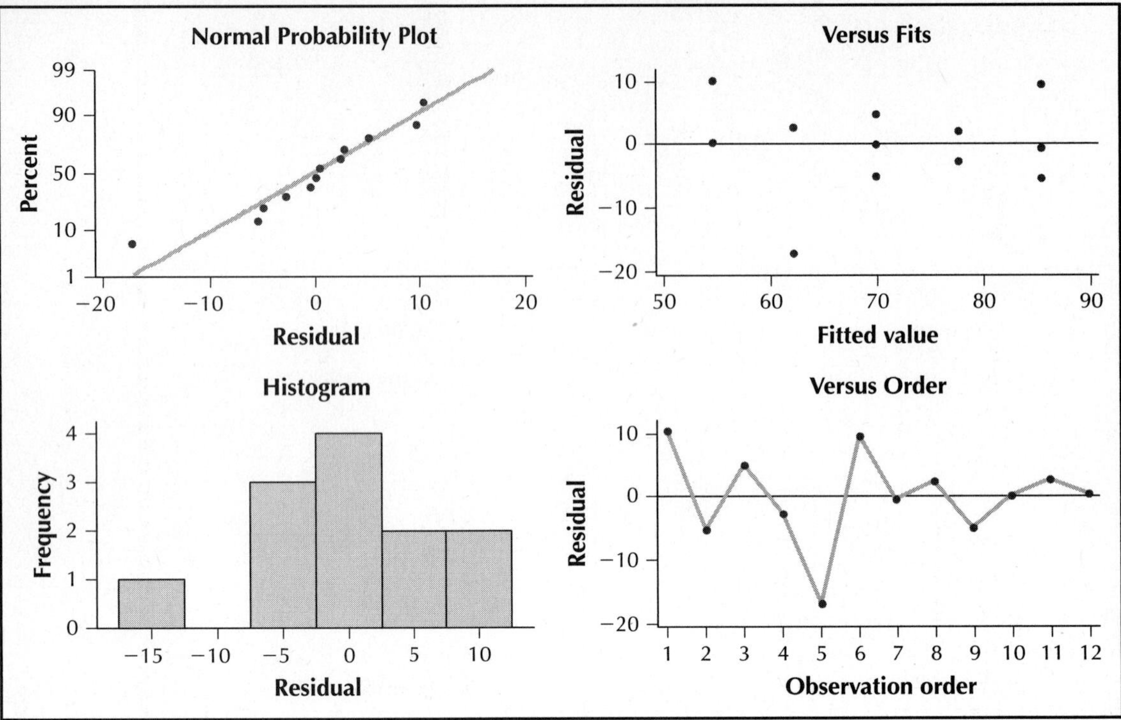

The scatterplot of the residuals contains no strong evidence of unhealthy patterns and the normal probability plot indicates no evidence of departures from normality in the residuals. Therefore, we conclude that the regression assumptions are verified.

(b) Since 0 does not lie in the confidence interval (0.4166, 1.1256), we would expect to reject the null hypothesis that $\beta_1 = 0$.

(c) $\text{df} = n - 2 = 12 - 2 = 10$, $\alpha = 0.05$, $t_{crit} = 2.228$

(d) From Exercise 30, $s_{b_1} = 0.1591$. Thus, $t_{data} = \dfrac{b_1}{s_{b_1}} = \dfrac{0.7711}{0.1591} \approx 4.847$

(e) $H_0 : \beta_1 = 0$. There is no relationship between *midterm exam* (x) and *overall grade* (y).
$H_a : \beta_1 \neq 0$. There is a linear relationship between *midterm exam* (x) and *overall grade* (y).
Reject H_0 if $t_{data} < -2.228$ or $t_{data} > 2.228$. Since $t_{data} > 2.228$, we reject H_0. There is evidence for a linear relationship between *midterm exam* (x) and *overall grade* (y).

47. **(a)**

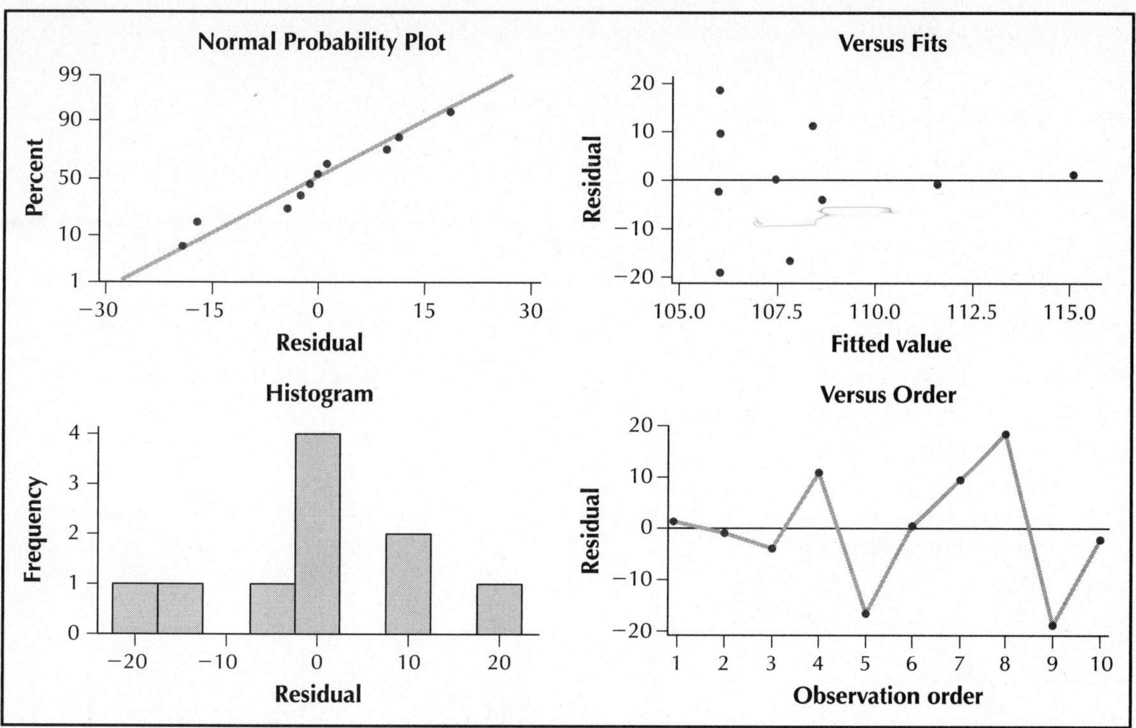

The residuals versus predicted values plot shows a funnel pattern.

(b) The funnel pattern in the residuals versus predicted values plot violates the constant variance assumption.

(c) No, because one of the regression assumptions is violated.

(d) Yes.

49. **(a)** Use part (c) and the identities $t_{data} = \dfrac{b_1}{s/\sqrt{\sum(x-\bar{x})^2}}$ and $b_1 = \dfrac{\sum(x-\bar{x})(y-\bar{y})}{\sum(x-\bar{x})^2}$. Since the five added points all have coordinates $(\bar{x}, \bar{y})$, the quantities $(x-\bar{x})$, $(x-\bar{x})^2$, and $(y-\bar{y})$ are all 0 for those five points and remain the same for the original points. Thus, b_1 and $\sqrt{\sum(x-\bar{x})^2}$ remain the same. From part (c) s decreases. Thus, t_{data} increases if b_1 is positive and decreases if b_1 is negative.

(b) Use 48 (c), (d), and the identity $r^2 = \dfrac{\text{SSR}}{\text{SST}}$. Since SSR and SST remain the same, r^2 remains the same.

(c) Use 48 (a), (b), and the identity $s = \sqrt{\dfrac{\text{SSE}}{n-2}}$. Since SSE remains the same and n increases, s decreases.

(d) Use (a) and the identity p-value $= 2 \cdot P(t > |t_{data}|)$. From (a), $|t_{data}|$ increases. Thus, the p-value decreases.

(e) Since we don't know what the new p-value will be, we don't know if p-value will decrease enough to change the conclusion from do not reject H_0 to reject H_0.

51. **(a)** From part (c), s decreases. By an argument similar to the argument given in 50 (c), $\sum(x-\bar{x})^2$ increases. Since $s_{b_1} = \dfrac{s}{\sqrt{\sum(x-\bar{x})^2}}$, s_{b_1} decreases. Since $t_{data} = \dfrac{b_1}{s_{b_1}}$ and b_1 remains the same, t_{data} increases if b_1 is positive and decreases if b_1 is negative.

(b) Use 50 (b), (c), and the identities $r^2 = \dfrac{\text{SSR}}{\text{SST}}$ and SSR = SST − SSE. Combining the identities gives us $r^2 = \dfrac{\text{SSR}}{\text{SST}} = \dfrac{\text{SST} - \text{SSE}}{\text{SST}} = 1 - \dfrac{\text{SSE}}{\text{SST}}$. From (b), SSE remains the same. From (c), SST increases. Thus, $\dfrac{\text{SSE}}{\text{SST}}$ decreases, so r^2 increases.

(c) Use 50 (a), (b), and the identity $s = \sqrt{\dfrac{\text{SSE}}{n-2}}$. Since SSE remains the same and n increases, s decreases.

(d) Use (a) and the identity p-value $= 2 \cdot P(t > |t_{data}|)$. From 50 (a), $|t_{data}|$ increases. Thus, the p-value decreases.

(e) Since the conclusion was already reject H_0 and the p-value decreases from (d), the conclusion will still be reject H_0.

53. (a) It can be shown that another equation for t_{data} is $t_{data} = r\sqrt{\frac{n-2}{1-r^2}}$.
From part (b), r^2 and therefore r both decrease. Thus, $1 - r^2$ increases. From Exercise 52 (a), n remains the same. Thus, $n - 2$ remains the same. Thus, t_{data} decreases.

(b) Use 52 (c), (d), and the identity $r^2 = \frac{SSR}{SST}$. From 52 (d), SSR decreases. From 52 (c), SST remains the same. Thus, r^2 decreases.

(c) Use 52 (a), (b), and the identity $s = \sqrt{\frac{SSE}{n-2}}$. Since SSE increases and n remains the same, s increases.

(d) Use (a) and the identities p-value $= 2 \cdot P(t > |t_{data}|)$ and $t_{data} = \frac{b_1}{s}$. Since b_1 and s are both positive, t_{data} is positive. From (a), $t_{data} = |t_{data}|$ decreases. Thus, the p-value increases.

(e) It depends on the new p-value.

55. (a)

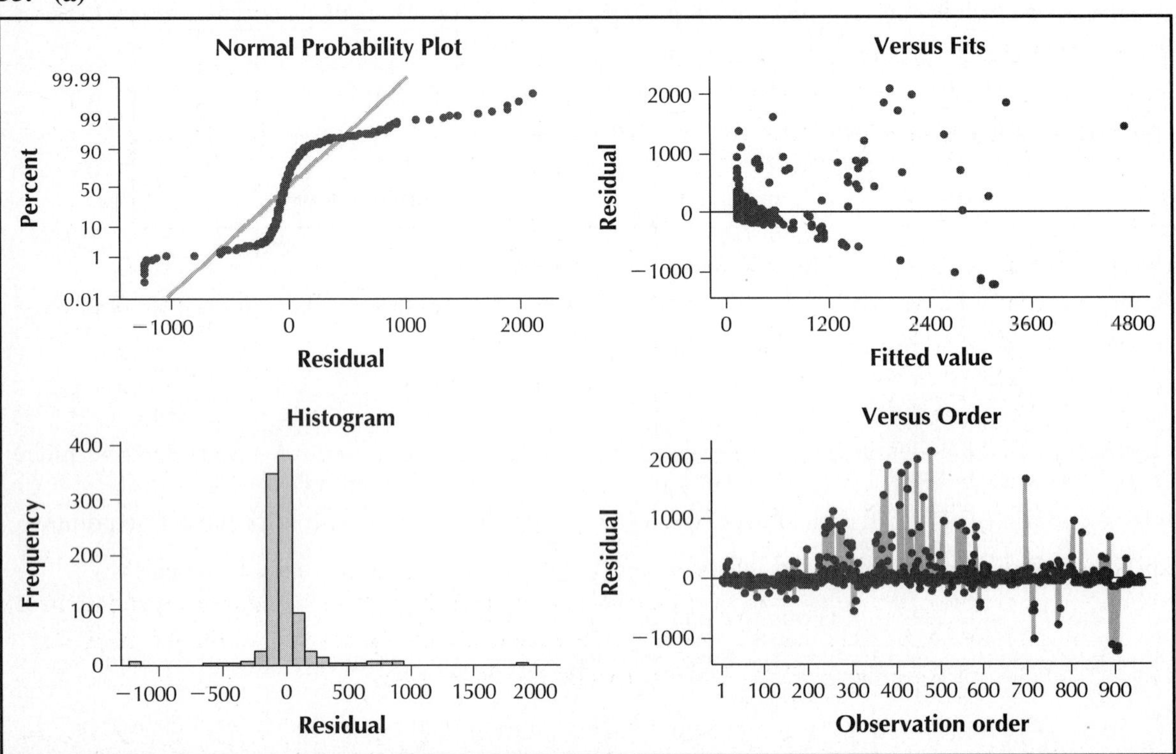

The scatterplot of the residuals contains no strong evidence of unhealthy patterns but the normal probability plot indicates evidence of departures from normality in the residuals. Therefore, we conclude that the regression assumptions are not fully verified.

(b) From Exercise 28, Section 13.2, $b_1 = 14.0842$, $s_{b_1} = 0.2701$, and df $= 959$. Thus, $t_{\alpha/2} = 1.984$. Therefore, $b_1 \pm t_{\alpha/2} \cdot s_{b_1} = 14.0842 \pm (1.984)(0.2701) = 14.0842 \pm 0.5359 = (13.5483, 14.6201)$.
We are 95% confident that the interval (13.5483, 14.6201) captures the population slope β_1 of the relationship between *fat per gram* and *calories per gram.*

(c) Since 0 does not lie in the confidence interval, we would expect to reject the null hypothesis that $\beta_1 = 0$.

(d) $H_0: \beta_1 = 0$. There is no relationship between *fat per gram* (x) and *calories per gram* (y).
$H_a: \beta_1 \neq 0$. There is a linear relationship between *fat per gram* (x) and *calories per gram* (y).
Reject H_0 if p-value < 0.05. From Exercise 28, Section 13.2, $t_{data} = 52.15$ and p-value ≈ 0. Since p-value < 0.05, we reject H_0. There is evidence for a linear relationship between *fat per gram* (x) and *calories per gram* (y).

57. **(a)** No.

(b) Positive relationship.

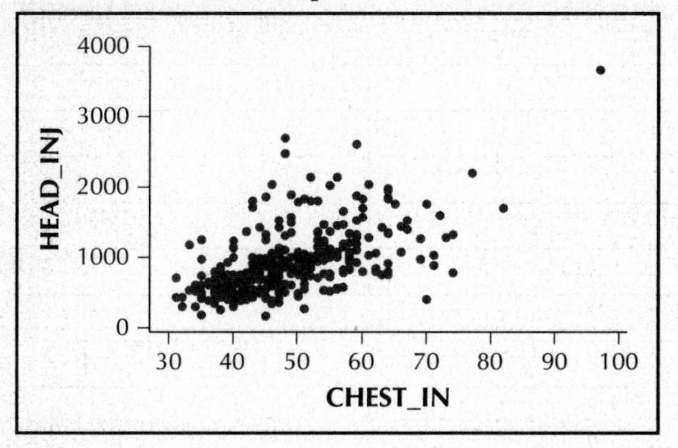

(c) Unclear. It is possible to have a head injury without having a chest injury or to have a chest injury without having a head injury, so it is unclear which variable should be the response variable and which variable should be the predictor variable.

59. **(a)** No.

(b) No apparent relationship between the variables.

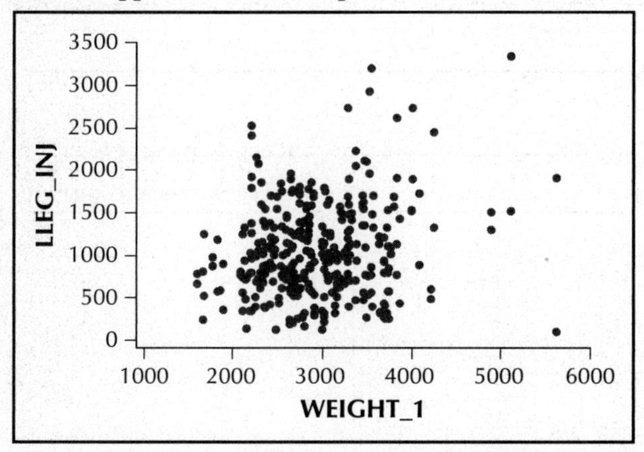

(c) Weight of the vehicles should be the predictor variable and the severity of the left leg injuries should be the response variable. It may be possible that the weight of the vehicle affects the severity of the left leg injury but the severity of the left leg injury will not affect the weight of the vehicle.

Chapter 13 Review

1. **(a)**

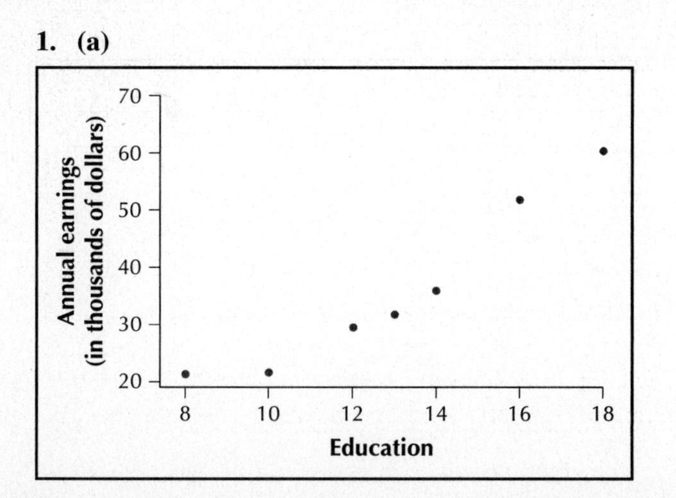

(b) As the number of years of education increases, the annual earnings increase.

(c) $b_1 = \dfrac{\sum xy - \left(\sum x\right)\left(\sum y\right)/n}{\sum x^2 - \left(\sum x\right)^2/n} = \dfrac{3436.7 - (91)(240.3)/7}{1253 - (91)^2/7} = \dfrac{312.8}{70} = 4.468571429 \approx 4.4686$

x	y	xy	x^2	y^2
8	18.6	148.8	64	345.96
10	18.9	189	100	357.21
12	27.3	327.6	144	745.29
13	29.7	386.1	169	882.09
14	34.2	478.8	196	1169.64
16	51.2	819.2	256	2621.44
18	60.4	1087.2	324	3648.16
$\sum x = 91$	$\sum y = 240.3$	$\sum xy = 3436.7$	$\sum x^2 = 1253$	$\sum y^2 = 9769.79$

Now, $\bar{x} = \dfrac{\sum x}{n} = \dfrac{91}{7} = 13$ and $\bar{y} = \dfrac{\sum y}{n} = \dfrac{240.3}{7} = 34.32857143$ giving,
$b_0 = \bar{y} - (b_1 \cdot \bar{x}) = 34.32857143 - (4.468571429)(13) = -23.76285714 \approx -23.76286$.
So $b_0 = -23.76286$, $b_1 = 4.4686$, and $\hat{y} = -23.76286 + 4.4686\,x$.

(d) $b_1 = 4.4686$ means that, for every additional year of education, the estimated annual earnings are increased by $4,468.60.

(e) $b_0 = -23.76286$ means that the estimated annual earnings of a person that has 0 years of education is −$23,762.86. This does not make sense in this data set because a person can't earn a negative amount of money.

(f)

x	y	Predicted Value $\hat{y} = -23.76286 + 4.4686\,x$	Residual $(y - y)$	(Residual)^2 $(y - \hat{y})^2$
8	18.6	11.98594	6.61406	43.74578968
10	18.9	20.92314	−2.02314	4.09309546
12	27.3	29.86034	−2.56034	6.555340916
13	29.7	34.32894	−4.62894	21.42708552
14	34.2	38.79754	−4.59754	21.13737405
16	51.2	47.73474	3.46526	12.00802687
18	60.4	56.67194	3.72806	13.89843136

$SSE = \sum (y - \hat{y})^2 = 122.8651439$, $s = \sqrt{\dfrac{SSE}{n-2}} = \sqrt{\dfrac{122.8651439}{7-2}} \approx 4.95712$.

If we know the number of years of education (x) that a person has, then our estimate of the person's annual earnings will typically differ from the person's actual earnings by $4957.12.

3. (a)

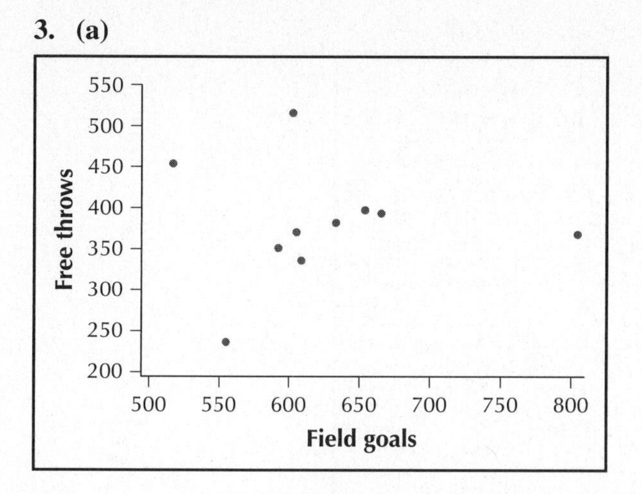

(b) As the number of field goals increases, the number of free throws stays the same.

(c)

x	y	xy	x^2	y^2
653	398	259,894	426,409	158,404
665	394	262,010	442,225	155,236
804	368	295,872	646,416	135,424
516	454	234,264	266,256	206,116
602	517	311,234	362,404	267,289
554	237	131,298	306,916	56,169
608	336	204,288	369,664	112,896
592	352	208,384	350,464	123,904
605	371	224,455	366,025	137,641
633	383	242,439	400,689	146,689
$\sum x = 6232$	$\sum y = 3810$	$\sum xy = 2,374,138$	$\sum x^2 = 3,937,468$	$\sum y^2 = 1,499,768$

$$b_1 = \frac{\sum xy - \left(\sum x\right)\left(\sum y\right)/n}{\sum x^2 - \left(\sum x\right)^2/n} = \frac{2,374,138 - (6232)(3810)/10}{3,937,468 - (6232)^2/10} = \frac{-254}{53,685.6} = -0.0047312501 \approx -0.0047$$

Now, $\bar{x} = \frac{\sum x}{n} = \frac{6232}{10} = 623.2$ and $\bar{y} = \frac{\sum y}{n} = \frac{3810}{10} = 381$ giving,

$b_0 = \bar{y} - (b_1 \cdot \bar{x}) = 381 - (-0.0047312501)(623.2) = 383.9485151 \approx 383.9485.$

So $b_0 = 383.9485$, $b_1 = -0.0047$, and $\hat{y} = 383.9485 - 0.0047x$.

(d) $b_1 = -0.0047$ means that, for every additional field goal, the estimated number of free throws decreases by 0.0047.

(e) $b_0 = 384$ means that the estimated number of free throws for a player who has 0 field goals is 384. This makes sense in this data set because it is possible for a person to have 384 free throws and 0 field goals.

(f)

x	y	Predicted Value $\hat{y} = 383.9485 - 0.0047x$	Residual $(y - \hat{y})$	$(\text{Residual})^2$ $(y - \hat{y})^2$
653	398	380.8794	17.1206	293.1149444
665	394	380.823	13.177	173.633329
804	368	380.1697	−12.1697	148.1015981
516	454	381.5233	72.4767	5,252.872043
602	517	381.1191	135.8809	18,463.61898
554	237	381.3447	−144.3447	20,835.39242
608	336	381.0909	−45.0909	2,033.189263
592	352	381.1661	−29.1661	850.6613892
605	371	381.105	−10.105	102.111025
633	383	380.9734	2.0266	4.10710756

$\text{SSE} = \sum(y - \hat{y})^2 = 48{,}156.8021$, $s = \sqrt{\dfrac{\text{SSE}}{n-2}} = \sqrt{\dfrac{48{,}156.8021}{10-2}} \approx 77.5861.$

If we know the number of field goals (x) that a player has, then our estimate of the player's free throws typically differs from the player's actual number of free throws by 77.5861.

5. **(a)** From Exercise 1 (f), SSE = 122.8651439. From Exercise 1 (c), $\sum y = 240.3$ and $\sum y^2 = 9769.79$. Thus,

$\text{SST} = \sum y^2 - \dfrac{\left(\sum y\right)^2}{n} = 9769.79 - \dfrac{240.3^2}{7} = 1520.634286$. Then, SSR = SST − SSE = 1520.634286 − 122.8651439 = 1397.769142.

(b) $r^2 = \dfrac{\text{SSR}}{\text{SST}} = \dfrac{1397.769142}{1520.634286} = 0.919201385$. Since r^2 is close to 1, the regression equation fits the data extremely well.

(c) From Exercise 1 (c), $b_1 = 4.4686$, which is positive, so r is positive.
Thus, $r = \sqrt{r^2} = \sqrt{0.919201385} = 0.9587499074$. Therefore, the variables are positively correlated.

7. **(a)** From Exercise 3 (f), SSE = 48,156.8021. From Exercise 3 (c), $\sum y = 3810$ and $\sum y^2 = 1{,}499{,}768$. Thus,

$\text{SST} = \sum y^2 - \dfrac{\left(\sum y\right)^2}{n} = 1{,}499{,}768 - \dfrac{3810^2}{10} = 48{,}158.$

Then, SSR = SST − SSE = 48,158 − 48,156.8021 = 1.1979.

(b) $r^2 = \dfrac{\text{SSR}}{\text{SST}} = \dfrac{1.1979}{48{,}158} = 0.00002487437186.$

Since r^2 is close to 0, the regression equation fits the data extremely poorly.

(c) From Exercise 3 (c), $b_1 = -0.0047$, which is negative, so r is negative.
Thus, $r = \sqrt{r^2} = -\sqrt{0.00002487437186} = -0.0049874214$. Therefore, the variables are not correlated.

9. **(a)** Calculating $\sum(x - \bar{x})^2$: From Exercise 1 (c), $\bar{x} = 13$.

x	$(x - \bar{x})^2$
8	$(8 - 13)^2 = (-5)^2 = 25$
10	$(10 - 13)^2 = (-3)^2 = 9$
12	$(12 - 13)^2 = (-1)^2 = 1$
13	$(13 - 13)^2 = (0)^2 = 0$
14	$(14 - 13)^2 = (1)^2 = 1$
16	$(16 - 13)^2 = (3)^2 = 9$
18	$(18 - 13)^2 = (5)^2 = 25$
	$\Sigma(x - \bar{x})^2 = 70$

From Exercise 1 (f), $s \approx 4.95712$. Thus, $s_{b_1} = \dfrac{s}{\sqrt{\sum(x - \bar{x})^2}} = \dfrac{4.95712}{\sqrt{70}} \approx 0.5925$. The typical error in using b_1 to estimate β_1 is $s_{b_1} = 0.5925$.

(b) df $= n - 2 = 7 - 2 = 5$, $t_{\alpha/2} = 2.571$

(c) From Exercise 1 (c), $b_1 = 4.4686$. Thus,
$b_1 \pm t_{\alpha/2}\, s_{b_1} = 4.4686 \pm (2.571)\,(0.5925) \approx 4.4686 \pm 1.5233 = (2.9453, 5.9919)$.
We are 95% confident that the interval (2.9453, 5.9919) captures the population slope β_1 of the relationship between *education* and *annual* earnings.

11. **(a)** Calculating $\sum(x - \bar{x})^2$: From Exercise 3(c), $\bar{x} = 623.2$.

x	$(x - \bar{x})^2$
653	$(653 - 623.2)^2 = (29.8)^2 = 888.04$
665	$(665 - 623.2)^2 = (41.8)^2 = 1{,}747.24$
804	$(804 - 623.2)^2 = (180.8)^2 = 32{,}688.64$
516	$(516 - 623.2)^2 = (-107.2)^2 = 11{,}491.84$
602	$(602 - 623.2)^2 = (-21.2)^2 = 449.44$
554	$(554 - 623.2)^2 = (-69.2)^2 = 4{,}788.64$
608	$(608 - 623.2)^2 = (-15.2)^2 = 231.04$
592	$(592 - 623.2)^2 = (-31.2)^2 = 973.44$
605	$(605 - 623.2)^2 = (-18.2)^2 = 331.24$
633	$(633 - 623.2)^2 = (9.8)^2 = 96.04$
	$\Sigma(x - \bar{x})^2 = 53{,}685.6$

$\sum(x - \bar{x})^2 = 53{,}685.6$. From Exercise 3 (f), $s \approx 77.5861$. Thus, $s_{b_1} = \dfrac{s}{\sqrt{\sum(x - \bar{x})^2}} = \dfrac{77.5861}{\sqrt{53{,}685.6}} \approx 0.3349$.

The typical error in using b_1 to estimate β_1 is $s_{b_1} = 0.3349$.

(b) df $= n - 2 = 10 - 2 = 8$, $t_{\alpha/2} = 2.306$

(c) From Exercise 3 (c), $b_1 = -0.0047$. Thus, $b_1 \pm t_{\alpha/2}\, s_{b_1} = -0.0047 \pm (2.306)\,(0.3349) \approx -0.0047 \pm 0.7723 = (-0.7770, 0.7676)$. We are 95% confident that the interval $(-0.7770, 0.7676)$ captures the population slope β_1 of the relationship between *field goals* and *free* throws.

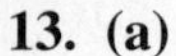

13. (a)

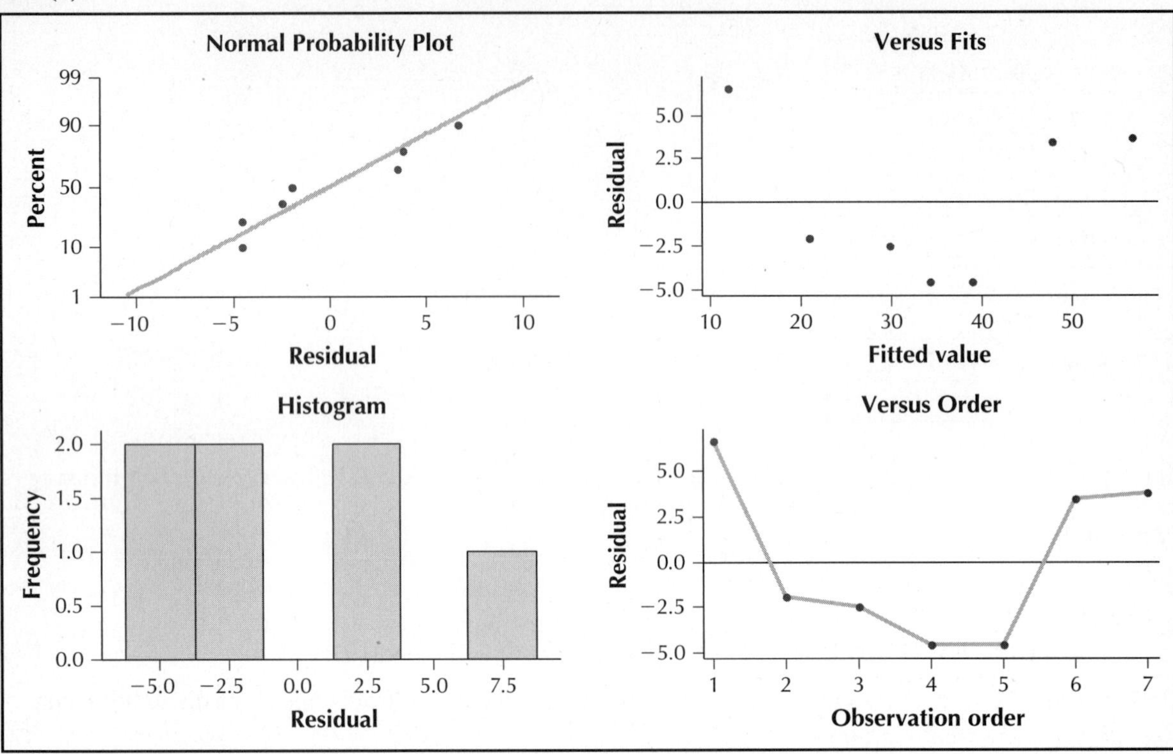

The scatterplot of the residuals contains evidence of an unhealthy pattern. Therefore, we conclude that the regression assumptions are not verified.

(b) Since 0 is not in the confidence interval (2.9453, 5.9919), we would expect to reject the null hypothesis that $\beta_1 = 0$.

(c) From Exercise 9, $s_{b_1} = 0.5925$. From Exercise 1 (c), $b_1 = 4.4686$. Thus, $t_{data} = \frac{b_1}{s_{b_1}} = \frac{4.4686}{0.5925} \approx 7.54$

(d) df $= n - 2 = 7 - 2 = 5$, p-value $= 2 \cdot P(t > |t_{data}|) = 2 \cdot P(t > |7.54|) = 0.0006499730713$

(e) $H_0 : \beta_1 = 0$. There is no relationship between *education* (x) and *annual earnings* (y).
$H_a : \beta_1 \neq 0$. There is a linear relationship between *education* (x) and *annual earnings* (y).
Reject H_0 if p-value < 0.05. Since p-value < 0.05, we reject H_0. There is evidence for a linear relationship between *education* (x) and *annual earnings* (y).

15. (a) Since 0 lies in the confidence interval $(-0.7770, 0.7676)$, we would expect to not reject the null hypothesis that $\beta_1 = 0$.

(b) df $= n - 2 = 10 - 2 = 8$, $\alpha = 0.05$, $t_{crit} = 2.306$

(c) From Exercise 11, $s_{b_1} = 0.3349$. From Exercise 3 (c), $b_1 = -0.0047$. Thus, $t_{data} = \frac{b_1}{s_{b_1}} = \frac{-0.0047}{0.3349} \approx -0.014$.

(d) $H_0 : \beta_1 = 0$. There is no relationship between *field goals* (x) and *free throws* (y).
$H_a : \beta_1 \neq 0$. There is a linear relationship between *field goals* (x) and *free throws* (y).
Reject H_0 if $t_{data} < -2.306$ or $t_{data} > 2.306$. Since t_{data} is not less than -2.306 and t_{data} is not greater than 2.306, we do not reject H_0. There is insufficient evidence for a linear relationship between *field goals* (x) and *free throws* (y).

Chapter 13 Quiz

1. False. The difference $(y - \hat{y})$ is called the *residual.*
2. True.
3. False. SSR measures the amount of improvement in the accuracy of our estimates when using the regression equation as compared to ignoring the x information.

4. y, y, x
5. predicted value of y, actual observed value of y
6. coefficient of determination
7. between 0 and 1 inclusive
8. SST = SSR + SSE
9. $n - 2$
10. **(a)**

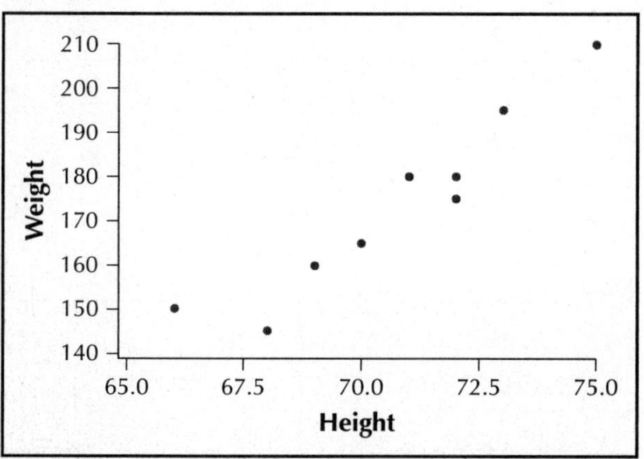

(b) As the height of a person increases, the weight of the person increases.

(c)

x	y	xy	x^2	y^2
66	150	9,900	4,356	22,500
68	145	9,860	4,624	21,025
69	160	11,040	4,761	25,600
70	165	11,550	4,900	27,225
70	165	11,550	4,900	27,225
71	180	12,780	5,041	32,400
72	175	12,600	5,184	30,625
72	180	12,960	5,184	32,400
73	195	14,235	5,329	38,025
75	210	15,750	5,625	44,100
$\sum x = 706$	$\sum y = 1725$	$\sum xy = 122{,}225$	$\sum x^2 = 49{,}904$	$\sum y^2 = 301{,}125$

$$b_1 = \frac{\sum xy - \left(\sum x\right)\left(\sum y\right)/n}{\sum x^2 - \left(\sum x\right)^2/n} = \frac{122{,}225 - (706)(1725)/10}{49{,}904 - (706)^2/10} = \frac{440}{60.4} = 7.284768212 \approx 7.2848$$

Now, $\bar{x} = \dfrac{\sum x}{n} = \dfrac{706}{10} = 70.6$ and $\bar{y} = \dfrac{\sum y}{n} = \dfrac{1725}{10} = 172.5$ giving, $b_0 = \bar{y} - (b_1 \cdot \bar{x}) =$ $172.5 - (7.284768212)(70.6) = -341.8046358 \approx -341.8046$.

So $b_0 = -341.8046$, $b_1 = 7.2848$, and $\hat{y} = -341.8046 + 7.2848\,x$.

11. (a)

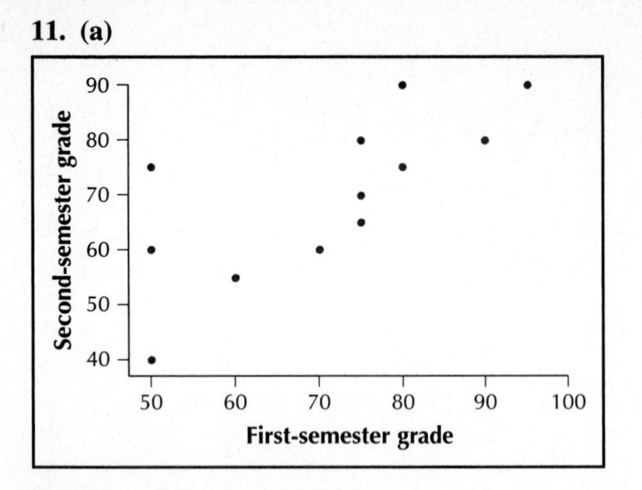

(b) As the first semester grade increases, the second semester grade increases.

(c)

x	y	xy	x^2	y^2
75	65	4,875	5,625	4,225
80	90	7,200	6,400	8,100
50	75	3,750	2,500	5,625
70	60	4,200	4,900	3,600
90	80	7,200	8,100	6,400
75	80	6,000	5,625	6,400
50	60	3,000	2,500	3,600
95	90	8,550	9,025	8,100
80	75	6,000	6,400	5,625
50	40	2,000	2,500	1,600
60	55	3,300	3,600	3,025
75	70	5,250	5,625	4,900
$\sum x = 850$	$\sum y = 840$	$\sum xy = 61{,}325$	$\sum x^2 = 62{,}800$	$\sum y^2 = 61{,}200$

$$b_1 = \frac{\sum xy - \left(\sum x\right)\left(\sum y\right)/n}{\sum x^2 - \left(\sum x\right)^2/n} = \frac{61{,}325 - (850)(840)/12}{62{,}800 - (850)^2/12} = \frac{1825}{2591.666667} = 0.7041800642 \approx 0.7042$$

Now, $\bar{x} = \frac{\sum x}{n} = \frac{850}{12} = 70.83333333$ and $\bar{y} = \frac{\sum y}{n} = \frac{840}{12} = 70$ giving, $b_0 = \bar{y} - (b_1 \cdot \bar{x}) =$ $70 - (0.7041800642)(70.83333333) = 20.12057879 \approx 20.1206$. So $b_0 = 20.1206$, $b_1 = 0.7042$, and $\hat{y} = 20.1206 + 0.7042\,x$.

12. (a)

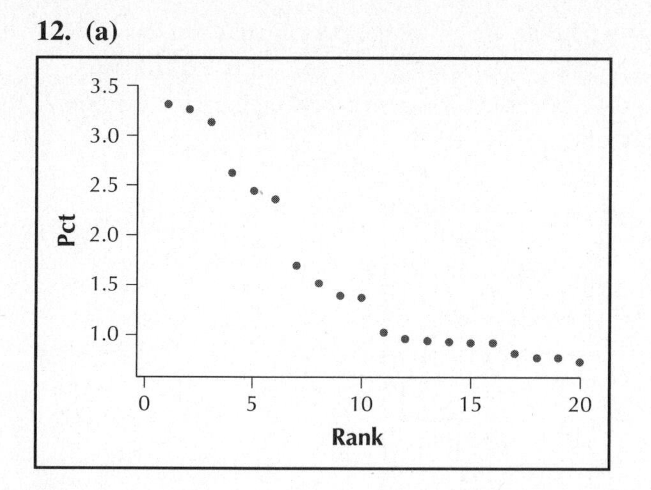

(b) As the rank of the boy's name increases the percent of boys with that name decreases.

(c)

x	y	xy	x^2	y^2
1	3.318	3.318	1	11.009142
2	3.271	6.542	4	10.699441
3	3.143	9.429	9	9.878449
4	2.629	10.516	16	6.911641
5	2.451	12.255	25	6.007401
6	2.363	14.178	36	5.583769
7	1.703	11.921	49	2.900209
8	1.523	12.184	64	2.319529
13	0.948	12.324	169	0.898704
9	1.404	12.636	81	1.971216
14	0.938	13.132	196	0.879844
10	1.380	13.8	100	1.9044
15	0.931	13.965	225	0.866761
11	1.035	11.385	121	1.071225
16	0.927	14.832	256	0.85329
12	0.974	11.688	144	0.948676
17	0.826	14.042	289	0.682276
18	0.780	14.04	324	0.6084
19	0.779	14.801	361	0.606841
20	0.736	14.72	400	0.541696
$\sum x = 210$	$\sum y = 32.059$	$\sum xy = 241.708$	$\sum x^2 = 2{,}870$	$\sum y^2 = 67.148931$

$$b_1 = \frac{\sum xy - (\sum x)(\sum y)/n}{\sum x^2 - (\sum x)^2/n} = \frac{241.708 - (210)(32.059)/20}{2{,}870 - (210)^2/20} = \frac{-94{,}9115}{665} = -0.1427240602 \approx -0.1427$$

Now, $\bar{x} = \frac{\sum x}{n} = \frac{210}{20} = 10.5$ and $\bar{y} = \frac{\sum y}{n} = \frac{32.059}{20} = 1.60295$ giving, $b_0 = \bar{y} - (b_1 \cdot \bar{x}) =$ $1.60295 - (-0.1427240602)(10.5) = 3.10155263 \approx 3.1016$. So $b_0 = 3.1016$, $b_1 = -0.1427$, and $\hat{y} = 3.1016 - 0.1427\,x$.

13. (a) $b_1 = 7.2848$ means that for every additional inch in height the weight of the person increases by 7.2848 pounds. **(b)** $b_0 = -341.8046$ means that the estimated weight of a person who is 0 inches tall is -341.8046 pounds. This does not make sense in this data set because it is impossible for a person to weigh a negative number of pounds and it is impossible for a person to be 0 inches tall.

(c)

x	y	Predicted Value $\hat{y} = -341.8046 + 7.2848x$	Residual $(y - \hat{y})$	$(\text{Residual})^2$ $(y - \hat{y})^2$
66	150	138.9922	11.0078	121.17166084
68	145	153.5618	−8.5618	73.30441924
69	160	160.8466	−0.8466	0.71673156
70	165	168.1314	−3.1314	9.80566596
70	165	168.1314	−3.1314	9.80566596
71	180	175.4162	4.5838	21.01122244
72	175	182.701	−7.701	59.305401
72	180	182.701	−2.701	7.295401
73	195	189.9858	5.0142	25.14220164
75	210	204.5544	5.4446	29.64366916

$\text{SSE} = \sum(y - \hat{y})^2 = 357.2020388$, $s = \sqrt{\frac{\text{SSE}}{n-2}} = \sqrt{\frac{357.2020388}{10-2}} \approx 6.68208$.

If we know the height (x) of a person, then our estimate of the person's weight typically differs from the person's actual weight by 6.68208 pounds.

14. (a) $b_1 = 0.7042$ means that, for every additional point of the first-semester grade, the second-semester grade increases by 0.7042 points. **(b)** $b_0 = 20.1206$ means that the estimated grade of a student with a grade of 0 in the first semester is 20.1206 . This does not make sense in this data set because a student with a grade of 0 in the first semester of accounting would not be eligible to take the second semester of accounting.

(c)

x	y	Predicted Value $\hat{y} = 20.1206 + 0.7042x$	Residual $(y - \hat{y})$	$(\text{Residual})^2$ $(y - \hat{y})^2$
75	65	72.9356	−7.9356	62.97374736
80	90	76.4566	13.5434	183.4236836
50	75	55.3306	19.6694	386.8852964
70	60	69.4146	−9.4146	88.63469316
90	80	83.4986	−3.4986	12.24020196
75	80	72.9356	7.0644	49.90574736
50	60	55.3306	4.6694	21.80329636
95	90	87.0196	2.9804	8.88278416
80	75	76.4566	−1.4566	2.12168356
50	40	55.3306	−15.3306	235.0272964
60	55	62.3726	−7.3726	54.35523076
75	70	72.9356	−2.9356	8.61774736

$\text{SSE} = \sum(y - \hat{y})^2 = 1114.871408$, $s = \sqrt{\frac{\text{SSE}}{n-2}} = \sqrt{\frac{1114.871408}{12-2}} \approx 10.5587$.

If we know the grade the student earned in the first semester of accounting (x), then our estimate of the student's grade in the second semester of accounting typically differs from the person's actual grade by 10.5587 points.

15. (a) $b_1 = -0.1427$ means that, for every increase of 1 in rank of a boy's name, the percent of boys with that name decreases by 0.1427. **(b)** $b_0 = 3.1016$ means that the estimated percent of boys with a name that has a rank of 0 is 3.1016%. This does not make sense in this data set because a name can't have a rank of 0.

(c)

x	y	**Predicted Value** $\hat{y} = 3.1016 - 0.1427x$	**Residual** $(y - \hat{y})$	**(Residual)^2** $(y - \hat{y})^2$
1	3.318	2.9589	0.3591	0.12895281
2	3.271	2.8162	0.4548	0.20684304
3	3.143	2.6735	0.4695	0.22043025
4	2.629	2.5308	0.0982	0.00964324
5	2.451	2.3881	0.0629	0.00395641
6	2.363	2.2454	0.1176	0.01382976
7	1.703	2.1027	−0.3997	0.15976009
8	1.523	1.96	−0.437	0.190969
13	0.948	1.2465	−0.2985	0.08910225
9	1.404	1.8173	−0.4133	0.17081689
14	0.938	1.1038	−0.1658	0.02748964
10	1.380	1.6746	−0.2946	0.08678916
15	0.931	0.9611	−0.031	0.00090601
11	1.035	1.5319	−0.4969	0.24690961
16	0.927	0.8184	0.1086	0.01179396
12	0.974	1.3892	−0.4152	0.17239104
17	0.826	0.6757	0.1503	0.02259009
18	0.780	0.533	0.247	0.061009
19	0.779	0.3903	0.3887	0.15108769
20	0.736	0.2476	0.4884	0.23853456

$\text{SSE} = \sum(y - \hat{y})^2 = 2.2138045$, $s = \sqrt{\frac{\text{SSE}}{n-2}} = \sqrt{\frac{2.2138045}{20-2}} \approx 0.3507$.

If we know the rank of a boy's name (x), then our estimate of the percent of boys with that name typically differs from the actual percent of boys with that name by 0.3507%.

16. (a) From Exercise 13, SSE = 357.2020388. From Exercise 10 (c), $\sum y = 1725$ and $\sum y^2 = 301{,}125$. Thus,

$\text{SST} = \sum y^2 - \frac{\left(\sum y\right)^2}{n} = 301{,}125 - \frac{1725^2}{10} = 3562.5$. Then, $\text{SSR} = \text{SST} - \text{SSE} = 3562.5 - 357.2020388 = 3205.297961$.

(b) $r^2 = \frac{\text{SSR}}{\text{SST}} = \frac{3205.297961}{3562.5} = 0.899732761$. Since r^2 is close to 1 the regression equation fits the data extremely well.

(c) From Exercise 10 (c), $b_1 = 7.4828$, which is positive, so r is positive. Thus, $r = \sqrt{r^2} = \sqrt{0.899732761} = 0.9485424403$. Therefore, the variables are positively correlated.

17. (a) From Exercise 14, SSE = 1114.871408. From Exercise 11(c), $\sum y = 840$ and $\sum y^2 = 61{,}200$. Thus,

$\text{SST} = \sum y^2 - \frac{\left(\sum y\right)^2}{n} = 61{,}200 - \frac{840^2}{12} = 2400$. Then, $\text{SSR} = \text{SST} - \text{SSE} = 2400 - 1114.871408 = 1285.128592$.

(b) $r^2 = \frac{\text{SSR}}{\text{SST}} = \frac{1285.128592}{2400} = 0.5354702467$. Thus, the regression equation fits the data moderately well.

(c) From Exercise 11(c), $b_1 = 0.7042$, which is positive, so r is positive. Thus, $r = \sqrt{r^2} = \sqrt{0.5354702467}$ $= 0.7317563253$. Therefore, the variables are positively correlated.

18. (a) From Exercise 15, SSE = 2.2138045. From Exercise 12 (c), $\sum y = 32.059$ and $\sum y^2 = 67.148931$.

Thus, $\text{SST} = \sum y^2 - \frac{\left(\sum y\right)^2}{n} = 67.148931 - \frac{32.059^2}{20} = 15.75995695$. Then, SSR = SST − SSE $= 15.75995695 - 2.2138045 = 13.54615245$.

(b) $r^2 = \frac{\text{SSR}}{\text{SST}} = \frac{13{,}54615245}{15.75995695} = 0.8595297876$. Since r^2 is close to 1, the regression equation fits the data extremely well.

(c) From Exercise 12 (c), $b_1 = -0.1427$, which is negative, so r is negative. Thus, $r = -\sqrt{r^2}$ $= -\sqrt{0.8595297876} = -0.9271082934$. Therefore, the variables are negatively correlated.

19. (a) Calculating $\sum(x - \bar{x})^2$: From Exercise 10(c), $\bar{x} = 70.6$.

x	$(x - \bar{x})^2$
66	$(66 - 70.6)^2 = (-4.6)^2 = 21.16$
68	$(68 - 70.6)^2 = (-2.6)^2 = 6.76$
69	$(69 - 70.6)^2 = (-1.6)^2 = 2.56$
70	$(70 - 70.6)^2 = (-0.6)^2 = 0.36$
70	$(70 - 70.6)^2 = (-0.6)^2 = 0.36$
71	$(71 - 70.6)^2 = (0.4)^2 = 0.16$
72	$(72 - 70.6)^2 = (1.4)^2 = 1.96$
72	$(72 - 70.6)^2 = (1.4)^2 = 1.96$
73	$(73 - 70.6)^2 = (2.4)^2 = 5.76$
75	$(75 - 70.6)^2 = (4.4)^2 = 19.36$
	$\Sigma(x - \bar{x})^2 = 60.4$

From Exercise 13 (c), $s \approx 6.68208$. Thus, $s_{b_1} = \frac{s}{\sqrt{\sum(x - \bar{x})^2}} = \frac{6.68208}{\sqrt{60.4}} \approx 0.8598$. The typical error in using b_1 to estimate β_1 is $s_{b_1} = 0.8598$.

(b) df $= n - 2 = 10 - 2 = 8$, $t_{\alpha/2} = 2.306$

(c) From Exercise 10 (c), $b_1 = 7.2848$. Thus, $b_1 \pm t_{\alpha/2}\, s_{b_1} = 7.2848 \pm (2.306)\,(0.8598) \approx 7.2848 \pm 1.9827$ $= (5.3021, 9.2675)$.

We are 95% confident that the interval (5.3021, 9.2675) captures the population slope β_1 of the relationship between *weight* and *height*.

20. (a) Calculating $\sum(x - \bar{x})^2$: From Exercise 11 (c), $\bar{x} = 70.83333333$

x	$(x - \bar{x})^2$
75	$(75 - 70.83333333)^2 = (4.16666667)^2 = 17.36111114$
80	$(80 - 70.83333333)^2 = (9.16666667)^2 = 84.02777784$
50	$(50 - 70.83333333)^2 = (-20.83333333)^2 = 434.0277776$
70	$(70 - 70.83333333)^2 = (-0.83333333)^2 = 0.6944444389$
90	$(90 - 70.83333333)^2 = (19.16666667)^2 = 367.3611112$
75	$(75 - 70.83333333)^2 = (4.16666667)^2 = 17.36111114$
50	$(50 - 70.83333333)^2 = (-20.83333333)^2 = 434.02777784$
95	$(95 - 70.83333333)^2 = (24.16666667)^2 = 584.0277779$

(Continued)

x	$(x-\bar{x})^2$
80	$(80 - 70.83333333)^2 = (9.16666667)^2 = 84.02777784$
50	$(50 - 70.83333333)^2 = (-20.83333333)^2 = 434.02777784$
60	$(60 - 70.83333333)^2 = (-10.83333333)^2 = 117.361111$
75	$(75 - 70.83333333)^2 = (4.16666667)^2 = 17.36111114$
	$\Sigma(x-\bar{x})^2 = 2591.666666$

From Exercise 14 (c), $s \approx 10.5587$. Thus, $s_{b_1} = \dfrac{s}{\sqrt{\sum(x-\bar{x})^2}} = \dfrac{10.5587}{\sqrt{2591.666666}} \approx 0.2074$. The typical error in using b_1 to estimate β_1 is $s_{b_1} = 0.2074$.

(b) df $= n - 2 = 12 - 2 = 10$, $t_{\alpha/2} = 2.228$

(c) From Exercise 11 (c), $b_1 = 0.7042$. Thus, $b_1 \pm t_{\alpha/2}\, s_{b_1} = 0.7042 \pm (2.228)(0.2074) \approx 0.7042 \pm 0.4621 = (0.2421, 1.1663)$. We are 95% confident that the interval (0.2421, 1.1663) captures the population slope β_1 of the relationship between *first-semester grade* and *second-semester grade.*

21. (a) Calculating $\sum(x-\bar{x})^2$: From Exercise 12 (c), $\bar{x} = 10.5$

x	$(x-\bar{x})^2$
1	$(1 - 10.5)^2 = (-9.5)^2 = 90.25$
2	$(2 - 10.5)^2 = (-8.5)^2 = 72.25$
3	$(3 - 10.5)^2 = (-7.5)^2 = 56.25$
4	$(4 - 10.5)^2 = (-6.5)^2 = 42.45$
5	$(5 - 10.5)^2 = (-5.5)^2 = 30.25$
6	$(6 - 10.5)^2 = (-4.5)^2 = 20.25$
7	$(7 - 10.5)^2 = (-3.5)^2 = 12.25$
8	$(8 - 10.5)^2 = (-2.5)^2 = 6.25$
13	$(13 - 10.5)^2 = (2.5)^2 = 6.25$
9	$(9 - 10.5)^2 = (-1.5)^2 = 2.25$
14	$(14 - 10.5)^2 = (-3.5)^2 = 12.25$
10	$(10 - 10.5)^2 = (-0.5)^2 = 0.25$
15	$(15 - 10.5)^2 = (4.5)^2 = 20.25$
11	$(11 - 10.5)^2 = (0.5)^2 = 0.25$
16	$(16 - 10.5)^2 = (5.5)^2 = 30.25$
12	$(12 - 10.5)^2 = (1.5)^2 = 2.25$
17	$(17 - 10.5)^2 = (6.5)^2 = 42.25$
18	$(18 - 10.5)^2 = (7.5)^2 = 56.25$
19	$(19 - 10.5)^2 = (8.5)^2 = 72.25$
20	$(20 - 10.5)^2 = (9.5)^2 = 90.25$
	$\Sigma(x-\bar{x})^2 = 665$

From Exercise 15 (c), $s \approx 0.3507$. Thus, $s_{b_1} = \dfrac{s}{\sqrt{\sum(x-\bar{x})^2}} = \dfrac{0.3507}{\sqrt{665}} \approx 0.0136$. The typical error in using b_1 to estimate β_1 is $s_{b_1} = 0.0136$.

(b) df $= n - 2 = 20 - 2 = 18$, $t_{\alpha/2} = 2.101$

(c) From Exercise 12 (c), $b_1 = -0.1427$. Thus,

$b_1 \pm t_{\alpha/2}\, s_{b_1} = -0.1427 \pm (2.101)(0.0136) \approx -0.1427 \pm 0.0286 = (-0.1713, -0.1141)$.

We are 95% confident that the interval $(-0.1713, -0.1141)$ captures the population slope β_1 of the relationship between *rank* and *pct*.

22. $H_0 : \beta_1 = 0$. There is no relationship between *height* (x) and *weight* (y).
$H_a : \beta_1 \neq 0$. There is a linear relationship between *height* (x) and *weight* (y).

Reject H_0 if p-value < 0.05. From Exercise 19, $s_{b_1} = 0.8598$. From Exercise 10 (c), $b_1 = 7.2848$. Thus, $t_{data} = \dfrac{b_1}{s_{b_1}} = \dfrac{7.2848}{0.8598} \approx 8.47$. df $= n - 2 = 10 - 2 = 8$, p-value $= 2 \cdot P(t > |t_{data}|) = 2 \cdot P(t > |8.47|) \approx 0$.

Since p-value < 0.05, we reject H_0. There is evidence for a linear relationship between *height* (x) and *weight* (y).

23. $H_0 : \beta_1 = 0$. There is no relationship between *first-semester grade* (x) and *second-semester grade* (y).
$H_a : \beta_1 \neq 0$. There is a linear relationship between *first-semester grade* (x) and *second-semester grade* (y).

Reject H_0 if p-value < 0.05. From Exercise 20, $s_{b_1} = 0.2074$. From Exercise 11 (c), $b_1 = 0.7042$. Thus, $t_{data} = \dfrac{b_1}{s_{b_1}} = \dfrac{0.7042}{0.2074} \approx 3.40$. df $= n - 2 = 12 - 2 = 10$, p-value $= 2 \cdot P(t > |t_{data}|) = 2 \cdot P(t > |3.40|) =$ 0.0067710724. Since p-value < 0.05, we reject H_0. There is evidence for a linear relationship between *first-semester grade* (x) and *second-semester grade* (y).

24. $H_0 : \beta_1 = 0$. There is no relationship between *rank* (x) and *pct* (y).
$H_a : \beta_1 \neq 0$. There is a linear relationship between *rank* (x) and *pct* (y).
df $= n - 2 = 20 - 2 = 18$, $\alpha = 0.05$, $t_{crit} = 2.101$.

Reject H_0 if $t_{data} < -2.101$ or $t_{data} > 2.101$. From Exercise 21, $s_{b_1} = 0.0136$. From Exercise 12 (c), $b_1 = -0.1427$. Thus, $t_{data} = \dfrac{b_1}{s_{b_1}} = \dfrac{-0.1427}{0.0136} \approx -10.49$. Since $t_{data} < -2.101$, we reject H_0. There is evidence for a linear relationship between *rank* (x) and *pct* (y).